Uranium for Nuclear Power

Related Titles

Nuclear Fuel Cycle Science and Engineering
(ISBN 978-0-85709-073-7)

Nuclear Decommissioning: Planning, Execution and International Experience
(ISBN 978-0-85709-115-4)

The Future of Nuclear Power
(ISBN 978-0-08044-489-5)

Woodhead Publishing Series in Energy: Number 93

Uranium for Nuclear Power

Resources, Mining and Transformation to Fuel

Edited by

Ian Hore-Lacy

World Nuclear Association, London, United Kingdom

ELSEVIER

AMSTERDAM • BOSTON • CAMBRIDGE • HEIDELBERG
LONDON • NEW YORK • OXFORD • PARIS • SAN DIEGO
SAN FRANCISCO • SINGAPORE • SYDNEY • TOKYO
Woodhead Publishing is an imprint of Elsevier

WP WOODHEAD PUBLISHING

Woodhead Publishing is an imprint of Elsevier
The Officers' Mess Business Centre, Royston Road, Duxford, CB22 4QH, UK
50 Hampshire Street, 5th Floor, Cambridge, MA 02139, USA
The Boulevard, Langford Lane, Kidlington, OX5 1GB, UK

British Library Cataloguing-in-Publication Data
A catalogue record for this book is available from the British Library.

Library of Congress Cataloging-in-Publication Data
A catalog record for this book is available from the Library of Congress.

ISBN: 978-0-08-100307-7 (print)
ISBN: 978-0-08-100333-6 (online)

For information on all Woodhead Publishing visit our website at https://www.elsevier.com/

Publisher: Joe Hayton
Acquisition Editor: Alex White
Editorial Project Manager: Alex White
Production Project Manager: Omer Mukthar
Designer: Mark Rogers

Typeset by MPS Limited, Chennai, India

Contents

List of contributors xi
Woodhead Publishing Series in Energy xiii

Part I Geology, Resources and Markets: Primary Uranium 1

1 Uranium for nuclear power: an introduction 3
Ian Hore-Lacy
1.1 Introduction and history 3
1.2 Energy density, other characteristics 4
1.3 Resource situation 4
1.4 Technological perspective 5
1.5 The electromobility frontier and methanol 6
1.6 Relationship with nonhydro renewables 7
1.7 Safety, regulation 8
1.8 Nonpower uses 9
1.9 Wastes, radiation, proliferation 9
1.10 Uranium in the future 10
1.11 Further information 10

2 Geology of uranium deposits 11
Patrice Bruneton and Michel Cuney
2.1 Introduction 11
2.2 Uranium provinces 11
2.3 Uranium minerals 13
2.4 Classification of uranium deposits 17
2.5 The IAEA UDEPO database 33
2.6 Unconventional resources and deposits in UDEPO 34
2.7 New uranium deposits (2004–2014) 38
2.8 Future trends 44
Acknowledgments 46
Books on uranium and general publications 46
Publications in scientific reviews 47
Main websites 48
References 48

3 Exploration for uranium **53**
Kurt Kyser
3.1 Introduction 53
3.2 Uranium and the nuclear fuel cycle 54
3.3 Drivers of uranium exploration 55
3.4 Prospectivity, explorability, and exploration targeting 57
3.5 Exploration techniques 60
3.6 Critical factors in deposit models 68
3.7 Drilling and evaluation 71
3.8 Synopsis 72
References 72

4 Uranium resources **77**
Robert Vance
4.1 Introduction 77
4.2 Resources and reserves 78
4.3 Global uranium resources as of January 1, 2013 85
4.4 Unconventional resources 91
4.5 International fuel reserves 92
4.6 United Nations Framework Classification 94
4.7 Conclusion 95
4.8 Sources for further information and advice 96
References 96

5 The uranium market, supply adjustments from secondary sources, and enrichment underfeeding **99**
Steve Kidd
5.1 Introduction 99
5.2 Uranium demand 99
5.3 Uranium supply 106
5.4 Costs of production 112
5.5 Uranium supply and demand to 2020 and beyond 114
5.6 The world uranium market and prices 118
5.7 Sources of further information and advice 121
References 121

Part II Mining and Alternative Fuel Sources **123**

6 Uranium mining (open cut and underground) and milling **125**
Peter H. Woods
6.1 Introduction 125
6.2 Open cut mining 126
6.3 Underground mining 130
6.4 Milling and extraction 135

6.5 Retreatment of mine tailings 145
6.6 Future trends 146
6.7 Sources of further information and advice 148
Acknowledgments 150
References 150

7 Introduction to uranium in situ recovery technology 157
Mark S. Pelizza and Craig S. Bartels
7.1 General description 157
7.2 Geology and hydrogeologic conditions 170
7.3 Environmental evaluations 180
7.4 Facilities 182
7.5 Processes 192
7.6 Waste management 201
7.7 Well drilling, installation, completion, and operation 203
7.8 Reclamation 209
References 210

8 Nuclear fuel from secondary supplies of uranium and plutonium 215
Jeff Combs
8.1 Introduction 215
8.2 Commercial inventories of natural and enriched uranium 218
8.3 Other natural and enriched inventories 222
8.4 HEU supplies 225
8.5 Recycled uranium and plutonium 228
8.6 Future of recycled uranium and plutonium 229
8.7 Enrichment of tails material 230
8.8 Underfeeding 234
8.9 Implications of generation IV nuclear reactors 235
8.10 Summary: The declining but continuing role of secondary supplies 236

9 Production of byproduct uranium and uranium from unconventional resources 239
Ian Hore-Lacy
9.1 Introduction 239
9.2 Uranium from phosphate deposits 239
9.3 Processing phosphates 241
9.4 Production potential 242
9.5 Uranium from rare earths deposits 243
9.6 Uranium from black shales 247
9.7 Uranium from seawater 249
9.8 Uranium from gold tailings 249
9.9 Uranium from coal ash 250
Sources and references 251

10 Thorium as a nuclear fuel **253**
Bradley S. Van Gosen and Harikrishnan Tulsidas
10.1 Introduction 253
10.2 Thorium fuel cycle 254
10.3 Previous work on the thorium fuel cycle 258
10.4 Current research and future possibilities 267
10.5 Thorium geology and resources 273
10.6 Current and recent thorium recovery 289
10.7 Disclaimer 290
References 290

Part III Conversion, Enrichment and Fuel Fabrication **297**

11 Conversion of natural uranium **299**
Andrew J. Oliver and Engin Özberk
11.1 Introduction 299
11.2 Conversion and needs 299
11.3 Conversion technologies 300
11.4 Current status of the conversion industry 314
11.5 Factors that impact converters 315
11.6 Conversion to UO_2 316
11.7 Potential future developments 318
References 318

12 Uranium enrichment **321**
Paul Harding
12.1 Introduction 321
12.2 How is uranium enriched? 324
12.3 Historic technologies 337
12.4 Future technologies 339
12.5 Quality control of uranium hexafluoride in enrichment 344
12.6 Management of tails 346
12.7 Experience enriching reprocessed uranium 349
References 351

13 Nuclear fuel fabrication **353**
Eileen Supko
13.1 Introduction 353
13.2 Fuel assembly design components 367
13.3 Current and future trends 374
13.4 Sources of further information and advice 378
References 379

Part IV Environmental and Social Issues **383**

14 Management for health, safety, environment, and community in uranium mining and processing **385**
Frank Harris
14.1 Framework for health, safety, environment, and community in uranium operation 385
14.2 HSEC across the uranium life cycle 385
14.3 Managing health in uranium operations 387
14.4 Managing safety in uranium operations 394
14.5 Managing environmental impacts in uranium operations 396
14.6 Managing community impacts in uranium operations 401
14.7 Integration of HSEC and conclusion 404

15 Safe and secure packaging and transport of uranium materials **405**
Dylan Hem
15.1 Introduction 405
15.2 Class 7 dangerous goods and regulatory practice/codes 405
15.3 Packaging of radioactive materials 406
15.4 Security measures for transport of nuclear materials 408
15.5 Current issues for the transport of nuclear materials 409
15.6 Transport of uranium concentrates and uranium hexafluoride 410
15.7 Transport and packaging of plutonium 411
15.8 Transport of fabricated fuel—uranium and MOX 412
15.9 Ships for MOX (and used fuel or high-level wastes) 412
15.10 Sources of further information 413
References 414

16 Uranium mine and mill remediation and reclamation **415**
Doug Collier
16.1 Introduction 415
16.2 Uranium mine and milling sites 419
16.3 Site remediation and reclamation 423
16.4 Risks addressed in rehabilitation 424
16.5 Tools used in rehabilitation 427
16.6 Funding for rehabilitation 432
16.7 Future trends 432
16.8 Conclusions 434
References 435

17 Interregional technical cooperation for education and training in the uranium industry **439**
Patricia Wieland
17.1 Introduction 439

17.2 Relevant interregional initiatives in education and training 440
17.3 Building competence in the uranium industry 445
17.4 Future trends 448
References 451

Index **453**

List of contributors

Craig S. Bartels Heathgate Resources Pty Ltd, Adelaide, SA, Australia

Patrice Bruneton Le Chalard, France

Doug Collier Australian Nuclear Science and Technology Organisation—ANSTO, Sydney, New South Wales, Australia

Jeff Combs The Ux Consulting Company, LLC, Roswell, GA, United States

Michel Cuney Université de Lorraine, GeoRessources-Centre National de la Recherche Scientifique-Centre de recherche sur la géologie des matières premières minérales et énergétiques, Vandoeuvre-lès-Nancy, France

Paul Harding URENCO Limited, Stoke Poges, United Kingdom

Frank Harris Rio Tinto Uranium, Brisbane, Queensland, Australia

Dylan Hem RMIT University, Melbourne, VIC, Australia

Ian Hore-Lacy World Nuclear Association, London, United Kingdom

Steve Kidd East Cliff Consulting, Bournemouth, United Kingdom

Kurt Kyser Department of Geological Sciences and Geological Engineering, Queen's University, Kingston, ON, Canada

Andrew J. Oliver Retired Cameco Corporation, Port Perry, ON, Canada

Engin Özberk Retired Cameco Corporation, Saskatoon, SK, Canada

Mark S. Pelizza M.S. Pelizza & Associates LLC, Plano, TX, United States

Eileen Supko Energy Resources International, Washington, DC, United States

Harikrishnan Tulsidas International Atomic Energy Agency, Vienna, Austria

Bradley S. Van Gosen US Geological Survey, Denver, CO, United States

Robert Vance Nuclear Energy Agency (NEA), Issy-les-Moulineaux, France

Patricia Wieland World Nuclear University, London, United Kingdom

Peter H. Woods International Atomic Energy Agency, Vienna, Austria

Woodhead Publishing Series in Energy

1. **Generating power at high efficiency: Combined cycle technology for sustainable energy production**
   Edited by Eric Jeffs
2. **Advanced separation techniques for nuclear fuel reprocessing and radioactive waste treatment**
   Edited by Kenneth L. Nash and Gregg J. Lumetta
3. **Bioalcohol production: Biochemical conversion of lignocellulosic biomass**
   Edited by Keith W. Waldron
4. **Understanding and mitigating ageing in nuclear power plants: Materials and operational aspects of plant life management (PLiM)**
   Edited by Philip G. Tipping
5. **Advanced power plant materials, design and technology**
   Edited by Dermot Roddy
6. **Stand-alone and hybrid wind energy systems: Technology, energy storage and applications**
   Edited by John K. Kaldellis
7. **Biodiesel science and technology: From soil to oil**
   Edited by Jan C. J. Bart, Natale Palmeri and Stefano Cavallaro
8. **Developments and innovation in carbon dioxide (CO_2) capture and storage technology Volume 1: Carbon dioxide (CO_2) capture, transport and industrial applications**
   Edited by M. Mercedes Maroto-Valer
9. **Geological repository systems for safe disposal of spent nuclear fuels and radioactive waste**
   Edited by Joonhong Ahn and Michael J. Apted
10. **Wind energy systems: Optimising design and construction for safe and reliable operation**
    Edited by John D. Sørensen and Jens N. Sørensen
11. **Solid oxide fuel cell technology: Principles, performance and operations**
    Edited by Kevin Huang and John Bannister Goodenough

12. **Handbook of advanced radioactive waste conditioning technologies**
Edited by Michael I. Ojovan

13. **Membranes for clean and renewable power applications**
Edited by Annarosa Gugliuzza and Angelo Basile

14. **Materials for energy efficiency and thermal comfort in buildings**
Edited by Matthew R. Hall

15. **Handbook of biofuels production: Processes and technologies**
Edited by Rafael Luque, Juan Campelo and James Clark

16. **Developments and innovation in carbon dioxide (CO_2) capture and storage technology Volume 2: Carbon dioxide (CO_2) storage and utilisation**
Edited by M. Mercedes Maroto-Valer

17. **Oxy-fuel combustion for power generation and carbon dioxide (CO_2) capture**
Edited by Ligang Zheng

18. **Small and micro combined heat and power (CHP) systems: Advanced design, performance, materials and applications**
Edited by Robert Beith

19. **Advances in clean hydrocarbon fuel processing: Science and technology**
Edited by M. Rashid Khan

20. **Modern gas turbine systems: High efficiency, low emission, fuel flexible power generation**
Edited by Peter Jansohn

21. **Concentrating solar power technology: Principles, developments and applications**
Edited by Keith Lovegrove and Wes Stein

22. **Nuclear corrosion science and engineering**
Edited by Damien Féron

23. **Power plant life management and performance improvement**
Edited by John E. Oakey

24. **Electrical drives for direct drive renewable energy systems**
Edited by Markus Mueller and Henk Polinder

25. **Advanced membrane science and technology for sustainable energy and environmental applications**
Edited by Angelo Basile and Suzana Pereira Nunes

26. **Irradiation embrittlement of reactor pressure vessels (RPVs) in nuclear power plants**
Edited by Naoki Soneda

27. **High temperature superconductors (HTS) for energy applications**
Edited by Ziad Melhem

28. **Infrastructure and methodologies for the justification of nuclear power programmes**
Edited by Agustín Alonso

29. **Waste to energy conversion technology**
Edited by Naomi B. Klinghoffer and Marco J. Castaldi

30. **Polymer electrolyte membrane and direct methanol fuel cell technology Volume 1: Fundamentals and performance of low temperature fuel cells**
Edited by Christoph Hartnig and Christina Roth

31. **Polymer electrolyte membrane and direct methanol fuel cell technology Volume 2: *In situ* characterization techniques for low temperature fuel cells**
Edited by Christoph Hartnig and Christina Roth

32. **Combined cycle systems for near-zero emission power generation**
Edited by Ashok D. Rao

33. **Modern earth buildings: Materials, engineering, construction and applications**
Edited by Matthew R. Hall, Rick Lindsay and Meror Krayenhoff

34. **Metropolitan sustainability: Understanding and improving the urban environment**
Edited by Frank Zeman

35. **Functional materials for sustainable energy applications**
Edited by John A. Kilner, Stephen J. Skinner, Stuart J. C. Irvine and Peter P. Edwards

36. **Nuclear decommissioning: Planning, execution and international experience**
Edited by Michele Laraia

37. **Nuclear fuel cycle science and engineering**
Edited by Ian Crossland

38. **Electricity transmission, distribution and storage systems**
Edited by Ziad Melhem

39. **Advances in biodiesel production: Processes and technologies**
Edited by Rafael Luque and Juan A. Melero

40. **Biomass combustion science, technology and engineering**
Edited by Lasse Rosendahl

41. **Ultra-supercritical coal power plants: Materials, technologies and optimisation**
Edited by Dongke Zhang

42. **Radionuclide behaviour in the natural environment: Science, implications and lessons for the nuclear industry**
Edited by Christophe Poinssot and Horst Geckeis

43. **Calcium and chemical looping technology for power generation and carbon dioxide (CO_2) capture: Solid oxygen- and CO_2-carriers**
Edited by Paul Fennell and E. J. Anthony

44. **Materials' ageing and degradation in light water reactors: Mechanisms, and management**
Edited by K. L. Murty

45. **Structural alloys for power plants: Operational challenges and high-temperature materials**
Edited by Amir Shirzadi and Susan Jackson

46. **Biolubricants: Science and technology**
Edited by Jan C. J. Bart, Emanuele Gucciardi and Stefano Cavallaro

47. **Advances in wind turbine blade design and materials**
Edited by Povl Brøndsted and Rogier P. L. Nijssen

48. **Radioactive waste management and contaminated site clean-up: Processes, technologies and international experience**
Edited by William E. Lee, Michael I. Ojovan, Carol M. Jantzen

49. **Probabilistic safety assessment for optimum nuclear power plant life management (PLiM): Theory and application of reliability analysis methods for major power plant components**
Edited by Gennadij V. Arkadov, Alexander F. Getman and Andrei N. Rodionov

50. **The coal handbook: Towards cleaner production Volume 1: Coal production**
Edited by Dave Osborne

51. **The coal handbook: Towards cleaner production Volume 2: Coal utilisation**
Edited by Dave Osborne

52. **The biogas handbook: Science, production and applications**
Edited by Arthur Wellinger, Jerry Murphy and David Baxter

53. **Advances in biorefineries: Biomass and waste supply chain exploitation**
Edited by Keith Waldron

54. **Geological storage of carbon dioxide (CO_2): Geoscience, technologies, environmental aspects and legal frameworks**
Edited by Jon Gluyas and Simon Mathias

55. **Handbook of membrane reactors Volume 1: Fundamental materials science, design and optimisation**
Edited by Angelo Basile

56. **Handbook of membrane reactors Volume 2: Reactor types and industrial applications**
Edited by Angelo Basile

57. **Alternative fuels and advanced vehicle technologies for improved environmental performance: Towards zero carbon transportation**
Edited by Richard Folkson

58. **Handbook of microalgal bioprocess engineering**
Edited by Christopher Lan and Bei Wang

59. **Fluidized bed technologies for near-zero emission combustion and gasification**
Edited by Fabrizio Scala

60. **Managing nuclear projects: A comprehensive management resource**
Edited by Jas Devgun

61. **Handbook of Process Integration (PI): Minimisation of energy and water use, waste and emissions**
Edited by Jiří J. Klemeš

62. **Coal power plant materials and life assessment**
Edited by Ahmed Shibli

63. **Advances in hydrogen production, storage and distribution**
Edited by Ahmed Basile and Adolfo Iulianelli

64. **Handbook of small modular nuclear reactors**
Edited by Mario D. Carelli and Dan T. Ingersoll

65. **Superconductors in the power grid: Materials and applications**
Edited by Christopher Rey

66. **Advances in thermal energy storage systems: Methods and applications**
Edited by Luisa F. Cabeza

67. **Advances in batteries for medium and large-scale energy storage**
Edited by Chris Menictas, Maria Skyllas-Kazacos and Tuti Mariana Lim

68. **Palladium membrane technology for hydrogen production, carbon capture and other applications**
Edited by Aggelos Doukelis, Kyriakos Panopoulos, Antonios Koumanakos and Emmanouil Kakaras

69. **Gasification for synthetic fuel production: Fundamentals, processes and applications**
Edited by Rafael Luque and James G. Speight

70. **Renewable heating and cooling: Technologies and applications**
Edited by Gerhard Stryi-Hipp

71. **Environmental remediation and restoration of contaminated nuclear and NORM sites**
Edited by Leo van Velzen

72. **Eco-friendly innovation in electricity networks**
Edited by Jean-Luc Bessede

73. **The 2011 Fukushima nuclear power plant accident: How and why it happened**
Edited by Yotaro Hatamura, Seiji Abe, Masao Fuchigami and Naoto Kasahara. Translated by Kenji Iino

74. **Lignocellulose biorefinery engineering: Principles and applications**
Edited by Hongzhang Chen

75. **Advances in membrane technologies for water treatment: Materials, processes and applications**
Edited by Angelo Basile, Alfredo Cassano and Navin Rastogi

76. **Membrane reactors for energy applications and basic chemical production**
Edited by Angelo Basile, Luisa Di Paola, Faisal Hai and Vincenzo Piemonte

77. **Pervaporation, vapour permeation and membrane distillation: Principles and applications**
Edited by Angelo Basile, Alberto Figoli and Mohamed Khayet

78. **Safe and secure transport and storage of radioactive materials**
Edited by Ken Sorenson

79. **Reprocessing and recycling of spent nuclear fuel**
Edited by Robin Taylor

80. **Advances in battery technologies for electric vehicles**
Edited by Bruno Scrosati, Jürgen Garche and Werner Tillmetz

81. **Rechargeable lithium batteries: From fundamentals to applications**
Edited by Alejandro A. Franco

82. **Calcium and chemical looping technology for power generation and carbon dioxide (CO_2) capture**
Edited by Paul Fennell and Ben Anthony

83. **Compendium of hydrogen energy Volume 1: Hydrogen production and purificiation**
Edited by Velu Subramani, Angelo Basile and T. Nejat Veziroglu

84. **Compendium of hydrogen energy Volume 2: Hydrogen storage, transmission, transportation and infrastructure**
Edited by Ram Gupta, Angelo Basile and T. Nejat Veziroglu

85. **Compendium of hydrogen energy Volume 3: Hydrogen energy conversion**
Edited by Frano Barbir, Angelo Basile and T. Nejat Veziroglu

86. **Compendium of hydrogen energy Volume 4: Hydrogen use, safety and the hydrogen economy**
Edited by Michael Ball, Angelo Basile and T. Nejat Veziroglu

87. **Advanced district heating and cooling (DHC) systems**
Edited by Robin Wiltshire

88. **Microbial electrochemical and fuel cells: Fundamentals and applications**
Edited by Keith Scott and Eileen Hao Yu

89. **Renewable heating and cooling: Technologies and applications**
Edited by Gerhard Stryi-Hipp

90. **Small modular reactors: Nuclear power fad or future?**
Edited by Daniel T. Ingersoll

91. **Fuel flexible energy generation: Solid, liquid and gaseous fuels**
Edited by John Oakey

92. **Offshore wind farms: Technologies, design and operation**
Edited by Chong Ng & Li Ran

93. **Uranium for nuclear power: Resources, mining and transformation to fuel**
Edited by Ian Hore-Lacy

94. **Biomass supply chains for bioenergy and biorefining**
Edited by Jens Bo Holm-Nielsen and Ehiaze Augustine Ehimen

95. **Sustainable energy from salinity gradients**
Edited by Andrea Cipollina and Giorgio Micale

96. **Membrane technologies for biorefining**
Edited by Alberto Figoli, Alfredo Cassano and Angelo Basile

Part I

Geology, Resources and Markets: Primary Uranium

Uranium for nuclear power: an introduction

1

Ian Hore-Lacy
World Nuclear Association, London, United Kingdom

1.1 Introduction and history

Compared with other mineral commodities, especially metals, whose utility has become evident by centuries of trial and error, the appreciation of uranium has developed from theories based in physics in the 1930s to exploit the unique energy density of uranium's transformation in nuclear fission. Initially this was in the crucible of a world war, but always beyond military uses was the promise of the "uranium boiler" canvassed in the second British MAUD (Military Application of Uranium Detonation) report in July 1941—the work of "one of the most effective scientific committees that ever existed."

Less than eight years after atomic bombs demonstrated the immense potential of uranium, the first precursor of today's power reactors and numerous naval reactors had started up in Idaho, and a year later, electricity was generated in Russia. The focus was now on safe, controlled, long-lasting, and economical machines to harness nuclear fission for reliable electricity supplies. The focus has remained there, with over 500 civil nuclear reactors notching up more than 16,000 reactor-years of operation to 2015 with remarkably few accidents—and even those accidents had far less adverse effects than feared.

Today nuclear power has a unique position in relation to national energy policies as the only well-proven technology able to be deployed anywhere that can provide continuous reliable supply of electricity on a large scale and without nearly any CO_2 emissions or air pollution. It is widely agreed that energy generally, and electricity in particular, must increasingly be produced with much lower carbon dioxide emissions than hitherto. And as one-third of the world's population aspires to enjoy the benefits of electricity that they have so far missed out on, the question of affordability looms larger than in the West, where it is by no means insignificant as a cost of living and an input to production, which must be competitive. Nuclear plants operate at low cost, and make electricity very affordable relative to any other low-carbon source.

Reliability is a key attribute of nuclear generation, and nuclear plants typically operate at near full capacity 24 h per day and year-round with only a pause for refueling every 18–24 months. This operation is irrespective of weather or season.

However, nuclear power is capital-intensive and this affects its ability to compete in liberalized electricity markets, particularly in competition with subsidized renewables and cheap gas, as elaborated next. Some long-term assurance of

Uranium for Nuclear Power. DOI: http://dx.doi.org/10.1016/B978-0-08-100307-7.00001-6

electricity sales at competitive prices with other sources apart from any subsidies on those is required (and without any subsidy beyond what is required to counter market distortions due to those sources).

But the clear message from practically every international authority and their reports is that nuclear power is essential for meeting the world's growing need for affordable, clean, and reliable electricity. There is no credible reason to not greatly increase its role in world electricity production, and for industrial heat including desalination.

1.2 Energy density, other characteristics

The single most remarkable characteristic of uranium is its energy density: the amount of energy that a single kilogram can yield. Even so, exactly how much energy depends on the technology used to liberate it. One fuel pellet the size of a fingertip produces as much energy as one tonne of coal, even in the least efficient reactor.

"Natural uranium" is that which is found in nature—mostly in the Earth's crust in a variety of geological environments. In most nuclear reactors, only about half of 1% of this is actually used, but even so it yields about 500 GJ/kg, about 20,000 times as much as black coal. In a fast neutron reactor, about 60 times this is achievable, and one day potentially more if technology and economics were pressed. But, in fact, uranium is fairly common and not a high-priced commodity—it has been less than $100/kg in recent years—so there is little incentive as of yet to push those boundaries from a resource perspective.

Like most other elements, uranium occurs as a mixture of isotopes, but with uranium, that fact is central to its use. Only one of the natural isotopes is directly usable, and that comprises only 0.7% of natural uranium. Hence, either power plants need to be designed accordingly, or the uranium needs to be enriched in that minor isotope, which is the subject of Chapters 11 and 12. In fact, the latter course of action accounts for 88% of the world's nuclear power reactors (and all naval reactors).

Also, like most metals, uranium occurs in a variety of chemical forms, though most is as a mixed oxide of UO_2 and UO_3, characteristically U_3O_8. Chapter 2 offers a deeper discussion.

The concentration of uranium in its geological settings can range up to about 20% in some Canadian deposits, though 2%U is generally called a high-grade ore. Geological occurrences where it is defined as "ore" (economically recoverable) range down to about 0.01%U.

1.3 Resource situation

Uranium is approximately as common in the Earth's crust as tin and zinc, and occurs in most rocks. Granites typically have up to 5 ppm U (which incidentally and at higher levels provides the heat for geothermal energy). Seawater contains a vast amount at 0.003 ppm U, which is recoverable, but not economical.

Our knowledge of what uranium (or anything else) is in the Earth's crust arises from mineral exploration activities (see Chapter 3), which are expensive and mostly undertaken by mining companies that have negotiated the right to mine what they find and quantify. Therefore, any figures published based on this refer only to known resources. Beyond geological theorizing, we have little idea of what there is beyond this.

The world's known resources related to cumulative exploration expenditure are shown in figure.

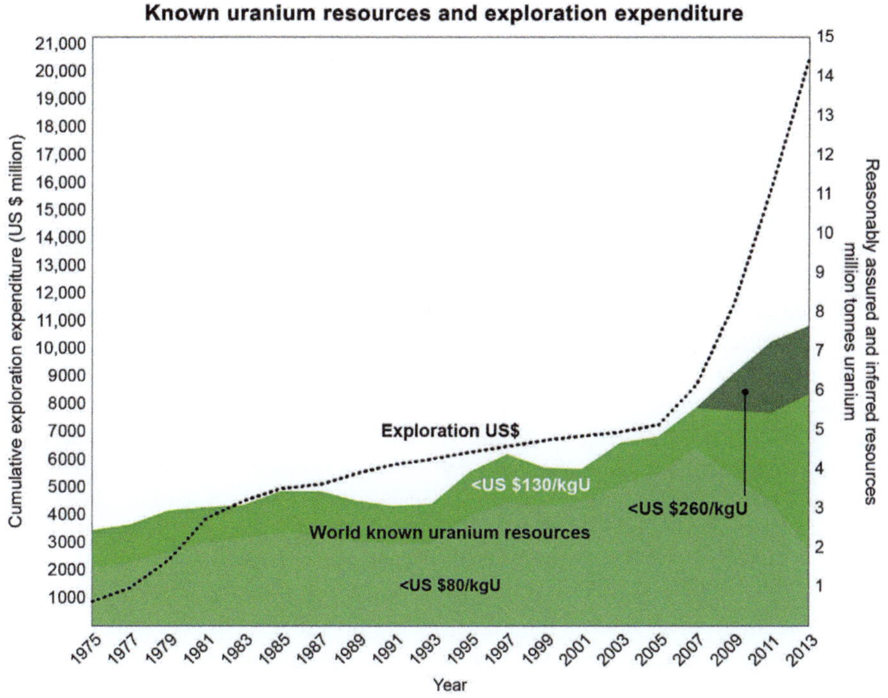

It can be seen how increased exploration leads to increased known resources. In 2013, there were 5.9 million tonnes of uranium known and recoverable at up to \$130/kgU, and a bit more at double that. The graph shows only historical figures, without allowing for inflation. Chapter 4 provides more detail, and Chapter 5 looks at the complexities of the uranium market.

1.4 Technological perspective

Mention has been made of reactor technologies to liberate uranium's energy in a controlled fashion. Prior to discussing these, it is necessary to outline a little basic physics. The objective in any nuclear reactor is to achieve criticality, which results in an ongoing chain reaction of nuclear fission, essentially of the uranium-235 (^{235}U) isotope's nucleus. This means that each ^{235}U atomic nucleus splits into two parts (fission products) and releases two or three neutrons in the process. These are "fast" neutrons with high energy. But to keep the chain reaction going, with one nuclear fission producing one more nuclear fission, the neutrons need to be slowed down to about one-ten thousandth of their velocity, by a moderator.

The most efficient moderators are graphite and heavy water. With a reactor employing these, natural uranium can be used as fuel. However, for cost and technical reasons, these are not usually preferred, and the main moderator used (in 88% of the world's power reactors and all naval ones) is ordinary "light" water, which doubles as coolant. But this means that the fuel needs to be enriched to at least 3% ^{235}U, and typically today is enriched to 4–5%. (Naval reactors use much higher enrichment to give a more compact unit.) So uranium enrichment plants are a major feature of the world's nuclear fuel cycle, as will be explained in Chapter 12.

It is also possible to have a reactor running with fast neutrons, though these are a little more complex, and bring plutonium more fully into the picture. Most generation IV nuclear reactors expected to be deployed from about 2030 are fast reactors.

Although a normal nuclear reactor is usually loaded with only uranium fuel, in fact about one-third of its energy output comes from plutonium. How so? Well, 99% of natural uranium is ^{238}U, which is said to be "fertile" rather than "fissile," and in a reactor, it captures some of those spare neutrons to become plutonium, principally ^{239}Pu. ^{239}Pu fissions in the same way as ^{235}U.

In a fast neutron reactor, plutonium has a greater role, and there is no moderator. Cooling the core requires a heat transfer fluid with minimal moderation, hence liquid metals such as sodium are used, and perhaps fluoride salts in the future.

To generate electricity, the physics described is applied in a plant designed to achieve high temperatures in the core to make steam that drive turbines. A coolant, most commonly water under considerable pressure, transfers the heat from fission in the core at a little over 300 °C to make steam either above the core or in separate steam generators. Over 400 of such reactors are operable for power generation, and another 180 or so smaller ones are used for naval propulsion.

About 240 "research" reactors are designed simply as neutron factories, and these are much smaller and operate at much lower temperatures. The neutrons are beamed out from the core for research, production of medical and industrial isotopes, or other purposes.

Since the 1940s, thorium has been considered as a possible source of nuclear fuel. Although it has no fissile isotope, it does produce fissile ^{233}U by neutron capture in a nuclear reactor, and this is an attractive source of energy, like ^{235}U and ^{239}Pu. However, some technical challenges remain for the time being and Chapter 10 describes the situation.

1.5 The electromobility frontier and methanol

While France has shown the world that well over half of the electricity for a major industrial country can economically be generated by nuclear power, a high proportion of electricity from nuclear plants in a system creates the need for either load-following to reduce output overnight, or some other application to compensate for temporary low system demand (in the absence of large-scale storage).

Increasing world interest in battery-powered cars and light commercial vehicles is helpful here, if batteries can be charged off-peak and overnight. This will have the effect of increasing the proportion of total electricity that can be generated by base-load plants, that is, on a continuous 24/7 basis, which will in most circumstances reduce the unit (per kilowatt-hour) cost of all electricity. If those base-load plants are nuclear rather than coal-fired, then automotive as well as the electricity usage is virtually emission-free.

Estimates vary concerning the potential effect of this electromobility on electricity demand, but a 20% increase seems a conservative estimate if there is widespread change from petrol/gasoline and diesel power to electric.

Another way nuclear power may become important for motor vehicles is making hydrogen which is used for converting carbon dioxide into methanol. This has about two thirds the energy density of gasoline and utilises present internal combustion engines. Methanol can be dehydrated to dimethyl ether (DME), a good diesel fuel.

1.6 Relationship with nonhydro renewables

Government policies in many countries demand that nonhydro renewables, mostly wind and solar power, should supply a significant proportion of electricity. Economic incentives are applied to achieve this, regardless of actual generating costs from those sources, and regardless of how well the supply from those sources over any 24 h may fit the normal demand pattern that it is vital for transmission operators to fully and reliably maintain supply.

The economic characteristics of any generating plant with high capital costs and low running costs—such as nuclear—mean that it is best run continuously, leaving other (eg, gas-fired) plants with lower capital costs and higher running costs to operate for only part of each day, picking up peak loads. If a significant amount of electricity from intermittent and unpredictable wind and solar sources is superimposed on this, so that output from other plants is curtailed, then costs of power from those plants increase. If this cost increase is added to the subsidies and other support for those disruptive renewables, then the effect on the ultimate consumer can be great. Of particular relevance here is that the economic virtues of nuclear power are not properly exploited.

But apart from economics, any major input of intermittent and unpredictable renewables means that major challenges arise for transmission operators to keep the grid stable and reliable, a role for which nuclear power has long been appreciated.

1.7 Safety, regulation

From the outset in mid-20th century, the energy density of nuclear power has made safety a high priority. This is achieved by engineering, both in the quality of containment and the physics of the reactor systems.[1] A considerable part of the engineering in any nuclear power plant is back-up provision in design and safety equipment, which has only ever been fully called upon twice in the history of nuclear power.[2]

To achieve optimum safety, nuclear plants[3] operate using a "defense-in-depth" approach, with multiple safety systems supplementing the natural features of the reactor core. Key aspects of the defense-in-depth approach include the following.

- High-quality design and construction
- Equipment that prevents operational disturbances or human failures and errors developing into problems
- Comprehensive monitoring and regular testing to detect equipment or operator failures
- Redundant and diverse systems to control damage to the fuel and prevent significant radioactive releases
- Provision to confine the effects of severe fuel damage (or any other problem) to the plant itself

These steps can be summed up as prevention, monitoring, and action (to mitigate consequences of failures).

Looked at functionally, the three basic safety functions in a nuclear reactor are to control reactivity, to cool the fuel, and to contain radioactive substances.

The main safety features of most reactors are inherent—negative temperature coefficient and negative void coefficient. The first means that beyond an optimal level, as the temperature increases, the efficiency of the reaction decreases (this in fact is used to control power levels in some new designs). The second means that if any steam has formed in the cooling water, there is a decrease in moderating effect so that fewer neutrons are able to cause fission and the reaction slows down automatically.

Regulation of nuclear facilities is undertaken nationally, but with a great deal of international collaboration. Reactor designs are subject to close scrutiny, as is construction and then operation.

Nuclear power is safe, with the best record of any major form of electricity generation.

[1] For instance, if water boils in the core of a light water reactor, moderation is reduced and the chain reaction slows. Or in a fast or high-temperature reactor, very high core temperatures slow the chain reaction.

[2] These two times are at Three Mile Island in 1979 and Fukushima Daiichi in 2011. No one was killed or seriously harmed in either accident. Chernobyl had a less-than-standard safety provision, and perhaps 50 deaths were attributed to the accident directly and indirectly.

[3] Once one might have added "in the Western world," but today high standards prevail in both the east and west.

1.8 Nonpower uses

Apart from charging batteries for cars and light commercial vehicles referred to previously, which have a potentially significant effect on how nuclear or other base-load provisions fit into the whole electricity demand picture, there are a number of other uses for nuclear energy fueled by uranium.

The most obvious of these is desalination of seawater or brackish groundwater to produce potable water for cities. This may use either electricity (ideally off-peak) or waste heat from electricity generation. Most desalination today uses fossil fuels, and thus contributes to increased levels of greenhouse gases. Total world capacity in 2013 was 80 million m^3/day (29,200 GL/year) of potable water in over 17,000 plants. A majority of these are in the Middle East and north Africa. Two-thirds of the world capacity is processing seawater, and one-third uses brackish artesian water.

Direct industrial uses of nuclear heat are also possible. The potential application of nuclear heat depends mainly on the temperature required. Most of today's reactors produce heat at a little over 300 °C; fast reactors go to 550 °C. With reactor output temperatures of up to 700 °C, there is a wide range of possible applications: at 900 °C, there are further possibilities, and at 950 °C, an important future application to hydrogen production opens up. About 20% of US energy consumption goes into process heat applications, compared with 35–40% into electricity. In this 15–20%, replacing fossil fuels with nuclear heat promises much in energy security, price stability, and reduced emissions.

1.9 Wastes, radiation, proliferation

No source of energy is without issues that, if not managed properly, are negatives. Coal has been a massive benefit to humanity since the Industrial Revolution, but has killed many in mine accidents, occupational health effects, and environmental pollution. Today, with those issues mostly under control, it is in disrepute because of its CO_2 emissions.

Nuclear power has at least the same potential as coal to benefit humanity, but only if its wastes continue to be properly contained and managed, if ionizing radiation arising directly and indirectly from it causes no harm, and if it clearly does not increase the risk of nuclear weapons proliferation.

Waste from nuclear power, considering the whole fuel cycle from uranium mining onward, range from innocuous to very hazardous. They are managed and if necessary contained accordingly. The high-level wastes comprising either used fuel or the fission products and actinides separated from it, need both cooling and shielding. This is straightforward, and has been done for more than half a century with no harm to anyone and no environmental effects. Long-term disposal of this waste is also technically straightforward—deep geological disposal being the universal choice once the radioactivity has decayed to less than 1% of its original level.

Ionizing radiation at low levels is ubiquitous and experienced by all of us constantly, and the nuclear industry, under the watchful eye of regulators, ensures that does not add significantly to natural background levels for neighbors, or reach more than a fraction of harmful levels for workers. The Chernobyl accident is the one occasion when workers (and firefighters) received harmful radiation doses. (In contrast, radioactive sources and accidents outside the civil nuclear industry have led to significant harm on a few occasions.) Chapter 16 addresses occupational health aspects of it.

Civil nuclear power arose by new application of the technology developed for bombs in World War II. The military links continue, but all one way: in every country that has both nuclear weapons and nuclear power, the weapons came first, and more recently, substantial amounts of uranium in military stockpiles have been diluted and used for electricity generation—about 10% of US electricity from 1993 to 2013 came from Russian military stockpiles. See also Chapter 8.

But at the level of routine uranium supply, international accounting and auditing procedures called safeguards are applied to ensure that any diversion from civil to military application is detected and high-level diplomatic pressure can be applied as a deterrent. Similar provisions apply regarding the use of equipment, and recently Iran's uranium enrichment program has been the subject of sanctions applied by the UN Security Council. Safeguards are applied under the Nuclear Non-Proliferation Treaty of 1970, to which practically all countries subscribe.[4]

Nuclear technology for making weapons is in the public domain, and no country is without enough uranium to make a few bombs. The Nuclear Non-Proliferation Treaty provides strong disincentives against weapons proliferation by any country beyond the nine that now have nuclear weapons.

1.10 Uranium in the future

There are a number of authoritative projections from the Organisation for Economic Co-operation and Development (OECD) and UN sources concerning the future of energy supplies in general and electricity in particular. Practically all of these show a considerably increased role for nuclear energy in meeting human needs much more widely than today. Therefore, uranium will remain a vital energy source, even if by mid-century much of it is essentially recycled, drawn from the huge stockpiles of depleted uranium arising from a century's enrichment activity.

The assembled contributions comprising this book set the scene for a much greater role for uranium in the decades ahead. Readers are urged to look beyond these pages to what is happening in more than 30 countries that are implementing plans and making heavy investments in nuclear power plants that depend on the world's abundant uranium.

1.11 Further information

World Nuclear Association: Information Library http://www.world-nuclear.org/Information-Library/

[4] The exceptions are India, Pakistan, Israel, and North Korea. Since 2008, India has effectively come under safeguards. Pakistan has its civil nuclear facilities under safeguards. Israel and North Korea have no civil nuclear power programs.

Geology of uranium deposits

2

Patrice Bruneton[1,*] *and Michel Cuney*[2]
[1]Le Chalard, France, [2]Université de Lorraine, GeoRessources-Centre National de la Recherche Scientifique-Centre de recherche sur la géologie des matières premières minérales et énergétiques, Vandoeuvre-lès-Nancy, France

2.1 Introduction

Uranium deposits formed at nearly all steps of the geological cycle, from high-grade metamorphic conditions (up to 850°C and 5–7 kbar) for the uranothorianite pyroxenite deposits from the Tranomaro area in the southern part of Madagascar, to plutonic, metasomatic, hydrothermal, basinal diagenetic, metamorphic, and volcanic to sedimentary and superficial environments with the calcretes (Dahlkamp, 1993, 2009; Cuney, 2009, 2010, 2011) (Fig. 2.1).

They formed from Neoarchean to Quaternary times, but none is expected to exist before about 3.1 Ga. Large deposits exist on all continents. Today, the largest defined resources are located in Australia, and are dominated by the huge Olympic Dam deposit with resources of more than 2 million tonnes of uranium. The current world uranium resource inventory (Reasonably Assured + Inferred Resources) was estimated in 2012 at 7.5 MtU (OECD/NEA-IAEA, 2014). 1555 deposits, each with estimated resources over 300 tU are listed in the International Atomic Energy Association (IAEA) World Distribution of Uranium Deposits (UDEPO) database. Among those, three types contain more than three-fourths of the worldwide uranium resources: (1) Proterozoic unconformity-related deposits, (2) iron oxide–copper–gold (IOCG) deposits, and (3) sandstone-hosted deposits.

Important past or current uranium production also comes from a variety of additional deposit types such as paleo quartz–pebble conglomerates, granite-related, volcanic-related, intrusive anatectic metamorphic, and metasomatites. Other types present either smaller resources such as the surficial, breccia pipe, and carbonate deposits, or very large, but low-grade resources (unconventional resources of the IAEA) with more than 7.6 Mt U (OECD/NEA-IAEA, 2014), such as sedimentary phosphates and black shales. Uranium enrichment in coal and lignite represents only potential resources.

2.2 Uranium provinces

The global distribution of mineral deposits shows that some areas concentrate large resources, whereas others are almost devoid of any resource (Laznicka, 1999; Goldfarb et al., 2001; Hronsky and Groves, 2008). One of the earliest and best definitions of a uranium province is the one proposed by Klepper and Wyant (1956): "Economic uranium

* retired

Uranium for Nuclear Power. DOI: http://dx.doi.org/10.1016/B978-0-08-100307-7.00002-8

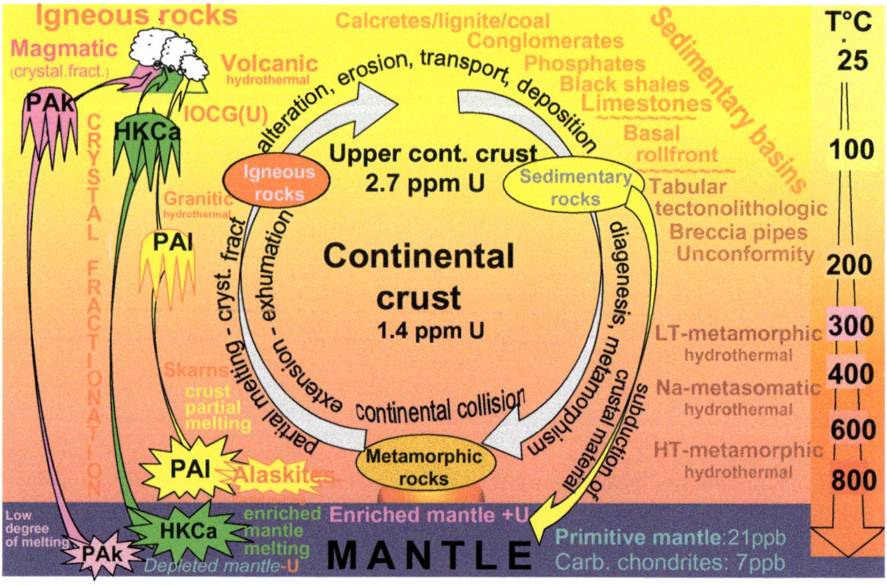

Figure 2.1 The extreme diversity of uranium deposits through the geological cycle.
Source: Modified from Cuney (2009, 2010, 2011).

deposits resulted from original inhomogeneities of uranium distribution in the Earth's crust that commonly persisted through long periods of time, and through a combination of orogenic, metamorphic, and sedimentary processes produced rocks with enriched uranium contents. The initial enriched uranium domain was successively remobilized and concentrated into new enrichments of one or more magnitudes above normal background forming uranium ore deposits."

Typically, a mature uranium province includes several types of uranium deposits formed at different periods of time as the result of extended uranium fractionation processes occurring during a complete geologic cycle or eventually during successive geological cycles and in which several types of uranium deposits can reach an economic value. In contrast, immature uranium provinces have a much shorter uranium fractionation cycle and mainly present one major type of uranium deposit, which may reach an economic to subeconomic value.

The geographic extent of a uranium province may vary greatly. Some provinces have a relatively limited geographic extension and may cover only a small part of a large country (such as the Athabasca–Thelon Uranium Province) or several small countries (such as the mid-European Uranium Province). However, some have a very large geographic extents and may cover large parts of continent and are called Super-Provinces (eg, the Columbia–Venezuela Phosphate Super-Province and the Karoo Basin Super-Province).

Many of the uranium provinces have very old histories, rooted in ancient geologic domains with progressive uranium fractionation starting during the Archean

Eon (such as the Witwatersrand Uranium Province in South Africa), or the Archean Eon and Paleoproterozoic Era (in the case of the Athabasca–Thelon Uranium Province in Canada and the Alligator Rivers Uranium Province in Australia). Others are young uranium provinces such as the Mid-European Uranium Province and the Central Asiatic Uranium Mega-Province, which have a history of uranium fractionation mainly starting during the Paleozoic Period.

Major world uranium provinces with their geological resources and production are listed in Table 2.1.

2.3 Uranium minerals

- Uranium in nature

Average uranium concentration in the continental crust is 1.7 ppm, reaching 2.7 ppm in its upper part. Uranium is typically enriched in the most silicic magmatic rocks such as granites and rhyolites (up to several tens to hundreds of ppm) and in sedimentary rocks rich in organic carbon (black shale and lignite) or phosphorus (phosphorites). Uranium concentration in seawater is 3.3 ppb, whereas in the fresh water of lakes and rivers, the concentration is typically only a few tenths of a part per billion.

Uranium occurs in rocks as its own minerals, as a substitute element in accessory minerals, adsorbed on Fe and Fe-Ti hydroxides-oxides and clay minerals adsorbed by organic matter, in crystal defects of major rock forming minerals, and dissolved in geologic fluids and fluid inclusions. Approximately 5% of all known minerals contain uranium as an essential structural constituent, which reflects the diversity of uranium minerals (Burns, 1999). For low uranium concentrations (1 to several ppm), most of the uranium is substituted in the structure of one or several accessory minerals such as apatite, zircon, monazite, xenotime, and titanite, which represent refractory uranium sources for mobilization by most types of geologic fluids. At higher uranium concentrations, it appears as a major constituent of some specific mineral species from which it can be more or less easily leached by geologic fluids and concentrated in deposits. Burns (1999), Finch and Murakami (1999), and Burns and Sigmond (2013) offer a comprehensive review of the structure and chemical formulas of uranium minerals.

Uranium minerals. Three categories are distinguished: (1) tetravalent U-Th minerals with minor other constituents, (2) hexavalent uranium minerals, and (3) Nb-Ta-Ti uranium oxides with U and Th as major constituents.

- *Tetravalent U-Th minerals* generally represent the U and/or Th ore minerals in most deposit types, but may also occur as accessory mineral phases disseminated in U-rich rocks. Uraninite (written as UO_2 but chemically UO_{2+x} (Janeczek and Ewing, 1992)) is the most common uranium ore mineral in most deposit types. It may occur as euhedral crystals generally with several weight percent ThO_2 and thousands of parts per million to several weight percent rare earth elements (REE), mostly in high-temperature deposits. It also occurs as colloform crystals with low REE contents and nearly no Th, but with

Table 2.1 Major world uranium provinces

Provinces	Deposit types	Geological resources UDEPO 2014 (tU)	Historical production to 2013 (tU)
South Mediterranean Super-Province (Morocco, Algeria, Tunisia, Libya, Egypt, Jordan, Syria, Iraq, Saudi Arabia, Israel, Turkey, Columbia, Venezuela)	Phosphate (Late Cretaceous, Early Tertiary)	> 7500.000	700
Northwest USA (Phosphoria Formation), Idaho, N. Utah, W. Wyoming, S.W. Montana	Phosphate (Permian)	7000.000	184
Northeast USA (Chattanooga Shale)	Black shale	5000.000	
Gawler Craton-Curnamona (Australia)	Polymetallic iron oxide breccia	2350.000	60.000
	Sandstone	68.000	7200
	Intrusive plutonic		
South/Central Sweden, Norway, Estonia	Black shale	1300.000	300
South-central Kazakhstan + Kyzylkum District (Uzbekistan)	Sandstone	840.000	172.000
		240.000	131.000
Florida (USA)	Phosphate (Miocene)	765.000	17.150
	Intrusive anatectic	720.000	92.000
Damara Belt (Namibia)	Metamorphite-skarn	5000	
	Surficial	205.000	8200
Thelon Basin–Athabasca Basin—(Canada)	Proterozoic unconformity	52.000	
	Proterozoic unconformity	880.000	308.000
	Intrusive anatectic		
Witwatersrand Basin (South Africa)	Paleo-QPCs	925.000	160.000
Tim Mersoi Basin (Niger/Algeria)	Sandstone	740.000	131.000
Alligator Rivers (Australia)	Proterozoic unconformity	450.000	106.000
Elkon District (Russia)	Metasomatite	345.000	
Grants Mineral Belt	Sandstone-Tabular	746.000	134.000
Colorado Plateau (USA)	Collapse breccia pipes	16.000	9000
Elliott Lake District (Canada)	QPC	465.000	140.000
Transbaikalia (Russia, Mongolia)	Volcanic-related	270.000	165.000
	Surficial paleovalley	100.000	1050
	Volcanic-related	78.000	535
Rocky Mountain and Intermontane Basins (USA)	Sandstone-roll front	255.000	87.000
	Granite-related	12.500	

Central Ukraine		Metasomatite	255.000	129.000
		Intrusive anatectic	15,000	
		Sandstone	14.000	
Karoo Formation (Botswana, Malawi, South Africa, Tanzania, Zambia)		Sandstone	340.500	3850
Kokshetau District (Kazakhstan)		Metamorphite	235.000	35.000
Illimaussaq (Denmark)		Intrusive plutonic	225.000	
Precaspian Region (Kazakhstan, Russia)		Phosphate	145.000	15.000
Itataia-Santa Quiteria (Brazil)		Metasomatite	122.000	
Mid-Europe Super-Province	France	Granite-related	70.000	65.000
	Germany	Granite-related	125.000	105.000
	Czech Republic	Granite-related	69.000	
	Spain, Portugal	Granite-related	65.000	9000
	Ronneburg District (Germany)	Black shale	200.000	115.000
	France, Germany, Czech Rep.	Sandstone		
	Spain	Phosphate	100.000	
Gulf Coast (USA-Mexico)		Sandstone-roll front	118.000	30.000
Lagoa Real (Brazil)		Metasomatite	101.000	1400
Cuddapah Basin (India)		Unconformity-related	19.000	
		Carbonate	100.000	
Western Australia		Surficial	100.000	
South-East China		Granite-related	110.000	n.a.
		Volcanic-related		
		Black shale		
North China		Sandstone		n.a.
Basin and Range (USA-Mexico)		Volcanic-related	14.000	
Singhbhum District (India)		Metamorphite	70.000	11.000
		Granite-related		
Hercynian–Andean Province (Peru, Bolivia, Argentina)		Metamorphite		
		Sandstone		
		Surficial		

n.a. = non available

several percent Ca, Si, and/or Zr, and is named pitchblende. When occurring as pulverulant microcrystals, this mineral is named "sooty pitchblende." Coffinite, $USiO_4$, is the second most common mineral in uranium ore deposits and occurs either as a primary ore mineral mostly in sandstone-hosted deposits or as a common alteration product of uraninite in most deposits. Uranium may also occur as ningyoite $(U,Ca,Ce)_2(PO_4)_2,1.5H_2O$) in some rollfront-type deposits and brannerite $(U,Ca,Ce)(Ti,Fe)_2O_6$ mainly in hydrothermal metasomatic deposits. More rarely, uranium occurs in uranothorianite $(Th,U)O_2$ in skarn deposits and uranothorite $(Th,U)SiO_4$ in deposits associated with peralkaline magmatism.

- *Nb-Ta-Ti oxides* form a wide family of minerals with the predominance of one or two of the cations Nb, Ta, and Ti and the presence of other cations such as REE′, Y, and Ca. Their uranium content is quite variable but may reach up to 18 wt% UO_2. The main minerals are uranmicrolite $(U,Ca,Ce)_2(\mathbf{Ta},Nb)_2O_6(OH,F)$, uranpyrochlore $(U,Ca,Ce)_2(\mathbf{Nb},Ta)_2O_6(OH,F)$, betafite $((Ca,Na,U)_2(\mathbf{Ti},Nb,Ta)_2O_6(OH,F))$, and euxenite $(Y,Ce,La,U)(Nb,Ti,Ta)_2(O,OH)_6$. They are commonly present in uranium deposits associated with peralkaline magmatism, but betafite may occur in intrusive anatectic deposits, as at Rössing. They represent refractory uranium sources and the cost of uranium extraction during ore processing is high.
- *Hexavalent uranium minerals* are highly colored and some fluoresce under ultraviolet light. More than 200 species exist. They can be deposited either as primary ore minerals such as carnotite $K_2(UO2)_2\ (VO4)_2,3H_2O$ in calcrete-type deposits, but more commonly, hexavalent uranium minerals represent alteration products of uranium oxides by oxidizing surface water or groundwater, mostly in the uppermost parts of deposits (Finch and Murakami, 1999) such as autunite $Ca(UO_2)_2(PO_4)_2,10H_2O$ and uranophane $Ca(UO_2)_2SiO_3(OH)_2,5H_2O$. For this reason, they are also commonly referred to as secondary uranium minerals.

Accessory minerals. These minerals occur in low abundance in most rocks and incorporate trivalent elements such as REE′ and Y; quadravalent elements such as U, Th, Zr, and Hf; and pentavalent elements such as Nb and Ta. They generally contain a few tens to several thousands ppm U, rarely up to several percent. They are highly concentrated in sedimentary placers and in fractionated peralkaline igneous rocks. A complete solid solution exists with Th^{4+} in uranothorianite and uranothorite may contain up to 30% UO_2 (Cuney and Friedrich, 1987). More limited substitution of U^{4+} occurs for Ca^{2+} in fluorapatite $(Ca)_5(SiO_4)_3(F,OH)$ and allanite $(Ce,Ca,Y,Th,U)_2(Al,Fe,Mg)_3(SiO_4)_3(OH)$, for Zr^{4+} (0.84 Å) in zircon $(Zr,U)SiO_4$, REE in rare earth fluorocarbonates such as bastnaesite, $(LREE,U)CO_3F$, and in phosphates such as monazite $(LREE, Th,U,Ca)[PO_4, SiO_4]$, and xenotime $(Y, HREE,U)PO_4$.

Adsorbtion. The adsorption capacity of uranium by some mineralsmay in some cases dominantly control its geochemistry, most particularly in the exogenous cycle. Iron, titanium oxides, and oxi-hydroxides are the most effective adsorbents of uranium under near-neutral pH conditions. The adsorption capacity of clayminerals for uranium is lower and varies with the surface charge increasing from kaolinite, to illite, to montmorillonite, and to interlayered clay minerals.

Associated with organic matter. The spatial connection between uranium and organic matter is often observed mostly in sedimentary but also in other rock types. It is related to the power of complexation and reduction of organic matter.

Uraniumin major rock forming minerals. This is only present at the ppb level because of its large ionic radius and high valency. When detected in higher concentrations, this reflects the presence of microinclusions of uranium-bearing accessory minerals or adsorption along cleavage planes or fractures.

Dissolved in geologic fluids and fluid inclusions. In most geologic fluids, uranium concentration is below ppb levels and reaches 3.3 ppb in seawater. Meteoric water in equilibrium with uranium oxides may reach 1–2 ppm U. The highest uranium concentrations, reaching several hundred ppm, have been measured in fluid inclusions corresponding to magmatic fluids and to fluids associated with the genesis of unconformity uranium deposits.

2.4 Classification of uranium deposits

Various global and regional classifications of uranium deposits have been published previously. These classifications generally follow two alternative approaches, focusing either on descriptive features of the mineralization, such as host rock type and orebody morphology (geological classification), or on genetic aspects (genetic classification). Other classifications considering the geotectonic position of deposits (tectono-lithologic classification) have also been published, in particular by Russian geologists for various regions of the world.

2.4.1 The IAEA geological classification of uranium deposits

In 2010, a working group was created by the IAEA to review the various existing classifications and propose a new or a modified classification that would be adopted internationally. The IAEA classification, used in particular in the 2012 Red Book (OECD/NEA-IAEA, 2012), dated back to 1993 and increasing availability of information about the deposits from Eastern countries, used extensively by Dahlkamp to establish his new classification (Dahlkamp, 2009), new publications and a great increase in company data following the increase of uranium prices starting in 2005 all provided a wealth of information on uranium deposit geology that was used to revise the classification.

The revised classification of uranium deposits was officially accepted in 2013 (Bruneton et al., 2014) and is presented in an IAEA report "Geological classification of uranium deposits and description of selected examples" with publication planned for 2016 (IAEA, 2016a). The revised classification is used in the 2014 version of the Red Book (OECD/NEA-IAEA, 2014).

Fifteen types of deposits have been retained in the new IAEA classification scheme. They are listed in order from deep primary magmatic deposits to sedimentary and surficial deposits following the geological cycle of Cuney (2010) illustrated in Fig. 2.1. The economic ranking used in the previous IAEA classifications has not been taken into account.

Most subtypes and classes defined by Dahlkamp (2009) have been retained with minor modifications and some new subtypes and classes have been created. Most deposit types are named by the host rocks except for types 3, 7, and 8 (which are related to structures) and type 5 (which is related to metasomatic alteration).

2.4.1.1 Detailed geological classification with types, subtypes, and classes

Type 1. Intrusive

1.1. Anatectic pegmatite-alaskite (Rossing, Namibia; Bancroft District, Canada)

1.2. Plutonic

Quartz monzonite (Bingham Canyon, USA; Chuquicamata, Chile)

Peralkaline complexes (Kvanefjeld, Greenland; Poços de Caldas, Brazil)

Carbonatite (Palabora, South Africa; Catalao, Brazil)

Type 2. Granite-related

2.1. Endogranitic (La Crouzille District, France; Xiazhuang District, China)

2.2. Perigranitic (Pribram Region, Czech Republic; Niederschlema-Alberoda, Germany)

Type 3. Polymetallic iron oxide breccia complex (Olympic Dam, Carrapateena, Australia)

Type 4. Volcanic-related

4.1. Structure-bound (Streltsov-Antei, Russian Federation; Kuriskova, Slovakia)

4.2. Strata-bound (Dornod N° 7 ore zone, Mongolia; Maureen, Australia)

4.3. Volcano-sedimentary (Anderson Mine, USA; Sierra Pintada District, Argentina)

Type 5. Metasomatite

5.1. Na-metasomatite

Granite-derived (Kirovograd District, Ukraine; Lagoa Real, Brazil; Coles Hill, USA)

Metasediments-metavolcanics-derived (Krivoy Rog District, Ukraine; Michelin, Canada)

5.2. K-metasomatite (Elkon District, Russian Federation)

5.3. Skarn (Mary Kathleen, Australia; Tranomaro, Madagascar)

Type 6. Metamorphite

6.1. Strata-bound (Forstau, Austria)

6.2. Structure-bound

Monometallic (Schwartzwalder, USA; Ace-Fay-Verna, Canada; Rozna, Czech Republic)

Polymetallic (Shinkolobwe, Democratic Republic of Congo; Port Radium, Canada; Jaduguda, India)

6.3. Marble-hosted phosphate (Itataia, Brazil; Nuottijarvi, Finland)

Type 7. Proterozoic unconformity

7.1. Unconformity-contact (Cigar Lake, Key Lake, McArthur River, Canada)

7.2. Basement-hosted (Jabiluka, Ranger, Australia; Eagle Point, Millennium, Canada)

7.3. Stratiform fracture-controlled (Lambapur, Chitrial, India)

Type 8. Collapsebreccia pipe (Arizona Strip, USA)

Type 9. Sandstone
9.1. Basal channel (Dalmatovskoye, Russian Federation; Beverley, Australia)
9.2. Tabular (Arlit District, Niger; Ambrosia Lake District, USA)
Continental fluvial, uranium associated with intrinsic reductant (Arlit type, Niger)
Continental fluvial, uranium associated with extrinsic humate/bitumen (Grants type, USA)
Continental fluvial vanadium-uranium (Salt Wash type, USA)
9.3. Rollfront (Wyoming Province, USA; Chu-Sarysu Basin, Kazakhstan)
Continental basin, uranium associated with intrinsic reductant (Wyoming type, USA)
Continental to marginal marine, uranium associated with intrinsic reductant (Chu-Saryisu type, Kazakhstan)
Marginal marine, uranium associated with extrinsic reductant (South Texas, USA)
9.4. Tectonic-lithologic (Lodève Basin, France; Franceville Basin, Gabon)
9.5. Mafic dykes/sills in Proterozoic sandstone (Westmoreland District, Australia; Matoush, Canada)

Type 10. Paleo-quartz-pebble conglomerate
10.1. U-dominant (Elliot Lake District, Canada)
10.2. Au-dominant (Witwatersrand Basin, South Africa)

Type 11. Surficial
11.1. Peat-bog (Kamushanovskoye, Kyrgyzstan; Flodelle Creek, USA)
11.2. Fluvial valley (Yeelirrie, Australia; Langer Heinrich, Namibia)
11.3. Lacustrine-playa (Lake Maitland and Lake Way, Australia)
11.4. Pedogenic and fracture fill (Beslet, Bulgaria)

Type 12. Lignite and coal
12.1. Stratiform (Koldzhat, Kazakhstan; Williston Basin, North Dakota, USA)
12.2. Fracture-controlled (Freital, Germany; Turakavak, Kyrgyzstan)

Type 13. Carbonate
13.1. Strata-bound (Tumalappalle, India)
13.2. Cataclastic (Mailuu-Suu, Kyrgyzstan; Todilto District, USA)
13.3. Paleokarst (Sanbaqi, China; Tyuya-Muyun, Kyrgyzstan)

Type 14. Phosphate
14.1. Organic phosphorite (Mangyshlak Peninsula, Kazakhstan; Ergeninsky Region, Russian Federation)
14.2. Minerochemical phosphorite (Oulad Abdoun Basin, Morocco; Florida Land Pebble Formation, USA)
14.3. Continental phosphates (Bakouma District, Central African Republic)

Type 15. Black shale
15.1. Stratiform (Haggan, Sweden; Chattanooga Shale Formation, USA)
15.2. Stockwork (Ronneburg District, Germany; Dzhantuar, Uzbekistan)

At the end of 2014, 1555 uranium deposits were recorded in the UDEPO database (Table 2.2).

Table 2.2 Number of deposits for each deposits type and subtype (UDEPO 2014)

	Type of deposit	Number of deposits	Subtype	Number of deposits
1	Intrusive	87	1.1. Anatectic	55
			1.2. Plutonic	31
2	Granite-related	132	2.1. Endogranitic	80
			2.2. Perigranitic	52
3	Polymetallic ironoxide breccia complex	16		16
4	Volcanic-related	120	4.1. Structure-bound	98
			4.2. Strata-bound	18
			4.3. Volcano-sedimentary	4
5	Metasomatite	76	5.1. Na-metasomatite	54
			5.2. K-metasomatite	17
			5.3. Skarn	5
6	Metamorphite	106	6.1. Strata-bound	5
			6.2. Structure-bound	92
			6.3. Marble-hosted	9
7	Proterozoic unconformity	85	7.1. Unconformity-contact	36
			7.2. Basement-hosted	44
			7.3. Stratiform fracture-controlled	5
8	Collapse breccia pipe	16		16
9	Sandstone	636	9.1. Basal channel	76
			9.2. Tabular	300
			9.3. Roll front	234
			9.4. Tectonic-lithologic	18
			9.5. Mafic dykes/sills	8
10	Paleo-QPC	70	10.1. U-dominant	26
			10.2. Au-dominant	44
11	Surficial	64	11.1. Peat-bog	2
			11.2. Fluvial valley	36
			11.3. Lacustrine-playa	23
			11.4. Pedogenic/fracturefill	3
12	Lignite-coal	35	12.1. Stratiform	33
			12.2. Fracture-controlled	2
13	Carbonate	10	13.1. Strata-bound	1
			13.2. Cataclastic	6
			13.3. Paleokarst	3
14	Phosphate	55	14.1. Organic phosphorite	7
			14.2. Phosphorite	43
			14.3. Continental phosphate	5
15	Black shale	47	15.1. Stratiform	28
			15.2. Stockwork	19
		1555		1555

2.4.1.2 Description of the uranium deposits types of the IAEA geological classification

2.4.1.2.1 Type 1. Intrusive deposits

Deposits included in this type are hosted in intrusive rocks of many different petrochemical compositions. Two main subtypes are recognized: (1) intrusive anatectic deposits associated with partial melting processes and contained in granite–pegmatite (Rössing and Husab, Namibia, and deposits in the Bancroft area, Canada) and (2) intrusive plutonic deposits related to magmatic differentiation processes and subdivided into three classes: quartz monzonite (Bingham Canyon, USA; Chuquicamata, Chile), peralkaline complexes (Kvanefjeld, Greenland; Pocos de Caldas, Brazil), and carbonatite (Palabora, South Africa; Catalao, Brazil).

2.4.1.2.2 Type 2. Granite-related deposits

Deposits related to granite include: (1) true veins composed of ore and gangue minerals in granite or adjacent (meta-) sedimentary rocks and (2) disseminated mineralization in granite as episyenite bodies. Uranium mineralization occurs within, at the contact or peripheral to the intrusion. In the Hercynian Belt of Europe and other parts of the world, these deposits are generally associated with peraluminous, two-mica granite complexes (leucogranites). Resources range from small to large and grades vary from low to high. Two subtypes are distinguished based on their spatial setting with respect to the pluton contact, endogranitic deposits within the granite and perigranitic deposits in the adjacent country rocks.

2.4.1.2.3 Type 3. Polymetallic ironoxide breccia complex deposits

They have been attributed to a broad category of ironoxide–copper–gold deposits from around the world. Olympic Dam (Gawler Craton, Australia) is the only known representative of this type with significant by-product uranium resources. The deposit contains the world's largest uranium resources with more than 2 MtU at 230 ppm. Deposits of this group occur in hematite-rich granite breccias (Olympic Dam) or metasedimentary-metavolcanic breccias (Salobo, Carajas, Brazil) and contain low-grade disseminated uranium in association with Cu-Au-Ag-REE.

2.4.1.2.4 Type 4. Volcanic-related deposits

They are located within or near volcanic calderas filled with mafic to felsic volcanic lavas or more commonly pyroclastic rocks and intercalated clastic sediments. Uranium mineralization is controlled by structures as veins and stockworks (structure-bound deposits), but may also be found as disseminations and impregnations in permeable flows and volcaniclastic sediments (strata-bound deposits). This mineralization occurs at several stratigraphic levels of the volcanic and sedimentary units and may extend deeply into the underlying basement. Uranium minerals (pitchblende, coffinite, U^{6+} minerals, and less commonly brannerite) are associated with Mo-sulfides and pyrite. Associated gangue minerals consist of violet-colored fluorite, carbonates, barite, and quartz. The most significant deposits are located within the Streltsovska Caldera in the Russian Federation. Other examples are known in China (Xiangshan District), Mongolia (Dornot and Gurvanbulag Districts), USA

(McDermitt Caldera), Peru (Macusani District), and Mexico (Peña Blanca District). Uncommon volcano-sedimentary deposits consist of peneconcordant, low-grade, carbonaceous lacustrine sediments with an important tuffaceous component (Anderson Mine, USA).

2.4.1.2.5 Type 5. Metasomatite deposits

They are generally confined to Precambrian shields (except Coles Hill, USA), in orogenic belts affected by intense Na-metasomatism (eg, Novokonstantinovskoye deposit, Ukraine; Lagoa Real, Brazil; Lianshanguan, China; Michelin, Canada) or K-metasomatism (Elkon District, Russian Federation), which produced K-feldspathized, albitized, or illitized rocks along deeply rooted fault systems. In Ukraine, these deposits are developed within a variety of rocks, including granite, migmatite, gneiss, and ferruginous quartzite, which produced albitite, aegirinite, and alkali-amphibolite. Principal uranium phases are uraninite, brannerite, and other Ti-U-bearing minerals, pitchblende, coffinite, and hexavalent uranium minerals. The resources range from medium to very large. Three subtypes of metasomatite deposits are distinguished based on precursor rock lithology and the type of metasomatism: Na-metasomatite, K-metasomatite, and skarn.

2.4.1.2.6 Type 6. Metamorphite deposits

These deposits consist of disseminations, impregnations, veins, and shear zones within metamorphic rocks of various ages with no relation to granitic intrusions. These deposits are highly variable in size, resources, and grades. Three subtypes are recognized: strata-bound deposits, which are uncommon (Forstau, Austria; Nuottijarvi and Lampinsaari, Finland); structure-bound deposits, which are well-represented (Schwartzwalder, USA; Ace-Fay-Verna, Canada; Kamyshevoye, Kazakhstan; Shinkolobwe, Democratic Republic of the Congo; Port Radium, Canada; Jaduguda, India); and marble-hosted phosphate deposits (Itataia, Brazil; Zaozernoye, Kazakhstan).

2.4.1.2.7 Type 7. Proterozoic unconformity deposits

Unconformity-related deposits are associated with and occur immediately below, above, or spanning an unconformable contact that separates Archean to Paleoproterozoic crystalline basement from overlying, redbed clastic sediments of Proterozoic age. In most cases, the basement rocks immediately below the unconformity are strongly hematized and clay altered, possibly a result of paleoweathering and/or diagenetic/hydrothermal alteration. Deposits consist of pods, veins, and semimassive replacements consisting mainly of pitchblende and they usually feature strong quartz dissolution. They are preferentially located in two major districts, the Athabasca Basin (Canada) and the Pine Creek Orogen (Australia). The Proterozoic unconformity deposits include three subtypes of variable importance: unconformity-contact deposits, which all occur in areas in the Athabasca Basin (Canada) such as Cigar Lake; basement-hosted deposits such as Jabiluka and Ranger (Australia), Eagle Point (Athabasca Basin) and Kiggavik (Thelon Basin, Canada) and stratiform structure-controlled deposits (Chitrial and Lambapur, Cuddapah Basin, India).

2.4.1.2.8 Type 8. Collapse breccia pipe deposits

Deposits in this group occur in sedimentary basins within cylindrical, vertical breccia pipes within sedimentary basins. The collapse breccia pipes contain down-dropped fragments from overlying lithological units filling karst dissolution cavities developed in the thick carbonate layers. Uranium occurs as primary pitchblende in the permeable breccia matrix, and in the arcuate, ring-fracture zone surrounding the pipe. Pitchblende is associated with numerous Cu, Fe, V, Zn, Pb, Ag, Mo, Ni, Co, As, and Sesulfidesor oxides. Type examples are located north and south of the Grand Canyon, USA. Resources are small to medium (300–2500 t) with relatively high grades (0.20–0.80%).

2.4.1.2.9 Type 9. Sandstone deposits

Sandstone-hosted uranium deposits occur in medium- to coarse-grained sandstones deposited in continental fluvial or marginal marine sedimentary environments. Volcanicash may represent a major uranium source within the sandstone in some regions (Niger; Lodève, France; Wyoming, USA). Uranium is precipitated by reduction processes caused by a variety of possible reducing agents within the sandstone, including carbonaceous material (detrital plant debris), sulfides (pyrite), ferromagnesian minerals (chlorite), bacterial activity, migrated fluids from underlying hydrocarbon reservoirs, and others. Sandstone deposits can be divided into five main subtypes with frequent transitional types between them.

9.1. Basal channel deposits consist of wide channels filled with thick, permeable alluvial-fluvial sediments. Uranium is predominantly associated with detrital plant debris, forming ore bodies that display, inplain view, an elongated lens or ribbon-like configuration and, in a sectionview, a lenticular or, more rarely, a rollshape. Individual deposits may range from several hundred to 20.000 tU, at 0.01–3%. Examples are the deposits of Dalmatovskoye (Transural Region) and Khiagdinskoye (Vitim District) in Russia and Beverley (Australia).

9.2. Tabular deposits consist of uranium disseminations that form irregularly shaped lenses within reduced sediments. The mineralized zones are oriented parallel to the depositional trend. Individual deposits may contain several hundred to 150.000 tU at 0.05–0.5%. Examples of deposits include Hamr-Stráz (Czech Republic), Akouta, Arlit, and Imouraren (Niger) and those of the Colorado Plateau (USA).

9.3. Rollfront deposits are mineralized zones with convex shape, oriented down the hydrologic gradient. They display diffuse boundaries with reduced sandstone on the down-gradient side and sharp contacts with oxidized sandstone on the up-gradient side. The mineralized zones are elongated and sinuous along strike and perpendicular to the direction of deposition and groundwater flow. Resources range from a few hundred to several thousand tonnes uranium, at 0.05–0.25%. Examples are Tortkuduk, Moynkum, Inkai, and Mynkuduk (Kazakhstan) and Crow Butte and Smith Ranch (USA).

9.4. Tectonic-lithologic deposits are discordant to strata. They occur in permeable fault zones and adjacent sandstone beds in reduced environments created by hydrocarbons and/or detrital organic matter. Uranium is precipitated in fracture or fault zones related to tectonic extension. Individual deposits contain a few hundred to 5000 tU at 0.1–0.5%. Examples include the deposits of the Lodève District (France) and Franceville Basin (Gabon).

9.5. Mafic dykes/sills in Proterozoic sandstones: Mineralization is associated with mafic dykes and sills that are concordant with or crosscut Proterozoic sandstone formations. Deposits may be subvertically oriented along dyke margins (Matoush, Otish Basin, Canada), or hosted within the dykes, or strata-bound within the sandstones along lithological contacts with mafic sills (Red Tree, Westmoreland District, Australia). Deposits are small to medium (300–10.000 tU) at 0.05–0.40%.

2.4.1.2.10 Type 10. Paleo-quartz-pebble conglomerate deposits

Detrital uranium oxide ores are found in quartz-pebble conglomerates (QPCs) deposited as basal units (Elliot Lake District, Canada) or intraformational conglomerates (Witwatersrand Basin, South Africa) in fluvial to lacustrine braided streams or lacustrine systems older than 2400–2300 Ma. The conglomerate matrix contains pyrite and gold, as well as other accessory, oxide, and sulfide detrital minerals. Two economic deposit subtypes are distinguished:

- U-dominant: with REE and Th (Elliot Lake District, Canada).
- Au-dominant: with uranium as a by-product (Witwatersrand Basin, South Africa) ± REE, Th.

2.4.1.2.11 Type 11. Surficial deposits

Surficial uranium deposits are broadly defined as young (Tertiary to recent), near-surface uranium concentrations in sediments and soils. The largest of the surficial uranium deposits are found in calcretes (Ca- and Mg-carbonates). These calcrete-hosted deposits mainly occur in valley-fill sediments along Tertiary drainage channels (Yeelirrie, Australia; Langer Heinrich, Namibia) and in lacustrine-playa sediments (Lake Way, Lake Maitland; Centipede, Australia) in areas of deeply weathered, uranium-rich granites. Carnotite is the main uraniferous mineral. Surficial uranium deposits also occur less commonly in peat bogs (Kamushanovskoye, Kyrgyzstan), and soils (Beslet, Bulgaria).

2.4.1.2.12 Type 12. Lignite-coal deposits

Elevated uranium content occurs in lignite or coal mixed with mineral detritus (silt, clay) and in immediately adjacent carbonaceous mud and silt/sandstone beds. The pyrite content is high. Lignite-coal seams are often interbedded or overlain by felsic pyroclastic rocks. Two subtypes are recognized: (1) stratiform lignite-coal deposits (North and South Dakota, USA; Ambassador, Australia) and (2) fractured-controlled lignite-coal deposits (Cave Hills and Slim Buttes, USA and Freital, Germany).

2.4.1.2.13 Type 13. Carbonate deposits

Deposits are hosted in carbonate rocks (limestone and dolostone). Mineralization can be syngenetic and strata-bound or more commonly structure-related within karsts, fractures, faults, and folds. Three types of carbonate-hosted uranium deposits are recognized: (1) strata-bound carbonate deposits (Tumalappalle, India),

(2) cataclastic carbonate deposits (Mailuu-Suu, Kyrgyzstan; Todilto District, USA), and (3) karst deposits (Bentou-Sanbaqi, China; Tyuya-Myuyun, Kyrgyzstan).

2.4.1.2.14 Type 14. Phosphate deposits

Phosphate deposits are principally represented by marine phosphorite of continental-shelf origin containing synsedimentary, stratiform, disseminated uranium in fine-grained apatite. Phosphorite deposits constitute large uranium resources (millions of tonnes), but at very low grade (0.005–0.015%). Uranium can be recovered as a by-product of phosphate production. Three subtypes of uranium-bearing phosphate deposits are identified: (1) minerochemical phosphorite deposits (Land Pebble District, Florida and Phosphoria Formation, Idaho and Montana, USA; Gantour, Morocco), (2) organic phosphorite deposits with argillaceous marine sediments enriched in uranium-rich fish remains (Melovoe, Mangyshlak District Kazakhstan; Ergeninsky Region, Russia), and (3) continental phosphate deposits only known in the Bakouma District (Central African Republic).

2.4.1.2.15 Type 15. Black shale deposits

Black shale-related uranium mineralization includes marine, organic-rich shale, and coal-rich pyritic shale, containing synsedimentary, disseminated uranium adsorbed onto organic material and clay minerals, and fracture-controlled mineralization within or adjacent to black shale horizons. Examples include the uraniferous alum shale in Sweden and Estonia, the Chattanooga shale (USA), the Chanziping deposit (China), and the Gera-Ronneburg deposit (Germany).

2.4.2 Genetic classification and formation of uranium deposits

The physical and chemical conditions prevailing during the transport and deposition of uranium, and the timing of these processes relative to the geological history of the area, are relatively well known for most deposit types. It is now possible to propose a classification of uranium deposits based on the geological, physical, and chemical conditions controlling their formation (Cuney, 2009, 2011; Skirrow et al., 2009).

Skirrow et al. (2009) proposed a genetic classification based on a mineral system approach that takes into account the mineralizing processes from crustal scale to deposit scale. Similarly to the geological cycle proposed by Cuney (2009, 2011) (Fig. 2.1), the mineral system classification considers the fractionation of uranium from the mantle to the crust and its progressive concentration and recycling via sedimentary, metamorphic, igneous, and fluid processes. The 15 deposits of the IAEA classification may fit within three end-member families of uranium mineralizing systems involving three end-member geological fluids (Fig. 2.2): (1) basin- and surface-related uranium systems with surface waters (meteoric and seawater evolving to diagenetic waters), (2) metamorphic-related uranium systems with metamorphic fluids, and (3) magmatic-related uranium systems with magmatic and magmatic-hydrothermal fluids. Hybrid systems between these end-members are also considered.

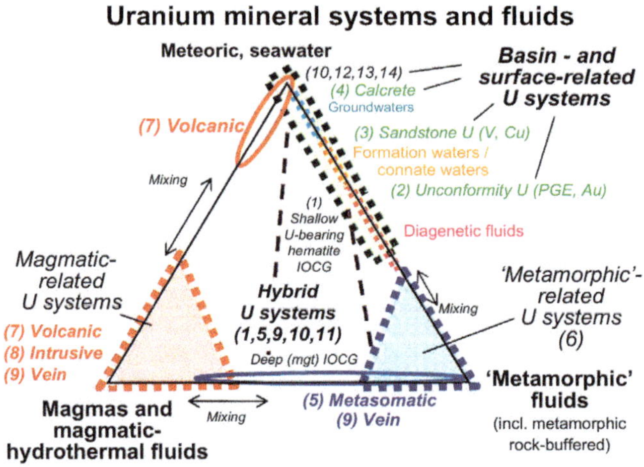

Figure 2.2 Schematic representation of the three end-members of uranium mineralizing systems and the three end-member fluid types (Skirrow et al., 2009). The numbers correspond to the deposit types from a former IAEA classification (IAEA, 1993).

The genetic classification proposed by Cuney (2009, 2011) is based on the dominant mechanism believed to be the main factor of primary uranium concentration for a given deposit. The different types of uranium deposits are presented from the highest to the lowest temperature of genesis and regrouped into five main types, each subdivided into several subtypes (Table 2.3).

2.4.2.1 Magmatic deposits (M)

Magmatic uranium deposits dominantly result from magmatic processes. Two contrasting subtypes are considered:

Crystal fractionation (MCF). These deposits result from extreme fractional crystallization. In peralkaline melts, U, Th, Zr, REE, and Nb are most soluble and thus are simultaneously and continuously enriched during magmatic fractionation (Linnen and Cuney, 2008). The type example is the Kvanefjeld deposit in Greenland (Sørensen et al., 1974). When the melt crystallizes, complex U-Th-Zr-REE-Nb silicates, oxides, and phosphates are formed, but these minerals are too refractory to allow for an economic recovery of uranium as the only product.

Related to partial melting (MPM). Such deposits result from low degrees of partial melting of epicontinental metasedimentary and/or acidic volcanic rocks. The uranium mineralization occurs as disseminated uraninite in granitic pegmatoids occurring within high-grade metamorphic and migmatite domainsn. The Rössing uranium deposit in Namibia is the type example (Berning et al., 1976; Cuney and Kyser, 2008).

Table 2.3 Genetic classification of uranium deposits

	Deposit type	Size	U (%)	Ages (Ga)	Type example	References
1—M	Magmatic					
MCF	Crystal fractionation	Small/ medium	0.1	2.7 to present	Ilimaussaq	Sorensen et al. (1974)
MPM	Partial melting	Large	×0.01	Neoarchean—1.8–1–0.5	Rössing	Berning et al. (1976)
2—H	Hydrothermal					
HV	Hydrothermal volcanic	Small/very large	0.05–0.2	Largest Jurassic–Cretaceous	Streltsovkoye	Chabiron et al. (2003)
HG	Hydrothermal granitic	Small/large	0.1–0.6	Permian, Cretaceous	Niederschlema	Golubev et al. (2000)
HD	Hydrothermal diagenetic					
HDia	• *Intraformational redox*					
HDiaTb	• *Tabular*	Small/large	0.1–0.2	Post-Silurian	Grants	Hansley and Spirakis (1992)
HDiaTl	• *Tectonolithologic*	Small/large	0.1–0.4	Post-Silurian	Lodève	Mathis et al. (1990)
HDiaCb	• *Collapse breccia*	Small	0.1–0.9	Jurassic	Colorado	Wenrich and Titley (2008)
HDbb	• *Basin/basement redox*	Small/very large	0.2–20	1.7–1.3	McArthur River	Jefferson et al. (2007)
HDir	• *Interformational redox*		0.1–0.4	2.0–1.5	Oklo	Gauthier-Lafaye (1986)
HMp	Hydrothermalmetamorphic	Small			Luiswishi	Eglinger et al. (2013)

(*Continued*)

Table 2.3 (Continued)

	Deposit type	Size	U (%)	Ages (Ga)	Type example	References
HMt	Hydrothermalmetasomatic					
HMtNa	• *Na-metasomatism*	Small/large	0.1–0.2	1.8–0.5	Michurinka	Cuney et al. (2012)
HMtK	• *K-metasomatism*	Large	0.1		Elkon district	Kazanskii (2004)
HMtSk	• *Skarns*	Small	0.1	1.8–1.5–1–0.5	Mary Kathleen	Maas et al. (1987)
3—MW	Meteoric water infiltration					
MWB	• *Basal-type*	Small/large	0.01–0.1	Tertiary–Quaternary	Blizzard	Boyle (1982)
MWRf	• *Roll front*	Small/very large	0.01–0.2	Jurassic–Cretaceous	Wyoming	Adams and Cramer (1985)
4—S	Synsedimentary					
SMs	• Mechanical sorting	Large	0.02–0.15	2.9–2.4	Witwatersrand	Frimmel et al. (2005)
SRt	• Redox trapping					
SRtM	• *Marine (black shales)*	Very large	0.005–0.3	Paleozoic–Mesozoic	Randstatt	Andersson et al. (1985)
SRtC	• *Continental (coal, lignite...)*	Small/medium	0.001–0.5	Paleozoic–Mesozoic	Kazakhstan	IAEA (2005)
SCcr	• Crystal-chemical/redox phosphates	Very large	0.005–0.03	Cretaceous, Mio-Pliocene	Morocco	IAEA (2001)
5—E	Evapotranspiration	Small/large	0.014–0.1	Tertiary–Quaternary	Yeelirrie	Carlisle et al. (1978)
6—O	Others	Medium/very large	0.01–0.04	Mesoproterozoic	Olympic Dam	Hitzman and Valenta (2005)

Source: From Cuney (2009, 2011).

2.4.2.2 Hydrothermal deposits (H)

Volcanic-hydrothermal systems (HV). These deposits consist mainly of uranium-bearing veins and disseminations formed in geothermal systems developed shortly after the eruption of felsic volcanic rocks. The main uranium districts are hosted by caldera complexes, filled with alternating mafic and felsic volcanic rocks, with subordinate sedimentary layers. Highly fractionated felsic volcanic units are generally predominant. Pyroclastic rocks represent particularly favorable source rocks because they are relatively permeable and a high proportion of the uranium is hosted by the glassy matrix, from which it can be easily mobilized during devitrification.

Peralkaline magmas represent the most efficient uranium source, as shown by the association of the largest uranium districts with this type of volcanism, such as the Streltsovskoye caldera (Transbaikalia, Russia) (Ishukova et al., 1991). High-K calc-alkaline metaluminous volcanic rocks are a less favorable uranium source because a significant but variable portion of the uranium in these rocks is trapped in accessory minerals (Leroy and George-Aniel, 1992). Highly peraluminous felsic volcanic rocks, mineralized in uranium, are only known from the Macusani district, Peru.

Granite-hydrothermal systems (GH). These deposits are related to post-magmatic, high-level hydrothermal circulation dominantly within two mica peraluminous leucogranites (Cuney et al., 1990). The largest uranium district of this type is the mid-European Variscan uranium province, which extends for more than 2000 km from Spain–Portugal to Bohemia. The mineralization may occur within the granites (intragranitic deposits) such as most of the deposits from the French Massif Central (Cathelineau et al., 1990; Turpin et al., 1990) or within the adjacent metamorphic rocks (peribatholitic deposits) as the deposits from the Erzgebirge.

Hydrothermal uranium deposits may also be associated with highly fractionated high-K calc-alkaline granites, but they are less common and generally of smaller size such as Hotagen (Sweden). However, the most significant example of the relation between high-K calc-alkaline magmatism and uranium mineralization is the Olympic Dam deposit, Australia, hosted by the Roxby Downs granite (Hitzman and Valenta, 2005).

Hydrothermal diagenetic systems (HD). These deposits form from the circulation of low-temperature basinal fluids of variable salinity through porous fluvial, lacustrine, deltaic to near-shore marine siliciclastic sediments, more rarely in limestone. This leads to the formation of a large variety of epigenetic uranium deposit types. The diagenetic fluids are brines that in some cases interact with infiltrating meteoric waters. Three main subtypes are distinguished according to the location of the redox control within the basin:

Intraformational redox control deposits (HDia)

Tabular intraformational redox-control deposits (HDiaTb) are transitional with synsedimentary deposits, because uranium minerals may begin to precipitate shortly after sedimentation and burial, but the majority of the mineralization is usually deposited during diagenesis. Ore bodies are conformable with the stratigraphy and form within permeable sandy layers, rich in detrital organic matter, intercalated between less permeable clay-rich horizons, generally at the margins of palaeochannels. In some sandstone, the reduction of uranium may be caused either by H_2S (biogenic and/or nonbiogenic) produced from the interaction of oxidized groundwater with pyrite in the aquifer or by reduced fluids derived from deep hydrocarbon reservoirs (Spirakis, 1996).

Tectonolithologic intraformational redox-control deposits (HDiaTl) are controlled both by tectonic structures and by paleochannels along the sedimentary bedding, as in the preceding deposit type. Typical examples are the deposits from the Lodève Basin in France (Mathis et al., 1990).

Diagenetic-hydrothermal collapse breccia deposits (HDiaCb) form near-vertical, 30–175 m in diameter, cylindrical pipes, located in flat-lying upper Paleozoic to Triassic marine platform sediments in the Grand Canyon region, USA. Solution of the Mississippian Redwall Limestone (0–210 m thick) has induced the collapse of the overlying strata up to a distance of 1000 m. The fluids associated with the mineralization are low temperature (80–173°C), saline (4–17 wt%. NaCl equiv) and oxidized diagenetic fluids (Wenrich and Titley, 2008).

Basin/basement redox-controlled deposits (HDbb) are usually termed unconformity-related deposits because they are generally located in the vicinity of an unconformity between metamorphic basement and undeformed clastic sedimentary cover. The basement comprises Archean granitic domes rimmed by uranium-rich and graphite-rich metamorphosed Paleoproterozoic epicontinental sediments intruded by granitic intrusions. Most of the preserved part of the basins consists of oxidized continental fluviatile, eolian to marginal marine, highly mature, siliclastic sediments. Uranium was transported by oxidized, acidic, and high-chlorinity basinal diagenetic brines (Derome et al., 2003, 2005). Uranium solubility was favored by their high f_{O2} and chlorinity (up to 6 molal) (Richard et al., 2012). The two main provinces hosting deposits of this type are the Athabasca Basin, Saskatchewan, Canada and the East Alligator River Uranium Province, Australia.

Interformational redox-controlled uranium deposits (HDir) are characterized by the location of the redox barrier, an oil generating shale formation above an oxidized sandstone formation. The best examples are the deposits of the Paleoproterozoic Franceville Basin in Gabon (Gauthier-Lafaye, 1986). All the uranium deposits of the Franceville basin (Oklo, Mounana, and Okelonbondo) are located in the upper part of the sandstone of the FA Formation, which has been reduced byhydrocarbons that migrated from the FB black shales (Gauthier-Lafaye, 1986). The diagenetic fluid is a highly saline brine (Mathieu et al., 2000).Uranium has precipitated from the diagenetic brines at the redox interface between the oxidized FA sandstones and FB pelites.

Hydrothermal metamorphic systems (HMp) correspond to uranium deposits formed during the circulation of metamorphic fluids in association with folding and/or thrusting of the rocks. Typical examples are the Mistamisk uranium veins of Labrador, Canada (Kish and Cuney, 1981) and uranium deposits of the African Copper Belt, as exemplified by the Kansanshi deposit (Eglinger et al., 2013). At higher temperature, the metamorphic origin of the deposits is rarely well constrained. Many of them are in fact metamorphosed uranium deposits.

2.4.2.3 Hydrothermal metasomatic systems (HM)

HM deposits include those associated with Na-metasomatism, K-metasomatism, and skarns.

Na-metasomatism related deposits (HMNa). Na-metasomatism can affect just the wall rocks of uranium veins, as at Beaverlodge, Canada, but more commonly affects a large volume of rocks, with mineralization occurring in the core of the alteration zones with the most advanced stage of fluid–rock interaction. Na-metasomatism is most commonly developed in felsic rocks to form albitites, but can also be developed in other rock types. The albitized zones may be developed along discrete structural zones within granitic batholiths or may extend regionally over several tens of kilometers along deep regional structures, as in central Ukraine (Belevtsev and Koval, 1968).The metasomatic fluids are believed to be derived from either or both a basinal brine and a magma, as proposed for the Valhalla uranium deposit in Queensland, Australia (Polito et al., 2007), or fluids that infiltrated from the surface, as in the central Ukraine uranium deposits (Cuney et al., 2012).

K-metasomatism related deposits (HMtK). These are only known from the huge Elkon district (Aldan Shield, Russia). The brannerite–gold mineralization is controlled by steeplydipping faults reactivated during the Mesozoic, along which pyrite–carbonate–K-feldspar alteration developed in Archean to Paleoproterozoic high-grade metamorphic rocks (Boitsov and Pilipenko, 1998).

Skarn-related deposits (HMSk). These are associated with metasomatic processes occurring during contact metamorphism between of carbonate rocks by intruding granite, or during infiltration of fluids into carbonate rocks during regional metamorphism. Therefore, they could also be classified as metamorphic deposits. The type example is the Mary Kathleen U-REE deposit in Australia hosted in Paleoproterozoic high-grade metamorphosed calc-silicate, mafic to intermediate igneous and sedimentary rocks.

Meteoric water infiltration-related deposits (MW) comprise basal-type (or sealed paleovalleys) and roll fronts. Uranium is leached from a granitic basement or from synsedimentary volcanic ash and precipitates by reaction with organic matter during groundwater percolation through permeable siliciclastic sediments.

Basal-type uranium deposits (MWB) are located in poorly consolidated, highly permeable, fluvial to lacustrine, organic matter-bearing gravels and sands deposited in paleovalleys incised in uranium-rich granitic basement and capped by basalt flows. They are named paleovalley or infiltration-type in Russian literature (eg, Vitim deposits, Transbaïkalia, Russia, Kondrat'eva et al., 2004).

Roll fronts deposits (Mrf) represent the best example of epigenetic uranium deposition at a redox interface (front). Hostrocks are younger than Ordovician and were deposited in fluvial to lacustrine environments in intermontane basins (eg, the Tertiary Powder River Basin in Wyoming, Adams and Cramer, 1985), or in channel, lagoonal, and beach-bar settings on marginal marine plains (South Texas, USA; Finch and Davis, 1985). The reductants may be either of synsedimentary origin (detrital plant debris), or epigenetic, by the infiltration of fluids derived from deep hydrocarbon reservoirs along faults (South Texas, USA; Goldhaber et al., 1978).

2.4.2.4 Synsedimentary deposits (S)

They are subdivided into three major types according to the main mechanism controlling uranium deposition during sedimentation.

Mechanical sorting-related synsedimentary deposits (SMs) are represented by QPCs. The earliest examples occur in the Dominion Group of South Africa (3086 ± 3 to 3076 ± 6 Ma; Frimmel and Minter, 2002), the latest ones are hosted by the Elliott Lake Group in Canada (between 2496 and 2215 Ma; Corfu and Andrews, 1986). Genetic models proposed for these deposits vary from a purely syngenetic origin (placer model) with a detrital deposition of uraninite in fluvial to fluvio-deltaic environments (Frimmel et al., 2005), to an epigenetic origin with uranium precipitating from hydrothermal fluids (Barnicoat et al., 1997) or a combination of both in the modified placer theory (Pretorius, 1961). The high Th contents and the magmatic signature of the REE in the uraninites support a detrital origin of the primary uranium concentration (Cuney, 2010).

Redox trapping in marine environments(SRtm) is represented by black shales. Large amounts of uranium associated with Mo, V, Ni, Zn, and Cu may be adsorbed on organic matter and clay minerals in organic- and pyrite-rich black shales. The largest deposits are the Cambro-Ordovician Alum shales from Sweden (Andersson et al., 1985), but the Silurian graptolitic black shales of the Ronneburg–Gera deposits in Germany are the only deposits of this type which have been mined to a significant extent because of their higher grades resulting from a combination of hydrothermal and supergene enrichments (Lange and Freyhoff, 1991).

Redox trapping in continental environments (SRtc) occurs in coal, lignite, peat bogs, swamps, anoxic lakes, and karst caverns (Morales et al., 1985). Uranium is deposited by adsorption on the organic material and/or by reduction of UO_2^{2+} by anaerobic bacterial activity.

Crystalchemical and redox trapping-related deposits (SCcr) are represented by sedimentary phosphates, which represent the largest low-grade uranium reservoir. The largest episode of phosphate deposition occurred during the late Cretaceous to Eocene Period (90–45 Ma) at a paleolatitude of 8–15°N, along the southern margin of the Tethys Ocean. The phosphorite belt extends from Turkey to Morocco and across the Atlantic Ocean to Colombia and Venezuela.

2.4.2.5 Evapotranspiration-related deposits (E)

Evapotranspiration-related deposits (E) are commonly called calcretes. Known examples form from the Tertiary to present time. They are hosted by surficial sediments, deposited in fluviatile, to lacustrine/playa systems. They are cemented by a variety of minerals: calcite, gypsum, dolomite, ferric oxide, strontianite, and halite. Uranyl vanadates deposited during fluctuation of the groundwater level and evapotranspiration in arid to semiarid climatic conditions. The most economically significant calcretes are referred to as valley calcretes by Carlisle et al. (1978). The largest calcrete deposits are known in Australia and Namibia.

2.4.2.6 *Other types (O)*

This category is reserved for deposits that either cannot be attributed to any of the preceding types or could belong to several types. The largest presently mined deposit in the world, the Olympic Dam iron oxide Cu-Au (U-Ag) deposit (IOCG deposit) in South Australia, is classified here, because too little information is available about the uranium mineralization process. A series of other deposits, such as Itataia in Brazil, need further research to be classified according to the present genetic scheme.

2.5 The IAEA UDEPO database

The IAEA UDEPO database (Fig. 2.3) was created in 1996 containing 582 deposits in 48 countries (IAEA, 1996). To be included in the database, a deposit needed a minimum resource of ≥500 tU at an average grade of at least 0.03%.

In 2009, IAEA-TECDOC-1629, *World Distribution of Uranium Deposits (UDEPO) with Uranium Deposit Classification* was published (IAEA, 2009), including 874 uranium deposits in 67 countries.

At the end of 2014, 1555 uranium deposits/districts in 76 countries were recorded in the UDEPO database (Table 2.2). The increase from 874 deposits listed in 2009 to 1555 deposits in 2014 is due not only to the discovery of about 100 new deposits over that time, but also to the compilation of deposits retrieved from the literature (more than 400) and to the inclusion of deposits considered

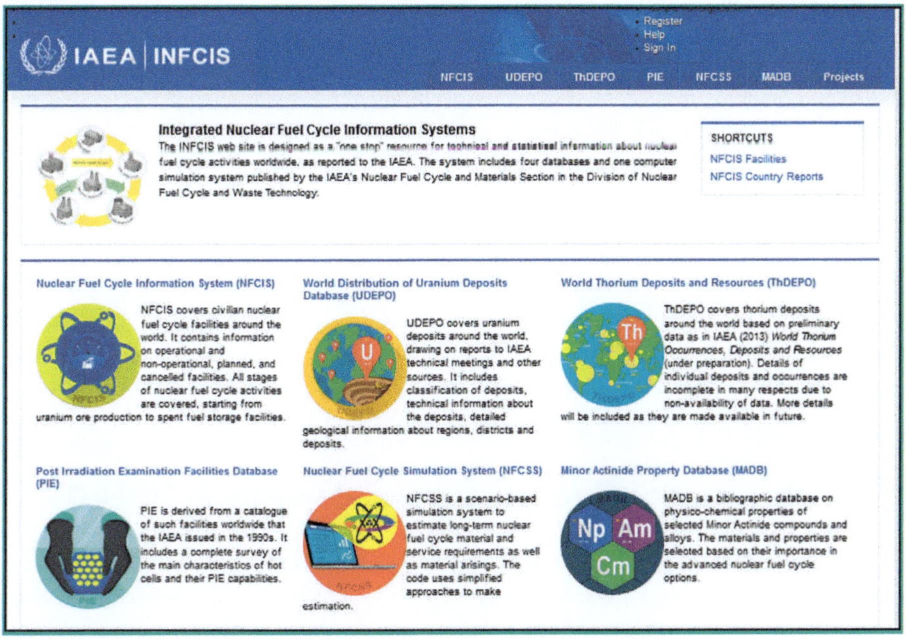

Figure 2.3 IAEA UDEPO website (http://www-nfcis.iaea.org/).

Table 2.4 Uranium resources by deposit types (UDEPO 2014)

Type of deposit		Number of deposits	UDEPO geological resources (tU)[a]
Type 1	Intrusive	87	1576.000
Type 2	Granite-related	132	462.000
Type 3	Polymetallic ironoxide breccia complex	15	2437.000
Type 4	Volcanic-related	120	640.000
Type 5	Metasomatite	76	1021.000
Type 6	Metamorphite	106	517.000
Type 7	Proterozoic unconformity	85	1405.000
Type 8	Collapse breccia pipe	16	16.000
Type 9	Sandstone	637	4254.000
Type 10	Paleo-QPC	70	1275.000
Type 11	Surficial	64	414.000
Type 12	Lignite-coal	35	7420.000
Type 13	Carbonate	10	93.000
Type 14	Phosphate	55	14,007.000
Type 15	Black shale	47	6799.000
		1555	**42,600.000**

[a]Numbers have been rounded to .000.

unconventional resources (120 deposits). In addition, the minimum resource size for inclusion in the database was reduced to 300 tU and there is no grade restriction, thus including low to very low-grade resources.

UDEPO is firstly a geological database. Resources correspond to "total resources" and include all resource categories plus any past production. Where a deposit has been estimated at several cutoff grades, the resources at the lowest cut-off grade are adopted. All data are in metric tonnes of uranium (tU). A total of 44 parameters are included in the database.

In 2014, total geological resources in UDEPO stand at 42.6 MtU (Table 2.4). The largest resources are unconventional resources contained within phosphate, black shale, coal-lignite, and polymetallic iron oxide breccia complex deposits (Table 2.5). Largest conventional resources are located within sandstone and Proterozoic unconformity deposits, which account for more than 75% of the world production in 2013.

A new document, *World Distribution of Uranium Deposits (UDEPO)—2015 Edition* is in progress and will be published in 2016 (IAEA, 2016b).

2.6 Unconventional resources and deposits in UDEPO

In the most recent Red Book (OECD/NEA-IAEA, 2014), conventional resources are defined as *"resources from which uranium is recoverable as a primary product, a co-product or an important by-product, while unconventional resources are*

Table 2.5 Uranium resources for deposit types by decreasing order of cumulative resources (UDEPO 2014)

	Type of deposit	Number of deposits	UDEPO resources (tU)[a]
1	Phosphate	55	14,007.000
2	Lignite-coal	35	7420.000
3	Black shale	47	6799.000
4	Sandstone	637	4254.000
5	Polymetallic ironoxide breccia complex	15	2437.000
6	Intrusive	87	1576.000
7	Proterozoic unconformity	85	1405.000
8	Paleo-QPC	70	1275.000
9	Metasomatite	76	1021.000
10	Volcanic-related	120	640.000
11	Metamorphite	106	517.000
12	Granite-related	132	462.000
13	Surficial	64	414.000
14	Carbonate	10	93.000
15	Collapse breccia pipe	16	16.000
		1555	**42,600.000**

[a]Numbers have been rounded to .000.

resources from which uranium is only recoverable as a minor by-product, such as uranium associated with phosphate rocks, nonferrous ores, carbonatite, black shale and lignite. Most of the unconventional uranium resources reported to date are associated with uranium in phosphate rocks, but other potential sources exist (eg, black shale and seawater)."

For the UDEPO database, which is firstly a geological database, there is no economic connotation. Unconventional resources/deposits are those of low to very low grade that are not or cannot be mined just for uranium.

In UDEPO 2014, 119 deposits are referred to as unconventional. Deposits like Olympic Dam are included in the unconventional resources, explaining some of the differences seen in these when compared with the Red Book data. Also, considered-as deposits are very low-grade resources present in deposits such as Kolmisoppi-Kuusilampi (Ni-Cu-Co-Zn mine) and Chuquicamata-Rodomiro Tomic (Cu mines).

Unconventional resources are found in several deposit types and subtypes (Tables 2.6 and 2.7):

- Type 1—Intrusive, subtype 1.2 Intrusive plutonic, with three classes, quartz monzonite, peralkaline complex, and carbonatite
- Type 3—Polymetallic iron oxide breccia complex
- Type 12—Lignite and coal
- Type 14—Phosphate
- Type 15—Black shale

Table 2.6 Deposits types, subtypes, and classes with unconventional resources (UDEPO 2014)

Deposit type	Deposit subtype	Deposit class	Number of deposits/ districts	Resources (tU)
1 Intrusive	1.2 Plutonic	1.2.1 Quartz monzonite	7	199.000
		1.2.2 Peralkaline complex	11	393.000
		1.2.3 Carbonatite	8	176.000
3 Polymetallic iron oxide breccia complex			10	2437.000
12 Lignite and coal	12.1 Stratiform		13	7411.000
	12.2 Structure-controlled		1	9000
14 Phosphate	14.1 Organic phosphorite		3	78.000
	14.2 Phosphorite		42	13,893.000
15 Black shale	15.1 Stratiform		23	6571.000
	15.2 Stockwork		1	228.000
			119	**30,920.000**

All these types also contain deposits that were mined in the past as conventional uranium deposits or could be mined in the future just for uranium due to their grades (U generally greater than 0.05%).

The largest unconventional resources in UDEPO are found in phosphate with 14 MtU. Using an average grade of 100 ppm U and world geological phosphate resources estimated to be 300 billion tons (USGS, 2011), uranium resources could be in the order of 30 Mt.

Very large uranium resources are also associated with black shales and coal-lignite deposits. Several hundred black shale and coal-lignite formations are known around the world, but only 38 deposits/districts are listed in UDEPO with uranium resources on the order of 7 Mt for each type.

For the polymetallic ironoxide breccia complex deposits, 90% of the resources are within a single deposit, Olympic Dam.

Large resources of uranium are also found in heavy mineral placers and paleo-placers with monazite, zircon, and xenotime. These resources are not included in UDEPO. However, most can be found in the IAEA ThDEPO database (Fig. 2.3). Very large resources are present in coal-lignite ash tailings and in South African gold mine tailings. Resources are on the order of 200.000–250.000 tU in 10 projects considered for extraction of uranium from gold tailings.

Table 2.7 **Largest deposits-districts with unconventional uranium resources >50.000 tU (UDEPO 2014)**

Deposit	Country	Type of deposit	Resource (tU)	Grade(% U)	Status
Phosphoria Formation	USA	Phosphate	7000.000	0.006−0.020	Dormant
Chattanooga Shale	USA	Black shale	5000.000	0.006	Dormant
Northern Great Plains	USA	Lignite-coal	5000.000	0.005	Dormant
Oulad Abdoum Basin	Morocco	Phosphate	3220.000	0.012	Operating (P)
Olympic Dam	Australia	Hematite breccia complex	2125.000	0.023	Operating (Cu-Au-Ag-U)
Meskala Basin	Morocco	Phosphate	2050.000	0.01	Dormant
Gantour Basin	Morocco	Phosphate	1210.000	0.015	Dormant
MMS Vicken	Sweden	Black shale	447.755	0.0147	Exploration
Haggan	Sweden	Black shale	308.000	0.0137	Exploration
East Florida District	USA	Phosphate	270.000	0.009	Dormant
Narke	Sweden	Black Shale	257.000	0.0175	Exploration
Ranstad	Sweden	Black shale	254.000	0.034	Dormant
Swab	Irak	Phosphate	245.000	0.007	Dormant
Central Florida District	USA	Phosphate	225.000	0.009	Dormant
Dwaima	Irak	Phosphate	193.000	0.007	Dormant
Carrapateena	Australia	Hematite breccia complex	184.000	0.0155	Development
NE Florida District	USA	Phosphate	180.000	0.009	Dormant
Marbat	Irak	Phosphate	148.000	0.007	Dormant
Lubin-Sieroszowice	Poland	Black shale	144.000	0.006	Dormant
Kvanefjeld	Greenland	Intrusive	134.750	0.022	Exploration
Nile Valley District	Egypt	Phosphate	118.500	0.007	Dormant
East Ebro Valley	Spain	Lignite-coal	102.000	0.017	Dormant
North Florida District	USA	Phosphate	90.000	0.009	Dormant
Al Shedeye-Eshidia	Jordan	Phosphate	83.000	0.007	Dormant
Springbok Flat	South Africa	Lignite-coal	82.000	0.042	Dormant
Twihinate	Morocco	Intrusive	75.000	0.025	Dormant
Catalao	Brazil	Intrusive	72.125	0.013	Operating(P-Nb)
Tummalappalle	India	Carbonate	72.000	0.04	Operating
Sorensen	Denmark	Intrusive	62.370	0.026	Development
Oued Eddahab Basin	Morocco	Phosphate	57.000	0.006	Dormant

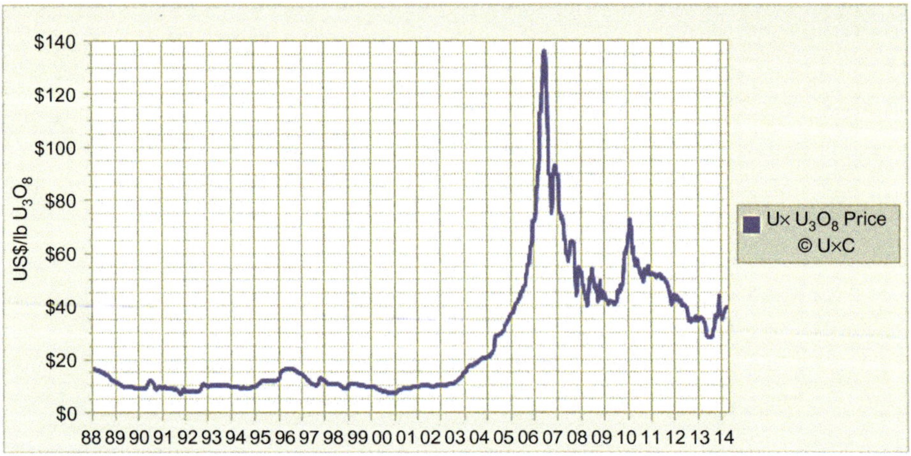

Figure 2.4 15 years U_3O_8 price.
Source: The Ux Consulting Company, LLC. http://www.uxc.com/.

2.7 New uranium deposits (2004–2014)

In 2004, following the sharp increase in uranium price (Fig. 2.4.), hundreds of junior companies began uranium exploration. Company portfolios included virtually any deposit, prospect, showing or anomaly known around the world, even those from very remote places. Many of these were drilled, resources were defined or confirmed, but only a few of the projects have reached the development stage today.

During the period from 2004 to 2014, about 300 uranium deposits were discovered or rediscovered, the majority in existing uranium provinces. Only a few of them (about 100) can be considered as "grassroots" discoveries.

2.7.1 Conventional deposits

Most significant "grassroots" discoveries are listed in Table 2.8 and the most significant confirmed or increased resources in existing prospects/deposits are listed in Table 2.9. For conventional resources, new deposits were found mainly in Namibia (intrusive anatectic and surficial), Peru (volcanic-related), Canada (Proterozoic unconformity), Niger (sandstone-tabular), Mongolia, and China (sandstone-roll front).

Type 1—Subtype 1.1. Intrusive anatectic: The largest discovery of the past 10 years is the Husab deposit (five orebodies) in Namibia within alaskite and containing more than 187.000 tU at 0.04%. The deposit is at the development stage. Other significant discoveries in Namibian alaskites are Ongolo, Oshively, and Onkelo. Also in Namibia, NI 43–101 resources have been defined in several deposits such as Z 20, Anomaly 18, Namibplass, Garnet Valley, and Ondjamba. Geological resources in 24 alaskite deposits are on the order of 720.000 tU. In addition, 205.000 tU have been defined in 15 surficial deposits. With geological resources in

Table 2.8 Major conventional uranium deposits discoveries (2004–2014)

Country	Deposit name	Resources (tU)	Grade (% U)	Deposit type	Deposit subtype	Operator
Namibia	Husab	187.920	0.04	Intrusive	Anatectic	China Guangdong Nuclear Power
Niger	Dasa 1-2-3	64.680	0.049	Sandstone	Tabular	Global Atomic Fuel Corp
Mongolia	Zoovch Ovoo	54.640	0.023	Sandstone	Roll front	AREVA
Canada	Triple R	40.611	1.28	Proterozoic unconformity	Basement-hosted	Fission Energy Corp
Canada	Phoenix	27.450	16.3	Proterozoic unconformity	Unconformity-contact	Denison Mines Corp
Australia	Four Miles East-West	27.048	0.28	Sandstone	Roll front	Quasar Resources
Canada	Roughrider	22.203	4.01	Proterozoic unconformity	Unconformity-contact	Rio Tinto
Mauritania	Ain SderOued FouleOum FerkikTenebdar	19.370	0.03	Surficial	Lacustrine-playa	Aura Energy Ltd
China	Bayinwula	10–25.000	0.01–0.05	Sandstone	Roll front	CNNC
Canada	North Shore	16.890	0.011	Intrusive	Anatectic	Uracan Resources
Australia	Blackbush	12.510	0.0195	Sandstone	Tabular	Uranium SA Ltd
Tanzania	Manyoni District	11.440	0.012	Surficial	Lacustrine-playa	Uranex NL
Peru	Kihitian	11.220	0.056	Volcanic-related	Structure-bound	Macusani Yellowcake Inc
Mongolia	Dulaan Ull	9.900	0.017	Sandstone	Roll front	AREVA
Canada	J-Zone	9.865	1.70	Proterozoic unconformity	Unconformity-contact	Denison Mines Corp

(Continued)

Table 2.8 **(Continued)**

Country	Deposit name	Resources (tU)	Grade (% U)	Deposit type	Deposit subtype	Operator
Namibia	Ongolo	8.780	0.037	Intrusive	Anatectic	Deep Yellow Ltd
Peru	Tantamaco	8.331	0.0186	Volcanic-related	Structure-bound	Azincourt Uranium Inc
Namibia	Oshively	7.970	0.0166	Intrusive	Anatectic	Bannerman Resources
Namibia	Onkelo	7.780	0.0165	Intrusive	Anatectic	Bannerman Resources
China	Sunjialang	5–10.000	0.01–0.05	Sandstone	Roll front	CNNC
Peru	Colibri 2-3	7.230	0.016	Volcanic-related	Structure-bound	Macusani Yellowcake Inc
Niger	Dajy	6.545	0.058	Sandstone	Tabular	Global Atomic Fuel Corp
Canada	J4-Ray	6.045	0.585	Metamorphite	Structure-bound	Kivallik Energy Corp
Namibia	Inca	5.160	0.0424	Metasomatite	Skarn	Deep Yellow Ltd

Table 2.9 New uranium resources in known deposits-districts

Country	Deposit name	Resources (tU)	Grade (% U)	Deposit type	Deposit subtype	Operator
Botswana	Gorgon Main-South-West	120.585	0.018	Sandstone	Tabular	A-Cap Resources
India	Tummalapalle	72.000	0.04	Carbonate	Strata-bound	UCIL
Tanzania	Nyota	55.135	0.025	Sandstone	Tabular	Uranium One
Namibia	Marenica	53.130	0.008	Surficial	Fluvialvalley	Marenica Energy Ltd
Botswana	Serule East-NW-West	46.430	0.017	Sandstone	Tabular	A-Cap Resources
Namibia	Z 20	46.274	0.03	Intrusive	Anatectic	Rossing Uranium Ltd
Namibia	Anomaly 18	40.570	0.01–0.05	Intrusive	Anatectic	Zhongue Resources Dvp
Namibia	Namibplass	30.415	0.013	Intrusive	Anatectic	Forsys Metal Corp
Australia	Ranger Deeps	29.483	0.242	Proterozoic unconformity	Basement-hosted	Energy Resources of Australia
Niger	Marianne-Maryline	19.890	0.13	Sandstone	Tabular	GoviEx Uranium Inc
Namibia	Garnet Valley	18.280	0.019	Intrusive	Anatectic	Extract Resources
Zambia	DibweDibwe East	14.340	0.025	Sandstone	Tabular	Denison Mines Corp
Niger	Isakanan	13.090	0.076	Sandstone	Tabular	Global Uranium Corp
Niger	Abakorum	12.790	0.17	Sandstone	Tabular	China National Nuclear Corp
Niger	Teguidda	12.760	0.20	Sandstone	Tabular	SOMINA
Namibia	Ondjamba	12.045	0.014	Intrusive	Anatectic	Bannerman Resources Ltd
Namibia	Anomaly 2–15	11.540	0.01–0.05	Intrusive	Anatectic	Zhongue Resources Dvp
Niger	Miriam	11.273	0.089	Sandstone	Tabular	GoviEx Uranium

(Continued)

Table 2.9 **(Continued)**

Country	Deposit name	Resources (tU)	Grade (% U)	Deposit type	Deposit subtype	Operator
Canada	Matoush	11.076	0.56	Sandstone	Mafic dyke	Strateco Resources
Namibia	Tubas Red Sand	10.935	0.0125	Surficial	Fluvial valley	Deep Yellow Ltd
Canada	Teasdale Lake Zone	10.822	0.0223	QPCs	U-dominant	Appia Energy Corp
Niger	Imca 25	10.000	0.12	Sandstone	Tabular	AREVA
Mauritania	A 238	9.010	0.02	Metasomatite	Na-metasomatite	Forte Energy NL
Spain	Alameda	8.124	0.039	Granite-related	Perigranitic	Berkeley Resources
Guinea	Firawa	7.890	0.0283	Metamorphite	Structure-bound	Forte Energy
Namibia	Aussinasis	6.960	0.02	Intrusive	Anatectic	Deep Yellow Ltd
Turkey	Temrezli	6.700	0.099	Sandstone	Tabular	Anatolia Energy Corp
Canada	GemicoBlock 3	6.262	0.017	QPC	U-dominant	Appia Energy Corp
Canada	GemicoBlock 10	5.977	0.039	QPC	U-dominant	Appia Energy Corp

excess of 900.000 tU, the Namib Desert is one of the largest uranium provinces in the world.

Type 2—Granite-related: Exploration in historical districts has taken place in Spain and China districts. In Spain, new resources have been defined in several perigranitic deposits and prospects. The district contains about 44.000 tU within 20 deposits.

Type 4—Volcanic-related: Several discoveries have been made in the volcanic Macusani District (Peru) such as Kihitian. At Macusani, the volcanic-related deposits are hosted in felsic peraluminous pyroclastic rocks. The district has 41.000 tU of low-grade geological resources in 13 deposits.

Types 5—Metasomatite: No important discoveries were related to this type of deposit but worthmentioning is the discovery of Inca (Namibia), an unusual skarn subtype deposit associated with the alaskites.

Types 6—Metamorphite: Only one deposit of this type, J4—Ray, was discovered in Nunavut (Canada), near the Lac Cinquante Deposit, whose resources have been significantly increased.

Type 7—Proterozoic unconformity: Several Proterozoic unconformity deposits were discovered in the Athabasca Basin (Canada) and surrounding areas, such as Triple R, Phoenix, Roughrider, and J-Zone. The Triple R, the most recent discovery, is a large, basement-hosted deposit situated outside the Athabasca Basin, along its southwestern margin. The Athabasca Basin has resources of 880.000 tU within 55 deposits. In Australia, the Ranger Deep deposit situated below the mined-out Ranger 3 deposit was confirmed and is presently being developed for underground mining.

Type 9—Sandstone: Important resources have been discovered or confirmed in the Tim Mersoi Basin (Niger). Daja and Dajy are new discoveries and Marianne-Marylin, Isakanan, and Abakorum hadresources confirmed or increased. Total geological resources of the Tin Mersoi Basin are 740.000 tU in 42 sandstone-tabular deposits.

The Frome Embayment (Australia) confirmed its potential with the discovery of Four Mile East and Four Mile West. Geological resources of the district near the Beverley mine are 68.000 tU in seven sandstone (basal channel, roll and tabular) deposits.

Two low-grade deposits in sandstone were discovered in the Sainshand Region of Mongolia (Zoovch Ovoo and Dulaan Ull) and two in the Erlian Basin, China (Bayinwula and Sunjialang). They are rollfront deposits in Mesozoic formations. The Zoovch Ovoo deposit has resources in excess of 50.000 tU.

In Kazakhstan, resources have been greatly increased and confirmed in known deposits. Geological resources are on the order of 850.000 t in about 30 deposits, some of them with resources larger than 100.000 tU (Inkai, Mynkuduk). They all correspond to sandstone deposits associated with stacked rollfronts.

The Karoo Formation in east Africa, running from South Africa to Tanzania through Zambia and Malawi, was the focus of exploration for numerous companies. Today, resources stand at 260.000 tU located within 25 deposits. They are all low-grade sandstone-tabular deposits with the largest one Nyota (Tanzania) containing 55.000 tU in several lenses.

Type 10—Paleo-QPC: Exploration drilling has restarted in the historical Elliott Lake District, not only for U, but also for REE. Seven explored deposits contain geological resources of 101.000 tU. The two largest ones are Banana Lake and the Eco Ridge Mine.

Type 11—Surficial: Two important new surficial deposit districts were delineated on the Reguibat Shield of Mauritania (19.400 tU) and the Manyoni District (11.450 tU) in Tanzania. Several deposits were discovered and/or confirmed in Western Australia, which contains 100.000 tU in 18 deposits and in Namibia with resources of more than 205.000 t in 15 deposits.

Type 13—Carbonate: A large unusual strata-bound deposit, Tummalappalle, has been defined in the dolomitic Vempalle Formation of Proterozoic age (India). Resources are on the order of 72.000 tU and the deposit is open in all directions.

2.7.2 Unconventional deposits and resources (Table 2.10)

*Type 1-Subtype 1.2—Intrusive plutonic (quartz monzonite, peralkaline complex, and carbonatites):*Resources were greatly increased at the Kvanefjeld REE-U-Zn project with 228.240 tU within three deposits. In Morocco, resources have been published for two carbonatite intrusions and one peralkaline complex. Twihinate, with 118.500 tU, is the largest.

Type 3—Polymetallic iron oxide breccia complex: Two new polymetallic iron oxide breccia complexes, Carrapateena and Prominent Hill, have been delineated on the Gawler Craton in South Australia. With Olympic Dam nearby, they are primarily copper deposits. Resources are in excess of 184.000 tU for Carrapateena.

*Type 14—Phosphate:*Chatham Rise (New Zealand), an offshore phosphorite deposit, has been delineated and is ready to be mined. The average grade (240 ppm U) is quite high for such a deposit.

Type 15—Black shale: Large polymetallic resources were defined in Sweden and in Canada. In Sweden, five deposits collectively contain resources of 1300.000 tU in the Scandinavian shales associated with V, Mo, Ni, and Zn.

2.8 Future trends

New conventional deposits: The greatest likelihood of discovering new deposits or increasing resources is first within large known uranium provinces:

- Athabasca Basin (Canada)
- Damara Belt (Namibia)
- Tin Mersoi Basin (Niger)
- Chu Saryssu, Syr Daria, and Kyzylkum districts (Kazakhstan, Uzbekistan)
- Grants Mineral Belt and Wyoming (USA)
- Elliott Lake District (Canada)
- Karoo Formation (East Africa)

Table 2.10 **Major unconventional uranium deposits and resources discovered/defined between 2004 and 2014**

Country	Deposit name	Resources (tU)	Grade (% U)	Deposit type	Deposit subtype	Operator
Sweden	MMS Vicken	447.755	0.0147	Black shale	Stratiform	Continental Precious Minerals
Sweden	Haggan	308.000	0.0133	Black shale	Stratiform	Aura Energy Ltd
Sweden	Narke	257.000	0.0175	Black shale	Stratiform	URU Metals Ltd
Denmark	Illimaussaq Kvanefjeld Sorensen	228.236	0.023	Intrusiveplutonic	Peralcaline complex	Greenland Minerals and Energy
Australia	Carrapateena	184.000	0.0155	Iron oxide breccia complex		Oz Minerals Ltd
Morocco	Twihinate	118.500	0.0212	Intrusive plutonic	Peralcaline complex	ONHYM
Morocco	Glibat Lafhouda	43.050	0.0425	Intrusive	Carbonatite	ONHYM
Canada	Buckton Zone	34.000	0.0011	Black shale	Stratiform	DNI Metals
Finland	Kolmisoppi Kuusilampi	27.000	0.0017	Black shale	Stratiform	Talvivaara Mining Company
Sweden	Marby	23.100	0.0115	Black shale	Stratiform	Aura Energy Ltd
New Zealand	Chatham Rise	12.250	0.024	Phosphate	Minerochemical phosphorite	Chatham Rock Phosphate Ltd

Other potential provinces:

- Frome Embayment (Australia)
- Southern Mongolia
- Chinese sedimentary basins
- Thelon Basin (Canada)
- Achala batholith(Argentina)
- Copper Belt (RDC and Zambia)
- Hoggar District (Algeria)
- Central Mineral Belt (Canada)
- Schist-Greywacke Group and associated two mica granites (Spain, Portugal).

Large portions of the world remain underexplored (northern Canada, Siberia, Brazil, Argentina, Mongolia, China, central Africa, Iran, and the Arabian Platform) containing potential targets for various types of deposits.

Most promising deposits for future planned production: Starting new mines is essentially driven by the price of uranium and increasingly, by societal acceptance. Thus, surficial deposits (western Australia, Namibia, Tanzania, Argentina), which are easy to mine and sandstone deposits where uranium is recovered by low-cost in Situ Recovery (ISR) methods (Wyoming, Mongolia, China) are the most likely to be mined in the near future. Several intrusive anatectic deposits (low grade, large volumes, open pits) are under development in Namibia (Husab, Valencia, Namibplass).

Unconventional deposits (co- and by-product): Very large, low-grade resources are present around the world but very little is known about the grades and resources of most of these deposits. Due to increasingly stringent environmental laws and with talk of comprehensive extraction for some new projects, extraction of uranium from very low-grade mineral deposits may bring additional uranium to the market such as phosphorites, black shales, peralkaline complexes, etc. Today, several polymetallic projects are being evaluated where uranium could be extracted as a coproduct: black shales from Sweden and Canada with Ni, Co, Zn, V, U, and Mo, or deposits in alkaline intrusions from Greenland (REE, Zn, Li, U, and Th). Uranium can also be extracted from formations such as porphyry copper deposits (Bingham Canyon, USA; Chuquicamata, Chile) or base metal deposits (Talvivaara, Sweden).

Acknowledgments

Large portions of the data presented in this chapter come from various documents published or being published by the IAEA. In particular, all numbers are extracted from the uranium deposits UDEPO database. The authors gratefully acknowledge the constructive comments of Gerard Zaluski for the improvement of the manuscript.

Books on uranium and general publications

Burns, P., Sigmond, G.E. 2013. Uranium: Cradle to Grave. Mineralogical Association of Canada, Short Course Series 43, 440p.

Committee on Uranium Mining in Virginia 2012. Uranium Mining in Virginia: Scientific, Technical, Environmental, Human Health and Safety and Regulatory Aspects of Uranium Mining and Processing in Virginia. http://www.nap.edu/catalog.php?record_id=13266, 345p.

Cuney, M., 2011. Uranium and thorium: The extreme diversity of the resources of the world's energy minerals. In Sinding-Larsen, R., Wellmer, F.-W. (Eds.), Non-Renewable Resource Issues: Geoscientific and Societal Challenges, International Year of Planet Earth, Springer, pp. 91–129.

Cuney, M. 2013. Uranium and thorium resources and sustainability of nuclear energy. In: Burns, P. (Ed.), Uranium: Cradle to Grave. Mineralogical Association of Canada, Short Course Series 43, pp. 417–438 (Chapter 15).

Cuney, M., Kyser, K., 2008. Recent and Not-So-Recent Developments in Uranium Deposits and Implications for Exploration. Mineralogical Association of Canada, Short Course Series 39, 257p.

Cuney, M., Kyser, K., 2015. Geology and Geochemistry of Uranium and Thorium Deposits. Mineralogical Association of Canada, Short Course Series 46, 320p.

Dahlkamp, F.J., 1993. Uranium Ore Deposits. Berlin, Springer-Verlag, 460p.

Dahlkamp, F.J., 2009. Uranium Deposits of the World—Asia. Berlin, Springer-Verlag, 493p.

Dahlkamp, F.J., 2010. Uranium Deposits of the World—USA and South America. Berlin, Springer-Verlag, 493p.

Publications in scientific reviews

Several international journals are publishing papers on ore deposit geology, geochemistry, and exploration. Regularly papers on uranium ore deposits are published in these journals. The main journals are the followings:

- *Mineralium Deposita*: the journal publishes since 1966 new observations, principles, and interpretations from the field of economic geology, including nonmetallic mineral deposits, experimental and applied geochemistry, with emphasis on mineral deposits.
- *Economic Geology*: the journal was first published in 1905. It publishes original papers in any segment of economic geology and studies of all classes of mineral deposits.
- *Ore Geology Review*: aims to familiarize all earth scientists with recent advances in a number of interconnected disciplines related to the study of, and search for, ore deposits. The reviews range from brief to longer contributions, but the journal preferentially publishes manuscripts that fill the niche between the commonly shorter journal articles and the comprehensive book coverages, and thus has a special appeal to many authors and readers.
- *Journal of Geochemical Exploration*: is dedicated to the publication of research studies that cover new developments in the application of analytical geochemistry and geoinformatics. Themes considered by the journal include geochemical exploration and the genesis of mineral deposits, environmental geochemistry and geology including metal migration and accumulation processes.

Main websites

- http://www.wise-uranium.org/indexu.html: WISE Uranium Project is part of World Information Service on Energy. It covers the health and environmental impacts of uranium mining and nuclear fuel production: Uranium Radiation and Health Effects, Nuclear Fuel Chain (Issues, Industry, Supply & Demand, Trade, Cost, Energy & CO_2, Impacts], Uranium Mining and Milling (Issues, Industry, Impacts, Tailings), Enrichment and Fuel Fabrication (Issues, Industry, Impacts), Depleted Uranium (Waste, Civilian and Military Use), Tailings Dam Safety, Phosphate.
- http://www.world-nuclear.org/: The World Nuclear Association is the international organization that promotes nuclear energy and supports the many companies that comprise the global nuclear industry: uranium mining, conversion, enrichment and fuel fabrication; reactor marketing; and nuclear engineering, construction, and waste management. WNA provides a global forum for sharing knowledge and insight on evolving industry developments; strengthen industry operational capabilities by advancing best practice internationally; speak authoritatively for the nuclear industry in key international forums that affect the policy and public environment in which the industry operates. WNA website provides information on nuclear energy with some 120 frequently updated information papers.
- http://www.mining.com/: A source for global mining news. Since launching in February 2011, the site has 1.1 million page views per month with over 500,000 subscribers.
- http://www.uxc.com/: The Ux Consulting Company (UxC) is a nuclear industry consulting company. UxC covers the full nuclear fuel cycle with special focus on market-related issues. UxC, founded in March 1994, publishes the Ux Weekly® and Market Outlook reports on uranium, enrichment, conversion, and fabrication as well as publishing the industry standard Ux Prices, referenced in many fuel contracts. UxC also prepares special reports on key topics of interest, as well as provides data services, such as nuclear fuel price indicator reporting, including support for the New York Mercantile Exchange (NYMEX) uranium future contract.

References

Adams, S.S., Cramer, R.T., 1985. Data-process-criteria model for roll-type uranium deposits. Geological Environments of Sandstone-Type Uranium Deposits. IAEA, Vienna, IAEA-TECDOC-328.

Andersson, A., Dahlman, B., Gee, D.G., Snäll, S., 1985. The Scandinavian Alum Shales: Sveriges Geologiska Undersoekning. Serie Ca. Avhandlingar och Uppsatser IA4, NR56.

Barnicoat, A.C., Henderson, I.H.C., Knipe, R.J., Yardley, B.W.D., Napier, R.W., Fox, N.P. C., et al., 1997. Hydrothermal gold mineralization in the Witwatersrand basin. Nature. 386, 820–824.

Belevtsev, Y.N., Koval, V.B., 1968. Genesis of uranium deposits associated with sodium metasomatism in crystalline rocks of shields. Geologicheskiy Zhurnal (Kiev.). 28, 3–17.

Berning, J., Cooke, R., Hiemstra, S.A., Hoffman, U., 1976. The Rössing uranium deposit, South West Africa. Econ. Geol. 71, 351–368.

Boitsov, V.E., Pilipenko, G.N., 1998. Gold and uranium in Mesozoic hydrothermal deposits of Central Aldan, Russia. Geologiya Rudnykh Mestorozhdenii. 40, 354–369.

Boyle, D.R., 1982. The formation of basal-type uranium deposits in South Central British Columbia. Econ. Geol. 77, 1176–1209.

Bruneton, P., CuneyM., Dahlkamp, F., Zaluski, G.,2014. IAEA geological classification of uranium deposits. URAM 2014, IAEA Vienna, to be published in 2016.

Burns, P.C.,1999. The crystal chemistry of uranium. In: Burns, P.C., Ewing, R.C. (Eds.), Uranium: Mineralogy, Geochemistry and the Environment. Mineralogical Society of America and Geochemical Society Reviews in Mineralogy and Geochemistry, vol. 38, pp. 23–90.

Burns, P.C., Sigmond, G.E., 2013. Uranium: Cradle to Grave. Mineralogical Association Canada Short Course Series 43, 440 p.

Carlisle, D., Merifield, P.M., Orme, A.R., Kohl, M.S., Kolker, O., 1978. The distribution of calcretes and gypcretes in southwestern United States and their uranium favorability based on a study of deposits in Western Australia and South West Africa (Namibia). US Deparment of Energy. Open File Rpt GJBX-29.

Cathelineau, M., Boiron, M.C., Holliger, P., Poty, B., 1990. Metallogenesis of the French part of the Variscan orogen. Part II: time-space relationships between U, Au and Sn-W ore deposition and geodynamic events—mineralogical and U-Pb data. Tectonophysics. 177, 59–79.

Chabiron, A., Cuney, M., Poty, B., 2003. Possible uranium sources for the largest uranium district associated with volcanism: the Steltsovka caldera (Transbaïkalia, Russia). Mineral. Dep. 38, 127–140.

Corfu, F., Andrews, A.J., 1986. A U–Pb age for mineralized Nipissing diabase, Gowganda, Ontario. Can. J. Earth Sci. 23, 107–109.

Cuney, M., 2009. The extreme diversity of uranium deposits. Mineral. Dep. 44, 3–9.

Cuney, M., 2010. Evolution of uranium fractionation processes through time: driving the secular variation of uranium deposit types. Econ. Geol. 105, 449–465.

Cuney, M., 2011. Uranium and thorium: the extreme diversity of the resources of the world's energy minerals. In: Sinding-Larsen, R., Wellmer, F.-W. (Eds.), Non-Renewable Resource Issues: Geoscientific and Societal Challenges, International Year of Planet Earth. Springer, Berlin, pp. 91–129.

Cuney, M., Friedrich, M., 1987. Physicochemical and crystal-chemical controls on accessory minerals paragenesis in granitoïds. Implications on uranium metallogenesis. Bull. Minéral. 110, 235–247.

Cuney, M., Kyser, K., 2008. Recent and Not-So-Recent Developments in Uranium Deposits and Implications for Exploration. Mineralogical Association of Canada, Short Course Series, 39, 257 p.

Cuney, M., Friedrich, M., Blumenfeld, P., Bourguignon, A., Boiron, M.-C., Vigneresse, J.-L., et al., 1990. Metallogenesis in the French part of the Variscan orogen. Part I. U-preconcentrations in the pre-Variscan and Variscan formations—a comparison with Sn, W and Au. Tectonophysics. 177, 39–57.

Cuney, M., Emetz, A., Mercadier, J., Mykchaylov, V., Shunko, V., Yuslenko, A., 2012. U deposits associated with sodium metasomatism from Central Ukraine: a review of some of the major deposits and genetic constraints. OGR. 44, 82–106.

Dahlkamp, F.J., 1993. Uranium Ore Deposits. Springer-Verlag, Berlin.

Dahlkamp, F.J., 2009. Uranium Deposits of the World–Asia. Springer-Verlag, Berlin.

Derome, D., Cuney, M., Cathelineau, M., Dubessy, J., Bruneton, P., Hubert, A., 2003. A detailed fluid inclusion study in silicified breccias from the Kombolgie sandstones (Northern Territory, Australia): application to the genesis of middle-Proterozoic unconformity-type uranium deposits. J. Geoch. Expl. 78–79, 525–530.

Derome, D., Cathelineau, M., Cuney, M., Fabre, C., Lhomme, T., Banks, D.A., 2005. Mixing of sodic and calcic brines and uranium deposition at McArthur river, Saskatchewan, Canada: a Raman and laser-induced breakdown spectroscopic study of fluid inclusions. Econ. Geol. 100, 1529–1545.

Eglinger, A., André-Mayer, A.-S., Vanderhaeghe, O., Mercadier, J., Cuney, M., Decrée, S., et al., 2013. Geochemical signatures of uranium oxides in the Lufilian belt: from unconformity-related to syn-metamorphic uranium deposits during the Pan-African orogenic cycle. OGR. 54, 197–213.

Finch, R., Murakami, T., 1999. Systematics and paragenesis of uranium minerals. In: Burns, P.C., Ewing, R.C. (Eds.), Uranium: Mineralogy, Geochemistry and the Environment, 38. Mineralogical Society of America and Geochemical Society Reviews in Mineralogy and Geochemistry, pp. 91–179.

Finch, W.I., Davis, J.F., 1985. Sandstone-type uranium deposits. An introduction. In: Geological Environments of Sandstone-Type Uranium Deposits. IAEA-TECDOC-38. Vienna, pp. 11–20.

Frimmel, H.E., Minter, W.E.L., 2002. Recent developments concerning the geological history and genesis of the witwatersrand gold deposits, South Africa. Soc. Econ. Geol. Spec. Pub. 9, 17–45.

Frimmel, H.E., Groves, D.I., Kirk, J., Ruiz J., Chesley J.,Minter W.E.L., 2005. The formation and preservation of the Witwatersrand Goldfields, the largest gold province in the world. In: Hedenquist, J.W., Thompson, J.F.H., Goldfarb, R.J., Richards, J.P. (Eds.), 100th Anniversary volume, Society of Economic Geologists, 9, pp. 769–797.

Fuchs, H., Hilger, W., 1989. Kiggavik (Lone Gull): an unconformity related uranium deposit in the Thelon Basin, Northwest Territories, Canada. Uranium Resources and Geology of North America In: Proceedings of a Technical Committee Meeting, Saskatoon, Saskatchewan: International Atomic energy Agency, Vienna IAEA-TECDOC.500 p.

Gauthier-Lafaye, F., 1986. Les gisements d'uranium du Gabon et les réacteurs d'Oklo. Modèle métallogénique de gites à fortes teneurs du Protérozoïque inférieur. Mém. Sci. Géol. Strasbourg Univ., 278.

Goldfarb, R.J., Groves, D.I., Gardoll, S., 2001. Orogenic gold and geologic time: a global synthesis. Ore Geol. Rev. 18, 1–75.

Goldhaber, M.B., Reynolds, R.L., Rye, R.O., 1978. Origin of a south texas roll-type uranium deposit: II. Sulfide petrology and sulfur isotope studies. Econ. Geol. 73, 1690–1705.

Golubev, V.N., Cuney, M., Poty, B., 2000. Phase composition and U-Pb isotopic systems of pitchblende of quartz-calcite-pitchblende veins at the Schlema-Alberoda deposit (Erzgebirge). Geol. Ore Deposits. 42, 761–773.

Hansley, P.L.O., Spirakis, C.S., 1992. Genesis of tabular uranium-vanadium deposits in the Morrison Formation, Colorado Plateau. Econ. Geol. 87, 352–365.

Hitzman, M.W., Valenta, R.K., 2005. Uranium in Iron Oxide–Copper–Gold (IOCG) systems. Econ. Geol. 100, 1657–1661.

Hronsky, J.M.A., Groves, D.I., 2008. Science of targeting: definition, strategies, targeting and performance measurement. Aust. J. Earth Sci. 55, 3–12.

IAEA, 1996. Guidebook to accompany IAEA map: World Distribution of Uranium Deposits. IAEA, Vienna.

IAEA, 2001. Analysis of Uranium Supply to 2050. STI-PUB-1104, IAEA, Vienna.

IAEA, 2009. World Distribution of Uranium Deposits (UDEPO) with Uranium Deposit Classification, IAEA-TECDOC-1629, 117 p.

IAEA, 2016a. Geological Classification of Uranium Deposits with Description of Selected Examples. IAEA, Vienna, In press.

IAEA, 2016b. World Distribution of Uranium Deposits (UDEPO)—2015 Edition. IAEA, Vienna, to be published in 2016.

Ishukova, L.P., Modnikov, I.S., Citchev, I.V., 1991. Uranium ore-forming systems of continental volcanism regions. Geol. Rudn. Mestorozhd. 3, 16–25 (in Russian).

Janeczek, J., Ewing, R.C., 1992. Structural formula of uraninite. J. Nucl. Mater. 190, 128–132.

Jefferson, C.W., Thomas, D.J., Gandhi, S.S., Ramaekers, P., Delaney, G., Brisbin, D., et al., 2007. Unconformity associated uranium deposits of the Athabasca Basin, Saskatchewan and Alberta. In: Jefferson, C.W., Delaney, G. (Eds.), EXTECH IV: Geology and Uranium EXploration TECHnology of the Proterozoic Athabasca Basin, Saskatchewan and Alberta, 588. Geological Survey of Canada Bulletin, pp. 23–67.

Kazanskii, V.I., 2004. The unique Central Aldan gold-uranium ore district, Russia. Geol. Rudn. Mestorozhd. 46, 195–211.

Kish, L., Cuney, M., 1981. Uraninite-albite veins from the Mistamisk Valley of the Labrador Trough, Québec. Mineral. Mag. 44, 471–483.

Klepper, M.R., Wyant, D.G., 1956. Uranium provinces. In: Page, L.R., Stocking, H.E., Smith, H.B., Contributions to the Geology of Uranium and Thorium by the United States Geological Survey and Atomic Energy Commission for the United nations International Conference on Peaceful Uses of Atomic Energy, Geneva, Switzerland, 1955, USGS Professional Paper 300, pp. 17–26.

Kondrat'eva, I.A., Maksimova, I.G., Nadyarnykh, G.I., 2004. Uranium distribution in ore-bearing rocks of the malinov deposit: evidence from fission radiography. Lithol. Mineral. Res. 39, 333–344.

Lange, G., Freyhoff, G., 1991. Geologie und Bergbau in der Uranlagerstätte Ronneburg/Thüringen. Erzmetal. 44, 264–269.

Laznicka, P., 1999. Quantitative relationships among giant deposits of metals. Mineral Deposits: Processes to Processing. Balkema, Rotterdam.

Leroy, J., George-Aniel, B., 1992. Volcanism and uranium mineralisations: the concept of source rock and concentration mechanism. J. Volcan. Geoth. Res. 50, 247–272.

Linnen, R.L., Cuney, M., 2008. Granite-related rare-element deposits and experimental constraints on Ta-Nb-W-Sn-Zr-Hf mineralization. In: Linnen, R.L., Samson, I.M. (Eds.), Rare-Element Geochemistry and Mineral Deposits, 17. Geological Association Canada, Short Course Notes, pp. 45–67.

Maas, R., Mcculloch, M.T., Campbell, I.H., Page, R.W., 1987. Sm–Nd isotope systematics in uranium–rare earth mineralisation at the Mary Kathleen uranium mine, Queensland. Econ. Geol. 82, 1805–1826.

Mathieu, R., Cuney, M., Cathelineau, M., 2000. Geochemistry of palaeofluids circulation in the Franceville basin and around Oklo natural nuclear reaction zones (Gabon). J. Geoch. Expl. 69–70, 245–249.

Mathis, V., Robert, J.-P., Saint Martin, J., 1990. Géologie et métallogénie des gisements d'uranium du bassin permien de Lodève (sud Massif central français). Chron. Rech. Min. 499, 31–40.

Min, M.Z., Luo, X.Z., Mao, S.L., 2002. The Saqisan Mine—a paleokarst uranium deposit, South China. OGR. 19, 79–93.

Morales, G., Nadal, P.J., Merino, J.L., Gasos, P., 1985. Uranium recovery as a byproduct from radioactive coal. Advances in Uranium Processing and Recovery from Non-conventional Sources. IAEA, Vienna, TC 491/16.

OECD IAEA, 2006. Uranium 2005: resources, production and demand. A Joint Report by the OECD Nuclear Energy Agency and the IAEA. OECD, Paris.

OECD IAEA, 2008. Uranium 2007: resources, production and demand. A Joint Report by the OECD Nuclear Energy Agency and the IAEA. OECD, Paris.

OECD/NEA-IAEA, 2012. Uranium 2011: resources, production and demand. A Joint Report by the OECD Nuclear Energy Agency and the IAEA. OECD, Paris.

OECD/NEA-IAEA, 2014. Uranium 2013: resources, production and demand. A Joint Report by the OECD Nuclear Energy Agency and the IAEA. OECD, Paris.

Page, R.W., 1983. Chronology of magmatism, skarn formation, and uranium mineralization, Mary Kathleen, Queensland. Aust. Econ. Geol. 78, 858–855.

Polito, P.A., Kyser, K.T., Stanley, C., 2007. The Proterozoic, albitite-hosted, Valhalla uranium deposit, Queensland, Australia, a description of the alteration assemblage associated with uranium mineralisation in diamond drill hole. Mineral Dep. 44, 11–40.

Pretorius, D.A., 1961. The nature of the Witwatersrand gold-uranium deposits. In: Wolf, K. H. (Ed.), Handbook of Stratabound and Stratiform Ore Deposits. Elsevier, Amsterdam, pp. 29–88.

Richard, A., Rozsypal, C., Mercadier, J., Banks, D.A., Cuney, M., Boiron, M.C., et al., 2012. Giant uranium deposits originate from exceptionally U-rich brines. Nat. Geosci. 5, 142–146.

Skirrow, R.G., Jaireth, S., Huston, D.L., Bastrakov, E.N., Schofield, A., Van Der Wielen, S. E., et al., 2009. Uranium mineral systems; processes, exploration criteria and a new deposit framework. Rec. Geosci. Aus., 44.

Sørensen, H., Rose-Hansen, J., Nielsen, B.L., Løvborg, L., Sørensen, E., Lundgaard, T., 1974. The uranium deposit at Kvanefjeld, the Ilímaussaq intrusion, South Greenland. Rpt Grønlands Geologiske Undersøgelse. 60, 54.

Spirakis, C.S., 1996. The roles of organic matter in the formation of uranium deposits in sedimentary rocks. Ore Geol. Rev. 11, 53–69.

Turpin, L., Leroy, J., Sheppard, S.M.F., 1990. Isotopic systematics (O, H, C, Sr, Nd) of superimposed barren and U-bearing hydrothermal systems in a granite, Massif Central, France. Chem. Geol. 88, 85–96.

USGS, 2011. USGS database of phosphates. Mineral Commodity Summaries. <http://minerals.usgs.gov/minerals/pubs/commodity/phosphate_rock/index.html#mcs>.

Wenrich, K.J., Titley, S.R., 2008. Uranium exploration for northern Arizona (USA) breccia pipes in the 21st century and consideration of genetic models. Arizona Geol. Soc. Digest. 22, 295–309.

Exploration for uranium

3

Kurt Kyser
Department of Geological Sciences and Geological Engineering,
Queen's University, Kingston, ON, Canada

3.1 Introduction

Uranium is unique among naturally occurring elements in that it is an energy-related material used almost exclusively for generating electricity. About 8% of the total energy consumed globally comes from nuclear power (Fig. 3.1) generated from 438 nuclear reactors in 31 countries. Globally, these reactors, plus 270 research reactors, require nearly 66,883 tons of uranium (tU) annually (WNA, 2015a). Even considering all existing reactors, the 68 reactors currently under construction in 13 countries, and those anticipated to be built, the need for uranium will increase by about 30% to 72,000–122,000 tU by the year 2035 (OECD/NEA-IAEA, 2014). One of the attractive aspects of nuclear power is that the energy generated from uranium has a minimal "carbon footprint" and offsets the additional emissions expected from the increase in energy consumption anticipated in the future (Pacala and Socolow, 2004), although the proportion of nuclear energy will probably not increase (Fig. 3.1). However, to meet the current and projected needs of the uranium industry, discovery of new deposits and development of new technologies for both exploration and processing are critical.

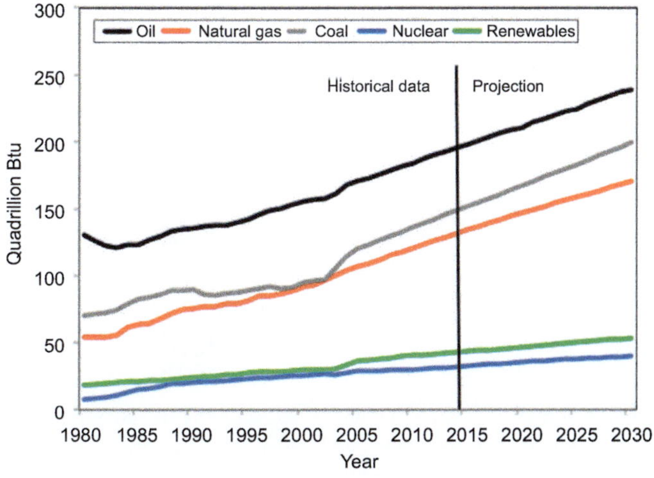

Figure 3.1 World marketed energy use by Fuel Type for 1980–2030.
Source: Data from EIA (2007) and OECD/NEA-IAEA (2010).

Uranium for Nuclear Power. DOI: http://dx.doi.org/10.1016/B978-0-08-100307-7.00003-X

3.2 Uranium and the nuclear fuel cycle

The uranium cycle is often referred to colloquially as the nuclear fuel cycle. In most renditions of the nuclear fuel cycle, the very first step involving exploration is often overlooked. This step is not only the first, it is critical because it begins the cycle and is the least trivial in the nuclear fuel cycle, as it cannot be engineered.

Uranium is a ubiquitous element that occurs everywhere, although the concentrations of uranium are normally low, averaging only 3 ppm in the upper continental crust (1.7 ppm in the entire crust) and 0.0032 mg/L in seawater (Clark et al., 1966). Uranium, like thorium and plutonium, is a member of the actinide series of elements. Uranium has an atomic number of 92 and has three naturally occurring radioactive isotopes, ^{234}U, ^{235}U, and ^{238}U, whose decay render uranium distinct from other commodities. The decay to radiogenic isotopes is one of the aspects exploited in the exploration for uranium deposits.

The most abundant isotopes of uranium, ^{235}U and ^{238}U, are commonly used in geological sciences as geochronometers because of their long half-lives and because they each decay to an isotope of Pb (Fig. 3.2). In deposit studies, the decay schemes

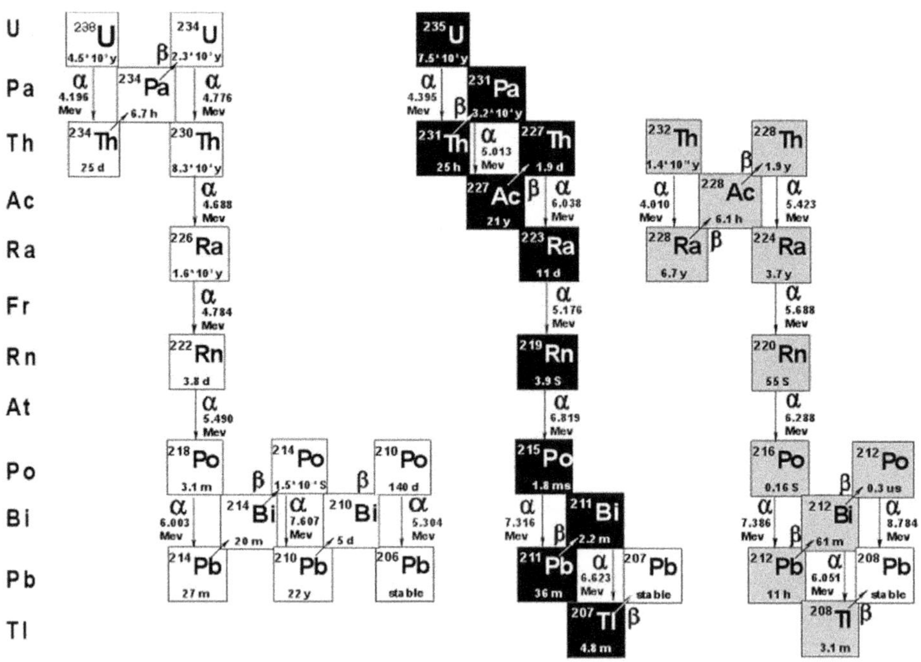

Figure 3.2 Decay schemes of ^{235}U, ^{238}U, and ^{232}Th showing the position of ^{234}U in the decay chain of ^{238}U, energies of the decay, and half-lives. The geochemical properties of the decay products are quite distinct, including production of radon gas. These will tend to separate from the ores in open systems and leak into the surrounding environment, where they can be used to indicate proximity to a deposit.
Source: From Cuney and Kyser (2009).

of ^{235}U and ^{238}U along with the isotopes of Pb are used to determine the age of uranium minerals, but only 0.7204% of natural uranium is ^{235}U. The isotope ^{234}U, which occurs as the decay product of ^{238}U and is used as a geochronometer as well, makes up only 0.0055% of natural uranium because its half-life of 246,000 years is 0.000055 times as long as the half-life of ^{238}U. However, the decays of ^{235}U and ^{238}U are complex, with many intermediate progeny (Fig. 3.2). These progeny decay to other progeny along the chain until ^{207}Pb or ^{206}Pb are produced from ^{235}U and ^{238}U, respectively (Fig. 3.2). Because of differences in the geochemical properties of the progeny elements produced, they can be differentially dispersed into the environment if the system is open, resulting in disequilibrium in the radioactivity of isotopes in the decay schemes. Some of the intermediate products such as Rn, Ra, and Bi can be used to indicate buried mineralization, and so are useful in exploration for deposits. The factors that affect the deposits (so they can be detected undercover) and the techniques used in exploration for uranium are the subjects of this chapter.

3.3 Drivers of uranium exploration

Two factors figure greatly in the exploration and exploitation of ore deposits, namely economic factors and the role of research and technology. Although awareness of economics has become more acute by most geologists because profit is fundamental to mining companies, effective analysis of economic factors is sometimes overlooked. As a metal commodity, uranium is distinguished by its use in energy generation, but also by its military weapon potential. The breakup of the former Soviet Union and changes in global politics afforded the release of research results on viable uranium deposits, such as volcanic-related uranium and metasomatite, that were formerly unavailable to the West, thereby expanding the potential importance of specific deposits that could be targeted (Cuney and Kyser, 2009).

The Organization for Economic Co-operation and Development (OECD) and the International Atomic Energy Association (IAEA) host the Uranium Group, which classifies uranium resources into Reasonably Assured Resources (RAR), Inferred Resources (IR), and Speculative Resources (SR). The first two categories are often combined into Identified Resources to reflect the realistic known resources recoverable from uranium deposits delineated by sufficient direct measurement to conduct prefeasibility and feasibility studies on projects. However, the true "resource" of a commodity is a function of the resources that can be recovered within a given price range (cf. Chapter 1). The current price is near <USD 130/kg U (USD 50/pound U_3O_8), with Identified Resources in 2013 at 5.903 million tU at <USD 130/kg. About 85% of RAR and IR are recoverable at <USD 130/kg, although there is a distinct lack of geographic diversity, with 64% of these resources located in just five countries. Australia has the greatest resources, followed by Kazakhstan, the Russian Federation, Canada, and Niger. However, the deposit types in each country are different, with the iron oxide–copper–gold (IOCG) deposit at Olympic Dam and unconformity-related deposits being the major uranium resources in Australia, sandstone-hosted deposits as the major uranium resources in Kazakhstan and Niger,

volcanic-hosted and a mix of other deposit types in the Russian Federation, and unconformity-related deposits as the exclusive resource of uranium in Canada. Thus, exploration strategies, which are based primarily on geology and metallogeny in the first stages of exploration, are different in each country. Most of these resources are related to sedimentary basins, which are areas of the world where exploration activities are most intense.

Uranium is a commodity whose value is determined by supply and demand. Uranium supply is divided into two categories, primary supply and secondary supply (Fig. 3.3). Primary supply includes newly mined and processed uranium, whereas secondary supply includes highly enriched uranium from dismantling of nuclear weapons, reprocessed uranium, mixed oxide fuels, and uranium from stockpiles (McMurray, 2006). Primary production of uranium has been less than reactor requirements since 1990 and secondary sources have made up the difference. Global uranium primary production nearly met reactor requirements in 2013 (Fig. 3.3). Primary sources typically require in excess of 10 years from discovery to production, so that after about the year 2020, the needs by commercial reactors will have to be met by additional primary uranium supplies, new technologies for more efficient exploitation of secondary sources, or both.

Exploration expenditures (in constant dollars) have an excellent correlation with uranium price, with the influence of a price increase on expenditures occurring 1–2 years after the price change (Fig. 3.4).

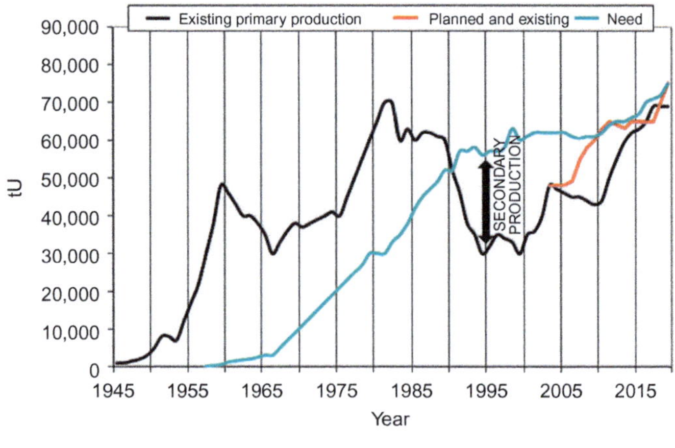

Figure 3.3 Primary uranium production (in tons of uranium), production + planned primary sources at <USD 130/kg and uranium needed for civilian use as a function of year, including estimates of these until 2020. Global primary production exceeded use of uranium until 1990. The difference between primary production and the uranium needed represents the inventory buildup prior to 1990 or uranium derived from secondary sources since 1990 (as indicated).

Source: Data from Vance et al. (2006), Price et al. (2006), Shatalov et al. (2006), OECD (2001, 2008), OECD/NEA-IAEA (2014), and WNA (2015a). Demand will be about 20% greater if military and civil naval ship needs are included.

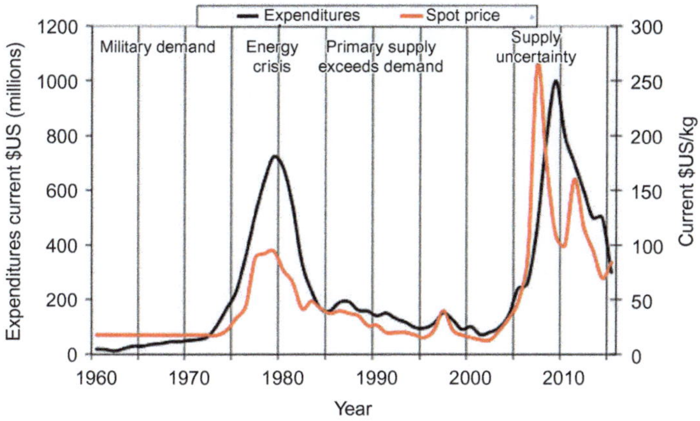

Figure 3.4 Spot market price of uranium in USD/kg for each year and expenditures in millions of USD for exploration of uranium deposits as a function of year since 1960. *Source:* Data from OECD (2001, 2008), OECD/NEA-IAEA (2014), and Price (2006).

Beginning in 1975, expenditures for exploration rose rapidly to a peak of USD 756 million in 1979 and then plummeted rapidly along with the spot price of uranium (Vance et al., 2006). This rise in expenditures and price resulted first from a combination of growth in military needs until the early 1960s, a period of time during which the cost of uranium extraction was not controlled by economic parameters, and then to feed the reactors since 1973. Since 1945, the total spent on uranium exploration has been about USD 15 billion, with the former USSR accounting for about 20% of this and Canada only 10% (Price, 2006). Estimates of the average historical expenditures aggregated across the exploration industry required to find unconformity-related deposits in the Athabasca Basin of Canada vary from USD 1.5 to 2.5 billion, or slightly more than USD 1/lb. Future discoveries will come at significantly higher costs as the impact of resource depletion and current exploration technology limitations are realized.

3.4 Prospectivity, explorability, and exploration targeting

Formulation of effective exploration strategies involves integrating (1) the prospectivity of the area, (2) the economic potential of the deposit, and (3) the explorability of the area. The prospectivity is determined by the geology and possibility of a particular type of deposit forming, also related to the metallogeny. The economic potential is related to the deposit type, grade, and size because this will determine the cost of mining and hence profit margin. The economics of deposits are such that grade is critical for deep deposits and less so for those near the surface. The explorability is the feasibility of exploring in that area, normally related to available infrastructure and politics of the host country. Countries that already mine uranium are usually the most favorable because they have a positive uranium mining policy, although this can change with new governments.

Most of the uranium in the world is supplied from deposits associated with sandstone-related environments; hence, these are among those prospective for exploration (Fig. 3.5). For example, in 2013, about 45% was by *in situ* leach extraction from sandstone-related deposits (ca. 12,500 tU/year), 10% was from conventional mining of sandstone-related deposits, and 13% of uranium production was from unconformity-related deposits (ca. 7000 tU/year)(OECD/NEA-IAEA, 2014). Therefore, the most lucrative targets for uranium exploration are those associated with sedimentary rocks, specifically those associated with Proterozoic unconformities and Phanerozoic clastic/marine basins.

In terms of mine types that have the greatest capacity potential, open pit mines generally have greater capacity than underground mines, and underground mines have greater capacity than mines where uranium is extracted by *in situ* leach technology. However, *in situ* leach technology dominates current production at 51%, underground mines supply about 24% of the global uranium production, and open pit mining methods supply about 18% of the uranium produced in 2014, with the remainder as a byproduct of other commodities (WNA, 2015b). The higher proportional contribution from *in situ* recovery (ISR) techniques reflects a substantial increase since 2008 and the low cost of mining using this technology. However, deposits near the surface, even at low grades, such as the Rössing deposit in Namibia with a grade of 0.03% uranium, can be significant economic sources of uranium.

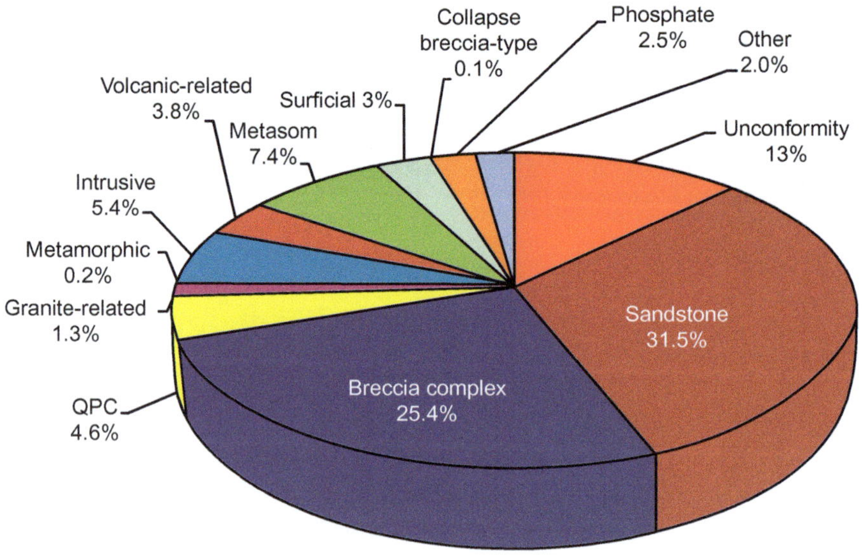

Figure 3.5 Percentage of the total production (59,370 tU) of uranium as a function of deposit type for 2013 (cf. Chapter 2).
Source: Data from OECD/NEA-IAEA (2014).

Metallogeny, which is the study of the relationship between geology and the genesis of ore deposits in space and time, is normally an important factor in formulating exploration strategies. Thus, areas with favorable geology for high-priority deposit types are desirable targets because they are prospective, although they may not be explorable. There are uranium mines operating in 21 countries; more than half of world production comes from just 10 mines that have average grades in excess of 0.10% uranium, with some Canadian mines such as McArthur River and Cigar Lake having average grades up to 25% uranium (Fig. 3.6). Countries where uranium mining is currently active normally have both prospectivity and explorability. Australia has the largest recoverable reserves of uranium at 29% of the global Identified Resources at <USD 130/kg U in 2013, in large part because of the Olympic Dam IOCG deposit and unconformity-related deposits. Kazakhstan has 11% of the global reserves, which are mainly sandstone-type deposits mined using

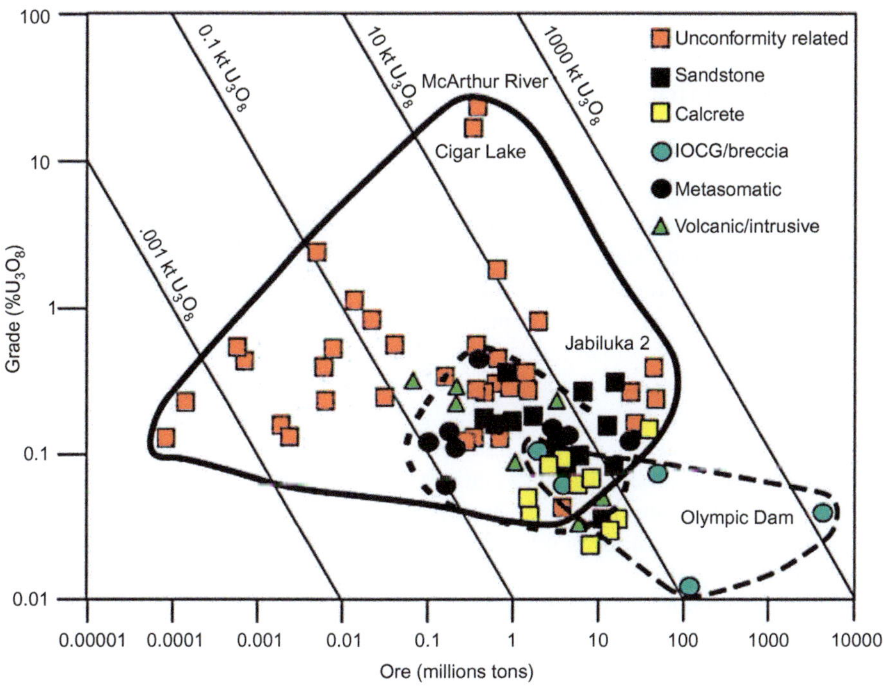

Figure 3.6 Grade vs tonnage for major types of uranium ore deposits. Unconformity-related deposits, encompassed by the solid line, have the highest grade and large reserves. Particularly high grades and reserves characterize the deposits at Cigar Lake and McArthur River in the Athabasca Basin, and large reserves also occur in the Jabiluka 2 deposit in Australia. Volcanic/intrusive, metasomatic, sandstone, and calcrete uranium deposits, indicated by the group surrounded by the short dashed line, have much lower grades and are smaller, as are IOCG/breccia related deposits, shown by group indicated by the long dashed line, except for Olympic Dam where uranium is a byproduct.
Source: From Cuney and Kyser (2009).

ISR methods; Russia has 9% with a large part coming from undeveloped K-metasomatites, sandstone, and volcanic-related deposits, and a lesser amount from intrusive related deposits (OECD/NEA-IAEA, 2014); and Canada has the next most plentiful at 8%, entirely from unconformity-related ore deposits.

Knowledge of the general properties of uranium, the nuclear power cycle, deposit types, mining methods, and the economics of uranium are all important to the uranium industry and hence to exploration strategies. To have a mine, discovery of ore bodies is necessary, which requires a knowledge base that must be constructed, expanded, and always refined. Most of the largest and most profitable mines are associated with sedimentary basins, particularly those mined using open pit methods. Finding these and other types of uranium deposits requires knowledge of how economics in the uranium industry and research results might affect exploration strategies, particularly as societal acceptance becomes an increasing challenge for the development of new deposits all over the world.

3.5 Exploration techniques

Exploration for uranium deposits, as with any type of geological search, requires the integration of regional geology, structural geology, geophysics, and geochemistry and must embrace new technologies and research results to be effective. Although "luck" and "serendipity" will always be factors of varying proportions, exploration must be more purposeful, especially to find deposits under cover. In formulating exploration strategies, an analogy might be someone looking for a treasure hidden in a desk drawer in a house in some unfamiliar city—geology and metallogeny can get you in the right city and neighborhood, geophysics can get you in the right house, and geochemistry can get you in the right room in that house. However, you still need to find the exact drawer.

The exploration process should initially address two issues: (1) is the terrain being considered prospective and (2) is the terrain explorable (Marlatt and Kyser, 2011)? Exploration geologists must continuously evaluate the political risk factors as well as the viability of technical applications as projects evolve. Analysis of the prospectivity is directed at ensuring that the exploration effort will be focused on the best area that has the potential to host economic uranium deposits. Thus, the metallogeny of the area—the status of the natural endowment of economic mineral deposits in this terrain—is a significant factor. The geological assessment of the uranium mineral endowment is a knowledge-based activity focused on understanding deposit models and the probability of their occurrence.

Exploration strategies for uranium deposits vary depending on which type of deposit is being sought. Some strategies are common to any type of deposit, such as property evaluation involving review of the metallogeny and evaluation of available data. Despite the diversity of deposit types, and therefore the diversity of exploration strategies, some general considerations apply to exploration for all types of uranium deposits.

3.5.1 Property evaluation

Exploration means first generating targets, ranging from the right area to be in, to the right spot on which to sample and drill. Thus, the potential property must be fully evaluated prior to staking or much financial investment. Many exploration geologists fail at property evaluation not because they have not assembled all the data required for evaluation, but because they either do not fully evaluate the data or they tend to rush to the conclusion that the property is positive, so that the project on the property must be promoted. Full property evaluation must be integrated and include the following:

1. Geology—Is the geology correct for the type of deposit being sought? This question includes the tectonic environment, the timing of geologic events, and an understanding of the metallogeny. For example, exploration for unconformity-related deposits is based firstly on Proterozoic redbed basins overlying basement complexes and source regions characterized by high uranium contents. There are nearly 200 basins globally that have these attributes, and about 15 of these have further attributes that make them prospective. Graphitic metasedimentary units within the basement complex are desirable, but certainly not necessary, and repeated brittle reactivations of ductile structures that offset the basal unconformity and were foci for fluid flow and ore deposition are required. The Athabasca and Thelon basins in Canada and the McArthur Basin in Australia have favorable proven metallogeny for unconformity-related deposits, whereas most other Proterozoic basins are unproven despite having similar geology. Most of them lack the metallogeny and critical data.
2. Land position availability—In areas of favorable geology and metallogeny, there must be land that can be acquired or used with secure tenure. Some countries have formidable land policies that can become a considerable financial burden, so knowledge of land policies must be part of the property evaluation.
3. Evaluation of historical exploration results, assessment data, and drill core data. These evaluations should be completed prior to securing a land position. During the mini-boom of uranium exploration in 2007 (Fig. 3.4), virtually all of the Athabasca Basin in Canada was claimed, despite much of it being almost impossible to explore. Apparently, many in exploration lacked the time or experience to evaluate their properties from government, company, and published results—these must be fully evaluated and researched to avoid re-inventing the wheel and to provide data to make intelligent exploration strategies.
4. Accessibility—Can the property be accessed by road or is a helicopter required? Assess what types of samples (outcrop, soils, till, vegetation, drill core) can be collected as part of the assessment.
5. Liabilities—What are the protocols for land staking and assessment requirements?

A decision to retain the property must be tempered with a complete assessment, and only then should effort be put into exploration. Otherwise, the property should be relinquished with minimal investment. The tendency of most companies is to retain properties well beyond the time they should because a decision was made to do so before the property was evaluated properly.

3.5.2 Exploration methods

Property evaluation is an ongoing and iterative process, but once all the available data are evaluated, exploration moves into a different phase wherein data lacking must be generated by the exploration team. A variety of methods is normally used for exploration, including a balance of geological mapping, remote sensing, geophysics, and geochemistry.

3.5.2.1 Geological mapping

Mapping is required to supplement what is available from government, industry, and academic sources, usually at a finer scale and concentrated on the property. In particular, mapping is focused on structures that are evident and what the structural evolution of the area is. Structures are controlling many types of uranium deposits related to hydrothermal fluid circulation. Structures have to be considered at different scales. Regional scale structures or lineaments are generally transcrustal and active over a long time period. Examples include the Arlit-In Azawa fault in Niger for the tabular deposits of the Arlit district, the Krivoi Rog and Kirovograd shear zones in central Ukraine for sodium metasomatic deposits, the Yuzhny fault zone for the deposits related to K-metasomatism of the Elkon district in Russia, and the P2 fault that hosts the McArthur River unconformity-related deposits in Canada. Individual deposits are controlled by localized second- or third-order structures. Structural results from mapping should be integrated with geophysical methods that identify possible structures at depth based on physical anomalies of the rocks. Although structure is critical in the formation of the deposits, the vast majority of structures with seemingly appropriate characteristics are not mineralized. Mapping is also used to characterize the cover, including Quaternary geology, soil character and type and character of vegetation.

3.5.2.2 Remote sensing

Sensing of the physical and chemical properties of the area remotely is often part of the property evaluation step. However, these data must be integrated into the exploration strategy at a variety of scales. Remote sensing is the science of acquiring, processing, and interpreting spectral information about the Earth's surface and recording interactions between matter and electromagnetic (EM) energy, mainly from satellites. Among the remote sensing data that should be integrated in property evaluation are digital elevation maps and light detection and ranging used to evaluate the geography of the area, radiometry as part of airborne geophysics packages or the National Oceanic and Atmospheric Administration—Advanced Very High Resolution Radiometer and Advanced Spaceborne Thermal Emission and Reflection Radiometer to identify radiometric anomalies within a few centimeters of the surface, and hyperspectral data from Landsat thematic mapper that identify types of vegetation, some mineralogy, and Fe contents (Bharti and Ramakrishnan, 2014). These techniques detect energy reflected and emitted from the Earth's surface from minerals, vegetation, soils, ice, water, and rocks, in selected wavelengths. For example, hyperspectral data have been used to characterize granite-type uranium deposits in South China (Zhang et al., 2014).

3.5.2.3 Geophysical methods

These vary with the type of uranium deposit because different aspects of the mineralization are targeted. The mainstay of exploration geophysics for uranium deposits is gamma ray radiometry and spectrometry used from airborne regional to local detailed surveys, down-hole logging, and on outcrops using handheld units (IAEA, 2013). The former is normally collected as part of airborne magnetic surveys with magnetic contrasts used to reveal the general geology. Airborne gamma ray spectrometry directly measures U, K, and Th in surficial material.

Magnetics are used to map basement lithologies in basin-hosted systems. Various EM techniques are used to image graphitic conductors that are occasionally associated with unconformity-related deposits, paleochannels in sandstone-hosted systems, and alteration halos in basin-related systems. For example, combined airborne EM and magnetic surveys over the Shea Creek area on the western side of the basin identified several conductive zones at depths of 700 m and greater that were drilled. Ground EM surveys are used routinely to follow up airborne surveys with deeper penetration through the sandstone, but they often provide detailed assessments of targets of interest. Both time and frequency domain systems are capable of deep penetration under conductive cover. Resistivity contrasts are used to identify areas of alteration around deposits because there are differences in porosity or mineralogy in the alteration halos. Gravity measurement reflects structures and alteration because of changes and breaks in lithology. Seismic techniques can reveal structures, areas of alteration, stratigraphy, and depth to basement.

Emanometry has been classically used for gas detection. Radon emanometry is based on the ability of radon (^{222}Rn), a gaseous progeny of from ^{238}U decay (Fig. 3.2), to migrate to the surface from buried mineralization (Reimer, 1985). This is facilitated through the pumping action of diurnal pressure variations and a high permeability of the cover strata. Most instrumentation relies on alpha-particle detection, although detection can also be done by measuring the gamma emission from Rn decay products, ^{214}Bi and ^{214}Pb, following adsorption of the radon onto activated charcoal or other substances. This technique has been used successfully for the delineation of the Tumas calcrete-type deposit, buried under a thin cover of calcrete or gypcrete duricrust, and the Husab alaskite-type deposit, covered by sand dunes (Corner et al., 2010a,b), both located in Namibia. Radon emanometry has also been used in uranium exploration for unconformity-related mineralization in the Athabasca Basin in Canada (Earle and Drever, 1983) and the Beharchuwa-Bokarda-Labed area of India (Banerjee et al., 2012). Helium resulting from the alpha decay of uranium has also been used occasionally in exploration with varying success (Clarke et al., 1983; Earle and Drever, 1983; Reimer, 1985).

Improved magnetotelluric methods have detected deep conductors and shallow alteration zones in the search for deep unconformity-related deposits in the Athabasca Basin (Powell et al., 2007; Tuncer et al., 2006). Clay-rich, quartz-corroded, quartz-arenite has relatively low resistivity, whereas quartz-rich silicified zones are characterized by high resistivity. Although expensive, three-dimensional seismic has been used to image details of basement topology, thereby locating more

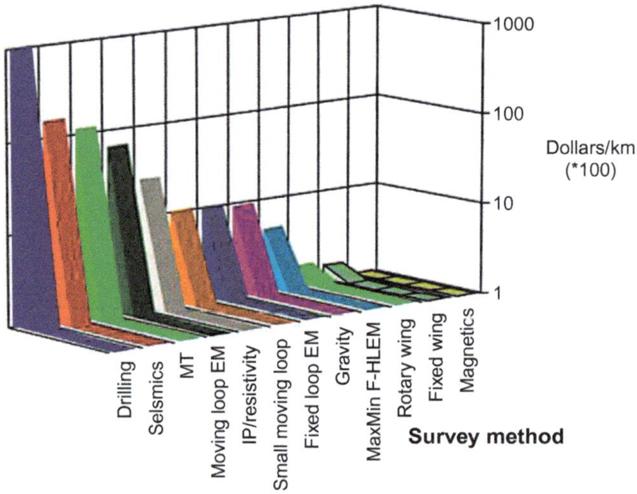

Figure 3.7 Estimated costs for various geophysical surveys used in uranium exploration relative to cost for drilling.
Source: From R.R. Koch presentation at PDAC Workshop—Giant Uranium Deposits: Exploration Guidelines, Models and Discovery Techniques (2006).

favorable areas for deep drilling (Hajnal et al., 2010; Wood et al., 2012) and to detect—directly or indirectly—uranium ore using borehole seismic (Cosma et al., 2006). In areas such as sedimentary basins where the deposits tend to be undercover, geophysical techniques are powerful tools to see physical anomalies undercover. However, it is their integration with the metallogeny and the geochemistry that forms a sum that is greater than the parts.

Prompt fission neutron (PFN) borehole logging technology directly measures the content of uranium in boreholes, overcoming the problem of disequilibrium, which limits the interpretation of uranium concentrations using gamma-logging tools (Penney, 2012). PFN uranium logging system has been recently used in the evaluation of the Four Mile deposits in South Australia. The effectiveness of these methods for uranium in organic-rich terrains, such as the Canadian Shield, is complicated by stable, soluble organo-uranium "complexes" that allow the metal to pass through organic-rich traps.

A substantial amount of exploration dollars are normally allocated to geophysical techniques in the hunt for uranium deposits to image possible mineralization or favorable environments for deposits. Although the costs of geophysics can be substantial, drilling is much more costly than almost all of the other techniques used in uranium exploration (Fig. 3.7).

3.5.2.4 Geochemical methods

Uranium deposits are geochemical anomalies and, as such, geochemistry must be an integral part of the exploration repertoire. Uranium itself is commonly used as a direct indicator of its deposits and this may be deflected using many different media. Amplifying the footprint of a deposit is the purpose of using geochemistry in exploration, thereby enabling detection at depth or traced near the surface (Cohen et al., 2010; Kelley et al., 2006). There are two distinct geochemical processes that expand the geochemical footprint of a deposit (Fig. 3.8): (1) primary dispersion, which provides information on alteration and primary element dispersion associated with ore emplacement, and (2) secondary dispersion, which provides clues about element migration from alteration and ore zones well after emplacement.

During primary dispersion (Fig. 3.8), components in the mineralizing fluids permeate into the country rock, which alters primary minerals and elevates the concentrations of "pathfinder" elements. These enrichments relative to background concentrations are sometimes evident in the lithogeochemistry up to several kilometers away from the uranium mineralization. Hydrothermal alteration minerals associated with mineralizing fluids are normally detected in drill core, outcrop, and from airborne surveys of the surface using spectroscopic techniques. Enrichments and zoning in pathfinder elements specific to an ore-forming fluid may be detected

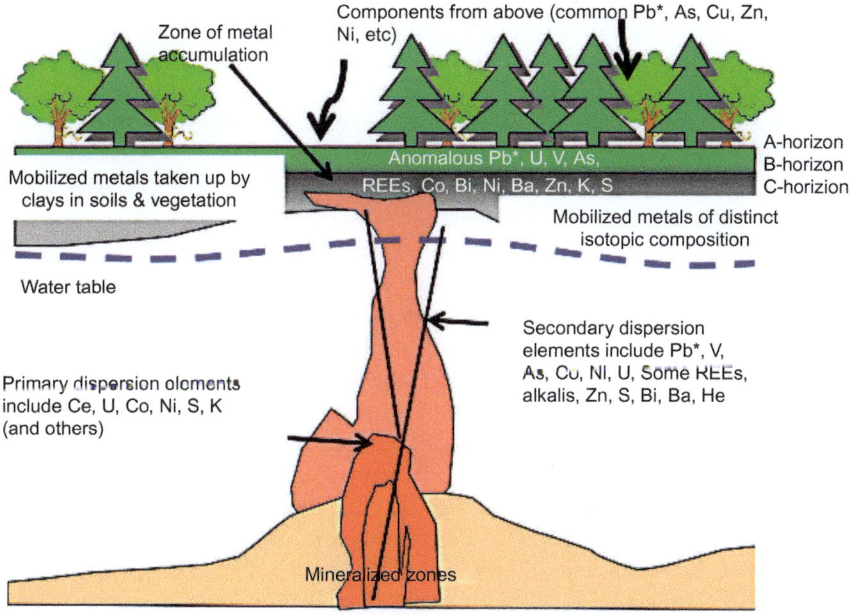

Figure 3.8 Diagram of buried unconformity-type uranium deposit showing elements associated with primary dispersion (synmineralization) and secondary dispersion (postmineralization). During secondary dispersion, elements such as Rare Earth Elements (REEs) are mobilized from the deposit and primary dispersion alteration halo and can be fixed by fracture fillings, clays, and Fe-Mn oxides in soils, and by vegetation. Also shown is the influence of components deposited from above because of anthropogenic activity. Exploration geochemistry targets both primary and secondary dispersion.
Source: After Cameron et al. (2004).

through element isotopic compositions that reflect the presence of ores and alteration halos such as H, Li, B, C, O, S, Cu, Zn, Tl, Mo, Pb, and U or in the trace-element compositions of specific minerals such as aluminum phosphate sulfates (APS) and clay minerals.

The levels of uranium concentration that can be considered anomalous vary according to the deposit type and its environment. In U-rich granites, significant anomalies should be above some tens of ppm, the granites having themselves a background uranium concentration of 10–30 ppm, whereas for unconformity-related deposits, anomalies of a few ppm in the sandstone may be meaningful because most of the sandstone has an average uranium content below 1 ppm.

In contrast to primary dispersion that is contemporaneous with the ore-forming process, secondary dispersion occurs subsequently via the mobilization of ore or alteration elements into the environment around the deposits. Tracing element migration in the near-surface environment involves understanding the secondary dispersion of elements, including those associated with anthropogenic activity that can mask the elemental signals coming from the ores (Fig. 3.8).

Because most ore deposits are electron-rich and contain elements in reduced forms, they are potential havens for microbes that can mobilize elements during secondary dispersion. Microbe-mobilized elements can involve aqueous or gaseous metal complexes, with the metals from the ore and the ligands from microbial waste products or from the decay products of dead microbes. Such complexes migrate to the surface, particularly along fractures and faults, become sorbed on clay and Fe-Mn oxide surfaces in soils, and make their way into the biosphere. These complexes have specific element and isotope signatures that can reflect the deposit at depth (Fig. 3.8). However, the controlling geochemical process by which elements migrate is fraught with uncertainty and requires additional research. Despite these uncertainties, surface media that have been used to trace secondary dispersion of uranium ores or alteration minerals include drainage sediment sampling (streams or lakes), water sampling (streams, lakes or wells), overburden sampling (soil or glacial tills) (McClenaghan et al., 2013; Sarala and Peuraniemi, 2007), boulder sampling and tracing (Earle et al., 1990), vegetation sampling or biogeochemistry (Dunn, 2007), and rock sampling. Later mobilization of uranium from unconformity-related deposits can be detected several hundred meters from the ore zone.

Mineral exploration in glaciated terrain has successfully used till geochemistry and indicator mineral methods for diamonds, gold, base metals, and other commodities that reflect secondary dispersion from outcrop. There is a paucity of studies using indicator minerals and till geochemistry for uranium exploration. A recent study on heavy minerals in till near the Kiggavik uranium deposit in Nunavut, Canada indicate that the highest metal contents are located directly to the west of the deposit in locally derived, basement-dominated grey till (Robinson et al., 2014). Till geochemistry exhibits a polymetallic dispersion signature, with anomalies in U, Bi, Mo, Au, Ag, Co, Cs, Pb, and W up to 1 km down-ice and up-ice of the Kiggavik Mineralized Zone. Elevated numbers of gold grains can be used as an indicator for uranium mineralization, as can Pb-rich apatites grains. However, no uranium minerals were observed, perhaps because they are in a very fine fraction. Till geochemistry surveys in glaciated areas may be a viable technique for uranium exploration.

An indirect guide to uranium mineralization may be reflected by elevated concentrations of other mobile elements such as Ba or Sr, and Pb isotopes and iron oxidation ratio (Ng et al., 2013). Exploration success will be enhanced if there are coincident anomalies in other elements (eg, Se, V, Mo, Pb, and Cu) that are typically enriched in some types of uranium deposits. Many types of uranium deposits are characterized by enrichment of one or more trace metals in addition to uranium. Although specific tectonic environments, structural settings, and lithologies are required for all uranium deposit types, none of these is a definitive indicator of mineralization because in most of these settings, structures and lithologies do not host deposits. Thus, they are required for the deposits to form, but are not definitive indicators of mineralization. In effect, the only definitive indicators are physical and geochemical contrasts, both in terms of the ores and the associated elements.

Mineralogy is an excellent exploration guide when specific alteration minerals are developed in association with ore deposit formation, many of them can be analyzed directly in the field by portable short-wave infrared (SWIR) spectrometers (Zhang et al., 2001) or using hyperspectral techniques (Lower et al., 2011). A typical example are the Mg-rich minerals (sudoïte, dravite), altered monazite and zircon, and APS associated with unconformity-related ore deposits, clay alteration around sandstone-hosted systems, and albitization associated with deposits related to sodium metasomatism. The mineralogy of the uranium in the potential source rocks also controls the formation of a deposit, and the mineralogy of the uranium is also critical for evaluating the processing costs for uranium extraction.

Chemical analysis of groundwater can be a useful strategy for regional exploration for uranium in reduced sediments in paleochannels, although multielement data are required. Lead isotopes, which originate from the ore and are mobilized by groundwaters, can be used to confirm the interpretations from the composition of the groundwaters. Paleochannels containing uranium deposits in South Australia have neutral, moderately saline groundwaters, whereas others are often saline and acidic waters that preferentially mobilize radium, thereby negating the use of radon or down-hole gamma-logging (Dickson & Giblin, 2007). If reduction of uranium by bacteria is an effective mechanism for formation of deposits in paleochannels, then microbially induced geochemical signals such as C and S isotopes or enhanced mobile metals should indicate favorable areas.

Groundwater samples collected from boreholes tens of meters from unconformity-related uranium mineralization have consistently high levels of uranium, radium, radon, and helium (Earle & Drever, 1983). Biogeochemistry of spruce twigs indicate that tree roots can extract anomalous uranium from groundwater that may reflect deposits at 300 m depth (Dunn, 1984).

Lakewater and sediment geochemistry and radiometric prospecting are significant tools in early regional exploration for uranium deposits in Canada (Cameron, 1980). This is because of the superior mobility of uranium in surface waters, which allows the element to disperse widely from its source. Interpretation of the radiometric and spectrometric data must take into account that measured radiations reflect the decay products of uranium.

Two sampling media in the drainage basins of lakes include organic-rich, center-lake sediments and surface waters. Waters have certain advantages over center-lake sediments, such as lower sampling and preparation costs. The pH of lakewater has minimal effect on the partitioning of uranium between organic-rich sediment and water over the pH range 5.0–7.4, but above pH 7.4 there is a marked increase in the uranium content of lake waters relative to organic sediments (Cameron, 1980). In glaciated terrain, such as the Canadian Shield, the development of anomalies in lakes is a two-stage process wherein U-rich detritus is transported down-ice from the mineralized source and then the metal is dispersed in solution from this detritus into the lakes. As a consequence, lake anomalies are most effectively followed up by boulder tracing.

3.6 Critical factors in deposit models

Once an area has been selected because of favorable geology, metallogeny must be considered in the evaluation process. The geology for the type of uranium deposit must be correct, which requires knowledge of both geology and deposit models. Associated with deposit models are certain key factors that include the timing of the mineralization, source of the uranium, transport mechanisms of the uranium, origin and character of the fluid, trapping mechanisms of the uranium, and controls on paleohydrology. All of these must be understood in exploration not only to find prospective areas, but also to eliminate areas that are unlikely to host mineralization.

3.6.1 Timing of mineralization

The temporal evolution of the area in relation to the timing of the mineralization must be assessed, but this is often overlooked or data are not available. Without an idea of both the relative and absolute ages of geologic events such as faulting and formation and alteration of the deposit and the host, exploration strategies cannot evolve beyond being prospector-driven (Marlatt and Kyser, 2011). For example, magmatic-related deposits form after significant differentiation of postorogenic parent magmas to allow residual melts and magmatic fluids to become enriched in uranium (Le Carlier de Veslud et al., 2000). In collision zones, uranium deposits require melting of the crust, which is facilitated by crustal thickening at an appropriate time during convergence. Metasomatic uranium deposits are associated with the uplift of an orogen, almost exclusively during the Proterozoic. Unconformity-related deposits, also exclusively Proterozoic, appear to require 75–100 million years after the basins form to allow the fluids the appropriate evolution in their chemistry and temperature to be able to leach and mobilize uranium (Alexandre et al., 2009). Hydrothermal processes associated with veins near granites are developed about 50 million years after the emplacement of the fertile granites. Effective exploration in any deposit requires detailed knowledge of the time–space relationships and deposit models (Cuney and Kyser, 2009).

Although determining the age of uranium deposits is not straightforward, recent advances in Inductively Coupled Plasma Mass Spectrometry (ICP-MS) and micro-sampling technologies (Chipley et al., 2007; Kotzer and Kyser, 1993, 1995; Ludwig, 1978, 1979; Ludwig et al., 1987), use of multiple decay systems such as both U-Pb and Sm-Nd, precise dating of associated gangue minerals of specific paragenesis, and integration of age data with other geologic factors render interpretation of geochronologic data easier to obtain and more meaningfully for exploration. The timing of the mineralizing process is required in exploration so that the geologic, chemical, and physical environment conducive to the mineralizing process at a critical time in the evolution of an environment can be realized.

Along with the actual age of the ores and, therefore, the critical time in Earth history that an effective mineralizing environment was present, the timing of events that have subsequently affected the ores can reveal when elements such as radiogenic Pb have been mobilized from the deposits and moved into the surrounding environment during secondary dispersion (Holk et al., 2003). These elements would elevate element concentrations in the surrounding environment, with gradients in concentrations and isotopes as vectors to the deposits.

3.6.2 Source, transport, and fixing of the uranium

The source of the uranium is often overlooked by many exploration geologists, but should be a factor in evaluating areas for potential mineralization. For many deposits, the source of the uranium is in units that have aberrantly high uranium contents, such as in volcanic glasses in the case of tabular deposits or in alkaline intrusions in the case of some magmatic-type deposits. Enrichment of uranium in the source region of basin-hosted deposits, available at the right time, certainly increases the probability that a deposit could form. Another critical aspect is the availability of uranium in the source region. For uranium enrichments in certain refractory minerals, the uranium is unlikely to be released unless those minerals become extensively damaged by radioactive decay. In the case of sandstone-hosted deposits, unstable volcanic glasses with high uranium contents make ideal sources because the uranium can be effectively mobilized.

Reduced carbon, particularly in the form of organic matter, is an effective reductant for fixing uranium, given that uranium is mobile in fluids as U^{6+} but immobile as U^{4+}. Exactly why carbon is such an effective reductant is unclear, especially given that there are alternatives such as sulfides and ferrous iron. Although the latter have been shown to be the likely reductants in the formation of some deposits, these deposits tend not to be as large or as high grade, relative to those where carbon is the reductant. In the case of calcrete-type deposits, redox plays a minimal role, except as a moderator of pH. For these deposits, solubility is most important.

3.6.3 Sources of fluids

Knowledge of the nature of the fluids involved in the ore generation process and manifestation of these fluids in appropriate environments are absolutely critical for refining exploration strategies. The critical factors involved in identifying the fluids include their temperature, pressure, oxygen fugacity, and chemical composition, aspects that are determined using a variety of techniques during deposit studies, including isotopic geothermometry (Kotzer and Kyser, 1995), fluid inclusions (Richard et al., 2013), and mineral equilibria (Langmuir, 1978). Without knowledge of the critical chemistry, temperature, and pressure required for generation of the ores, it is difficult to identify the correct environment in which to explore. For example, most highly evolved alkaline intrusives do not host uranium deposits because they do not evolve to concentrate uranium in their differentiates. Most areas of albitization do not host deposits presumably because the fluids were not carrying uranium or there was no trap. There are areas in the Athabasca Basin with the appropriate geology, structure, and alteration conducive to the ore-forming process, yet they are apparently devoid of any significant unconformity-related mineralization. Most sandstone-hosted deposits occur in meanders of paleostreams where organic detritus could accumulate, but most such areas do not host ore. Understanding why areas that should have ore do not, requires knowledge of the physical, chemical, and temporal characteristics of the fluids required to form and preserve various types of uranium deposits.

Fluids that produce many types of uranium deposits can produce significant alteration zones around the uranium mineralization. Clay minerals are ubiquitous up to hundreds of meters from hydrothermal uranium mineralization, and often there is zoning in the type of alteration minerals involved. The presence of reductants such as degraded graphite or organic matter with the alteration zones are among the major indicators of an environment conducive to uranium enrichment. Exploration techniques that exploit these features include airborne and ground geophysics, surface geochemistry, and clay typology.

3.6.4 Sedimentology

Most uranium deposits in sedimentary rocks are associated with geochemical provinces enriched in U and Th or with U-rich intrusives or volcanic rocks, although the deposits may be separated by tens of kilometers from these source rocks. However, weak regional U and Th anomalies in sediments containing uranium deposits may be present. Geochemical detection of uranium deposits in sandstone-type deposits depends on the geochemical behavior of uranium and pathfinder elements (Rose and Wright, 1980). Uranium is dispersed under oxidizing conditions, but is immobile under reducing conditions where accumulations can occur. In addition, adsorption on Fe oxides and certain types of organic matter also limits dispersion unless high concentrations of CO_3^{2-} or other complexers are present. Thorium accompanies uranium in most high-temperature plutonic processes, but the two elements are separated under oxidizing conditions.

Uranium ore deposits in the Grants Mineral Belt, New Mexico occur in fluvial sandstones in the Jurassic Morrison Formation, where uranium is concentrated by dark grey to black humate derived from decaying vegetation. The ores vary greatly in size and shape, generally occur in clusters, and often are difficult targets for drilling. Exploration is done primarily by drilling, delineating favorable ground on a wide spacing, and then using closely spaced drilling in mineralized areas. Criteria for favorable areas include the presence of a host sandstone, anomalous uranium contents, dark color of host rock, presence of carbonaceous matter, and position of an area with respect to mineralized trends (Fitch, 1979). Possible pathfinder elements associated with uranium in sandstone-type deposits include S, V, Mo, Se, As, He, Rn, and other radioactive decay products and some deposits have anomalies in Cu, Ag, Cr, Pb, Zn, Ni, Co, Re, Be, P, Mn, and REEs. In addition, Pb, S, and C isotopes, and textures of Fe and Ti oxides can also be vectors to ore.

In basin-related deposits, sedimentology is of outmost importance to locate the favorable aquitards within a basin. Coarse siliciclastic continental to near-shore sediments with detrital organic matter, confined between siltstone aquitards, are priority targets for tabular and roll front-type uranium deposits. Specific organic-rich sedimentary systems such as black shales or the Mulga Rock deposit in west Australia (Fewster, 2009) exemplify the importance of basin architecture, paleo-environment, and redox potential related to sedimentation. The importance of sedimentology and basin architecture has also been discussed for unconformity-related deposits (Hiatt et al., 2003; Hiatt and Kyser, 2007).

3.7 Drilling and evaluation

After targets have been identified by integrating favorable geophysical and geochemical results, the next stage of exploration is drilling. Most alteration halos associated with deposits are barren. Directional drilling allows several intersections to be made from a single pilot hole, reducing drilling costs and improving target precision. Core orientation methods are also used to better understand the structural controls on mineralization.

Drilling results can reveal much about the third dimension of a property. Given that drill hole samples represent an actual physical sample of the subsurface and are among the costliest aspect of exploration, much effort should be invested to understand the details of their character, mineralogy, geochemistry, and relationship to the geological, geophysical, and geochemical results. Core or cuttings are logged to reveal lithology and stratigraphy. Fracture densities, structure, and sedimentology must be documented to integrate with the geological and geophysical data. Mineralogy using SWIR or hyperspectral techniques should be an integral part, as this will reveal the clay minerals associated with the mineralizing system relative to background. Geochemical samples are collected, normally from the bottom of each row of the core box, and these are composited. Although often avoided, fractures and petrologically anomalous samples should also be collected, with the former being compared to surface geochemical results. In addition, down-hole EM, radiometrics, and neutron tomography should be done if possible. At each step, the property should continually be re-evaluated so that informed decisions about next steps can be made.

3.8 Synopsis

The unique properties of uranium as a radioactive element make it both useful for generating energy and contentious for its weapons potential. Given the existing and projected demand for uranium, new resources must be found and new technologies to find them and use them more efficiently must be developed.

Formulation of effective exploration strategies involves evaluating the prospectivity of an area, the economic potential of the deposit, and the explorability of the area. Countries where uranium mining is currently active tend to have both prospectivity and explorability. The nuclear power cycle requires discovery of ore bodies, which requires a knowledge base that must be constructed, expanded, and always refined. Most of the largest and most profitable mines are associated with sedimentary basins, particularly those mined using open pit methods.

Exploration strategies vary depending on which type of deposit is being sought, property evaluation, local exploration, production rules and risks, and market conditions for selection of the area to be explored. Exploration means first generating targets, ranging from the right area to be in, to the right spot on which to sample and drill. Thus, the potential property must be fully evaluated prior to financial investment. Full property evaluation should include an understanding of the geology, metallogeny, availability, accessibility, and liability. The techniques used to evaluate a property and complete effective exploration integrate geologic mapping, remote sensing, geophysics, and geochemistry with deposit models and the critical aspects of the models.

New technologies will continue to be developed to help minimize the luck and serendipity aspects of exploration. New Geographic Information System (GIS) capabilities are continually expanding, as is new software to display and aid in the interpretation of data. In geophysics, better resolution; refined models for EM, magnetics, and gravity; and less expensive seismic techniques are being developed. Lower detection limits, *in situ* analyses to minimize sample preparation, refined deposit models, and isotope tracing are at the frontier in geochemistry. However, regardless of the methods, success will always depend on understanding basic geology and metallogeny, and doing proper property evaluation.

References

Alexandre, P., Kyser, K., Jiricka, D., 2009. Critical geochemical and mineralogical factors for the formation of unconformity-related uranium deposits; comparison between barren and mineralized systems in the Athabasca Basin, Canada. Econ. Geol. 104, 413–435.

Banerjee, R., Deshpande, M.S.M., Roy, M.K., Maithani, P.B., 2012. Radon emanometry in uranium exploration: a case study of Beharchuwa-Bokarda-Labed area, Janjgir-Champa and Korba Districts, Chhattisgarh, India. Gondwana Geol. Mag. Spec. 13, 37–44.

Bharti, R., Ramakrishnan, D. 2014. Uraniferous calcrete mapping using hyperspectral remote sensing. In: IGARSS 2014–2014 IEEE International Geoscience and Remote Sensing Symposium, IEEE, 13–18 July 2014, Piscataway, NJ, USA, pp. 2902–2905.

Cameron, E.M. 1980. Geochemical exploration for uranium in northern lakes; geochemical exploration for uranium. In: Geochemical exploration for uranium; a part of The Geological Society of America, 92nd annual meeting, 5–8 November 1979, San Diego, CA, USA, pp. 13, 221–250.

Cameron, E.M., Hamilton, S.M., Leybourne, M.I., Hall, G.E.M., McClenaghan, M.B., 2004. Finding deeply buried deposits using geochemistry. Geochem. Expl. Environ. Anal. 4, 7–32.

Chipley, D., Polito, P.A., Kyser, T.K., 2007. Measurement of U-Pb ages of uraninite and davidite by laser ablation-HR-ICP-MS. Am. Mineral. 92, 1925–1935.

Clark, S.P., Jr., Peterman, Z.E., Heier, K.S., 1966. Abundances of uranium, thorium, and potassium. In: Clark, Jr., S.P. (Ed.), Handbook of Physical Constants, revised edition. Geol. Soc. Am., Memoir 97, pp. 521–541.

Clarke, W.B., Top, Z., Eismont, W.C., 1983. Helium isotope and tritium content of lakes and uranium exploration in the NEA/IAEA Athabasca test area. Paper—Geological Survey of Canada 82–11, pp. 139–146.

Cohen, D.R., Kelley, D.L., Anand, R., Coker, W.B., 2010. Major advances in exploration geochemistry, 1998–2007. Geochem.—Expl. Environ. Anal. 10, 3–16.

Cornell, J.C., 2006. Changing role of secondary supply in the global uranium market. Uranium Production and Raw Materials for the Nuclear Fuel Cycle–Supply and Demand, Economics, the Environment and Energy Security. IAEA, Vienna, IAEA-CN-128.

Corner, B., Osiyuk, V., Lytvyniuk, S., Kuchmin, A., Verran, D., 2010a. Radon emanometry case studies in Namibia: the Spitzkoppe and Tumas uranium deposits. In: 11th SAGA Biennial Technical Meeting and Exhibition. Session 3 C—Minerals exploration. <http://earthdoc.eage.org/publication/>.

Corner, B., Osiyuk, V., Lytvyniuk, S., Kuchmin, A., Verran, D., 2010b. A Radon emanometry case study of the Rössing South Deposit, Namibia. In: 11th SAGA Biennial Technical Meeting and Exhibition. Session 3 C—Minerals exploration. <http://earthdoc.eage.org/publication/>.

Cosma, C., Enescu, N., Powell, B., Wood, G., 2006. Structural mapping for uranium exploration by Borehole Seismic. EAGE Near Surface, 2006.

Cuney, M., Kyser, K., 2009. Recent and not-so-recent developments in uranium deposits and implications for exploration. In: Mineralogical Association of Canada Short Course Series Vol. 39. Mineralogical Association of Canada, Quebec, 258pp.

Dickson, B.L., Giblin, A.M., 2007. Effective exploration for uranium in South Australian palaeochannels; geology of uranium deposits. In: Australia's Uranium Conference 2006, Adelaide, South Australia, Australia, 10–11 July 2006, pp. 116, 50–54.

Dunn, C.E., 1984. Geochemical studies in the Athabasca test area, Saskatchewan. CIM Bull. 77, 72 (1974).

Dunn, C.E., 2007. Biogeochemistry in Mineral Exploration. Elsevier, Amsterdam, The Netherlands, 456pp.

Earle, S.A.M., Drever, G.L., 1983. Hydrogeochemical exploration for uranium within the Athabasca Basin, northern Saskatchewan. J. Geochem. Explor. 19, 57–73.

Earle, S., McGill, B., Murphy, J., 1990. Glacial boulder lithogeochemistry: an effective new uranium exploration technique in the Athabasca Basin, Saskatchewan. In: Beck, L.S., Harper, C.T. (Eds.), Modern Exploration Techniques; Proceedings of a Symposium Held in Regina, 20–21 November 1989, Saskatchewan Geological Society Special Publication No.10.

EIA, 2007. International Energy Outlook. Department of Energy, Washington, DC, <http://www.eia.doe.gov/oiaf/ieo/index.html>.

Fewster, M., 2009. Sandstone hosted uranium mineralisation associated with the Mulga Rock Deposits. In: Publication Series—Australasian Institute of Mining and Metallurgy 5/2009, 101.

Fitch, D.C., 1979. Exploration for uranium deposits in Grants mineral belt, New Mexico. In: Grants Uranium Region Symposium, Albuquerque, NM, USA, 14–16 May 1979, pp. 63, 688.

Hajnal, Z., White, D.J., Takacs, E., Gyorfi, I., Annesley, I.R., Wood, G., et al., 2010. Application of modern 2-D and 3-D seismic-reflection techniques for uranium exploration in the Athabasca Basin. Can. J. Earth Sci. 47, 761–782.

Hiatt, E.E., Kyser, T.K., 2007. Sequence stratigraphy, hydrostratigraphy, and mineralizing fluid flow in the Proterozoic Manitou Falls Formation, eastern Athabasca Basin, Saskatchewan. Bull. Geol. Surv. Can. 4, 489–506.

Hiatt, E.E., Kyser, K., Dalrymple, R.W., 2003. Relationships among sedimentology, stratigraphy, and diagenesis in the Proterozoic Thelon Basin, Nunavut, Canada: implications for paleoaquifers and sedimentary-hosted mineral deposits. J. Geochem. Explor. 80, 221–240.

Holk, G.J., Kyser, T.K., Chipley, D., Hiatt, E.E., Marlatt, J., 2003. Mobile Pb-isotopes in Proterozoic sedimentary basins as guides for exploration of uranium deposits. J. Geochem. Explor. 80, 297–320.

IAEA, 2013. Advances in Airborne and Ground Geophysical Methods for Uranium Exploration. Nuclear Energy Series NF-T-1.5, 0401-STI/PUB/1558, 58p.

Kelley, D.L., Kelley, K.D., Coker, W.B., Caughlin, B., Doherty, M.E., 2006. Beyond the obvious limits of ore deposits: the use of mineralogical, geochemical, and biological features for the remote detection of mineralization. Econ. Geol. 101, 729–752.

Kotzer, T.G., Kyser, T.K., 1993. O, U, and Pb isotopic and chemical variations in uraninite—implications for determining the temporal and fluid history of ancient terrains. Am. Mineral. 78, 1262–1274.

Kotzer, T.G., Kyser, T.K., 1995. Petrogenesis of the Proterozoic Athabasca Basin, Northern Saskatchewan, Canada, and its relation to diagenesis, hydrothermal uranium mineralization and paleohydrogeology (Vol 120, Pg 45, 1995). Chem. Geol. 120, 45–89.

Langmuir, D., 1978. Uranium solution-mineral equilibria at low temperatures with applications to sedimentary ore deposits. Short Course Handb. 3, 17–55.

Le Carlier de Veslud, C., Cuney, M., Royer, J.J., Floc'h, J.P., Ameglio, L., Alexandrov, P., et al., 2000. Relationships between granitoids and mineral deposits; three-dimensional modelling of the Variscan Limousin Province (NW French Massif Central). Trans. R. Soc. Edinb. Earth Sci. 91 (Parts 1–2), 285–301.

Lower, C., Ehrig, K., Macmillan, E., Kittler, P., Leibezeit, V., Yang, K., 2011. Quantitative clay mineralogy for a calcrete-hosted uranium deposit; innovative application of existing technology on an unprecedented scale. In: Publication Series—Australasian Institute of Mining and Metallurgy 10/2011, pp. 173–180.

Ludwig, K.R., 1978. Uranium-daughter migration and U/Pb isotope apparent ages of uranium ores, Shirley Basin, Wyoming. Econ. Geol. 73, 29–49.

Ludwig, K.R., 1979. Age of uranium mineralization in the Gas Hills and Crooks Gap districts, Wyoming, as indicated by U-Pb isotope apparent ages. Econ. Geol. 74, 1654–1668.

Ludwig, K.R., Grauch, R.I., Nutt, C.J., Nash, J.T., Frishman, D., Simmons, K.R., 1987. Age of uranium mineralization at the Jabiluka and Ranger deposits, Northern Territory, Australia; new U-Pb isotope evidence. Econ. Geol. 82, 857–874.

Marlatt, J.L., Kyser, T.K., 2011. Paradigmatic shifts in the uranium exploration process; knowledge brokers and the Athabasca Basin learning curve. SEG Newsl. 84 (1), 17–23.

McClenaghan, M.B., Plouffe, A., McMartin, I., Campbell, J.E., Spirito, W.A., Paulen, R.C., et al., 2013. Till sampling and geochemical analytical protocols used by the Geological Survey of Canada. Geochem. Expl. Environ. Anal. 13, 285–301.

McMurray, J.M., 2006. Worldwide uranium resources and production capacity—the future of the industry. Uranium Production and Raw Materials for the Nuclear Fuel Cycle – Supply and Demand, Economics, the Environment and Energy Security. IAEA-CN-128. IAEA, Vienna.

Ng, R., Alexandre, P., Kyser, K., Cloutier, J., Abdu, Y.A., Hawthorne, F.C., 2013. Oxidation state of iron in alteration minerals associated with sandstone-hosted unconformity-related uranium deposits and apparently barren alteration systems in the Athabasca Basin, Canada: implications for exploration. J. Geochem. Explor. 130, 22–43.

OECD, 2001. Assessment of uranium deposit types and resources; a worldwide perspective; proceedings of the International Atomic Energy Agency, Technical Committee meeting; OECD Nuclear Energy Agency 1997, IAEA-Tecdoc. International Atomic Energy Agency, Vienna, International, p. 253.

OECD, 2008. Uranium 2007: Resources, Production and Demand. Nuclear Energy Agency. Organization for Economic Co-operation and Development, Paris, France.

OECD/NEA-IAEA, 2010. Uranium 2009; Resources, Production and Demand, Joint Report—OECD Nuclear Energy Agency and the International Atomic Energy Agency. Organisation for Economic Co-operation and Development, Paris, France.

OECD/NEA-IAEA, 2014. Uranium 2014: Resources, Production and Demand, 2014 Red Book. OECD, No. 7209, Paris, France, 504 p.

Pacala, S., Socolow, R., 2004. Stabilization wedges: solving the climate problem for the next 50 years with current technologies. Science. 305, 968–972.

Penney, R., 2012. Australian sandstone-hosted uranium deposits. Applied Earth Science. 121, 65–75.

Powell, B., Wood, G., Bzdel, L., 2007. Advances in geophysical exploration for uranium deposits in the Athabasca Basin. In: Milkereit, B. (Ed.), Proceedings of Exploration 07: Fifth Decennial International Conference on Mineral Exploration, Toronto, Canada, pp. 771–790.

Price, R.R., 2006. An analysis of historical data on uranium exploration expenditures and price. Uranium Production and Raw Materials for the Nuclear Fuel Cycle—Supply and Demand, Economics, the Environment and Energy Security. IAEA, Vienna, IAEA-CN-128/1P.

Price, R.R., Barthel, F., Blaise, J.-R., 2006. Forty years of uranium resources, production and demand in perspective. In: Facts and Opinions, NEA News 24.1, 4–6.

Reimer, G.M., 1985. Gaseous emanations associated with sandstone-type uranium deposits. In: Finch, WID., James, F. (Ed.), Geological Environments of Sandstone-type Uranium Deposits. IAEA-TECDOC-328. Int. At. Energy Agency, Vienna, Austria, pp. 335–346.

Richard, A., Cauzid, J., Cathelineau, M., Boiron, M.C., Mercadier, J., Cuney, M., 2013. Synchrotron XRF and XANES investigation of uranium speciation and element distribution in fluid inclusions from unconformity-related uranium deposits. Geofluids. 13, 101–111.

Robinson, S.V.J., Paulen, R.C., Jefferson, C.W., McClenaghan, M.B., Layton-Matthews, D., Quirt, D., et al., 2014. Till geochemical signatures of the Kiggavik uranium deposit, Nunavut, Open-File Report 7550—Geological Survey of Canada. Geological Survey of Canada, Calgary, AB, Canada, 168p.

Rose, A.W., Wright, R.J., 1980. Geochemical exploration models for sedimentary uranium deposits; geochemical exploration for uranium. Geochemical exploration for uranium; a part of The Geological Society of America. In: 92nd Annual Meeting, San Diego, CA, USA, 5–8 November 1979, pp. 13, 153–179.

Sarala, P., Peuraniemi, V., 2007. Exploration using till geochemistry and heavy minerals in the ribbed moraine area of southern Finnish Lapland. Geochem. Expl. Environ. Anal. 7, 195–205.

Shatalov, V.V., Takhanov, A.V., Boldyrev, V.A., Knyazev, O.I., 2006. Analysis of uranium world resources and ways of their extension. Uranium Production and Raw Materials for the Nuclear Fuel Cycle–Supply and Demand, Economics, the Environment and Energy Security. IAEA, Vienna, IAEA-CN-128.

Tuncer, V., Unsworth, M.J., Siripunvaraporn, W., Craven, J.A., 2006. Exploration for unconformity-type uranium deposits with audiomagnetotelluric data: a case study from the McArthur River mine, Saskatchewan, Canada. Geophysics. 71, B201–B209.

Vance, R.E., Price, R.R., Barthel, F., 2006. Recent activities of the joint Nuclear Energy Agency (NEA)/International Atomic Energy Agency (IAEA) Uranium Group. Uranium Production and Raw Materials for the Nuclear Fuel Cycle–Supply and Demand, Economics, the Environment and Energy Security. IAEA, Vienna, IAEA-CN-128.

WNA, 2015a. The Nuclear Fuel Cycle. World Nuclear Association, <http://www.world-nuclear.org/info/inf03.html>.

WNA, 2015b. World Uranium Mining Production. World Nuclear Association, <http://www.world-nuclear.org/info/Nuclear-Fuel-Cycle/Mining-of-Uranium/World-Uranium-Mining-Production/>.

Wood, G., O'Dowd, C., Cosma, C., Enescu, N., 2012. An interpretation of surface and borehole seismic surveys for mine planning at the Millennium uranium deposit, northern Saskatchewan, Canada. Geophysics. 77, 203–212.

Zhang, G., Wasyliuk, K., Pan, Y., 2001. The characterization and quantitative analysis of clay minerals in the Athabasca Basin, Saskatchewan; application of shortwave infrared reflectance spectroscopy. Can. Mineral. 39 (Part 5), 1347–1363.

Zhang, J.-L., Wang, J.-H., Zhou, M., Huang, Y.-J., Wu, D., 2014. Aerial visible-thermal infrared hyperspectral feature extraction technology and its application to object identification. IOP Conf. Ser. Earth Environ. Sci. 17, 1–5.

Uranium resources

4

Robert Vance
Nuclear Energy Agency (NEA), Issy-les-Moulineaux, France

1. The opinions expressed and arguments employed herein are those of the author and do not necessarily reflect the official views of the OECD/NEA or of the governments of its member countries.

4.1 Introduction

Nuclear energy offers opportunities for diversifying energy supply and ensuring long-term security, since nuclear power plants (NPPs) provide a largely domestic supply of energy. The main advantages of nuclear power for energy security are the high energy density of uranium—one tonne of uranium is the energy equivalent of 14,000 to 23,000 tonnes of coal, if used in light water reactors with a once-through cycle (NEA/IAEA, 2014)—combined with the geographically diverse distribution of uranium resources and the ease with which strategic stockpiles of nuclear fuel can be maintained. Because nuclear fuel costs represent only 5–7% of the total cost of generating electricity, uranium price volatility is not as significant an issue for NPP owners and operators as it is for fossil fuel alternatives. This chapter focuses on global uranium resources, the processes used to compile and monitor the resource base, and production cost estimates for the starting point in the production of nuclear fuel.

4.1.1 History

First separated from the mineral pitchblende while mining for silver in St. Joachimsthal (Jáchymov) in 1789, and initially valued because it gave glass a brilliant yellow color and green fluorescence, the radioactive properties of uranium were not confirmed until 1895. This stimulated further research, particularly in the medical field, and interest was intensified during World War II when the discovery of the power of fission properties was applied to military uses. In the mid-1960s, interest in uranium increased further as the raw material component in fuel for NPPs to produce base-load electricity.

With the rising importance of its use in nuclear fuel, combined with the long operational lifetime of NPPs, government interest grew in compiling information on uranium resources. In early 1965, the Steering Committee of the European Nuclear Energy Agency (ENEA) set up a study group on the long-term role of nuclear energy in Western Europe and the group decided that documenting world estimates of uranium (and thorium) resources, initially in the Western world, would form the

Uranium for Nuclear Power. DOI: http://dx.doi.org/10.1016/B978-0-08-100307-7.00004-1

basis of its work. The first report (ENEA, 1965) classified resources as either Reasonably Assured or Possible Additional Resources in three production cost categories: USD 5–10, 10–15, and 15–30/lb U_3O_8.

4.1.2 The Red Book

Since the first publication of the aforementioned report, the ENEA and its successor, the Organisation for Economic Co-operation and Development's (OECD's) Nuclear Energy Agency (NEA), has continued producing reports on uranium resources, production, and demand, roughly every 2 years, initially with the informal cooperation of the International Atomic Energy Association (IAEA) and beginning in the mid-1990s with the formal participation of the IAEA's member states. As the majority of the editions in the series have featured a red cover, the publication became commonly known as the "Red Book." Over the course of its history, the Red Book has become a recognized authoritative source of government-sponsored information on uranium resources, and over 100 countries have contributed data to the 25 Red Books published to date. The most recent edition in this series, *Uranium 2014: Resources, Production and Demand*, was released in September 2014 (NEA/IAEA, 2014). A compilation edition entitled *Forty Years of Uranium Resources, Production and Demand in Perspective: "The Red Book Retrospective"* (NEA, 2006) summarized information in the Red Book series between 1965 and 2003, describing and analyzing the evolution of the commodity and its market and drawing conclusions from its well-documented history. Because these publications are a time series of comprehensive, global assessments of what is known about the uranium resource base, they are the primary references used in this chapter.

4.2 Resources and reserves

In recent years, there has been a move toward strict methods and standards for reporting mineral resources and reserves to provide investors and governments with assurance that amounts reported by companies are based on standard methods and are approved by certified third party experts. Although several national and international organizations have published codes and standards to govern reserve and resource reporting (eg, the Australasian Code for Reporting Exploration Results; Mineral Resources and Ore Reserves, commonly known as the JORC code; and the South African Code for Reporting of Exploration Results, Mineral Resources and Mineral Reserves, commonly known as the SAMREC code), they are broadly consistent. The basics are captured by the Canadian Institute of Minerals, Metallurgy and Petroleum following the NI 43-101 code (CIM, 2014), as outlined next (italics added for emphasis).

> A mineral resource is a concentration or occurrence of solid material of economic interest in or on the Earth's crust in such form, grade or quality and quantity that there are *reasonable prospects for eventual economic extraction.* The location,

quantity, grade or quality, continuity and other geological characteristics of a mineral resource are known, estimated or interpreted from specific geological evidence and knowledge, including direct sampling. A mineral resource is further subdivided into measured, indicated and inferred categories, in order of decreasing confidence of the estimates based on the amount of drilling and sampling conducted.

A mineral reserve, on the other hand, is the *economically mineable part of a measured and/or indicated mineral resource.* A reserve declaration includes diluting materials and allowances for losses, which may occur when the material is mined or extracted. It is supported by pre-feasibility and feasibility studies that demonstrate, at the time of reporting, extraction could reasonably be justified. At the highest level of confidence, a proven mineral reserve includes an assessment of modifying factors, including but not limited to mining, processing, metallurgy, infrastructure, economics, marketing, legal, environmental, social and governmental factors. It is the kind of detailed analysis that a mining company must conduct prior to making a decision to develop a mine to extract the well-defined reserves in the deposit(s) of interest.

While in practice the differences are more specific and detailed, in essence a mineral resource has the potential for economic extraction at some time, whereas a mineral reserve is sufficiently well defined and it is understood that under the economic conditions prevailing at the time of the declaration, the mineral(s) of interest can be mined economically. Only resources defined to the highest level of confidence (measured and indicated) have the potential to be declared a mineral reserve. Other categories of resources typically require more sampling and analyses prior to making a case for mine development. Because the purpose of the Red Book is to define the amount of uranium of economic interest that could be available for use in a global fleet of long-lived NPPs (60 years or more), uranium resource figures, and not reserves, are reported in the Red Book.

4.2.1 Conventional and unconventional uranium resources

Uranium is relatively common in the Earth's crust and has been recovered in a variety of forms and settings, ranging from very low-grade to very high-grade ores that have been mined using a number of methods. Because the focus of uranium mining has been principally on higher-grade sources of ore where uranium is the primary mining target, resources reported in the Red Book have been divided, somewhat arbitrarily, into conventional and unconventional resources. Conventional resources are those from which uranium is recoverable as a primary product, coproduct, or an important byproduct, whereas unconventional resources refer to those from which uranium is only recoverable as a minor byproduct, although uranium resources are huge.

Examples of conventional resources include those from which uranium minerals are extracted and processed exclusively, from very high-grade deposits accessed by underground mines in Canada (eg, McArthur River and Cigar Lake) to much lower-grade ores mined by open pit in Australia (eg, Ranger) and Namibia (eg, Rössing) and in situ leach ((ISL), sometimes referred to as in situ recovery) amenable deposits in Kazakhstan (eg, Katko). Conventional resources also include multimineral deposits where uranium is extracted as an important byproduct, such as the Olympic Dam mine in Australia, where copper is the primary mining target.

Examples of unconventional resources include low-grade uranium occurrences in phosphate rocks that are mined principally to produce phosphoric acid for the production of fertilizer, rare earth elements (REEs), nonferrous ore, carbonatite, black shale, lignite, and even the very low-grade background of uranium in seawater.

The majority of the unconventional uranium resources reported in the Red Book are associated with uranium occurrences in phosphate rocks. Although none of these unconventional resources is currently being used for uranium supply, phosphate rocks and black shales have been sources of uranium in the past and interest has not completely waned (see Section 4.4). Unconventional resources represent significant amounts of uranium that could potentially be utilized, but the focus of this chapter is on conventional uranium resources that will, in all likelihood, be the source of the majority of uranium mined in the coming decades.

4.2.2 Resource classification

Uranium resources in the Red Book are classified by a two-dimensional system of geological certainty and production costs. This classification scheme resulted from efforts to combine resource estimates from a number of different countries into harmonized global figures (Fig. 4.1), with amounts reported as recoverable tonnes of uranium metal (tU). The horizontal axis of Fig. 4.1 represents the geological

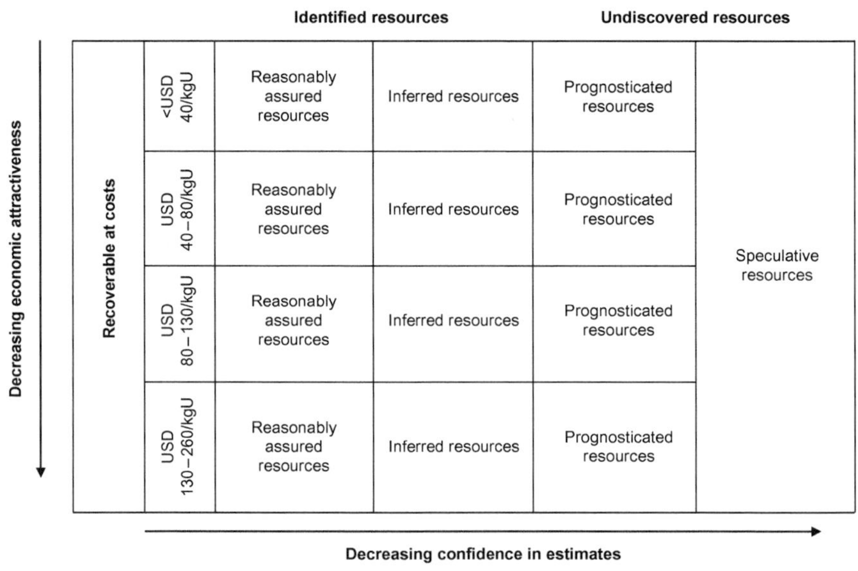

Figure 4.1 Red Book uranium classification scheme (NEA/IAEA, 2014).

confidence of the estimates, from those best defined in terms of amount of drilling, geochemical analysis, and modeling, referred to as either Reasonably Assured Resources (RAR) or Inferred Resources (IR). Collectively, these two categories comprise what is known as Identified Resources. Undiscovered Resources, on the other hand, refer to resources that are expected to exist based on knowledge of previously discovered deposits and regional geological mapping. They are subdivided into two categories: (1) Prognosticated Resources (PR) refer to those expected to exist in known uranium provinces, generally supported by some direct evidence, whereas (2) Speculative Resources (SR) refer to those expected to exist in geological provinces that may host uranium deposits. Methods of calculating tonnages of Undiscovered Resources have been available for some time (IAEA, 1992). Undiscovered Resources provide an indication of possible resource availability according to currently understood deposit models in areas that have been the subject of regional geological mapping. Both PR and SR require significant exploration efforts before their existence can be confirmed and grades and tonnages can be defined.

Because the goal of the Red Book is to provide an assessment of global uranium resources in a particular year, and the information collected comes from a number of countries that, in relative isolation, developed national uranium resource classification schemes, compromises had to be made to accommodate these well-functioning classification schemes into a global system. This was particularly important after the dissolution of the Soviet Union and the direct participation in the Red Book exercise by a number of countries from the former Soviet Union. A great deal of effort was made to incorporate regional classification systems under the global umbrella portrayed in Fig. 4.1, while respecting the national classification scheme definitions.

Fig. 4.2 shows the correlation of national classification schemes with the Red Book system. Although not always a seamless fit, the Red Book classification scheme allows major uranium resources from a variety of jurisdictions around the world to be classified in one system.

An important aspect of Red Book resource classification is the cost of production (vertical axis, Fig. 4.1). No other energy commodity is classified in this way. Production cost categories were incorporated into the first edition of the Red Book (ENEA, 1965) and are integral to the system. Governments and utilities planning for the installation of long-lived nuclear reactors (50–60 years or more) benefit from the assurance that potential future raw material resources will be available at reasonable costs. Currently, RAR, IR, and SR are subdivided into four production cost categories: <USD 40, 40–80, 80–130, and 130–260/kgU (roughly equivalent to <USD 15, 15–30, 30–50, and 50–100/lb U_3O_8). The highest cost category (130–260/kgU) was added to the Red Book in 2009 because high case scenario projections of growth in nuclear generating capacity indicated that uranium requirements could exceed identified global resources available at costs of less than USD 130/kgU (the currency conversion of national production costs into USD is based on the average exchange rate for the month of June in the year of data collection).

	Identified resources			Undiscovered resources		
NEA/IAEA	Reasonably assured		Inferred	Prognosticated	Speculative	
Australia (JORC)	Demonstrated		Inferred	Undiscovered		
	Measured	Indicated				
Canada (NRCan; NI 43–101)	Measured	Indicated	Inferred	Prognosticated	Speculative	
United States (DOE)	Reasonably assured		Estimated additional		Speculative	
Russian Federation, Kazakhstan, Ukraine, Uzbekistan	A+B	C1	C2	P1	P2	P3

Figure 4.2 Approximate correlation of terms used in major resources classification systems (NEA/IAEA, 2014).

When estimating the cost of production, countries are asked to take into account the following:

- The direct costs of mining, transporting, and processing the uranium ore
- The costs of associated environmental and waste management during and after mining
- The costs of maintaining nonoperating production units, where applicable
- In the case of ongoing projects, those capital costs that remain nonamortized
- The capital cost of providing new production units, where applicable, including the cost of financing
- Indirect costs such as office overheads, taxes, and royalties, where applicable
- Future exploration and development costs for further ore delineation until the ore is ready to be mined
- Sunk costs are not normally taken into consideration

All countries are asked to provide estimates of recoverable uranium resources in the RAR and IR categories. In other words, quantities are reported as recoverable uranium from mineable ore, as opposed to quantities of uranium contained in mineable ore (in situ) that do not take into account mining and milling losses. If a

Table 4.1 Conversion factors applied to in situ resources in cases where estimates of mining and milling losses are not provided (NEA/IAEA, 2014)

Mining and milling method	Overall recovery factor (%)
Open pit mining with conventional milling	80
Underground mining with conventional milling	75
ISL (acid)	75
ISL (alkaline)	70
Heap leaching	70
Block and stope leaching	75
Coproduct or byproduct	65
Unspecified method	75

country reports in situ resources and does not provide recovery factors, a recovery factor is assigned to determine recoverable resources (Table 4.1).

4.2.3 Resource dynamics

In a market-based system, quantities of RAR, IR, or PR will vary to some extent with fluctuations in uranium prices. Mining companies may update resource figures on a quarterly or semiannual basis to assess impacts of changing prices and other factors. This underscores an important point concerning uranium resource figures published in the Red Book: the figures are estimates from a specific point in time, in essence snapshots of a dynamic system. The resource base includes deposits that might or might not be mined because of political constraints, public opposition, unfavorable economics, or other reasons. Equally, resources not currently included in the resource base could become the source of ore for large production facilities in the future, given the limits of the geographical coverage of uranium exploration and the current understanding of uranium deposit formation.

Uranium is a relatively common element of the Earth's crust, and periods of increased investment in uranium exploration have always led to the discovery of new resources of economic interest, although doing the required amount of drilling and geochemical analyses to adequately document the deposit(s) takes time. Uranium exploration requires a great deal of expertise and investment to locate deposits of interest and do the work required to develop confidence in the estimated grade and size, processing methods and costs to the point that investment decisions can be made.

Data on exploration expenditures from 13 leading resource countries analyzed in the Red Book Retrospective (NEA, 2006) show that the historic cost of discovery during the early phases of uranium exploration averaged less than USD 2/kgU (not inflation adjusted). While it can be argued that all the easy-to-find, near-surface mineable deposits have already been discovered, unexpected breakthroughs can lead to new discoveries and new exploration targets. History shows that new

Table 4.2 Uranium resource balances in Red Book history (expressed in inflation-adjusted 2013 USD equivalent; PPI, 2015)

	tU	2013 USD cost
RAR 1965 (in situ)	1,020,000	USD 95/lb U_3O_8
Total mined: 1965–2013	2,700,000	
Balance	(−1,680,000)	
RAR 2013 (recoverable)	4,585,000	USD 100/lb U_3O_8
Added	**6,265,000**	
RAR 1993 (recoverable)	1,424,000	USD 137/lb U_3O_8
Total mined: 1993–2013	865,000	
Balance	(559,000)	
RAR 2013 (recoverable)	4,585,000	USD 100/lb U_3O_8
Added	**4,026,000**	

exploration methods and the development of new geological models have led to new discoveries. For example, only 25% of the more than 500,000 tU of the uranium resources identified in the Athabasca Basin of northern Saskatchewan (Canada) by 2000 were discovered during the initial phase of surface-based exploration. A sustained second phase, based on advances in deep penetrating geophysics and new geological models, was required to discover the remaining 75% identified by 2000 (MacDonald, 2001).

When exploration activity has increased, typically stimulated by increased commodity prices, so too has the resource base, despite continuous drawdown through mine production. Although cost categories and aspects of the classification scheme have periodically been changed throughout the history of the Red Book series, simple comparisons illustrate this point (Table 4.2).

In 1965, identified in situ RAR (excluding the former Soviet Union) amounted to the equivalent of about 1,020,000 tU at less than USD 15/lb (ENEA, 1965), or about USD 95/lb U_3O_8 in inflation-adjusted 2013 USD (PPI, 2015). In 2013, recoverable RAR at an estimated cost of less than USD 260/kgU (USD 100/lb U_3O_8) amounted to more than 4,585,000 tU, despite resource depletion of 2,700,000 tU by global mining since 1965 (NEA/IAEA, 2014). This amounts to a net gain of over 6 million tU at comparable costs of recovery in less than 50 years.

In 1993, when Red Book coverage included countries in the former Soviet Union for the first time, recoverable RAR available at USD 80/kgU (about USD 137/lb U_3O_8 in inflation-adjusted 2013 USD) amounted to 1,424,000 tU. In 2013, 4,585,000 tU were estimated to be available at USD 260/kgU (USD 100/lb U_3O_8), despite depletion by mining of over 865,000 tU from 1993 to 2013. This amounts to a net global gain of over 4 million tU at comparable costs of recovery in 20 years.

Clearly, uranium resources are being added at a greater rate through exploration than resources are being depleted through mining. Moreover, this strong positive balance exists despite a lengthy period of low prices between 1983 and 2003, when secondary supplies met a significant share of global demand (roughly 35–40%)

and prospects for global nuclear power expansion were low. Exploration activity during this period was also low, as prices were not high enough to stimulate little more than deposit extension activities.

Each time uranium demand was expected to increase (for military purposes in World War II and when civil nuclear power was projected to grow significantly in the 1970s and 2000s), uranium was considered in short supply and regarded as a potential limiting factor to such growth. However, in each period that prices increased as demand increased, exploration activity was stimulated and more uranium resources than were required were identified. Much of the excess material mined during these periods remains an important source supply today.

In the early years of uranium exploration activity, new discoveries were necessary to build up the resource base. In the most recent cycle of increased demand, exploration, and resources in the mid-2000s, only a small portion of the increases were related to new discoveries. The majority were the result of reevaluations of previously identified resources and exploration in the vicinity of known deposits. Recent exploration activities remain concentrated in areas with potential for exceptionally rich deposits (such as very high-grade unconformity-related deposits) and low-cost ISL amenable sandstone deposits. Should favorable market conditions and increased prices once again stimulate increased exploration activity, additional discoveries can be expected, as was the case during past periods of heightened exploration activity. This could include successful exploration activity in areas not previously explored ("greenfield" areas). Uranium is a nonrenewable resource. However, there is no indication over its short exploration and production history that the total amount of mineable uranium in the Earth's crust has been defined.

4.3 Global uranium resources as of January 1, 2013

As of January 1, 2013, total Identified Resources (RAR and IR) amounted to over 5,900,000 tU in the <USD 130/kgU (<USD 50/lb U_3O_8) cost category and over 7,600,000 tU in the highest cost category (<USD 260/kgU or <USD 100/lb U_3O_8). Both totals represent increases since the previous 2011 edition of the Red Book of 10.8% and 7.6%, respectively.

Although total Identified Resources increased in both the higher cost categories over the 2 years between Red Book reports, the <USD 80/kgU (<USD 30/lb U_3O_8) cost category declined by 36%, owing principally to increased mining costs. However, the lowest cost category (<USD 40/kgU or <USD 15/lb U_3O_8) changed little.

Overall, the majority of resource increases are the result of reevaluations of previously Identified Resources and additions to known deposits to extend the life-of-mine or expand production capacity at existing mining facilities, particularly in Australia, Canada, the People's Republic of China, Greenland, Kazakhstan, the Russian Federation, South Africa, Tanzania, and Zambia, with the most significant increases in the Czech Republic (a result of reclassification of deposits previously considered inaccessible), Greenland, and Mongolia. Despite relatively low uranium prices during this 2-year period, continued investment and associated exploration efforts resulted in the identification of additional resources of economic interest.

Along with the increase in total Identified Resources, reductions were reported in some countries owing to reevaluations (mainly in Botswana) and depletion by mining (eg, Namibia and Niger). A notable exception to a general decline of low-cost resources in recent years is Kazakhstan, where exploration and deposit extensions resulted in an 88% increase in low-cost Identified Resources (<USD 40/kgU) from 47,400 tU in 2011 to 89,300 tU in 2013. Globally, it is likely that resources in this lowest cost category are actually more abundant than reported, because some countries have indicated that estimates of low-cost resource availability are either not possible (Australia) or confidential (the United States).

Australia dominates the world's uranium resources with 29% of the total Identified Resources (<USD 130/kgU) and 24% of the highest cost category (<USD 260/kgU). Kazakhstan hosts 12% in both cost categories, with all other countries having less than a 10% share. Globally, 15 countries (Fig. 4.3) have more than a 1% share of the world's Identified Resources available at costs of <USD 130/kgU and 16 countries have more than a 1% share in the high-cost category.

RAR recoverable at <USD 40/kgU, the best-defined and most economically attractive category, increased slightly by 13,500 tU (2.7%) mainly as a result of exploration near existing production centers in Canada, China, and Kazakhstan. However, the most significant change in RAR was a decrease of over 800,000 tU (39.9%) in the <USD 80/kgU category, a result of a continued trend of higher production costs. The <USD 130/kgU and <USD 260/kgU categories experienced modest increases of 7% and 4.8%, respectively (compared to 2011), mainly as a result of the reevaluation and transfer of resources into higher cost categories. Increases in the overall total RAR were largely due to additions in Australia, Canada, the People's Republic of China, the Czech Republic, India, Mongolia, the Russian Federation, and South Africa.

As of January 1, 2013, Undiscovered Resources (PR and SR) amounted to about 7,700,000 tU, a significant decline from the 10,400,000 tU reported in 2011, primarily because the United States withdrew estimates produced in the 1980s to reevaluate the historic data using modern methods. Moreover, it is important to note that in some cases, including major producing countries with large identified resource inventories, undiscovered resource estimates are either not reported (eg, Australia) or have not been updated since the early 1990s (eg, Canada).

4.3.1 Widespread and variable

The geographical distribution of uranium resources helps reduce the risk of disruption of the raw material component of fuel supply. Identified Resources are documented in countries as diverse as Australia, Canada, Kazakhstan, Namibia, Niger, the Russian Federation, South Africa, and the United States (Fig. 4.3). Since 2009, Kazakhstan has emerged as the dominant producer, and based on the amount of low-cost resources reported, appears destined to be a significant producer for years to come.

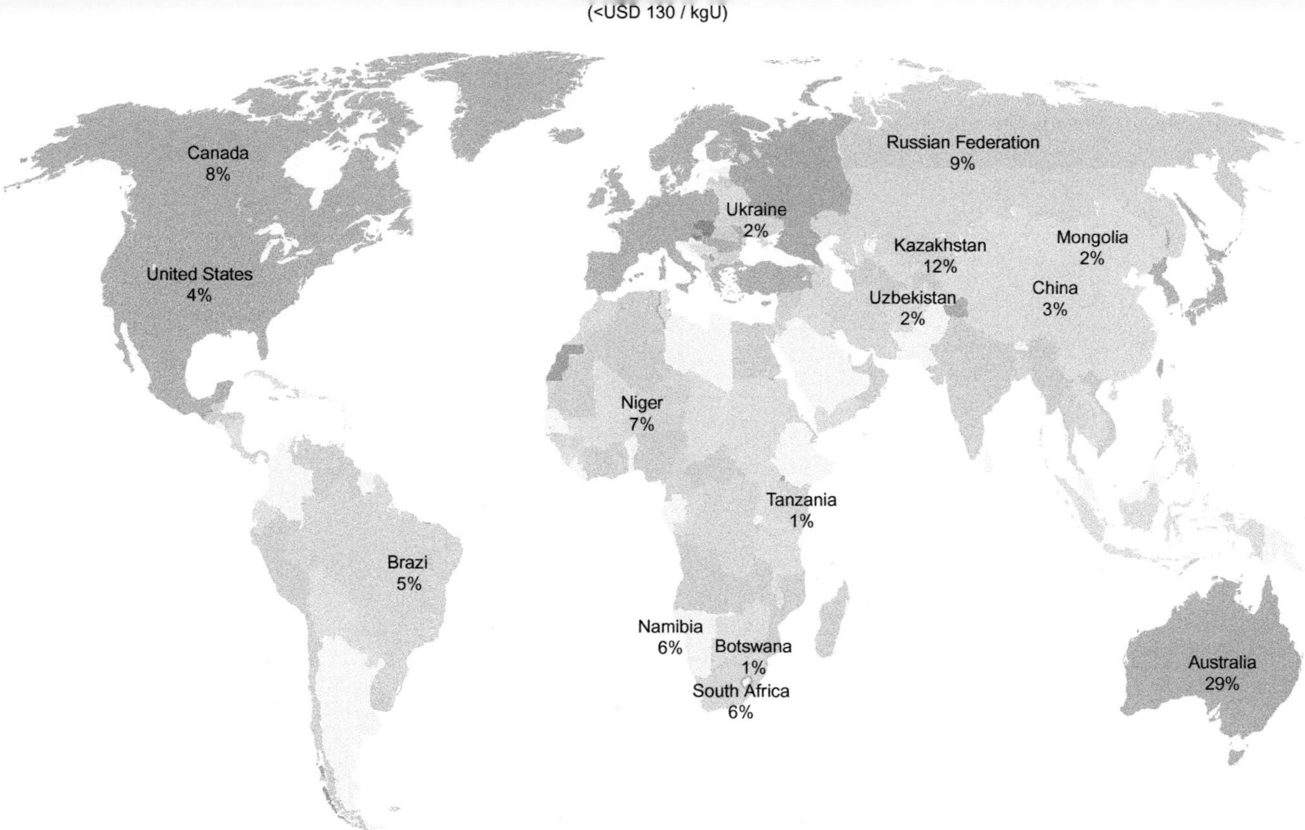

Figure 4.3 Global distribution of identified resources (NEA/IAEA 2014).
The global distribution of identified resources among the 15 countries with more than a 1% share of the total global identified resources available at costs of <USD 130/kgU illustrates the widespread distribution of these resources. Together, these 15 countries are endowed with 97% of the global identified resource base in this cost category (the remaining 3% are distributed among another 21 countries). The widespread distribution of uranium resources is an important aspect of nuclear energy in terms of security of energy supply.

In terms of resource endowments, some important distinctions should be noted. For example, although Australia hosts the largest uranium resource base, its national total is dominated by one deposit, Olympic Dam. Of the recoverable resources listed for the seven existing and planned production centers in Australia, Olympic Dam represents 90% of the total (NEA/IAEA, 2014). Because Olympic Dam is primarily a copper mine, with coproducts uranium, gold, and silver, the future of the mine will not necessarily be determined by the uranium market outlook. Indeed, the average uranium grade of the deposit is very low −0.023% U. Moreover, the significant capital investments needed to adopt the mining technology required to access all of this massive multimineral deposit have been deferred owing to a general decline in commodity demand, including copper. Recovering uranium in a more cost-effective manner by heap leaching is being investigated. Nonetheless, because the deposit is massive and well defined, interest in its development will remain. Reports indicate that when mine expansion was under consideration, despite drilling conducted continuously over 2 years, the full extent of the deposit could not be defined.

In contrast, the large, low-grade (0.035−0.07% U) resource base in Kazakhstan is principally (80%) amenable to extraction by low-cost ISL. With strong government support of the industry and a number of international partners and investors, it is likely that the resource endowment in Kazakhstan will form the bulk of global uranium production for the foreseeable future (currently about 40% of uranium mined in the world is produced in Kazakhstan). As noted previously, recent exploration efforts have added to national uranium resource totals, notably in the low-cost category. Until now, most production is from deposits discovered in the 1970s, and much promising terrain is only now being explored.

In Namibia and Niger, significant resource endowments have been identified, particularly in recent years, but the majority of these high-tonnage, low-grade deposits (0.01−0.05% U in Namibia and 0.01−0.02% U in Niger) are mainly accessible through the relatively high-cost open pit and underground mining methods. Both governments support the expansion of the uranium mining industry but the characteristics of the resource base limit, to some extent, mining that can be conducted when uranium prices are low, particularly when the deposits are located in areas with limited water resources. Conventional mining of low-grade deposits also means that the resulting significant waste rock and tailing accumulations require long-term management.

Although resources have been identified in a number of regions in Canada, all uranium production today comes from unconformity-related deposits in the Athabasca Basin of northern Saskatchewan, where resources tend to be high tonnage and high grade to very high grade, orders of magnitude greater than deposits mined elsewhere (eg, the average grade of the Cigar Lake and McArthur River deposits is between 11% and 14% U). The largest deposits are mined by innovative underground mining methods, although open pit mining has been undertaken and is expected to be conducted in mines under development. Relatively low volumes of wastes are produced because of the combination of high-grade ore and underground mining practices.

Unconformity-related deposits like these are understandably a focus of ongoing exploration, with several promising discoveries being made in the Athabasca Basin in recent years (eg, Patterson Lake, Wheeler River, and Shea Creek). Government support for exploration efforts, in the form of high-resolution airborne surveys and partnerships with industry and universities to investigate new exploration methods and models of deposit formation (eg, Jefferson et al., 2007), suggest that additional high-grade discoveries of significance are forthcoming. Companies operating the mines have successfully developed innovative mining techniques to safely access these technically challenging deposits. Despite these challenges and the significant investment required, mining costs per tonne of product are relatively low in these types of deposits, suggesting that mining will continue well into the foreseeable future.

4.3.2 How much is enough?

At the 2012 level of uranium requirements (61,600 tU), declining somewhat in recent years because of the idling of all NPPs in Japan following the accident at the Fukushima Daiichi NPP, currently Identified Resources are sufficient for over 120 years of supply for the global nuclear power fleet (NEA/IAEA, 2014). The overall increase in the identified resource base (including high-cost resources) from 2011 to 2013 alone added over 8 years of supply. This seems more than adequate for future nuclear power requirements, especially considering the growth in global uranium resource figures over the past 50 years, despite the significant tonnages extracted by mining, as previously outlined. Currently identified resources of other energy resources do not have such a lengthy time horizon (eg, oil 50 years, natural gas 55 years; BP, 2014). Moreover, fast neutron reactors are anticipated to be in operation by mid-century and the ability of this type of reactor to produce more fuel than it consumes promises to reduce uranium demand significantly.

Exploitation of the entire conventional resource base (including PR and SR) would increase the lifespan of the uranium resource base to nearly 300 years, although significant exploration and development would be required to advance these Undiscovered Resources into more definitive categories.

There are different perspectives on what constitutes an adequate uranium resource base. For miners, 20 years or more is typically all that is required for resource identification sufficient to recover capital expenditures and to operate profitably over the planned lifetime of a mine. On the other hand, utilities and governments planning to deploy NPPs want assurance of raw material supply for a much longer time horizon—60 years or more (lifetime extensions to 80 years of operation are currently under consideration). The concept of scarcity is important for miners to help maintain market prices at a level that would more than cover full life cycle mining costs, whereas utilities prefer to have a much larger resource base for assurance of reliable long-term supply capabilities.

As noted previously, every period in which uranium requirements were expected to increase has been characterized by price increases that stimulated exploration activity that in turn led to the identification (and often the mining of) more uranium than required. There is little reason to expect that this will not be the case once again if future uranium requirements are projected to increase.

4.3.3 Is that all there is?

Sufficient resources have already been identified to support the full lifespan of established nuclear capacity for electricity generation. Given the limited maturity and geographical coverage of uranium exploration worldwide, there remains considerable potential for discovering new resources of economic interest.

In the mid-2000s, the uranium industry emerged from a two-decade period of low prices, during which prices were too low to stimulate much investment in exploration. Prices began to rise in 2003 on heightened expectations of growth in nuclear power and associated uranium demand (dubbed by some the "nuclear renaissance"). The price rise, while dramatic, was cut short by the financial crisis in 2008 and the Fukushima accident in 2011. This brief period of intense exploration activity, stimulated by the rapid rise in spot prices through to mid-2007, is nonetheless instructive since it added almost 58,000 tU in Identified Resources available at <130 kg/U and over 2,600,000 tU to the highest cost category (<260 kg/U). At the peak of activity, some 400 junior exploration companies were involved and although the focus was on extensions to known deposits, some greenfield exploration resulted in discoveries. Although the cycle was short, it had the same outcome as previous bursts of exploration; increased prices resulted in increased exploration activity that led to the identification of additional resources.

Unlike other metals that have been in demand for centuries, uranium has been of interest for only a matter of decades. Significant nonmilitary demand did not arise until the 1970s and increased prices in that period resulted in the identification of more than enough uranium resources to fuel several decades of supply, before declining uranium prices dampened exploration considerably in the early 1980s.

As noted by MacDonald (2001), the significant investment in uranium exploration during the 1970–1982 exploration cycle would have been efficient in discovering exposed uranium deposits, due to the ease of detecting radioactivity. However, in a second round of exploration, new techniques and deposit models led to the discovery of large, deeply buried deposits. Not all prospective regions in the world have been the subject of a second round of intensive knowledge and technology-driven exploration. This experience suggests that the largest proportion of future uranium resources will be discovered in the more advanced phases of exploration. In addition, it is by no means certain that all possible deposit types have been identified. Any estimate of world uranium potential made 45 years ago would have missed the entire deposit class of unconformity deposits, simply because geologists did not know of its existence.

Over USD 25 billion has been invested in global exploration and mine development activities through to 2013 (NEA/IAEA, 2014). However, the answer to the question "have we found it all?" remains an emphatic "no" (McMurray, 2005). The list of potential exploration targets remains long. All that is needed is another round of exploration investment and experienced people to use it wisely to establish an even more geographically diverse resource base.

4.4 Unconventional resources

In addition to conventional uranium resources, there are unconventional sources of uranium, some of which have contributed to uranium production in the past and could be used to supplement uranium supply given appropriate market signals and technological development. By definition, these sources are low-grade, and uranium is only recoverable as a minor byproduct.

For example, uranium was extracted from phosphate rocks in Florida and Belgium until low uranium prices in the early 1990s forced the closure of these operations. Past estimates of uranium contained in phosphate rocks point to a possible upper global amount of 22 million tU (De Voto and Stevens, 1979), although additional exploration effort will be required to confirm this figure. With strong demand and higher prices, extraction of uranium from phosphates during the production of fertilizer may once again become attractive, particularly with the development of new initiatives to reduce the uranium content of fertilizers and more efficient extraction technology.

When uranium prices were higher prior to the Fukushima accident, fertilizer producers in the United States were reportedly considering the production of uranium as a byproduct of their operations using proven technology (solvent extraction). In Brazil, efforts are being directed at licensing the Santa Quitéria mine that will produce both uranium and phosphate in a single integrated process, with first production expected in 2016 at an annual production capability of 970 tU (NEA/IAEA, 2014). A noteworthy aspect of this initiative is the relatively high-grade of the uranium (0.08% U). Because of this high average grade (twice the upper end of most other phosphate deposits), Brazil includes uranium contained in this deposit in its conventional uranium resource total. Technological advances could also lead to increased production, even during times of low uranium prices, as shown by recent successful demonstration plant tests of uranium production using an ion exchange process at a fertilizer plant in the United States (PhosEnergy, 2014, 2015).

Uranium has also been produced in the past from low-grade black shale deposits and uranium resources in black shales have been reported in recent years by Finland and Sweden. In Sweden, this large, low-grade (0.013% U) resource is estimated to amount to over 95,000 tU (NEA/IAEA, 2014). Uranium recovery from coal ash and tailings, in particular gold tailings in South Africa, has also been under consideration since the last peak in uranium prices in 2007 (South Africa includes resources in gold mine tailings and coal and shale hosted deposits in the RAR and Inferred conventional resource categories). Uranium contained in REE deposits, such as the large Kvanefjeld deposit in Greenland, is also a potential source of uranium as a byproduct. Although the main mining target in the Kvanefjeld project area deposits is REEs, JORC-compliant resource estimates indicate that uranium amounts to over 220,000 tU (Greenland Minerals and Energy, 2015).

Finally, no uranium resource survey is complete without mention of the 4.5 billion tU contained in seawater. This enormous, very low-grade (3–4 ppb) constantly replenished potential source of uranium cannot be ignored despite the significant challenges associated with recovering sufficient quantities for viable commercial extraction. In the 1980s, Japanese researchers showed that extraction of uranium from seawater is feasible. Laboratory-scale efforts led to cost projections approximately 5–10 times the cost of conventionally mined uranium with optimistic assumptions about the scale of the operation and the reuse of collection materials. Recent efforts in the United States are exploring methods to improve the economics of extraction, although this work was initiated to determine the lowest cost possible for extraction from seawater to inform fuel cycle decisions, and is not commercially oriented. Efforts have reportedly reduced extraction costs significantly with further potential improvements to be tested (Ferguson, 2012). Despite this progress, recovery costs remain well above conventional uranium mining costs.

Adding unconventional resources (eg, uranium in phosphate rocks, black shales, REE deposits, etc.) have the potential to extend the longevity of fuel supply. However, most resource estimates are not produced using the same degree of standardization applied to conventional resources, and none of these potentially extensive resources has been brought into production for over 20 years. The extraction of uranium from seawater could potentially extend raw material fuel supply indefinitely, albeit at a much higher cost than other resources, but doing so will require significant effort, innovation, and investment to improve the economics and practicality of the extraction process.

4.5 International fuel reserves

4.5.1 Rationale

With rising global energy needs, nonproliferation, and waste concerns, the IAEA (with the assistance of some member states) developed a number of proposals aimed at strengthening nonproliferation by establishing multilateral enrichment and fuel supply centers. Specific proposals for reserves of enriched uranium are considered important for assuring countries with nuclear power, or those planning to develop nuclear power, that fuel supply will not be stopped for noncommercial reasons should the country remain in compliance with nonproliferation commitments. Fuel reserves also limit potential proliferation risks by providing an alternative to the development of sensitive domestic nuclear technologies, such as uranium enrichment and nuclear fuel reprocessing.

Although a number of proposals have been made in recent years, only one such reserve has been established to date. The proposed fuel reserves are designed to not affect the functioning of the commercial market for nuclear fuel. Because the amounts of uranium in each of the proposed reserves do not amount to much more than 900 tU equivalent, and one will be stocked by blending down highly enriched uranium (HEU), unless used frequently the fuel reserves are likely to have a minimal market impact on a global industry that has been producing between 55,000 and 59,000 tU annually in recent years.

4.5.2 Status

In March 2010, the IAEA and the Russian Federation signed an agreement that led to the establishment in 2011 of a low-enriched uranium (LEU) reserve under IAEA auspices at the International Uranium Enrichment Centre in Angarsk (the Russian Federation). A reserve of 120 t LEU with enrichment levels up to 4.95% was then made available to IAEA member states. Under IAEA safeguards, this reserve (equivalent to two full loads of fuel for a typical 1000 MWe reactor) will be accessible by IAEA member states in good standing that are unable to procure LEU on the open market for political reasons. The LEU is to be made available for nuclear power generation at market prices and proceeds from sales will be used to replenish the LEU stock. The Russian Federation is paying LEU storage, maintenance, safety, security, and safeguards costs. The reserve is not intended to distort the commercial market, but rather to reinforce existing market mechanisms of member states. Decisions on the release of materials will be made by the IAEA Director General.

In December 2010, the IAEA Board of Governors authorized the IAEA Director General to establish a LEU bank (owned and operated by the IAEA) to serve as a supply of last resort for nuclear power generation. The IAEA reserve is a back-up mechanism to the commercial market in the event that an eligible member state's supply of LEU is disrupted and cannot be restored by commercial transactions, state-to-state arrangements, or by other such means. The goal is to have sufficient LEU in the bank to meet the fuel fabrication needs of a single 1000 MWe light water reactor reload. Donors have pledged about USD 125 million and EUR 25 million toward the initial operational expenses, the purchase and delivery of the LEU to the host state, and operation of the facility for at least 10 years. The fuel is to be made available to eligible member states for fuel fabrication at prevailing market prices and the proceeds used to replenish the fuel in the LEU bank. Kazakhstan is the only member state that formally expressed interest in hosting this LEU bank and, after the IAEA assessed two sites offered as potential locations, agreements were signed in June 2015 to establish the LEU bank at the Ulba metallurgical plant in Oskemen (northeastern Kazakhstan) and transit fuel through the Russian Federation (IAEA, 2015). The IAEA LEU Bank is expected to be fully operational in 2017.

In March 2011, the IAEA Board of Governors approved a proposal for nuclear fuel assurance led by the United Kingdom. This initiative was designed to provide assurance that a commercial contract for nuclear fuel cannot be interrupted for non-commercial reasons. No stockpile of fuel was involved and contractual agreements between supplier and recipient states were instead proposed. As a response to this initiative, Germany proposed the establishment of an enrichment plant administered by the IAEA, referred to as the Multilateral Enrichment Sanctuary Project (MESP). This proposal foresees the construction of one or more enrichment facilities under the exclusive supervision of the IAEA. The MESP is designed to allow independent access to nuclear fuel cycle services, complementing other proposals concerning assurances of nuclear fuel supply. The proposed enrichment company could also establish and maintain a buffer stock or a physical reserve of nuclear fuel available on notice to the Director General of the IAEA for additional assurance.

In August 2011, the United States Department of Energy announced that the American Assured Nuclear Fuel Supply had been established to secure 230 t of LEU, sufficient for six reloads of an average 1000 MWe reactor. The LEU has been produced by blending down 17.4 t of HEU declared surplus to military needs, including an additional 60 t of LEU that will be sold to pay for production costs. The administrator of the program, the National Nuclear Safety Administration, will make the LEU available for use in NPPs in nations that are not pursuing uranium enrichment and reprocessing technologies. Qualifying countries will have access to the fuel at the current market price only in the event of an emergency that disrupts the normal flow of fuel supply.

4.6 United Nations Framework Classification

In 1997, the United Nations Economic Commission for Europe (UNECE) developed the first iteration of the United Nations Framework Classification for Fossil Energy and Mineral Reserves and Resources (UNFC). The stated principal function of the UNFC was to make national classification terms comparable internationally to allow reserves/resources of solid fuels and mineral commodities to be classified in a uniform international system (UNECE, 1997). Development of this type of classification system was expected to enhance communication at an international level, provide a better understanding and firmer knowledge of the reserves/resources available, and make investments in fuels and minerals safer and more attractive. The scheme developed combines geological, feasibility, and economic factors in a three-dimensional model. Because frequently used terms (eg, reserves and resources) can have different meanings in different countries, precisely defined numerical codes were developed to avoid any potential confusion.

In 2004, a second iteration of the UNFC was released (UNECE, 2004), after the system was further refined and extended to incorporate and harmonize all extractable energy commodities (eg, petroleum, coal, and uranium). UNECE proposed that the UNFC be implemented worldwide and accordingly developed related standards, issued guidelines, and held regional seminars to harmonize national classification systems with the UNFC. The main elements of UNFC-1997 were retained; that is, the three-dimensional numerical codes for economics (three subdivisions), feasibility (three subdivisions), and geology (four subdivisions).

In 2009, what was considered a stronger and simpler third iteration of the UNFC was released (UNECE, 2009) that retained the main elements of the system (numerical designation of each class in the main categories of economic and social viability, field project status and feasibility (now expanded to four subdivisions) and geological knowledge). UNECE recommended that the UNFC either be applied directly or used as a harmonizing tool. It was aligned with other widely used classification systems in extractive industries to facilitate mapping between systems to the extent possible. UNFC-2009 is designed to meet applications pertaining to energy and mineral studies, resource management functions, corporate business processes, and financial reporting standards.

The UNFC is now the only internationally applicable scheme for the classification and reporting of fossil energy and mineral reserves and resources. If not applied directly, it nevertheless serves as an umbrella classification system under which all existing classification schemes can be mapped. In the case of the classification scheme used for reporting uranium resources in the Red Book, mapping has recently been conducted to the highest level possible and a bridging document is currently in preparation to explain the relationship between these two classification schemes. The UNFC covers the total resource base, including quantities that are not currently economic, as well as the "unrecoverable" part of individual deposits. The Red Book classification scheme is designed to report resources on a national and international basis and, unlike other systems, incorporates estimated production cost as an integral part of the scheme. Despite these basic differences, it is nonetheless possible to align the systems to show the interrelatedness of the two systems and delineate uranium deposits in an international system.

4.7 Conclusion

Uranium is a common element of the Earth's crust. Currently, Identified Resources are sufficient to fuel the existing global nuclear fleet for 120 years. With sufficiently high prices to motivate the required work, Undiscovered Resources (those not yet discovered but expected to exist based on regional geological data and mapping) can be expected to be converted into Identified Resources. The result could increase uranium supply to almost 300 years for the existing fleet of NPPs exclusively using a once-through fuel cycle, though significant exploration and development would be required. Given the limited maturity and geographical coverage of worldwide uranium exploration, combined with the potential for the discovery of new uranium deposit types, there remains considerable potential for discovery of additional resources of economic interest. Given the expected development of fast neutron reactors, these conclusions on the length of fuel supply are conservative.

The existing uranium resource base is not a limitation to the development of new nuclear generating capacity. Erroneous assumptions about uranium resource figures used to make such claims often result from treating uranium resource figures published in the Red Book as estimates of the absolute amount of uranium available in the Earth's crust. This is not the case. The uranium resource figures published at any given time are epistemological and simply a reflection of the amount of exploration that has taken place and the price of uranium. Although there is a finite limit to the amount of mineable uranium in the Earth's crust, there is no indication that the limit has been approached after only 60 years of sporadic exploration effort. In fact, resources have been added at a greater rate than they have been mined.

Unconventional resources (eg, uranium in phosphate rocks) have the potential to greatly extend the time horizon of raw material requirements for nuclear fuel supply, although utilization of these resources will rely on a combination of technical developments and higher uranium prices. While recovery of uranium from seawater could extend supplies for millennia, significant technological development will be needed to recover uranium at costs competitive with conventional mining methods.

Identified uranium resources are widely distributed and growing. Maintaining strategic stockpiles of fuel is relatively easy and because the cost of uranium represents only a small percentage of the total cost of generating electricity, uranium price volatility is not as significant a factor as that of fuel prices for electricity generating facilities powered by fossil fuels. Such features enhance security of energy supply for countries with nuclear energy in their electricity generation mix.

4.8 Sources for further information and advice

Australasian Code for Reporting Exploration Results, Mineral Resources and Ore Reserves, www.jorc.org/

Committee for Mineral Reserves International Reporting Standards, www.crirsco.com/welcome.asp

South African Code for Reporting of Exploration Results, Mineral Resources and Mineral Reserves, www.samcode.co.za/

References

BP, 2014. Statistical Review of World Energy. <www.bp.com/content/dam/bp/pdf/Energy-economics/statistical-review-2014/BP-statistical-review-of-world-energy-2014-full-report.pdf>.

CIM, 2014. CIM Definition Standards for Mineral Resources and Mineral Reserves. Canadian Institute of Mining, Metallurgy and Petroleum, online article dated 10 May 2014. <http://web.cim.org/standards/MenuPage.cfm?sections=177&menu=178>.

Devoto, R.H., Stevens, N.R., 1979. Uraniferous phosphate resources and technology and economics of uranium recovery from phosphate resources, United States and Free World, Rep. GJBX-110 (79), Vols. 3, US Department of Energy, Washington, DC.

ENEA, 1965. World Uranium and Thorium Resources. European Nuclear Energy Agency, Organisation for Economic Co-operation and Development (OECD), Paris.

Ferguson, W., 2012. Record haul of uranium harvested from seawater. New Scientist, online article dated 22 August 2012. <https://www.newscientist.com/article/dn22201-record-haul-of-uranium-harvested-from-seawater/>.

Greenland Minerals and Energy, 2015. Maiden Ore Reserves of 108 Million Tonnes for Kvanefjeld Rare Earth—Uranium Project. Company Announcement Australia <http://www.ggg.gl/docs/ASX-announcements/Ore_reserves_announcement.pdf>.

IAEA, 1992. Methods for the Estimation and Economic Evaluation of Undiscovered Uranium Endowment and Resources: Instruction Manual. International Atomic Energy Agency, Vienna, STI/DOC/10/344.

Jefferson, C.W., Thomas, D.J., Gandhi, S.S., Ramaekers, P., Delaney, G., Brisbin, D., et al. 2007. EXTECH IV: Geology and Uranium EXploration TECHnology of the Proterozoic Athabasca Basin, Saskatchewan and Alberta; by Jefferson, C.W., Delaney, G. (Eds.), Geological Survey of Canada, Bulletin No. 588, Natural Resources Canada, Ottawa.

MacDonald, C, 2001. Rocks to reactors: uranium exploration and the market. In: Proceedings of the World Nuclear Association Symposium, London.

McMurray, J M, 2005. Worldwide uranium resources and production capacity—the future of the industry: uranium production and raw materials for the nuclear fuel cycle—supply and demand, economics, the environment and energy security. In: Proceedings of an International Symposium, Vienna, IAEA-CN-128.

NEA, 2006. Forty Years of Uranium Resources, Production and Demand, the Red Book Retrospective. OECD, Paris.

NEA/IAEA, 2014. Uranium 2014: Resources, Production and Demand. OECD, Paris.

PhosEnergy, 2014. Demonstration Plant Operations Progressing. <www.phosenergy.com/2014/05/may-2014-newsletter/>.

PhosEnergy, 2015. Demonstration Plant and PFS Show Robustness of Phosenergy Process. <www.phosenergy.com/wp-content/uploads/2015/01/PhosEnergy-Newsletter-December-14_Draft-V4.pdf>.

PPI, 2015. Producer Price Index for All Commodities. Economic Research, Federal Reserve Bank of St. Louis. <https://research.stlouisfed.org/fred2/series/PPIACO>.

UNECE, 1997. United Nations International Framework Classification for Reserves/Resources—Solid Fuels and Mineral Commodities of 1997 (UNFC-1997) <www.unece.org/energy/se/unfc_fc_sf.html>.

UNECE, 2004. United Nations Framework Classification for Fossil Energy and Mineral Resources 2004 (UNFC-2004) <www.unece.org/energywelcome/areas-of-work/unfc-and-resource-classification/areas-of-work/energyseunfc-gen/united-nations framework-classification-for-fossil energy-and-mineral-resources-2004-unfc-2004.html>.

UNECE, 2009. United Nations Framework Classification for Fossil Energy and Mineral Reserves and Resources 2009 (UNFC-2009) <www.unece.org/energy/se/unfc_2009.html>.

The uranium market, supply adjustments from secondary sources, and enrichment underfeeding

5

Steve Kidd
East Cliff Consulting, Bournemouth, United Kingdom

5.1 Introduction

All mineral commodity markets tend to be cyclical, that is, prices rise and fall substantially over the years, but with these fluctuations are superimposed on long-term movements in real prices. In the uranium market, high prices in the late 1970s gave way to very low prices in the 1990s, the spot prices being below the cost of production for most mines. Prices recovered strongly starting in 2003 and there was a significant spike in prices that peaked in 2007. Subsequent to this, prices have fallen back sharply but today are well ahead of the depressed level of the 1990s.

The obvious way to examine the uranium market is to consider firstly the demand for uranium (Section 5.2) and then to examine how that can be satisfied by both primary and secondary supply (Section 5.3). The costs of producing uranium can then be considered (Section 5.4), as well as the longer-term outlook for supply and demand (Section 5.5). Finally, we can consider in more detail how the world uranium market works and prices have moved in the past and will likely do so in the future (Section 5.6).

5.2 Uranium demand

There are various indicators of demand for uranium in common use within the nuclear fuel market, such as reactor requirements, utility procurements, and uncommitted and uncovered demand.

Reactor requirements can be calculated knowing the nuclear generating capacity of reactors in operation together with various data about their operations and fuel cycle characteristics. This provides a good measure of how much fissile material and fuel cycle services will be required to prepare the fuel to be physically loaded into reactors in a given year in the future and can be calculated for the past. The World Nuclear Association (WNA) and other bodies base their demand forecasts on such calculations, using data provided by operators and from other sources.

Uranium for Nuclear Power. DOI: http://dx.doi.org/10.1016/B978-0-08-100307-7.00005-3

Preparing a forecast of reactor requirements is a two-stage process. First, there must be an assessment of nuclear generating capacity (Section 5.2.1), which requires taking a view about national nuclear programs (Section 5.2.2). The second stage is to consider how reactors will actually use the fuel (Section 5.2.3), which involves closely considering the enrichment sector (Section 5.2.4). It is then possible to produce estimates of uranium demand going forward.

5.2.1 Nuclear generating capacity

There are essentially two main aspects to forecasting nuclear generating capacity: the outlook for the continued operation of existing plants and the prospects for the construction of new reactors. Although many of the same factors influence each aspect, there are significant differences in how these are likely to influence demand in particular year.

Most of today's 430 or so nuclear reactors around the world started up during the 1980s and 1990s. The normal operating lifetime of most types of reactors, for planning or licensing purposes, was generally considered to be 40 years, so this would imply that a significant number of presently operating plants would be scheduled for retirement during the period after 2020. Unless there is a substantial upturn in construction within the next 10 years or so, is conceivable that the majority of reactors likely to be operating in 2030 (and even in 2040) are already in use today. Thus, forecasting the operating lifetimes of current reactors is as important in estimating future capacity as is assessing the prospects for the construction of new reactors.

How long existing reactors will in fact remain in operation depends on a number of factors that vary from country to country. The most important of these are the licensing procedures applying to life extensions, public acceptance, and the economic attractiveness of continued operation. The last will depend partly on the state of the electricity market in which the reactor is operating, the price for which the plant's output can be sold, the types of electricity supply contract that are permitted, the availability of capital for construction of replacement generating capacity, etc. Environmental considerations (eg, the avoidance of carbon dioxide emissions) may also influence reactor lifetimes in the future.

Operating reactors are subject to continuous upgrading and replacement of components, as well as rigorous licensing and inspection regimes. It is expected that many, if not most, reactors could make a technical case for operating beyond 40 years and will therefore apply for relicensing. In principle, extending the lifetime of existing nuclear plants should normally be economically attractive. Nuclear power is characterized by high initial capital costs and low fuel costs, with operations and maintenance (O&M) costs varying according to operator efficiencies and regulatory practices. For well-managed plants with low O&M costs, the cost of producing electricity should be very competitive.

Life extensions, however, may be only one side of the coin. Nothing guarantees reactors will even operate for their nominal 40-year lifetime if operating costs are too high, if they encounter licensing problems, or if they run up against subsidized renewable energy with preferential grid access. Even if operating costs are not too high, a closure decision may come because a plant requires major additional capital expenditure to keep it in operation (eg, steam generator replacement). The cost of servicing the additional capital, added to existing costs, may make the plant uneconomic. There have been examples of this happening in the United States in 2013–14. There is also the possibility of political interference in licensing decisions, as happened in Germany after the Fukushima accident in 2011.

A final factor to consider when discussing existing plants is the potential available for uprating their capacity by capital expenditure on the plant, such as modifying the steam generators and/or replacing the turbine generator set. Several countries have already benefited from this, notably Finland, Germany, Spain, Sweden, Switzerland, and the United States, but it may represent an economic way of generating more power in many others. For example, in the United States, some reactors are now uprating their power output by anything up to 20% along with plans to seek extensions for total operating lives of 60 years.

Concerning the prospects for new nuclear plants, there are a number of global influences that are important, such as security of energy supply, environmental aspects, electricity market restructuring, and the strength of the opposition to nuclear power. The world's energy situation is endlessly debated and nuclear power is an important element within this, particularly as its possible role in curbing greenhouse gas emissions is emphasized.

On the other hand, there are currently few firm plans for new reactors in the developed industrial economies of Western Europe and North America. This partly reflects the relatively slow electricity demand growth and the large existing nuclear capacity already established in many of these countries. It is also the result of the economics of new nuclear power construction having become less competitive on a short-term basis in comparison with some other generating options notably natural gas. Historically, public opposition to continued nuclear expansion has also frustrated plans for additional capacity in some cases.

Over the past few years, however, there has been some revival of interest in building new nuclear power plants (NPPs), even in countries currently without them. There are now more reactors under construction (over 60) around the world than was the case 5 or 10 years ago and the center of gravity of the nuclear world has moved toward Asia, where both China and India have large nuclear power ambitions.

5.2.2 National nuclear programs

The past experience and future possibilities of nuclear power vary considerably by country. The only sensible way to determine where world nuclear generating capacity is heading is to examine the position on a country-by-country basis.

The United States currently has the largest number of reactors in service (99 with generating capacity of 98.7 GWe) with five under construction. The key question is how long the currently operating units can stay in service. Although most will eventually achieve licenses to operate for 60 years, economic pressures from cheap natural gas and from subsidized renewable energy are now bearing down on much of the fleet.

In France (58 reactors, 63.1 GWe), the government is now developing a new energy strategy that may result in a significant reduction in the nuclear share of power generation from the current 70% plus, with 50% by 2025 mentioned as a target. There is only one unit under construction.

In Japan, all the 43 reactors operable units (40.5 GWe) are currently still shut down following the Fukushima accident, and the restart schedule is a major question. There are various scenarios for this, with the more optimistic seeing more than half of the 43 returning to service over the next few years.

Russia currently has 34 reactors in operation (25.3 GWe) and the nuclear program has now at last got back on track with nine units currently under construction. Some older units will, however, likely close by 2030.

Nuclear power is now expanding rapidly in China, which now has 26 reactors in operation (23.1 GWe) with 23 units under construction. The program has, however, slowed down somewhat post-Fukushima but still has the potential to overtake the United States during the 2020s as the largest nuclear plant operator.

The Korean nuclear program is still expanding rapidly, with 23 units reactors now in service (21.7 GWe) and four under construction. As with China, the reactors are relatively new and none are expected to shut prior to 2030.

In Canada (19 reactors 13.5 GWe), building new units has been ruled out for now, so the only question is how long the existing units will last. It is expected that two nuclear stations will be refurbished to allow extended operation but one may close around 2020.

Although the recent political tension complicates matters, the nuclear sector is important to the Ukraine (15 reactors, 13.2 GWe). All reactors are expected to run beyond 2030 and there is the possibility of completion of further units at some stage.

In Germany, there are still nine reactors operating (12.0 GWe), but it seems highly unlikely that the nuclear phase-out plan will be reversed and the last reactor will therefore close in 2022. Only a seemingly incredible degree of political change will alter this.

The situation in the United Kingdom (16 units currently operating, 10.0 GWe) is entirely different and quite distinctive. Fourteen of the reactors are older advanced gas-cooled reactors that have received operating life extensions but all likely to close prior to 2030. There is now an active new build program with three separate consortia planning 11 new units.

Reviewing the experience of the 10 countries operating the largest number of current reactors (indeed 342 out of the current 437 in the world) provides a flavor of the complex processes at work. Beyond these 10, we can note that India currently has 21 smaller reactors in operation (5.3 GWe) but has ambitious nuclear plans and has six larger units under construction. Given its size, it could eventually become a very significant player in the world nuclear market (and thus for uranium). Nuclear power is now spreading to some new counties and both Belarus and the United Arab Emirates currently have first reactors under construction.

Table 5.1 Nuclear generating capacity, 2015–30, GWe

	2015	2020	2025	2030
United States	98.7	104.2	105.2	107.1
France	63.1	63.0	64.6	64.6
Japan	40.4	29.8	29.8	32.4
Russia	25.3	26.4	34.9	45.2
China	23.1	58.3	93.1	128.1
Korea	21.7	30.5	33.2	38.5
Canada	13.5	14.2	14.4	13.9
Ukraine	13.2	15.0	16.9	18.1
Germany	12.0	8.1	0	0
United Kingdom	10.0	8.7	10.2	14.7
Others	59.8	75.2	88.8	111.8
Total	380.8	433.4	491.1	574.4

Source: 2015 are actuals, 2020–30 from The World Nuclear Association (2013).

Indeed, the Middle East (and conceivably also Southeast Asia) is a possible region where nuclear power could expand in the future.

Table 5.1 shows the WNA's reference scenario for nuclear generating capacity in the period to 2030. It shows the world total expanding by 50% between 2015 and 2030. This is despite some of the Japanese reactors not returning to service and Germany exiting nuclear power. The main growth is clearly coming from China, backed by Russia and Korea. There are also more new nuclear countries included in the "others," for example, Turkey and Vietnam.

This report was published in 2013 and may now be subject to some downward revision in light of some of the previous comments about specific countries.

5.2.3 Important reactor operating factors

The amount of uranium required to operate a reactor is not a simple function of the amount of electricity produced. Changes in the way a reactor is operated and fueled can have a significant effect on its consumption of uranium. Therefore, to make an accurate forecast of uranium consumption, it is necessary to know the present operating characteristics of reactors and the fuel management policies of utilities, and to be able to estimate future trends in these.

Because nuclear power generation entered commercial use on a large scale, there has been a general trend toward improving the efficiency with which nuclear fuel is utilized in reactors. This has been reflected in a steady decline in the amount of uranium used per unit of electricity produced (specific uranium consumption). This trend has slowed in recent years as most of the easier gains have already been achieved, whereas a shift to longer operating cycles in some reactors (beyond the normal 12 months) has tended to actually increase specific uranium consumption.

Reactor operation and fuel management are complex topics, with each reactor requiring detailed planning and careful management to ensure its efficient operation in a way that meets the requirements of the utility. Three factors that are highly relevant to uranium demand are the load factor, the enrichment level, and the fuel burnup.

The load factor (or capacity factor) of a reactor is the actual power generated during the year expressed as a percentage of the power that would have been generated if it had operated at full rated power continuously throughout the year. Most reactors have scheduled outages, at annual or longer intervals, for refueling and maintenance. These normally limit achieved load factors to less than 100%. In addition, unplanned outages or power reductions due to technical failures or other reasons may further reduce load factors. Thus, load factor is a key indicator of reactor performance. Most utilities aim to achieve levels of over 85%, and some regularly achieve over 90%.

The load factor is one of the main influences on uranium requirements, as there is a nearly linear relationship between load factor and fuel requirements (ie, a 10% increase in load factor results in an increase of about 10% in uranium requirements, if other parameters are unchanged). Forecasts therefore have to make assumptions about which load factors will be achieved.

Most reactors were originally designed to operate with an annual outage for refueling, maintenance, and regulatory inspection. However, it is technically possible for utilities to operate reactors with longer cycle lengths, with outages every 15, 18, or even 24 months. A longer cycle length should allow a shorter overall annual average outage time, thus helping to increase load factors and reduce outage costs. In years with no refueling outage, a load factor of 100% can be achieved.

The main disadvantage is higher fuel costs, as longer cycles usually require higher enrichment levels—in other words, the proportion of ^{235}U in the fuel has to be increased from the formerly usual 3% to up toward 5% (see next). The tendency has been for this to rise steadily over time and this has been an important factor in pushing up uranium requirements.

An associated factor is the fuel burnup. The burnup of nuclear fuel is normally expressed as the amount of energy extracted per unit mass of fuel, in units such as gigawatt days per tonne of uranium (GWd/tU). Increases in burnup reduce the number of fresh fuel assemblies that need to be loaded (and indeed the number of spent fuel assemblies to be stored or reprocessed) so higher burnups therefore result in potential cost savings for the utility. However, increases in burnup usually require increased enrichment levels in the fuel assemblies, which increases the uranium required.

Considering the three factors together, they usually tend to cancel each other out in determining uranium requirements from a given amount of nuclear generating capacity. Load factors and enrichment levels have been increasing, which have pushed up uranium demand, but so has the fuel burnup, which has been tending to decrease it.

5.2.4 Interplay with the uranium enrichment sector

Beyond the enrichment level of the fuel (see the preceding), there is a further very important element in uranium demand related to the enrichment sector. This comes about because uranium and enrichment are (within a certain range) effectively substitutes for each other (see also Chapters 8 and 12).

Uranium enrichment involves increasing the proportion of the fissile ^{235}U isotope from its level of 0.711% in natural uranium to the level required for the reactor fuel, typically in the range of 3–5%. The process also produces tails material, which is depleted in ^{235}U content. The ^{235}U assay of the tails can be varied. Using a lower tails assay means that less natural uranium is needed but increases the requirement for enrichment effort, measured in separative work units (SWUs), and vice versa. Indeed, the required enriched uranium can be produced by a widely varied combination of natural uranium and enrichment services.

The selection of tails assay therefore has a significant impact on the requirements for both uranium and enrichment. The primary factor in determining the tails assay in a commercial order is the relative prices paid for uranium and enrichment. A rise in the price of uranium will make lower tails assays attractive (unless offset by a rise in the price of enrichment), and vice versa. At any price level, an optimum tails assay can be calculated that will minimize the cost of producing a given quantity of enriched uranium. So uranium and enrichment can certainly be regarded, to some extent at least, as substitutes for each other, but enrichment capacity may limit this.

Uranium demand is therefore very sensitive to the level of both uranium and enrichment prices. If uranium prices rise sharply with enrichment prices unchanged, the optimum tails assay can fall significantly and push down uranium demand as less uranium becomes needed, with more enrichment services utilized.

5.2.5 Estimates of total demand for uranium

Bringing together nuclear generating capacity and the various other factors, a forecast of uranium requirements can be produced. Table 5.2 shows the view of the WNA, corresponding to Table 5.1.

Demand is expected to grow by 45% over the period 2015–30, slightly below the growth of nuclear generating capacity (Table 5.1). The same countries are responsible for the growth, notably China, which is itself responsible for over half of the increase in demand.

Table 5.2 Uranium requirements, 2015–30, tonnes U

	2015	2020	2025	2030
United States	18,692	19,194	19,950	19,532
France	9230	9239	9505	9505
Japan	2549	4445	4445	4814
Russia	4206	5227	6052	6035
China	8161	12,688	18,737	24,709
Korea	5022	5526	5999	7002
Canada	1784	2073	1889	1820
Ukraine	2366	2681	3119	3331
Germany	1889	1268	0	0
United Kingdom	1738	1738	2530	2061
Others	11,246	13,536	16,981	18,641
Total	66,883	77,615	89,207	97,450

Source: The World Nuclear Association (2013).

5.3 Uranium supply

The supply of uranium can be conveniently divided into primary and secondary supply. It is very common in metal and general commodity markets for a substantial proportion of demand to be satisfied not by primary production but by secondary supplies. In the world uranium market, various categories of secondary supplies take on a crucial role.

Secondary supplies may be defined as all materials other than primary production sourced to satisfy reactor requirements. Secondary supply sources include inventories, stockpile drawdowns, and use of recycled materials of various types. In the widest sense, secondary supplies may be regarded as previous uranium production, returned or reassigned to the commercial nuclear fuel market.

Uranium is completely different from coal, gas, and oil as a fuel for thermal power generating plants. With the "fossil fuels," the fuel is burnt, giving off a variety of gases (that today are widely becoming subject to stringent environmental regulation) and leaving (in the case of coal) some residual ash in the furnace (that can be utilized in applications such as lightweight concrete).With uranium, however, we can still today physically identify virtually all the world production of roughly 2.6 million tonnes that has occurred since there was the first "uranium rush" (for military purposes at the time) after 1945.

This would not be so important if there were no obvious uses today (or in the future) for all of this historic production. But this is very far from being the case. Even elementary analyses of the world uranium market stress the importance both historically (and may be going forward too) of the so-called secondary supplies. The origins of these are bound up in the historical uranium production and they both have been (and will be in the future) very important elements in the market. They effectively compete with fresh primary uranium production. Looking at the future, some of the possibilities for utilizing these secondary supplies in a big way depend on technical developments in nuclear reactors. However, even today at the current stage of commercial development, there are several additional options to allow historical uranium production to play an even larger part of the market. This is arguably an important factor in the future because both China and India, the two largest and most dynamic future sources of future uranium demand growth, are both committed to a "closed fuel cycle," where secondary supply sources of uranium are likely to play a significant part and provide at least 25% more energy from each tonne of mined uranium.

5.3.1 Primary supply from mines

Table 5.1 shows the development of world uranium production by country in the period 2008–13. In the 5 years prior to 2008, annual world uranium production was close to the 40,000 tU per annum level. However, as we will see in Section 5.6, spot uranium prices rose sharply in the 2003–07 period and this caused renewed interest in the uranium sector. Substantial future shortages of uranium

Table 5.3 **World uranium production, 2008–13, tonnes U**

	2008	2009	2010	2011	2012	2013
Australia	8430	7982	5900	5983	6991	6350
Brazil	330	345	148	265	231	250
Canada	9000	10,173	9783	9145	8999	9363
China[a]	769	750	827	1500	1500	1500
Czech Republic	263	258	254	229	228	220
France	5	8	7	6	3	3
Germany	0	0	0	52	50	50
India[a]	271	290	400	400	385	400
Kazakhstan	8521	14,020	17,803	19,451	21,317	22,500
Malawi	0	104	670	846	1101	1131
Namibia	4366	4626	4496	3258	4495	4316
Niger	3032	3243	4198	4351	4667	4520
Pakistan[a]	45	50	45	45	45	45
Romania[a]	77	75	77	77	90	90
Russia	3521	3564	3562	2993	2872	3135
South Africa	655	563	583	582	465	540
Ukraine[a]	800	840	850	890	960	925
United States	1430	1453	1660	1537	1595	1849
Uzbekistan	2338	2429	2400	3000	3000	2420
Total	43,853	50,772	53,663	54,610	58,994	59,607

[a]Estimate.
Source: WNA website and East Cliff Consulting (2015).

were expected and literally hundreds of "junior" uranium companies suddenly appeared, seeking to raise funding to explore for, and then hopefully develop, uranium deposits. With such a substantial price rise, it is not surprising that production responded, as shown in Table 5.3.

However, production responded in a way that was not generally expected. What is remarkable about the 2008–13 period is that nearly all of the increase in production (which rose by almost 50% from the 2003–07 level) occurred in just one country, namely Kazakhstan. Production in most of the uranium producing countries, both big and small, remained almost constant over these years. Australia went down a little, whereas China and Niger rose, but otherwise there was little change. Indeed, excepting Kazakhstan, the world production position was very much in line with reactor requirements over this period, which themselves were almost constant.

There were some other notable developments during 2008–13. Arguably the most significant was Paladin's success in developing two mines in Africa. Of all the junior uranium companies, Paladin was the only one to develop significant production during this period—all the other major projects stalled somewhere. The development of Langer Heinrich in Namibia has so far not led to a significant increase in Namibia's uranium output (as Rio Tinto's Rössing mine has experienced lower production) but with the development of the China-owned Husab mine, this will likely soon change. Paladin's other mine, Kayelekera in Malawi, was the first uranium mine in that country, but has recently been put on a care-and-maintenance status, owing to the unfavorable market conditions.

It is, however, the expansion of production in Kazakhstan that has been the truly remarkable feature of the recent period. This has been accomplished by the rapid build-up of production from a large number of largely joint venture *in situ* leaching mines. To a great extent, this expansion is the mirror image of the Chinese nuclear growth.

Looking forward, the decline in the uranium price after 2007 has meant that most significant production expansion projects have been postponed or even effectively abandoned. The market has been heavily impacted by the Fukushima events in 2011. Prior to this, uranium producers were expanding or preparing to expand for projected uranium shortfalls in the coming years. Following Fukushima, nuclear plans in Japan and around the globe faced increased uncertainty that eventually led to reduced demand projections and thus many mining projects were delayed or deferred indefinitely. The reduced demand affected the uranium junior mining sector in particular. Many small firms found themselves unable to secure funding to develop their new mines, and this again led to delays and cancellations of a number of projects in the development pipeline.

Of the bigger projects, the Olympic Dam expansion in Australia, Imouraren in Niger, Trekkopje in Namibia, and a few other prospects in Africa have been put on hold. With no current mines beginning to run out of reserves, the immediate prospect is for world production to remain fairly stable at the 57,500–60,000 tU per annum level over the next few years.

5.3.2 Historical trends in meeting uranium demand

Fig. 5.1 shows that historically there has only been a weak relationship between the commercial demand for uranium through reactor requirements and uranium production. In the whole of the period up to the late 1980s, the level of annual production was in excess of demand. For the period since then up to very recently, the level of demand was well in excess of primary supply.

During the period from 1945 to about 1966, any generation of electricity from nuclear fuel was purely incidental to the nuclear arms race. After the detonation of the first atomic bombs over Japan in 1945, uranium production rose rapidly during the 1950s to satisfy the military requirement for highly enriched uranium (HEU) and plutonium. Nevertheless, uranium demand from the military programs fell sharply from 1960 onward and in response, world production fell by 40% by the mid-1960s.

Following a lot of experimentation with various reactor types and some early pilot programs, civil nuclear power picked up in a big way from the late 1960s onward. Uranium production picked up again as reactor orders expanded and many new mines were quickly brought into production, largely underwritten by long-term contracts agreed with electricity utilities in North America, Japan, and Western Europe. Nevertheless, production peaked again in 1980 and then declined sharply, as expectations for the growth of nuclear power turned out to have been rather

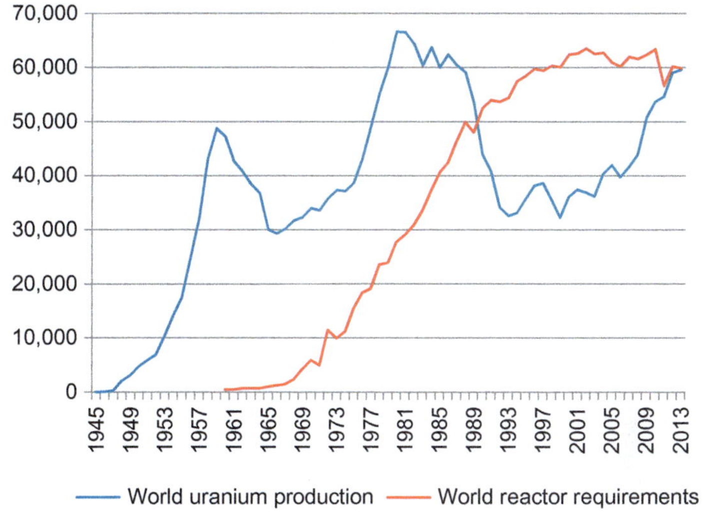

Figure 5.1 World uranium production and reactor requirements, 1945–2013, tonnes U.
Source: East Cliff Consulting (2015).

higher than was eventually to be fulfilled. Some commentators blame most of this on the Three Mile Island accident in the United States in 1979, but the story was much more complex than this. Other significant factors were slow rates of growth of world electricity demand and unexpected problems in building reactors on time and on budget.

The next period, from the point where demand rose above supply in 1989 until 2003, was an era dominated by secondary supplies and severe and prolonged difficulties for the uranium producers. Despite the abundance of the secondary supplies, it was always recognized that they could only last so long, at least with the present level of proven reactor technology. There was a strong market reaction in 2003 resulting from the perception that additional primary production was needed to support a renewed belief that there was once again going to be rapid growth in nuclear power. Concerns about climate change had mounted and nuclear power was seen as an ideal way to counter the build-up of greenhouse gases in the atmosphere. Many countries began to look again at building large numbers of reactors and the big developing countries, China and India, began to assert their role in the nuclear world. The uranium mining sector had been battered by bad economic conditions for many years, but it suddenly became exciting and dynamic once again. Therefore, production rose sharply once again, essentially on the back of declining and finite secondary supplies and the expectations of nuclear growth.

Yet despite all the talk about the "nuclear renaissance," uranium demand plateaued from the mid-1990s over the next 15 years. For one reason or another, nuclear plans have been dashed in many countries and it has become realized that it will take a longer time for many reactor plans to be realized. Commissioning of new reactors has been offset since 2003 by the closure of some units dating from the early days of civil nuclear power. This has been reinforced by the impact of the Fukushima accident in 2011, which has greatly constrained uranium demand from Japan and encouraged "political" reactor closures in others. It has become increasingly clear that uranium production has run ahead of itself, and it almost reached the same level as (diminished by Fukushima) reactor requirements in 2011. In addition, there are abundant secondary supplies remaining in the market.

One feature of uranium history is the extent to which national nuclear programs have sought their own secure supplies by encouraging or investing in production. For example, the French nuclear sector invested heavily in its former colonies in Africa (Gabon and Niger) and more recently in Canada and Kazakhstan. Russia promoted uranium production in Eastern Europe prior to 1990 and since then in Kazakhstan. Japan gave secure (and financially generous) contracts to producers in Australia and Namibia. In more recent times, China has adopted a three-pronged approach to securing its supplies, namely indigenous uranium production, taking equity in foreign production (eg, Husab in Namibia), and open market purchases. On the other hand, US uranium production has collapsed from previous levels and operators have replaced this with contracts signed with producers in many other countries, notably Canada.

5.3.3 *Secondary supplies*

Uranium that has been mined and held as inventory for a period of time before it is used is the simplest form of secondary supply, but accounts nowadays for only a small portion of the total supply picture. Military programs have been responsible for creating large quantities of HEU, with ^{235}U assays in excess of 90% and quantities of separated military plutonium. In both cases, substantial amounts have now been declared surplus to governmental requirements by both the United States and Russia and are thus available for potential use within the civil sector.

Thus, of the secondary sources, the most significant has been the low-enriched uranium (LEU) derived from Russian HEU being delivered to the United States under the so-called HEU Agreement. From both a nonproliferation and security of supply standpoint, this was an important program for the United States and provided significant revenues each year to Russia. During the lifetime of the agreement, from 1995 to 2013, some 500 tonnes of HEU (equivalent to 150,000 tonnes of natural uranium) were down-blended for civil use.

The other important contributors to secondary supply to consider are essentially fourfold—the recycling of materials from the reprocessing of used nuclear fuel, the actions of the US Department of Energy (DOE), the strategies of the enrichment companies, and the particular position of Russia.

Recycling plutonium in mixed oxide (MOX) fuel and using reprocessed uranium as a substitute for natural uranium has saved about 3000 tonnes of natural uranium per annum in the past and this is likely to continue at a similar rate. Nearly all of this is currently happening in France, but Japan, China, and India (and may be a few other countries) are likely to be saving some natural uranium in this way by 2030. On the other hand, France's contribution may diminish owing to reactor closures.

The US DOE has long been an important source of secondary supply to the uranium market. Prior to recent years, the biggest contribution came from the gift of almost 30,000 tU of uranium made to the United States Enrichment Corporation (USEC) when it was privatized. This had a huge impact on the market in the late 1990s and the early years of this century. Today's contribution comes from a DOE inventory management plan that (subject to some legal challenge today) is set to contribute about 3000 tU per annum up to 2020. Beyond then, the rate is likely to slow somewhat to perhaps 2000 tU up to 2030.

That uranium and enrichment services are essentially competitors was discussed in Section 5.2.4. When uranium prices rose rapidly (as in 2003–07), enrichment suppliers could "underfeed" their plants with natural uranium and use more SWUs, thus creating an additional secondary supply of uranium. Enrichment contracts that are being signed today reflect the newer price relationship (so there is arguably less incentive to underfeed) but another possibility has opened up.

With the closure of the old gaseous diffusion enrichment plants, the SWU price has fallen sharply and there is also excess supply capacity (stemming from Fukushima and the slower than expected rise in world demand). In certain economic circumstances, it can be worthwhile to reenrich stockpiles of depleted uranium tails stored at enrichment plants and create fresh uranium, at either the natural 0.71% ^{235}U assay or further enriched. Given the huge volume of depleted uranium available worldwide (approaching 2 million tonnes), there is substantial potential for reenrichment. The main economic limitation today is the ^{235}U assay of the stockpile. But the substantial quantities with assays of 0.30% and above (dating from the earlier ages of uranium involving military use or, alternatively, for civil use when uranium was relatively cheap and tails assays at enrichment plants relatively high by today's standards) make reenrichment economically viable when there is spare enrichment capacity.

The balance between underfeeding and reenrichment depends on the following. For underfeeding, the key is the tails assay specified in the current enrichment contracts. Whereas for reenrichment, it is the assay of the depleted uranium in storage. Going forward, both these actions at enrichment plants in Europe and the United States will likely add about 3000 tU per annum to uranium supply as a secondary source in the period to 2030.

Russia has historically contributed a significant element to secondary supplies. The largest element has been the down-blended HEU mentioned previously and after the contract has ended, some could be used internally. However, there has been a substantial additional element coming from running down any stockpiles of natural uranium and from either underfeeding or reenriching depleted uranium at Russian enrichment plants. In recent years, assessments of the supply–demand balance in Russia suggest that this additional supply is equivalent to about 5000 tU of natural uranium a year. Going forward to 2030, it is believed that this source of secondary supply will continue at a similar level.

5.3.4 Historical trends in meeting uranium demand according to category of secondary supply

Looking at the history in Fig. 5.1, the main sources of the secondary supplies that influenced the market were essentially fourfold. The reprocessed material and the activities of the enrichment companies had less significance at this time.

The first was what had turned out to be the overproduction of uranium for civil purposes in the period from 1966 to 1980. There was an obvious overhang of inventories from 1989, despite the fact that uranium demand carried on growing (but at a lower rate) up until about 2000. Many power utilities had signed uranium contracts in anticipation of building additional nuclear plants and honoring these resulted in a substantial build-up of utilities' inventories. As these were being run down, mines either closed or cut production. Production carried on falling up until 1993. This was accentuated by the break-up of the former Soviet Union in 1989, as uranium production also plunged there.

Nevertheless, this region became the source of the second significant element in secondary supplies. Significant quantities of natural or LEU became available from the former Soviet Union (from inventory) on Western markets. Therefore, utilities were able to satisfy much of their requirements without recourse to new production. Obviously relying on such inventories can only go on so long, but expectations that the downturn in uranium production and poor prices would end quite quickly were dashed by the third element in secondary supplies.

This was the arrival on Western markets of the down-blended HEU from Russia in the mid-1990s. The flows of material from the East meant that markets that were formerly almost completely segregated reached the first stage of integration, although the material was effectively going one way only.

Finally, the US DOE actions have undoubtedly had an influence on the market. In terms of volume, they have been a lot smaller than the Russian secondary supplies, but have been important in deflating market prices at crucial times.

5.4 Costs of production

Uranium is not scarce in a geological sense and the identified world resources are more than adequate to fuel any conceivable nuclear programs for many years in the future. Whether a deposit is developed into an operating mine depends largely on its costs of production. Although some countries may develop uranium mines for security of supply reasons, even though the costs are high, the world uranium market is competitive and buyers generally seek their supplies from those companies offering them the lowest prices. If the producers are going to offer competitive prices and also earn some return on capital, they must identify and develop the lowest cost mines first.

5.4.1 Cash costs of operating mines

For operating mines, the capital costs have already been sunk and so whether they operate or not depends on covering what are known as the cash costs of production.

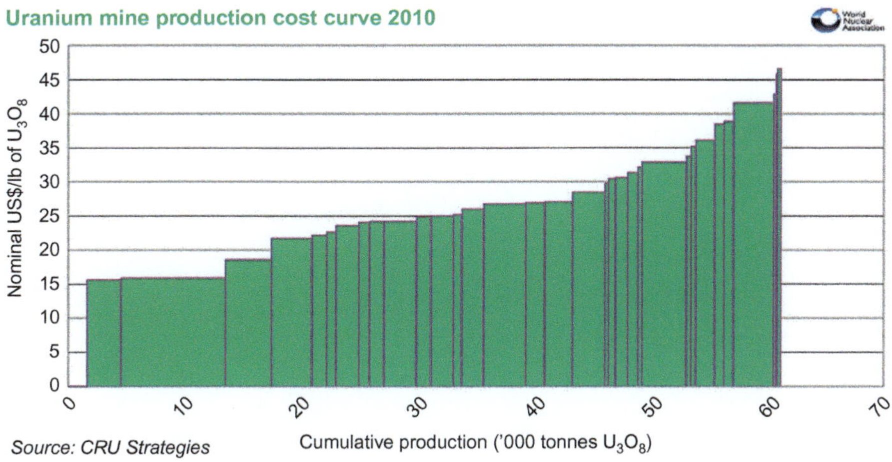

Figure 5.2 Uranium mine production cost curve 2010—cash costs.
Source: WNA website, originally sourced from CRU Strategies.

These include all the operating costs of a mine, including labor, raw materials, fuel, maintenance of machinery, and royalties/taxes paid to the local authorities. They are essentially the costs that must be covered if a mine is to break even financially and should be just sufficient to allow it to stay in business. However, if the revenue only covers the cash costs, it will not provide any return on capital invested by the mine owner.

Fig. 5.2 shows a uranium production cost curve for 2010. These are the cash costs and show the cumulative production available at various cost levels. For example, for 20,000 tonnes of production, the marginal cost (required to produce the 20,000th tonne of output) is around US $20 per pound and for 60,000 tonnes of output, around US $40 per pound. This suggests that producers need a price of US $40 per pound to cover even their cash costs at the current level of world uranium production. Any on the lower end of the cost curve will be earning some profit, but the curve escalates sharply over 60,000 tonnes per annum, indicating that there is little additional output available at low costs.

5.4.2 Full costs of planned and prospective mines

Fig. 5.2 suggests that if a higher level of uranium production is needed, it is essential to invest in new mines. These take time to pass through regulatory hurdles and to commence production and require a considerable input of capital expenditure. The magnitude of this depends on the type of mine and the grade of the ore, but for a large mine, it can amount to several billion US dollars. However, companies will only invest in new mines if they can achieve prices that cover not only the cash costs of production but also the capital investment costs, and include a rate of return for the investor. Fig. 5.3 shows a cost curve for uranium based on full costs. This shows that at a production level of 60,000 tonnes per annum, the full cost of

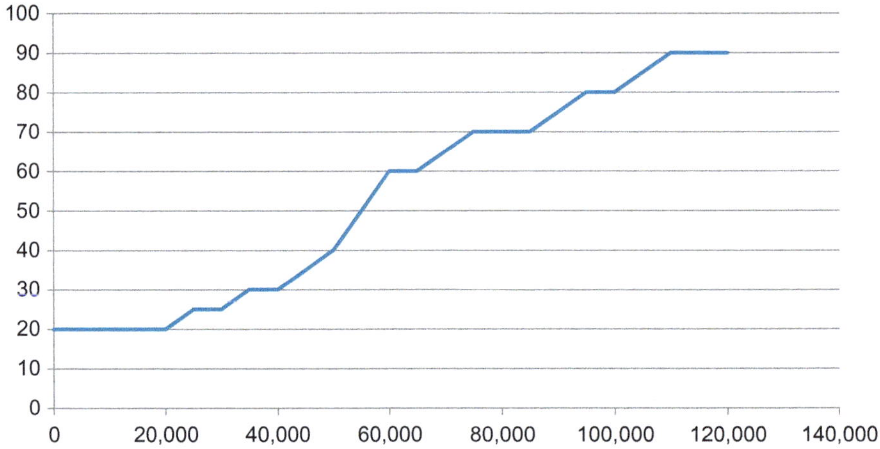

Figure 5.3 Uranium mine production cost curve 2014—full costs.
Source: East Cliff Consulting (2015).

production of the marginal mine is about US $60 per pound. To stimulate more production above this level, a price of US $60 per pound and above will be required. For example, if world production needs to rise to 80,000 tonnes per annum, a price of US $70 per pound will be required. Below this, the mine may cover its cash costs, but not its capital costs nor a return on capital. Hence, there would be no incentive to invest in the mine.

This is the argument used to justify higher prices when more production is needed. The price offered must cover not just the cash costs but also the full costs of the mine or the investor will effectively be losing money by investing in it.

5.4.3 Future trends

It is apparent that mining costs have escalated steeply in recent years, owing mainly perhaps to the resources boom to satisfy Chinese demand for raw materials. Uranium mines have not escaped this and if a higher level of world production is needed, the prices required need to be above those quoted in the spot market during 2013 and 2014 of US $30–40 per pound. There are currently two major mines coming into production in the 2015–20 time frame, namely Cigar Lake in Canada and Husab in Namibia, but it is difficult to see further mines beyond these unless prices are higher. This explains why so many development projects have been put on hold, as they are essentially waiting for a requirement for new production.

5.5 Uranium supply and demand to 2020 and beyond

It is possible to make a simple assessment of how supply and demand will pan out in the period to 2030 by plotting the likely numbers on a graph. This has a number of obvious limitations, but is attempted in Fig. 5.4.

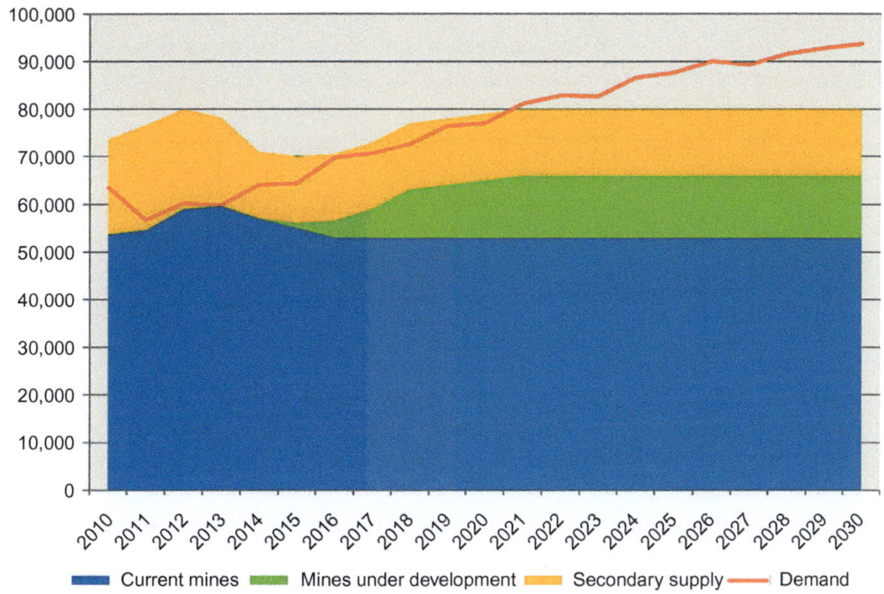

Figure 5.4 Supply and demand for uranium, 2010–30, tonnes U.
Source: East Cliff Consulting (2015).

The supply shows the mines currently in operation, those currently under development, and an assessment of future secondary supplies. Those currently in operation lose some output over the next few years, as economic factors begin to bite, but most stay in operation to 2030. The new mines are Cigar Lake and Husab, plus a few small operations. The secondary supplies are a persistent element in supply, remaining at 14,000 tonnes per annum all the way up to 2030.

The result is that, owing to rising demand, a deficit arrives in the 2020s and more primary production is needed. On the other hand, supply is ahead of demand in the period prior to 2020, substantially so in 2010–15, implying that inventory building has been taking place. If these are run down to add to the secondary supplies level after 2020, new mines may not be required into even further in the future.

This analysis is rather simplistic, as it ignores many market complications (eg, it can be argued that there is no global market, only a few ill-connected segments) while it cannot assess the dynamic response of supply and demand to price movements. On the other hand, it offers at least some perspective on when more production may be required, assuming that the demand outlook is correct.

5.5.1 Impact of prospective new technologies

There is little in the history of uranium's production and use that suggests a sudden step up in utilization either through the identification of a new, previously untried use of uranium or via the proving of a new nuclear technology. The uses today are much the same as in the early days of the nuclear industry immediately after World War II. Although it is not inconceivable that another arms race will happen and require a large input of fresh uranium, it is not currently very likely. While the nuclear powers (even France and the United Kingdom) possess huge destructive power in their nuclear arsenals, it seems that increased pressure for complete nuclear disarmament may lead to stronger moves toward the opposite. Despite the conflict in Ukraine and concerns over Iran and North Korea, it is surely more likely that additional military nuclear materials will be declared surplus to requirements (and therefore potentially available to the civil sector) than another rush to produce more weapons.

In terms of using uranium in other applications, the minor previous uses for depleted uranium because of its ready availability and substantial density (eg, in the tips of nonnuclear armaments or as counterweights in aircraft) seem unlikely to return in any major way. The amount of uranium used in creating the radioisotopes used in medicine, industry, and agriculture in small reactors is minor and will always remain so.

Applications of nuclear reactors for their great heat-providing properties beyond nuclear electricity are already well proven but are so far limited in application. Nuclear-powered ships, with the exception of some Russian nuclear-powered ice-breakers, are confined to the military sector. Whenever the price of oil rises sharply, there are calls to investigate once again the possibility of developing huge nuclear-powered bulk carriers and container ships, but this has never looked a highly likely route for greater uranium use. Seawater desalination and district heating look more likely future possibilities, particularly the former if the Middle East eventually gets into nuclear power in a big way. However, it will take a long time before a significant amount of demand for uranium comes from either of these—certainly not this side of 2030.

A final area worth noting is the possibility of using reactors that operate at higher temperatures to process heat, displacing fossil fuels. This is a serious consideration with high-temperature gas-cooled rectors, such as the program now underway with prototype reactors under construction in China. Again significant uranium demand is a long way in the future. These reactors, fundamentally different from the predominant light water reactors (LWRs) of today, are really within the same time frame as the generation IV reactors (see next), which will have very different fueling to today's models. It is unlikely that these will have much impact on the market this side of 2030 (at the earliest).

Indeed, concrete progress in nuclear technology development has, in reality, been relatively slow throughout the world (compared with other industries) and today's reactors under construction are clearly evolutionary versions of those of the late 1950s. There has been no major shift in the market, such as the introduction of jet airliners in the civil aviation industry. The prime reason for this is that the growth of the nuclear industry stalled very badly from the late 1970s onward and, despite hopes of a renaissance in the early years of this century, has so far still been unable to fulfill the most expansive hopes of its promoters.

5.5.2 Alternative fuel cycles

Nuclear technology is constantly evolving, largely in an incremental fashion. For instance, most of the world's operating reactors are referred to as generation II. These reactors use the heat created by ^{235}U and ^{239}Pu fission, moderated mostly either by light or heavy water, to raise steam for power generation. Generation III reactors are an evolutionary development of the generation II reactors aimed at enhancing the safety and economics of nuclear generation and further optimizing the operability and flexibility of nuclear power reactors. Essentially, the improved efficiency of fuel use by generation III reactors stems from their expected higher burnup capacities as well as the increased possibility to use recycled fissile materials from used fuels and continuous improvements in fuel assembly design. The greater efficiencies achieved in the nuclear steam supply system and the steam turbines further contribute to improved fuel efficiency.

More innovative, though still R&D level, improvement such as the use of ceramics in fuel claddings and nitride fuels could lead to yet higher burnups and greater utilization of the energy present in the fuel. Higher capacity factors, thermal efficiencies, and greater safety margins in the event of loss of cooling are other advantages. These possible advances in fuel claddings will have to undergo extensive safety demonstrations and trials prior to finding acceptance by the safety authorities and utilities.

Another technological development influencing specific uranium demand is the gradual increase of recycling and the consequent use of MOX fuel. Recycling could increase primarily as a contribution to meeting the long-term challenge of used fuel management. Various countries (France, the United Kingdom, the United States, Japan, Russia, India, and China) have developed recycling R&D programs and/or even industrial capabilities (France, the United Kingdom, and Russia) and more countries are starting to use recycling services in view of an improved sustainability of the nuclear energy system. The recycling of separated U and Pu (using MOX) reduces the specific uranium demand by about 30%.

The nuclear technology that has the greatest potential to change the future nuclear scene toward even more sustainable nuclear energy generation is the introduction of fast neutron reactors (FNRs), which is included in the generation IV reactor group. In conventional reactors, neutrons emitted by ^{235}U atoms after their fission have to be slowed by a moderator before they can trigger a new fission leading to a sustainable ^{235}U fission reaction. These are known as thermal neutrons. As the name implies, FNRs use neutrons unimpeded by the presence of a moderator. Because uranium does not fission sufficiently with fast neutrons, the fast reactor works with a plutonium (^{239}Pu) core and, optionally for breeding of additional fissile material, a blanket of ^{238}U around the core. The excess neutrons emitted by the ^{239}Pu convert the ^{238}U into ^{239}Pu and thereby breeds new fuel.

"Refueling" a fast reactor requires only 1–2% of the uranium needed with a conventional reactor. Supplies of ^{238}U are copious, as it is estimated that at up to 2 million tonnes exist in the form of depleted uranium at enrichment plants and other potential supplies exist in the form of used nuclear fuel. The large-scale deployment of fast reactors would greatly reduce the demand for new uranium supplies.

Fast reactors have been in existence since the 1950s and have accumulated over 400 reactor-years of operating experience. Initially developed as a means of economizing on the use of uranium resources that were thought to be in short supply, the discovery and exploitation of new uranium resources and the decline in uranium prices coupled with technical problems with the first fast reactors undermined the relative economics of fast reactors compared with conventional reactors. Only two fast reactors have been developed with primarily commercial objectives in mind, the French Superphenix, a 1.2 GW reactor that operated from 1985 to 1998, and the Russian BN Series.

Given the availability of uranium resources, it seems unlikely that serious concerns about future uranium supplies will be sufficient to justify a widespread development of fast reactors. Probably a more pressing rationale for the development of fast reactors post-2030 will be the difficulty of finding acceptable solutions for the storage of used nuclear fuel. Fast reactors have the capacity to "burn" actinides contained in the used fuel and thereby transform most of the long-lived actinides that make the permanent storage of used fuel so challenging into a smaller volume of shorter lived fission products that requires secure storage for a thousand years or so rather than the many thousands of years for used fuel from conventional reactors. In addition, the introduction of fast reactors in synergy with light water reactors furthers the multiple recycling of Pu in these light water reactors and thus gradually contributes to a more resource-efficient use of the mined uranium without requiring transition to a full fast reactor park.

5.6 The world uranium market and prices

The world uranium market is characterized by a relatively small number of buyers and sellers, the preponderance of long-term contracts, and some lack of market liquidity. Building NPPs and developing uranium mines are both very long-term businesses. It takes many years to get a reactor into operation but once this happens, they tend to operate for a long time. Uranium mines are similar. Hence, there are good reasons to explain why reactor operators find it best to sign long-term supply contracts, whereas the producers like these for the security they grant to invest.

5.6.1 Trading of uranium

Nuclear fuel trade is dominated by the long-term market. However, in recent times, the spot market has become a bit more active, but this has been catalyzed by external market players (hedge funds, uranium funds, etc.). The spot market provides an important function in yielding a price indicator, but a relatively small function in terms of satisfying reactor needs. For deliveries under long-term uranium contracts, there are two prevalent pricing mechanisms: specified pricing and market-related pricing.

Specified pricing is a fixed price, a series of fixed prices, or a base price plus adjustment for inflation to the date of delivery. The adjustment mechanism is usually either a combination of published indexes or a fixed annual percentage rate. Fixed/base escalated is the most predictable form of pricing and is attractive to both buyers and sellers in terms of managing cash flow and budgets. In a fixed-price contract, the parties simply agree to hard numbers that are fixed in the contract.

The base price agreed by the parties is the price that is reflected in the reported long-term price indicators that are published by market analysts. The long-term price has typically traded at a premium to the spot price—an average of about 10% since 1996.

Market-related pricing, on the other hand, is based on the uranium market price at or near the time of delivery, and/or some other published market index, such as the average US import price. In most instances, the price is the market price less a discount (or plus a premium). Market-related price mechanisms nearly always include a floor price, below which the contract price may not fall. The floor, which protects the seller, is usually either a base adjusted for inflation, a fixed price, or a production cost-related mechanism. Market-related price mechanisms also usually include a ceiling price above which the contract price may not rise. The ceiling, which protects the buyer, is usually a base adjusted for inflation or a fixed price.

Market-related pricing has a much greater level of uncertainty, but gives both parties the opportunity to gain value depending on their respective views of the market. Typically, in a market-related contract, annual delivery volumes will reference a single or combination of price indices.

5.6.2 Price trends and outlook

Fig. 5.5 shows the basic history of the spot uranium price. Contract prices have tended to be a little higher but the trends are the same.

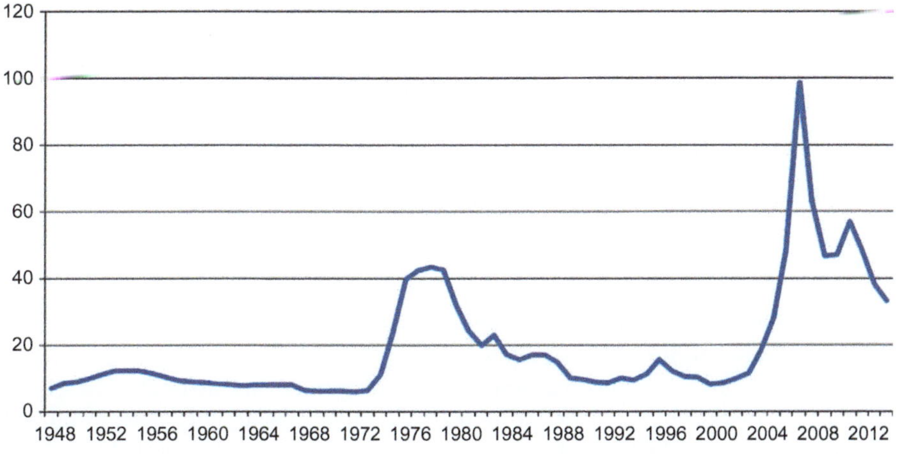

Figure 5.5 Uranium prices 1948–2014, annual averages, S per pound U_3O_8.
Source: Various market indicators from East Cliff Consulting (2015).

During the early days of the uranium business (1945–66), the US government essentially controlled what there was in terms of a uranium market via its purchase for military purposes and the prices paid ranged from $8 to $12 per pound. The United States also had a monopoly over uranium enrichment and it was only in the late 1960s, when the first commercial market price indicators emerged, that anything resembling a commercial uranium market appeared. During the next period (1966–89), the price was initially depressed at only $6 per pound but there was a price spike starting in 1974 caused by looming potential shortages of uranium following the volume of new reactor orders at the time. Prices rose rapidly and stood at over $40 per pound in the 1977–79 period. This escalation has often been blamed on a degree of cartel activity in the market, but although there were agreements between producers to allocate markets in the early 1970s, by the time of the price escalation it is clear that deeper market fundamentals were also at work. The price undoubtedly overshot and this became clear as reactor orders melted away in several countries. Prices then fell back sharply to $10 per pound by 1989.

The remarkable feature of the price during the next period (1989–2003) was its stability with only relatively minor fluctuations around the $10 per pound level. The key to this was the abundance of inventory available, backed by the arrival of supplies from the former Soviet Union. Although these suffered limitations via trade restrictions in both the United States and Western Europe, the additional supply clearly depressed the price and kept it low. There was a minor spike in the price during 1996 but this was effectively killed off by the deal whereby 500 tonnes of surplus Russian HEU (equivalent to 150,000 tU of natural uranium) was to be down-blended and introduced to the US market.

This period ended in 2003, as it became apparent that the uranium production sector was very depressed after a battering from low prices for many years and higher prices were needed to encourage new mines. As in the 1970s, we can subsequently see that the price overshot again, and this time it was partly financial sector players who were responsible. They became active in the uranium market once they believed that shortages could ensue from the mooted "nuclear renaissance," but prices of over $100 per pound during 2007 (not dissimilar to the 1980s peak in real inflation-adjusted terms) did not last long. Therefore, as with the earlier period in the 1970s, there was a price boom and bust. There was a recovery from the general decline in prices during 2010–11, which was arguably caused mainly by the arrival of significant Chinese buying in the market (to fuel their rapidly expanding nuclear power program). Then Fukushima killed demand and prices have subsequently continued to slide to the $30–40 per pound level during 2014.

Where the price will go next is always the subject of some debate. The market is currently (early 2015) oversupplied with uranium and there has been a substantial build-up of inventories. The enrichment companies are also adding to secondary supply as they have the spare capacity to both underfeed and reenrich. The outlook for prices is therefore that they are unlikely to move substantially upward until the excess supply is worked off. The general consensus is that this probably will not happen until at least the early 2020s. By then, reactor requirements should be markedly higher and the additional supply from Cigar Lake and Husab absorbed.

At this point, new primary uranium production may be needed, which will require a higher price if the necessary investments are to be made.

5.7 Sources of further information and advice

Two good public websites have information about the uranium business and the world market.

1. The Ux Consulting Company—http://www.uxc.com/
2. TradeTech—http://www.uranium.info/

References

East Cliff Consulting, 2015. The fifth age of uranium (private client report).
The World Nuclear Association, 2013. The global nuclear fuel market—supply and demand 2013–2030.

Part II

Mining and Alternative Fuel Sources

Uranium mining (open cut and underground) and milling

6

Peter H. Woods
International Atomic Energy Agency, Vienna, Austria

6.1 Introduction

6.1.1 Historic mining for pigment and radium (to 1939)

Uranium minerals were known, without a detailed knowledge of their elemental composition, at least from 1565 in the Saxonian Ore Mountains (Erzgebirge) in Germany and Bohemia (modern Czech Republic) (Dahlkamp, 1993). Uranium was identified in minerals from this region by Martin Klaproth, although he did not manage to produce the element as a metal (Dahlkamp, 1993). Later, the utility of uranium to color glass and ceramic glazes was recognized, a minor use for uranium widely used from the 1830s to 1940s (Strahan, 2001) that continues on a small scale to the present day. In the early decades, uranium ore was recovered as a byproduct of other mines in Saxony, Bohemia and Cornwall in the United Kingdom, and later as the main product in Joachimsthal, Bohemia (Dahlkamp, 1993). Some production was recorded in the United States and Sweden (Dahlkamp, 1993). When the medical uses of radium were discovered in the early decades of the 20th century, uranium and radium mining followed discoveries in Katanga (now the Democratic Republic of the Congo), Canada, Portugal, Uzbekistan, and Australia (Dahlkamp, 1993; McKay and Miezitis, 2001).

6.1.2 First extensive uranium mining and production booms (World War II to 1980)

Exploration for uranium began in earnest toward the end of World War II (1939–1945), initially for the supply of uranium to military programs to develop and manufacture atomic weapons and then for submarine and marine power plants. This was followed by domestic, civilian demand in the second half of the 1950s as nuclear power plants (NPPs) for electricity were developed. Mining developed rapidly on both sides of the Iron Curtain in the Soviet Union and allied nations, particularly East Germany and Czechoslovakia, and in the United States and its allies, notably Canada and Australia. France and some francophone countries in Africa also became significant producers.

The production of uranium from mines rose from approximately 407 tonnes in 1945 to about 46,000 tU in 1960 (OECD-NEA, 2006). The production rate eased to

Uranium for Nuclear Power. DOI: http://dx.doi.org/10.1016/B978-0-08-100307-7.00006-5

about 31,500 tU in 1965, prior to rising again to peak at nearly 65,000 tU in 1980 (OECD-NEA, 2006). The top five producing nations at 10-year intervals from 1950 are shown in Table 6.1.

6.1.3 Transition (1980–1995)

After many years of increasing production, as the pace of construction of NPPs and therefore uranium demand slowed, both the price of uranium and total production commenced a prolonged decline from 1980. By 1990, the mining output fell below the consumption rate, with the additional nuclear fuel coming from stockpiles and other secondary sources. Production fell to about 31,500 tU in 1994. A detailed exploration of the reasons for the fall in production is beyond the scope of this chapter, but has been discussed in chapters "Uranium Resources" and "The Uranium Market, Supply Adjustments from Secondary Sources and Enrichment Underfeeding."

6.1.4 Modern era

For the purposes of this account, the modern uranium mining era commenced after the low production of 1995. By this time, there had been a major shakeout in the industry. The Iron Curtain came down, commencing with the reunification of Germany in 1991, and uranium production there was quickly shut down, followed by closures or significant reductions in many other Eastern Bloc nations. In Western nations, many historic mines had been mined out or were nearing the end of their production. Thus, production had been taken up by mines established under more modern environmental and social expectations, led by Canada and Australia (which became the leading producers). Production remained significant in Russia, Namibia, and Niger. After a significant reduction in production, uranium mining using the in situ leach (ISL) or in situ recovery (ISR) technique (see chapter: Introduction to Uranium In Situ Recovery Technology) grew rapidly in Kazakhstan, with Kazakhstan becoming the world's leading uranium producer in 2009 and extending that lead up to the time of writing in 2015.

6.2 Open cut mining

Open cut mining is employed to access ore that occurs at or near the ground surface or close enough to the surface that removal of barren or waste rock (overburden) to access the ore is viable. Access to an open cut mine, often called an open cut or open pit, is via one or more ramps constructed at a suitable gradient and width to allow access by mining equipment. The design of the pit walls is strongly dependent on the geotechnical strength of the soil, sediment, or rock material. In general, steeper pits are more economically favorable, but the possibility of wall failure must be taken into account; the same geotechnical considerations as any open cut

Table 6.1 **Leading uranium producing countries and production (tU) 1950–2010 at 10-year intervals**

Rank	1950		1960		1970		1980		1990		2000		2010	
1	DR Congo	2220	USA	13,568	USA	9928	USA	16,811	USSR	14,500	Canada	10,683	Kazakh.	17,803
2	GDR	1224	Canada	9807	USSR	8300	USSR	15,700	Canada	8729	Aust.	7579	Canada	9775
3	USSR	417	GDR	5356	GDR	6389	Canada	7150	Aust.	3519	Niger	2914	Aust.	5918
4	USA	353	S. Africa	4930	Canada	3520	S. Africa	6146	USA	3420	Russ. Fed.	2760	Namibia	4503
5	Czechoslovakia	281	USSR	3800	S. Africa	3167	GDR	5245	Namibia	3221	Namibia	2715	Niger	4197
World Total		4804		46 290		37,736		69,692		50,026		35,755		54,670
% by top 5		94%		81%		83%		73%		67%		75%		77%

Notes: DR Congo, Democratic Republic of the Congo; GDR, German Democratic Republic; USSR, Union of Soviet Socialist Republics; USA, United States of America; S. Africa, Republic of South Africa; Aust., Australia; Russ. Fed., Russian Federation; Kazakh., Kazakhstan.
Sources: for 1950–2000, OECD-NEA (2006); for 2010, OECD-NEA/IAEA (2012).

Figure 6.1 Rössing open pit uranium mine, Namibia, April 2012.
Source: Photograph P. Woods.

mine apply. In general, the higher the grade and the larger the deposit, open pit mining is economically employed to greater depths than would be considered for lower-grade or smaller deposits. Hence, a large open pit for uranium may be over a kilometer across and hundreds of meters deep, for example,

- Rössing mine, Namibia, mined 1976–present: 3000 m long, 1500 m wide, and up to 390 m deep (Rössing Uranium, 2014, see Fig. 6.1)
- Lichtenberg open pit in Saxony, Germany, mined 1956–1976, now backfilled: 2000 m long, 1000 m wide, and up to 240 m deep (Wismut, 2014)
- Ranger Pit 3 in Australia, mined 1997–2012 and currently being backfilled: 1150 m long, 800 m wide, and up to 275 m deep (Wines et al., 2013)

Open pits must also be protected from flooding from nearby catchments or streams, and may also require control of groundwater inflows.

Over the history of particular deposits, open pit mines have been developed where underground mines formerly existed, and open pit mining may be followed by underground mining to access deeper ore.

6.2.1 Soft rock open cut mining

Rock properties are a major determinant of mining method. Where the hosting rocks are relatively friable (low geotechnical strength), rock material may be removed by scraper technology, where the equipment excavates, transports, and dumps material.

Figure 6.2 Drill-and-blast (left; Langer Heinrich mine, Namibia) and excavation (right; Rössing mine, Namibia) at hard rock open cut mines.
Source: Photographs: (left) P. Waggitt, reproduced with permission, (right) A. Hanly, reproduced with permission.

Generally, only a thin layer of material is removed with each pass. Alternatively, material can be loosened and moved by the blade of a bulldozer and pushed into a heap. The heap is then either loaded onto a truck or conveyor belt for removal, or in some cases mixed with water and made into a slurry that is pumped away.

For stronger material, ripping by large tines mounted on a grader or bulldozer may be used to break up the material before it is removed.

Alternatively, "free digging" material can be dug with an excavator without any additional assistance. The excavated material is placed in a truck or conveyor belt for removal. Selective mining of ore is proposed to be by a "surface miner" that removes and breaks up a relatively shallow strip of ore for direct loading onto a truck, and produces a "precrushed" product for further processing (Toro Energy Limited, 2015).

6.2.2 Hard rock open cut mining

Material with higher strength must be broken up or fragmented prior to excavation and removal. The most common method is drill-and-blast. A drill rig, or group of rigs, prepares holes, usually vertical, at a predetermined spacing and pattern, which are then loaded with explosives, usually a powder, pellets, or a packaged form. The explosives are detonated in a defined sequence designed to fragment the rock ready for excavation and placement onto a truck or conveyor for removal from the open pit (Fig. 6.2).

6.2.3 Trends and alternatives

Advancements in mining machinery have meant deeper and deeper open pits have been viable over the last century. Small open pits dug by hand are known historically. The optimum size of mining, rock, and ore hauling equipment is largely determined by the required rates of mining, the length of mine life, and the distances waste rock and ore must be hauled. These trends are apparent in uranium mines as much as any other industrial-scale mines and, with the exception of additional dust control to reduce radiation exposure where required, general mining industry planning and machinery are applied at open pit uranium mines.

Old tailings can sometimes be recovered by hydraulic mining or sluicing, where high-pressure water jets are used to both break up the tailings into slurry that can then be pumped for further processing (see Section 6.5.1).

6.3 Underground mining

Underground mining is employed to access ore that occurs below the ground surface. Hand-dug underground mines for minerals other than uranium are known since antiquity. Underground uranium mines, or mines with a significant uranium byproduct, have all been undertaken in the era of mechanized mining.

6.3.1 Access options

Access to an underground uranium orebody is by one or more means, and the design depends on the topography of the ground surface, the depth of the orebody, rock strength conditions, and possible requirements for road access. The three main access methods are shafts, adits, and declines.

Shafts are vertical openings typically a few meters in diameter or width. Shafts may be tens or hundreds of meters deep, or over a thousand meters in some cases. Shafts may start at the surface or be created between different levels within an underground mine. Shafts are typically lined with concrete (or historically with timber); although in strong, competent rock they may be unlined. A combination of lined and unlined shafts is possible in changing rock strength conditions. Lining or grouting may also be required to control the inflow of groundwater, even if rock strength would otherwise allow an unlined shaft. Materials and miners are transported in cages or large containers supported on cables operated by a winch beneath a headframe (Fig. 6.3).

Adits are near horizontal openings, typically from the side of a hill or mountain to access an orebody at a similar elevation. Again, they are typically a few meters in the vertical and horizontal dimensions, although in early mines the openings might only be 1.5–2 m in size. A railway may be installed in an audit to transport material and miners, or access may require using wheeled vehicles. Conveyors systems might also be installed in an adit for the removal of ore and waste rock. As with shafts, lining with concrete, steel, or timber may be required if potential collapse is a risk at the particular site. An adit may also act as a drain for groundwater that enters a mine, in which case it would slope gently upward from the entrance.

Declines are inclined ramps, often in the form of a spiral, constructed to allow vehicular access to the orebody. The entrance to a decline is typically called a

Figure 6.3 Left: Headframe at Dolni Rozinka uranium mine, Czech Republic, with some timber supports awaiting transport underground; right: headframes at shafts 1 and 2, Cigar Lake uranium mine, Canada.
Source: Photographs: left, P. Woods; right, Cameco; reproduced with permission.

portal. Declines are typically no steeper than a 15% slope. Most of a decline is wide enough only for the largest vehicle that uses it, so passing spaces must be provided at various intervals to allow two-way traffic. Again, lining may be required, depending on rock strength conditions.

These access options are often used concurrently: typically, a mine can have multiple shafts with one or more adits or declines, depending on the size and situation of the underground workings that require access. Where a single access point is used, particular attention to emergency access is required, should that access become blocked, such as via a ventilation shaft (see Section 6.3.3).

6.3.2 Mining options

Individual mining options are diverse, and new techniques are developed as technology advances or specifically to meet particular ore geometry or characteristics not economically amenable to established techniques. Four commonly used techniques are described here; the interested reader is referred to the extensive mining literature or to descriptions of individual mines for more information.

Further to the access from the surface, mining areas are accessed by horizontal or inclined tunnels and internal shafts, and small railways may be constructed or material moved in specialized underground trucks. In large, long-lived mines, the system of tunnels may be hundreds of kilometers long in total. The mining first described here presumes competent rock material; for an example of specialized mining in weaker material; refer to the end of this Section.

Where ore occurs in relatively narrow veins, small scale, labor-intensive techniques may be required. Usually drilling and blasting is required to break up the ore for excavation and removal. This type of mining may proceed from the base of the orebody upward, following the slope of the ore, from above proceeding downward, or a combination. Pillars of unmined ore may be required to hold the mining void open, or full removal may be possible. The void left behind may remain open, be allowed to collapse, or be partly or completely refilled with waste rock or tailings from an associated milling facility.

A current example of a small underground uranium mine using narrow vein mining techniques is Dolni Rozinka in the Czech Republic, which extends to more than 1000 m below the ground surface (Michálek, 2011) and there are many historic examples around the world. In larger mines, several areas of mining may be active at once to provide sufficient quantities of ore for further processing.

If the ore mass is larger, a typical technique used is called stoping. There are a number of variations, but it is used where the orebody can be mined in a series of vertical masses that might be tens or more meters high and several to tens of meters wide. Access tunnels are made and the selected portion of orebody is drilled and explosives placed. After blasting, the broken ore is removed from the base of the blasted area for crushing and processing. Previously mined stopes may be refilled with waste rock or tailings, often with the addition of cement. Once the refilled stope has gained sufficient strength, the adjacent segment or segments of the orebody can be mined. Again, multiple stopes may be mined simultaneously, or be in different stages simultaneously, to provide the required rate of ore for further processing.

A current example of a large underground mine using stoping is the Olympic Dam copper–uranium–gold mine in South Australia (Robertson et al., 2013; Webb, 1998; Oddie and Pascoe, 2005).

Where the orebody is relatively horizontal and extensive, ore is typically removed in galleries, leaving sufficient columns to prevent collapse—this is called "room and pillar" mining. When an area has been mined to the fullest extent possible using this method, the space can either be backfilled to allow mining of the pillars, or the pillars can be progressively removed to allow collapse of the mined-out area. An example of an underground uranium mine using this method with backfilling is COMINAK's d'Akouta mine, Niger (Bibert, 1982; Areva, 2015).

Some of the high-grade uranium mines in Canada operate in relatively unconsolidated sediments. The Athabasca Basin is at a high latitude and ground freezing can be used to isolate the ore from the surrounding highly hydraulically conductive, saturated sediments that would otherwise cause large-volume groundwater inflows and possibly collapsing conditions (Newman et al., 2011). Ore is extracted by the "raise bore" technique, such as with the McArthur River mine (Beattie and Davis, 2002; Jamieson, 2002). Tunnels are first constructed above and below the orebody to be mined. A relatively narrow vertical hole (bore) is drilled between the two levels. The raise borer, essentially a cutting disk, is rotated raised upward starting at the lower level, increasing the diameter of the bore to (for example) 3 m. The extracted ore is removed from below by a remote-controlled bucket loader and delivered for scanning of ore grade and either further milled underground and pumped as a slurry to the surface, or hoisted above ground for further milling (Beattie and Davis, 2002). At the newer Cigar Lake uranium mine, a variation of this technique is being used called "jet boring." Here, the orebody is artificially frozen, and rather than a mechanical device, rotating high-pressure jets of water are used to both loosen the ore and to create a slurry that can be collected and pumped away for processing

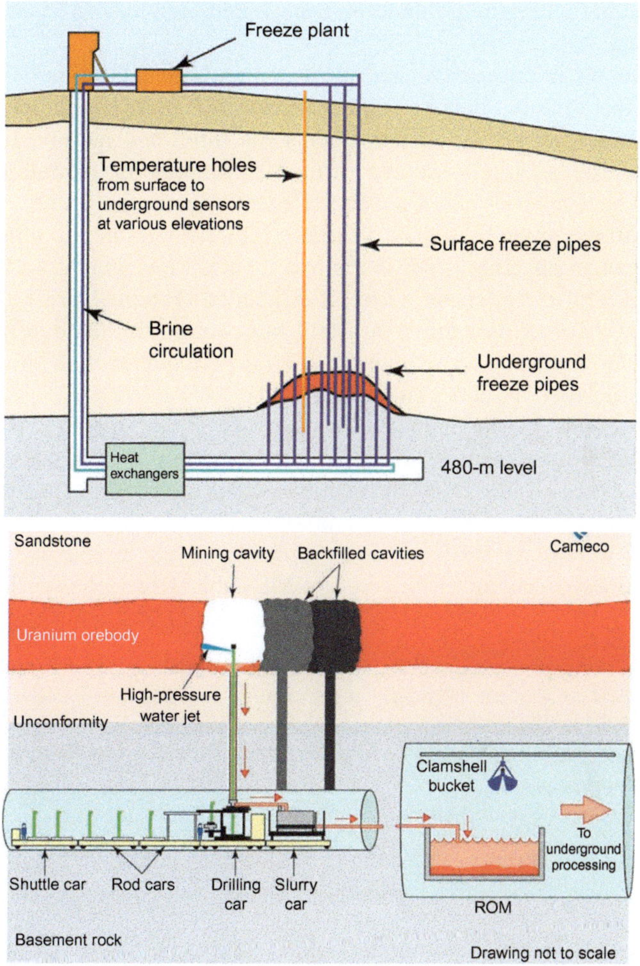

Figure 6.4 Ground freezing and jet boring mining technique, Cigar Lake, Canada.
Source: Courtesy Cameco, reproduced with permission.

(Schmitke, 2004; Edwards, 2004; Fig. 6.4). Only a tunnel beneath the ore is required. Mined voids are filled with a concrete mixture to allow later mining of adjacent ore.

6.3.3 Ventilation considerations

Ventilation is always important for underground mines, and this requirement must be a special consideration for uranium mines, or mines with significant uranium associated with them. This has been recognized from the early years of uranium mining (Westfield et al., 1958), although those authors acknowledged that at that time, "Too many underground uranium mines are operating with inadequate natural ventilation" (p. 10).

Uranium itself is radioactive, but most of the radioactivity associated with uranium ore is from the radioactive decay products. Of these, radon gas and its decay products (which may remain airborne for some time) and radionuclides associated with dust must be taken into account. A modern account of ventilation requirements is presented by Gherghel and De Souza (2008). Ventilation is forced by fans and controlled within a mine by special ventilation shafts and internal mine connections, as well as through general access shafts and tunnels. Special doors and vents can be opened or closed to direct air where it is needed. Personal protective equipment such as a respirator of dust mask may still be required for some tasks such as drilling. Computer modeling is required, undertaken together with in-mine measurements (eg, Bracke et al., 2006; El-Fawal, 2011).

As an example, at the high-grade McArthur River operation in Canada, the primary ventilation system is driven by surface exhaust fans on shaft no. 2 that draw fresh air (heated in winter) through two other shafts (Cameco, 2009). The ventilation is required to control diesel emissions and blasting gases, as at nearly all underground mines, as well as radon and long-lived radioactive dust (LLRD). General ventilation principles followed at McArthur River consist of the following (Cameco, 2009):

- Single pass ventilation used in areas with a significant radon or LLRD potential
- Capture and containment of radon at the source location (drill collars, radon bearing sumps, backfill holes, etc.) through the use of secondary ducting
- Equipment operator stations will be located in fresh air and upstream of potential radiation sources whenever possible
- Avoiding the short-circuiting of air from fresh air to exhaust air

Other documented recent case studies include the Indian Jaduguda uranium mine (Panigrahi et al., 2005) and at Olympic Dam in Australia (Uggalla, 1999; Bloss, 2009).

Occupational radiation protection in the mining and processing of raw materials is the subject of one of the International Atomic Energy Association (IAEA)'s main safety guide series documents (IAEA, 2004). Even with relatively low ore grades, along with silica dust and diesel fumes, radiation protection requirements may be one of the main drivers of ventilation at an underground uranium mine, and similar efforts may be required in mines not currently producing uranium as a byproduct, but with significant uranium present in the orebody.

6.3.4 Identification and placement of waste and mineralized material

The uranium grade of material in a pit is generally known in advance from a conceptual or computerized model of the orebody and waste rock, and by testing of some or all blast holes (where used) by analysis of samples, geophysical probing (using natural gamma), or both or by surface gamma scanning. Calibration of

Figure 6.5 Discriminator in action at Ranger uranium mine, Australia.
Source: Dr. J. Kvasnicka, Radiation Dosimetry Systems, Adelaide, Australia, reproduced with permission.

gamma loggers is important and can be difficult, particularly if grades are low and disequilibrium between uranium and the decay products responsible for the gamma is present (eg, Princep and Owers, 2012).

It is common for broken rock to be passed under a discriminator, that is, one or a set of gamma ray detectors that have been calibrated for local conditions. The discriminators provide an estimate of the grade of ore in that truck, so the truck can be directed to dump its load directly into a crusher or to the relevant stockpile of ore grade, mineralized material but below current ore grade, or clean waste rock (Fig. 6.5, cf. Tietzel et al., 2013 at Ranger Mine, Australia).

Rock may also need to be specifically placed to control dust generation, the generation of acidic or otherwise potentially toxic water runoff, or seepage to groundwater. Other criteria are required to separate potentially problematic rock, and a carefully considered plan to manage it must be prepared and implemented (eg, INAP, 2014). Historically, with poor quality water from early uranium mines designed without adequate environmental safeguards, the major problematic contaminants have not always been radioactive; rather, other elements and ions such as heavy metals (eg, by acid rock drainage) or salinity have more often been of most concern[1].

6.4 Milling and extraction

The milling and extraction of uranium from uranium ore is a mixed physical and chemical process. A generalized flow chart of the major steps is outlines in Fig. 6.6. A compendium of uranium extraction technology was published by the

[1] Committee on Uranium Mining in Virginia; Committee on Earth Resources; National Research Council (2011)

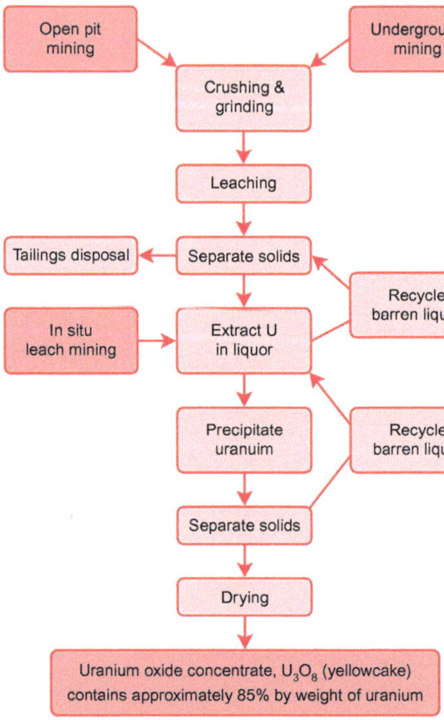

Figure 6.6 Generalized flow sheet for the processing of uranium ore. Not all variants of processing are shown.
Source: World Nuclear Association, reproduced with permission.

IAEA (1993) and earlier overviews by Seidel (1980) and the IAEA (1971). The more recent reviews by Edwards and Oliver (2000), Gupta and Singh (2003, Ch. 3 & 4), Lunt et al. (2007), Taylor (2011), Mishra et al. (2013), and Schnell (2014) provide useful summaries of uranium processing; Mishra et al. (2013) and Edwards and Oliver (2000) give more depth, whereas the others are useful but briefer practitioners' overviews.

6.4.1 Conventional crushing and grinding (comminution)

In most instances, uranium ore as produced by a mine—run-of-mine (ROM) ore—contains too much coarse material for further processing by comminution. Crushing (producing fragments typically of a few centimeters size) and grinding (producing sand or finer sized particles) are generally employed. Both are mechanical processes; crushing is generally a dry process, whereas grinding is typically a wet process. Because of the dust generated by crushing, dust control by water sprays or other means may be required for occupational health and visibility reasons.

Figure 6.7 Rod mill with steel grinding rods ready to load, Namibia.
Source: Photo: Rössing Uranium Mine, used with permission.

Note that "mill" and "milling" may be used to refer to just the comminution process of a mineral processing plant or to the plant as a whole, including comminution (or even in the absence of comminution), extraction, and purification.

The sizes of crushing and grinding plants must be matched to the type of ore (hardness, size of ROM ore, clay content, etc.) and the throughput of the processing plant. The purpose of crushing and grinding is to liberate or at least expose the surfaces of uranium minerals to allow the next stage of extraction. The number and design of mills and the optimum fineness of the grind will vary according many mineralogical, physical, metallurgical, and commercial factors. Finer grinding is typically required for alkaline leaching projects compared to acid leaching (see Section 6.4.3). Grinding is undertaken in large, rotating drums and may be enacted by the tumbling of rock pieces themselves (autogenous grinding) or with the addition of steel rods, steel balls, or rock pebbles to facilitate the grinding process (see Fig. 6.7).

Other factors can affect the design of a crushing and grinding circuit. For some clay ores (and other reasons), crushing may not be effective and the broken ore is sent immediately to a grinding process (eg, Rabbit Lake in Canada in the 1980s; Edwards, 2012, p. 72).

6.4.2 Ore beneficiation (including ablation, radiometric, and gravimetric sorting)

Ore beneficiation refers to the selection and collection of higher-grade ore fragments or rejection of lower-grade fragments from ROM ore. The upgraded ore will have a higher grade, and therefore require a smaller-scale processing plant, perhaps with different technology, compared to ROM ore. Ore beneficiation is only worthwhile if the majority of the uranium is retained and the majority of the mass is rejected.

In the earliest times of uranium mining hand sorting was employed, based on visual appearance or simple gamma scanning. In more recent times, mechanical sorting based on physical, mineralogical, or radiometric characteristics are employed.

At the Cluff Lake uranium mine and mill in Canada, in phase 1 operations high-grade ore was fed to a gravity concentration plant with jigs and vibrating tables. The concentrate averaged over 30% U (Schnell and Corpus, 2000). The gravity concentrator rejects were stored and retreated later (see Section 6.5.1).

Flotation (sometimes spelled floatation) separates mineral grains that respond differently when air bubbles are forced through a suspension in water with chemical additives, causing certain minerals to rise with the froth, from which they can be collected in a concentrated form. It was used in some Canadian uranium mines in the 1980s (eg, Muthuswami et al., 1983) and has been investigated in India (eg, Singh et al., 2001). It is used at the Olympic Dam copper–uranium–gold mine to separate sulfidic copper-bearing minerals (Alexander and Wigley, 2003) and only incidentally for uranium minerals; uranium is recovered from the reject stream of the flotation circuit. Some recent investigation of the technique for application to multimetallic ores containing uranium is described by Kurkov and Shatalov (2010).

Shatalov et al. (2001) report that the automated radiometric ore separating was widely adopted in former Eastern Bloc countries starting in 1955. Variants and improvements up to 2000 are discussed. Radiometric sorting has been used in recent times in Ukraine (OECD-NEA/IAEA, 2014, p. 426). Radiometric sorting trials at Ranger mine in Australia and Rössing in Namibia are reported by Schnell (2014), who also comments that "[G]ravity separation has been applied to uranium ores in the past with some success, but this was associated with radiation issues" (cf. Section 6.5.1, where it was used with very high-grade ore). Lund et al. (2007) mention radiometric sorters in use historically in Australia and South Africa.

A form of upgrading by physical means, rotary scrubbers, at the Langer Heinrich uranium mine in Namibia is described by Marsh (2014), who states that the rejected oversize "Barren Solids will contain 40–50% of the solids mass but only 5–10% of the uranium in the ROM feed."

When the price of uranium was high in the mid-2000s, there was relatively low interest in upgrading, rather treatment of low-grade ore was considered feasible and pursued (cf. Lund et al., 2007). However, with lower prices since 2010, more experimentation is being reported. For example, ablation can remove uranium-rich mineral crusts from some sandstone ores (Coates et al., 2014) with some ore types; Scriven (2014) cites the ablation technique resulting in rejection of 90–95% of the unprocessed ore mass but with a loss of only 5–10% of the uranium originally present. Another process under development (Becker et al., 2015) is for certain low-grade, surficial ores. It can reportedly increase the ore grade by a factor of 30 times or more without the use of chemicals, producing an inert waste and providing a leach feed suitable for acid leaching, although details of this second process were not yet released at that time. To date, neither has been undertaken at a commercial scale.

6.4.3 Acid and alkaline uranium extraction and purification

Uranium as an element can occur in four oxidation states, of which only U^{4+} and U^{6+} (tetravalent and hexavalent, or U(IV) and U(VI)) are geochemically and mineralogically important (Dahlkamp, 1993). In nature, it most commonly occurs as tetravalent uranium in the subsurface, although hexavalent uranium may be present under oxidizing conditions prevailing near the surface (Dahlkamp, 1993).

The following generalized descriptions of acidic and alkaline leaching, solvent, and ion exchange, precipitation, drying, and packing are largely based on Mishra et al. (2013), Taylor (2011), Lund et al. (2007), Edwards and Oliver (2000), and IAEA (1993). Specific variations or examples are cited for certain case studies or as the main information sources for a particular aspect of uranium extraction.

Hexavalent uranium is much more soluble than tetravalent uranium. Where uranium is already in the hexavalent form, it can usually be readily dissolved in dilute acid or carbonate (alkaline) solutions. A principal purpose of uranium extraction is to convert uranium from insoluble tetravalent to soluble hexavalent forms via a process of oxidation. Uranium oxides (such as pitchblende and uraninite) and silicates (such as coffinite) dissolve quite readily in oxidizing acid or carbonate solutions. As dissolution in carbonate solutions is typically slower than in acid solutions, requires finer grinding, and may be less complete, the acidic processing route is most commonly used, except where a carbonate content of more than 1–3% makes make acid consumption uneconomic and then alkaline leaching is used. Alkaline leaching may require elevated temperatures and possibly elevated pressure. The Dolni Rozinka operation in the Czech Republic, for example, uses alkaline leaching at atmospheric pressure and a temperature of 80°C to achieve a recovery factor of around 93% (Toman and Jezova, 2007); at Langer Heinrich in Malawi, >75°C is used for their alkaline leach circuit (Marsh, 2014).

Acidic leaching is usually completed with sulfuric acid, largely based on its relatively low price in commercial quantities and at industrial grade. Bhargava et al. (2015) review the acidic leaching of uraninite. Nitric and hydrochloric acids can be used, but are generally more expensive and may cause more serious environmental pollution; high chloride levels from hydrochloric acid can also cause difficulties later in the processing. The amount of sulfuric acid that must be added varies from a few to tens of kilograms per tonne of ore (15–50 kg/t ore; Taylor, 2011). Higher amounts of sulfuric acid are required for "refractory" uranium-bearing minerals such as brannerite, a uranium calcium titanium iron oxide, or uranium-bearing zircon, or in the presence of alkaline minerals such as carbonates. An economic tradeoff between uranium recovery and acid consumption means there is usually some residual uranium in the ore after treatment, especially in resistant mineral forms such as brannerite.

Acidic leaching is typically at a pH of between 1.5 and 2.0, which will also bring into solution other elements and compounds, such as iron, calcium, and magnesium. Alkaline leaching may leach vanadium from some ores; in fact, a project may produce both uranium and vanadium as endproducts, which was common in early post-World War II uranium mining in the United States from carnotite ores, and is still practiced at the White Mesa Mill. However, it is mostly considered a nuisance element in current mines. Other ions or species that may be problematic and must be removed later in the processing include arsenic, molybdenum, zirconium, and silica. Typically, the ground ore slurry is mixed with the acidic or alkaline extractant in leach tanks (pachuchas). These are either mechanically or pneumatically agitated to give good contact between the solid and liquid phases, with heating if required. Residence times are usually several hours to a day for acidic leach, although up to 4 days in some cases. Leaching residence time is typically over 1 and up to 4 days for alkaline leach.

Maintaining a suitable oxidizing environment during acid leaching is critical to high yields of uranium. Oxidizing agents are typically added, with agents such as pyrolusite (MnO_2), oxygen gas, Caro's acid (H_2SO_5), hydrogen peroxide, or chlorate (ClO_3^-). Unless free oxygen is present, the actual dissolution of uranium oxide is typically by the ferric (Fe^{3+}) ion, which is reduced to ferrous (Fe^{2+}) in the process; in this common case, the purpose of the other oxidants is to return the ferrous iron to the ferric state to sustain the dissolution of uranium.

Once the uranium is in an aqueous solution, be it acidic or alkaline, it is normal to further purify and concentrate the uranium before it is packaged. Note that the uranium is not normally present as a metal ion (cation), but rather as part of an anionic complex, typically with sulfate or carbonate. There are two routes used commercially to preferentially remove and concentrate uranium from a loaded (also called pregnant) solution:

- Solvent exchange (SX)
- Ion exchange (IX)

Solvent exchange is a separation, purification, and recovery technology based on the differing solubilities of compounds in two different immiscible liquids, usually water and an organic solvent such as kerosene with added amines. It is used not only for uranium recovery but also for other commodities such as copper, nickel, cobalt, and rare earth elements. Solvent extraction recovery of uranium can only be used in conjunction with acid leach solutions.

A pregnant solution and barren solvent are mixed, resulting in selective transfer of uranium into the solvent, leaving most other dissolved metals behind. The mixture is allowed to settle, with the now loaded solvent floating above the now barren aqueous solution. The barren aqueous solution is then returned to the extraction stage for reuse, possibly after cleaning and "topping up" of acid and oxidant.

The loaded solvent is then separately mixed with a new aqueous solution of a different composition. Under these different conditions, the uranium then transfers to the new aqueous solution, and the now barren solvent is recycled to extract more uranium from the original pregnant solution. At each step, the concentration of uranium increases and the amount of unwanted impurities decreases. Considerable initial testing and ongoing refinement is required to implement SX extraction of uranium (eg, Bartsch and Hall, 2011).

Ion exchange is the exchange of ions from a solution with ions from solid organic resins. The process is reversible, enabling the target ion (a uranium complex) to be recovered and the resin to be regenerated. Ion exchange is an important process for uranium recovery in both acid and alkaline leaching. The uranium is present in a pregnant solution in anionic forms, complexed with ions such as carbonate or sulfate, rather than metallic ions with a positive charge (cations). Hence, uranium IX is an anionic process.

IX resins are typically small polymeric beads of approximately 0.5 mm diameter. Pregnant solution is passed through tanks of resin, where uranium is selectively absorbed onto the resin as an anionic complex. In a variant of this, pulp (finely ground ore and pregnant solution after leaching) is physically mixed with resin, and then the resin is later separated out by screening. This is called the resin-in-pulp (RIP) technique and is used for uranium at the Kayelekera mine in Malawi (Marsh and Hladun, 2010) and the Rozna mine in the Czech Republic (Toman and Jezova, 2007). The RIP technique for uranium was not widely employed around the turn of the century, but additional interest has been shown in this technique in the last decade (Brown et al., 2010; Sreenivas and Rajan, 2013).

Loaded resin is then washed with a different aqueous solution, causing the uranium to leave the resin and enter the solution, a process called elution. The concentration of uranium in the elutate is much higher than it was in the original pregnant solution, with fewer impurities. The beads are then reconditioned (regenerated) with a third aqueous solution to prepare them to absorb more uranium and reused.

The large Rössing uranium mine in Namibia uses both solvent extraction and ion exchange, but other contemporary mines use one or the other.

Uranium can be precipitated from solution over a wide range of pH, depending upon the solution type and the precipitant used. A number of different precipitants have been effective, including hydrogen peroxide, ammonia, magnesia, magnesium hydroxide, and sodium hydroxide. The type of precipitating reagent chosen is influenced by factors such as the purity of the feed solution, the product specifications demanded, the relative reagent cost, and the possible environmental impact of the reagent. Recent trends are away from the use of ammonia, which can cause additional environmental liabilities and is hard to remove from wastewater streams.

The final moisture removal and drying of the uranium precipitate varies among operations, often after a rinse with good quality water. Horizontal belt filters (continuous), batch filters, or centrifuges may be employed. Final drying equipment and the temperature employed are also variable. Typically at larger mines, calcining (roasting) at approximately 700°C is used to provide a final uranium oxide product, usually U_3O_8 (strictly $UO_3.UO_2.UO_3$), which has a theoretical uranium metal content of 85%. Elsewhere, the final product contains peroxide, magnesium, sodium, or ammonium as well as oxygen and has a somewhat lower content of uranium metal.

Under certain circumstances, it may be possible to precipitate uranium of sufficiently good quality direct from the first pregnant solution without using SX or IX to purify and concentrate the uranium first (Direct Precipitation, eg, Van Tonder, 2012; Becis et al., 2014). In other circumstances, a two-stage precipitation may be necessary to achieve the required purity standards, even after SX or IX, as is the case at Langer Heinrich mine in Namibia (Marsh, 2014).

Uranium ore concentrate (UOC) is a general term to describe the final product of uranium mining and milling, although the term "yellowcake" is still loosely used. In fact, UOC may be yellow (eg, ammonium diuranate, uranyl peroxide), dark green-black, grey (U_3O_8), (Fig. 6.8), and sometimes orange.

The remnants of treated ore after leaching are termed tails or tailings; depending on the efficiency of the extraction technique used, these may contain from 40% (rarely) to less than 5% of the original uranium, typically in harder-to-leach mineral forms (see also Section 6.5).

The final product is drummed for transport to the customer in individually identified quality 200-L (44-gallon) drums sealed with all appropriate labeling. Loaded drums weigh around 300 kg (more in the case of U_3O_8). Typically, drums are

Figure 6.8 Left; taking a subsample of drummed UOC (uranyl peroxide); right, laboratory sample of calcined UOC (U_3O_8)
Source: Photos: Heathgate Resources (left), Rössing Uranium Mine (right), used with permission.

Figure 6.9 UOC drummed, sealed, labeled, and secured in a sea container prior to transport. *Source*: Photograph P. Woods.

loaded into sea containers for transport by road, rail, and sea, as required for transport to reach the customer (Fig. 6.9). Strict accounting of the quantity of UOC per drum and the number of drums per sea container are required both for commercial purposes (UOC is typically worth more USD 50–100/kg in the last decade, and was more than double that around the 2007 price peak), and to fulfil national and international safeguards requirements.

6.4.4 Heap leaching options

Heap leaching is a variation of uranium ore processing applicable to open cut or underground uranium mines. The technique as currently applied was largely adapted from copper (and to a lesser extent gold) heap leaching operations (Schnell, 2014; see also Van Zyl and Caldwell, 2013).

In heap leaching, crushed ore is placed up to several meters thick on relatively flat pads (but with an engineered slope for correct internal drainage) with an impervious base such as a robust plastic liner or bitumen-coated clay. Historically, a less sophisticated leap leach uranium extraction was undertaken with little preparation of ore other than crushing, such as in Argentina (Castillo, 2004) and Hungary (Banik et al., 2011). More recent operations, beginning in Brazil (see Gomiero, 2008), have undertaken agglomeration of more finely crushed ore before it is placed on the pads. The technique of agglomeration—the preparation of stable pellets of ore that allow an optimal combination of contact of ore with the leaching fluid and rate of drainage of that fluid—and its importance in heap leach operations are given in some detail by Bouffard (2008) and Dhawan et al. (2013).

The extracting fluid, acid (eg, at Caetité, Brazil (Gomiero, 2008) and at Somaïr, Niger (Durupt, 2009)) or alkaline (eg, Trekkopje, Namibia, (Schnell, 2014)) is spread on the heaps by lines of drippers or sprinklers. The solution passes through the heap, dissolving uranium and other elements or ions as with conventional leaching described previously. The leach solution is collected under gravity, sometimes with the assistance of drains previously installed on the impervious base, then processed conventionally to extract the uranium. Used leach solution is returned to the optimal pH and chemical composition and reused.

The heap leaching option is often investigated as the only option for low-grade uranium deposits (Trekkopje, Namibia, currently on care-and-maintenance) or to supplement production from lower-grade ores from mines with conventional treatment circuits (Somaïr, Niger); several other mines have or are investigating this processing option.

There are several variations to details of stacking and leaching arrangements. Pads may be "on-off," where leached-out material is removed from the pad for disposal elsewhere and replaced by fresh material. A single-layer pad may be left in place for possible reshaping and covering for closure or in some cases one or more additional layers of fresh material may be added on leached-out material. The arrangement of the mining solution movement can be varied, with multiple leach cycles with recycled solution before the solution is removed for product recovery, or passage of fresh mining solution through partially-leached pads (or segments of pads) prior to reuse through newly stacked material.

Under the right circumstances, heap leach is an attractive low-capital operating cost option for miners because it avoids the grinding stage of ore preparation, leaching in tanks, solid–liquid separation, and a large tailings facility. However, it requires a large area of suitable land and is not suitable for all orebodies. Careful assessment and extensive, progressive testing are required (eg, Robertson and Van Staden, 2009), with the scaling-up of tests from bench scale to columns of increasing height and field trials of small heaps to refine and prove the viability of the technique for a particular project. The distribution of uranium minerals in the ore is important, as the ore particles are relatively coarse, and the surface area available to react with the leaching solution is relatively low. Maintaining heap stability and uniform distribution of leaching solution through the heap requires good operating technique, and testing of ore in laboratory and columns should continue even at successfully operating projects. Reviewing experience at Caetité in Brazil, Gomiero (2008) stated, "Although heap leaching, in principle, is a very simple process, [Caetité's] operation has shown that attention to details is critical for the project success as a whole." His paper describes various improvements incorporated into that operation, a feature of all long-lived mines.

Leaching in the heaps may be coincidentally or deliberately accelerated and facilitated by biological action, where oxidation of sulfides facilitated by bacteria or other microorganisms provides some of the acid required.

Uranium can potentially be extracted from the leaching solution from heap leach projects with a different primary product or products. In recent times, this was investigated and constructed at the Talvivaara base metal mine in Finland, which uses bio-heap leaching to recover nickel, zinc, copper, and cobalt (Pekkala et al., 2012), although mining was suspended in 2014 before any uranium was produced.

Copper heap leaching operations, particularly in South America, also have the potential to produce modest quantities of uranium byproduct.

6.4.5 Block leaching (underground)

Underground block leaching is a process where ore is broken by blasting underground but not transported to the surface (or elsewhere underground) for further treatment. Rather, leaching solution is percolated through the broken ore in place and collected to be pumped to the surface for the recovery of the uranium.

The Ingul'skaya mine in Ukraine is reported to currently operate a block leaching complex, allowing it to significantly lower production costs (UX Consulting, 2011). Historically block leach[2] was used at the Königstein uranium mine in Germany, from 1984 to 1990, causing challenges to the final rehabilitation of the mine, which involves flooding the underground workings (Jenk et al., 2014).

6.5 Retreatment of mine tailings

6.5.1 Historic examples

Tailings from historic mining operations are sometimes remined and reprocessed to extract the original product that was incompletely extracted during the first treatment, additional products, or both when circumstances and economics allow. This can have the additional environmental benefit of depositing the retreated tailings in a more modern and secure tailings management facility, compared to the often very basic tailings dams of many historic mining operations. However, appropriate good management and precautions are required to ensure an environmental benefit from retreatment (cf. Fleming et al., 2010; Liefferink and Liefferink, 2015).

Tailings may be excavated conventionally and trucked, conveyed to a treatment plant, or recovered hydraulically. In hydraulic recovery, tailings are broken up and formed into slurry by high-pressure water jets, and the resulting slurry pumped away for treatment (Ford et al., 1987 describes both options).

Low-grade uranium occurs with gold in the Witwatersrand area of South Africa. Large historic tailings piles from over 100 years of gold mining occur in the area. During the 1970–1990 period, as well as plants extracting uranium as a byproduct of direct gold mining and treatment, three significant tailings retreatment operations in the East Rand, Klerksdorp, and Welkom areas were established (Ford, 1993; Ford et al., 1987).

In the United States, retreatment of some 550,000 tonnes of earlier uranium mill tailings was undertaken at Naturita, with recovery of both uranium and vanadium (Scheffel, 1981).

[2] The process is called ISL by the cited German authors and others, but in general the term is reserved as a synonym for the solution mining technique ISR (see chapter: Introduction to Uranium In Situ Recovery Technology) and is not preferred.

In Canada, phase 1 tailings from the Cluff Lake uranium mill were later recovered and retreated to recover uranium alone (gravity tailings, 1983–1984) and gold and uranium (from leach tailings, 1985–1986). The Cluff Lake tailings were particularly high in uranium by world standards, with gravity tailings containing over 1% U, and leach tails 58 g/t gold and 0.3–1% U (Schnell and Corpus, 2000).

6.5.2 Current considerations

In South Africa, 32% of the total identified uranium resources are in tailings facilities associated with historic Witwatersrand Basin gold (and some uranium) mining, that is, approximately 75,000 tU (OECD-NEA/IAEA, 2014). Historically there were several metallurgical plants that extracted uranium as a byproduct of gold production, with a diminished but steady production continuing to the present day. During the high uranium price in the mid-2000s, the potential to recover uranium held in historic tailings dams was reevaluated. One gold recovery plant for remined tailings had a uranium circuit added, and according to the owner AngloGold Ashanti (2014, p. 72), its "uranium plant at Mine Waste Solutions (MWS) was successfully completed with the first deliveries in the fourth quarter of the year."[3]

The large volumes of historic gold tailings in this part of South Africa are likely to attract further interest, depending perhaps on the success of the MWS plant, or as other technologies are proved (cf. Fleming et al., 2010).

Other mine tailings facilities around the world have been and may again be of interest, both from former uranium mines or from mines where uranium is present in significant concentrations and quantities in tailings from other commodity mining. This includes parts of Central Asia, where early uranium mining sometimes left considerable uranium behind, and remining, reprocessing, and better disposal of the retreated tailings could reduce current environmental liabilities (Jakubick, 2009, pers. comm.). In such cases, it may be justified to extract the uranium (and other products, if feasible) to "subsidize" the cost of making historic tailings dams safe; however, if the economics and other aspects including approvals and social acceptance are positive, other projects around the world may be established in the future.

6.6 Future trends

6.6.1 Technology developments

The mining industry is marked by many technology developments and breakthroughs over the last centuries. Mechanization, now combined with increased geotechnical understanding and the integration of traditional geological and mining engineering with geometallurgical, economic, and environmental considerations (now including improved computerized techniques), has allowed previously unexploitable mineral deposits to be successfully mined. This brings with it new

[3] This same company produces uranium as a byproduct of its main primary ore treatment plant.

challenges, as some mines take on bigger footprints and may bring greater volumes of problematic materials, such as those leading to acid drainage to the surface. This trend of innovation is expected to continue, and may allow remining of formerly "mined-out" deposits or their remains (eg, tailings, mineralized waste rock). In the case of uranium, a breakthrough in the extraction of uranium from seawater could create an upper cost limit for the production of uranium against which terrestrial production from mines would be compared and limited.

The use of remote-controlled mining equipment is likely to increase in the future, either to minimize the radiation exposure of workers or as an efficiency measure. This may allow mining in adverse temperature and humidity conditions in deep mines that are currently problematic for human miners or access to narrow veins or ore in soft rock formations that are currently neither safe nor economic to mine.

Other developments that could affect the demand for mined uranium are discussed in other chapters.

6.6.2 Considering the end at the beginning

Bitter experience with legacies from the early decades of uranium mining—and certainly, from most medium- to large-scale mining—has emphasized the need for a safe and secure transition from mining to an acceptable postmining land use. Many historical or "legacy" sites have been known (notably in the former Soviet Union and its Eastern European associated countries, but also in the United States, Australia, western Europe, and elsewhere) and an extensive literature exists. A key lesson learned is that a life cycle planning approach can prevent future problematical legacies from occurring (Woods et al., 2015). The International Council on Mining and Metals, in its *Planning for Integrated Mine Closure: Toolkit* (ICMM, 2008), strongly makes the case that closure planning begins in the earliest exploration stage. Early planning is conceptual, and as a project progresses through design, feasibility, approvals, and construction into operation, closure planning becomes progressively more detailed so that appropriate funds can be put aside. Effective closure planning engages all affected and the appropriate interested parties, giving better support by stakeholders, and allowing the proponent and regulator to derisk and improve the accuracy of closure cost estimates. Detailed plans will include milestones, methodologies to achieve these, monitoring, and validation processes (ICMM, 2008).

An important and encouraging trend is the "mainstreaming" of environmental and social aspects, which can be expressed as: environmental and social aspects are increasingly becoming a standard part of planning, regulating, mining, and milling of uranium, both in theory and in practice (Woods et al., 2015). The adoption of responsibility for social, environmental, and closure performance by operational management is a positive development, and allows environmental staff to provide specialist assistance to operational departments, rather than existing outside (and possibly ignored by) the main planning and operations of a mine. At the same time, environmental and social specialists are more effective when they have an appreciation of the technical and financial aspects of mining projects (Woods, 2015). This life cycle planning approach is seen as a way to prevent or at least minimize future liabilities (see also chapters: Management for Health, Safety, Environment and Community (HSEC) in Uranium Mining and Processing and Uranium Mine and Mill Remediation and Reclamation). Examples of how uranium mining has evolved to effectively manage impacts, with case studies contrasting old and new practices and outcomes, are highlighted by OECD-NEA (2014).

6.6.3 Other trends and discussion

The current overall picture of uranium supply is one where the closure of existing mines due to resource depletion is expected to be offset by the opening of new mines for the medium term (OECD-NEA/IAEA, 2014). Over 2013–2015, a number of uranium mines were put on care-and-maintenance and several projects of different sizes were postponed, but new production countered these closures. In the longer term, post-2020 or so, additional resources are expected to be required other than currently operating, on care-and-maintenance, planned or proposed mines to meet reasonably foreseeable demand. The 2014 Red Book (OECD-NEA/IAEA, 2014) stated:

> *... the uranium resource base described in this publication is more than adequate to meet projected requirements for the foreseeable future. The challenge is to continue developing safe and environmentally responsible mining operations to bring the required quantities of uranium to the market in a timely fashion.*

In general, and as with other metals, it is expected that many future mines will open with lower average grades than the mines of the last three decades, as most shallow deposits of higher grade are considered to have been already discovered and mined or planned to be mined. However, this is partially countered by ongoing (deep) high-grade discoveries in the Athabasca Basin in Canada (eg, Wheatley, 2014). The potential to effectively and economically upgrade the ore from low-grade deposits will have an impact on which deposits are mined, as other higher-grade resources become depleted and are (potentially) only partially replaced by resources of similar grade.

6.7 Sources of further information and advice

6.7.1 Books, proceedings, and technical reports

Anderson, C., Dunne, R., Uhrie, J. (Eds), 2014. Mineral Processing & Extractive Metallurgy. Society for Mining, Metallurgy & Exploration.

ALTA metallurgical conference series, http://www.altamet.com.au.

IAEA, in press. Proc. International Symposium on Uranium Raw Material for the Nuclear Fuel Cycle: Exploration, Mining, Production, Supply and Demand, Economics and Environmental Issues (URAM-2014). http://www-pub.iaea.org/iaea-meetings/46085/URAM-2014.

IAEA, 2014. Proc. International Symposium on Uranium Raw Material for the Nuclear Fuel Cycle: Exploration, Mining, Production, Supply and Demand, Economics and Environmental Issues (URAM-2009). http://www-pub.iaea.org/books/IAEABooks/8496/Uranium-Raw-Material-for-the-Nuclear-Fuel-Cycle-Exploration-Mining-Production-Supply-and-Demand-Economics-and-Environmental-Issues-URAM-2009-Proceedings-of-an-International-Symposium-Vienna-Austria-22-26-June-2009 and http://www-pub.iaea.org/books/IAEABooks/10751/Uranium-Raw-Material-for-the-Nuclear-Fuel-Cycle-Exploration-Mining-Production-Supply-and-Demand-Economics-and-Environmental-Issues-URAM-2009-Proceedings-of-an-International-Symposium-Vienna-Austria-22-26-June-2009.

IAEA, 2006. Proc. International Symposium, Uranium Production and Raw Materials for the Nuclear Fuel Cycle—Supply and Demand, Economics, the Environment and Energy Security. http://www-pub.iaea.org/MTCD/Publications/PDF/Pub1259_web.pdf.

Живов, В. Л., Бойцов, А. В., Шумилин, М. В. 2012. (Jivov, V.L., Boytsov, A.V., Shumilin, M.V.) Уран: геология, добыча, экономика (Uranium: geology, production, economy). Москва: Атомредметзолото (RIS «VIMS», Moscow) (in Russian).

OECD-NEA/IAEA (Organisation for Economic Cooperation and Development – Nuclear Energy Agency/International Atomic Energy Agency) (1965–2014), Uranium [year]: Resources, Production and Demand. Organization for Economic Cooperation and Development, Paris. ("Red Book" report series, biennial or annual, ongoing).

OECD-NEA (Organisation for Economic Cooperation and Development – Nuclear Energy Agency) (2006), Forty Years of Uranium Resources, Production and Demand in Perspective: "The Red Book Retrospective." Organization for Economic Cooperation and Development, Paris.

Uranium 2010 "The future is U," Proc. 3rd Int. Conf. on Uranium, 40th Ann. Hydrometallurgy Meeting, Saskatoon, Canada, 15–18 October 2010 (2 Volumes). http://legacy.metsoc.org/u2010/index.asp.

Uranium 2000 Proc. International Symposium on the Process Metallurgy of Uranium: 9–15 September 2000, Saskatoon, Canada.

6.7.2 Websites

6.7.2.1 International organizations

International Atomic Energy Agency https://www.iaea.org/

Organisation for Economic Co-operation and Development (OECD) Nuclear Energy Agency http://www.oecd-nea.org/

6.7.2.2 *Industry groups (world and some major producing countries, alphabetical order)*

Canadian Nuclear Association https://cna.ca/technology/energy/uranium-mining/

Namibian Uranium Association http://namibianuranium.org/

Minerals Council of Australia http://www.aua.org.au/resources/uranium

Uranium Producers of America http://www.theupa.org/

World Nuclear Association http://www.world-nuclear.org/

6.7.2.3 *Top seven uranium producing companies (alphabetical order, as at 2015)*

Areva Mining Business Group http://www.areva.com/EN/operations-635/mining-uranium-production-yellowcake-exploration-mining-milling.html

ARMZ Uranium Holding Company (includes Uranium One) http://www.armz.ru/eng/

BHP Billiton (Olympic Dam copper–uranium–gold-silver mine) http://www.bhpbilliton.com/home/businesses/copper/Pages/default.aspx

Cameco http://www.cameco.com/

Kazatomprom http://www.kazatomprom.kz/en/

Navoi Mining and Metallurgical Combinat http://www.ngmk.uz/

Rio Tinto (Rössing and Ranger uranium mines) http://www.riotinto.com/ http://www.rossing.com/ http://www.energyres.com.au/

6.7.2.4 *Nongovernment organization (may be philosophically opposed to uranium mining and milling)*

WISE (World Information Service on Energy) Uranium Project http://www.wise-uranium.org/

Acknowledgments

An extensive effort has been made to acknowledge the information sources, but some information is from conversations, talks, or unpublished manuscripts that do not lend themselves to formal citation. Particular acknowledgment is due, however, to an unpublished compilation of information from Geoscience Australia provided as an information source for an internal IAEA training document and available to the author.

References

Alexander, D.J., Wigley, P. 2003. Flotation circuit analysis at WMC Ltd Olympic Dam operation. In: 8th Mill Operators' Conference, Townsville, Queensland, Australia, 21–23 July 2003. Australasian Institute of Mining and Metallurgy, Melbourne, Australia, pp. 41–51.

AngloGold Ashanti, 2014. Integrated Report 2014.

Areva, 2015. COMINAK, operator of the largest underground uranium mine. <http://www.areva.com/EN/operations-602/cominak-operator-of-the-largest-underground-uranium-mine.html> (accessed March 2015).

Banik, J., Csoevari, M., Nemeth, G. 2011. Uranium ore mining and the remediation of the site in Hungary. In: The Uranium Mining Remediation Exchange Group (UMREG) Selected Papers 1995–2007. International Atomic Energy Agency, Vienna, Austria, STI/PUB/1524, pp. 125–137.

Bartsch, P., Hall, S. 2011. Solvent extraction of uranium—towards good practice in design, operation and management. In: Proc. Metallurgical Plant Design and Operating Strategies (MetPlant 2011), Perth, Australia, 8–9 August 2011. Australasian Institute of Mining and Metallurgy, Melbourne, Australia, pp. 468–479.

Beattie, D., Davis, T. 2002. High-grade uranium mining at McArthur River, Saskatchewan, Canada. In: Underground Operators' Conference, Townsville, Queensland, Australia, 29–31 July 2002. Australasian Institute of Mining and Metallurgy, Melbourne, Australia, pp. 11–19.

Becis, A., Guettaf, H., Ferhat, K., Semaoune, T., Hanou, K., Ferrad, F., et al. 2014. Elaboration of uranium ore concentrate by direct precipitation. In: Proc. International Symposium on Uranium Raw Material for the Nuclear Fuel Cycle: Exploration, Mining, Production, Supply and Demand, Economics and Environmental Issues (URAM-2009), Vienna, Austria, 22–26 June 2009, IAEA-TECDOC-1739, Appendices, Session 4.

Becker, G.S.M., Hill, M.P., Johnson, G.D. 2015. U-pgrade™ a technological breakthrough for surficial uranium ores. In: ALTA 2015, Perth, Western Australia, 23–30 May 2015.

Bhargava, S.K., Ram, R., Pownceby, M., Grocott, S., Ring, B., Tardio, J., et al., 2015. A review of acid leaching of uraninite. Hydrometallurgy 151, 10–24.

Bibert, F.-X. 1982. Récupération des piliers dans une exploitation sédimentaire d'uranium par remblayage cimente en pate, Mine d'Akouta – République du Niger. Industrie Minérale, Décembre 1982, 645–651 (reprinted 2001 <http://www.bibert.fr/Niger/Remblayage.pdf>).

Bloss, M., 2009. Increasing throughput at Olympic Dam by effective management of the mine bottleneck. Min. Technol. 118 (1), 33–46.

Bouffard, S.C., 2008. Agglomeration for heap leaching: equipment design, agglomerate quality control, and impact on the heap leach process. Miner. Eng. 21 (15), 1115–1125.

Bracke, G., Alkan, H., Müller, W. 2006. Modelling underground ventilation networks and radon flow for radiological protection using VUMA. In: Merkel, B.J., Hasche-Berger, A. (Eds.), Uranium in the Environment, Conference Proceedings 4th International Conference on Uranium Mining and Hydrogeology 2005, Freiberg, Germany. Springer, Berlin and Heidelberg, pp. 593–599.

Brown, J., Goode, J.R., Fleming, C. 2010. The re-emergence of resin in pulp with strong base resins as a low cost, technically viable process for the recovery of uranium. In: 42nd Annual Meeting of the Canadian Mineral Processors, Ottawa, Ontario, Canada, 19–21 January 2010 (as reprinted, SGS Mineral Services Technical Paper 2010–10).

Cameco 2009. McArthur River Operation Northern Saskatchewan, Canada, NI 43–101 Technical Report. Cameco Corporation, Canada.

Castillo, A. 2004. Leaching method: changes in Sierra Pintada mine. In: Proc. Tech Mtg, Recent Developments in Uranium Resources and Production with Emphasis on In Situ Leach Mining, Beijing, 18–23 September 2002, IAEA-TECDOC-1396, pp. 223–229.

Coates, J.A., Scriven, D.H., Coates, C., Coates, E. 2014. Devices, systems, and methods for processing heterogeneous materials. U.S. Patent No. 8,646,705. U.S. Patent and Trademark Office, Washington, DC.

Committee on Uranium Mining in Virginia; Committee on Earth Resources; National Research Council, 2011. Uranium Mining in Virginia: Scientific, Technical, Environmental, Human Health and Safety, and Regulatory Aspects of Uranium Mining and Processing in Virginia. National Academies Press (USA), Washington, DC (Chapter 6, Potential Environmental Effects of Uranium Mining, Processing, and Reclamation). Available from: <http://www.ncbi.nlm.nih.gov/books/NBK201052/>.

Dahlkamp, F.J., 1993. Uranium Ore Deposits. Springer-Verlag, Berlin and Heidelberg.

Dhawan, N., Safarzadeh, M.S., Miller, J.D., Moats, M.S., Rajamani, R.K., 2013. Crushed ore agglomeration and its control for heap leach operations. Miner. Eng. 41, 53–70.

Durupt, N. 2009. Heap leaching of low grade uranium ores at Somaïr. In: ALTA 2009 Uranium Conference, Perth, Western Australia, 25–30 May 2009.

Edwards, C., 2012. Uranium milling in northern Saskatchewan's Athabasca Basin: past, present and future. CIM J. 3 (2), 69–94.

Edwards, C.R., 2004. The Cigar Lake project—mining, ore handling and milling. CIM Bull. 97 (1078), 105–114.

Edwards, C.R., Oliver, A.J., 2000. Uranium processing: a review of current methods and technology. JOM. 52 (9), 12–20.

El-Fawal, M.M., 2011. Mathematical modelling for radon prediction and ventilation air cleaning system requirements in underground mines. J. Am. Sci. 7 (2), 389–402.

Fleming, C.A., Brown, J.A., Botha, M. 2010. An economic and environmental case for re-processing gold tailings in South Africa. In: 42nd Annual Meeting of the Canadian Mineral Processors, Ottawa, Ontario, Canada, 19–21 January 2010 (as reprinted, SGS Minerals Technical Paper 2010–03).

Ford, M.A., 1993. Uranium in South Africa. J. South. Afr. Inst. Min. Metall. 93 (2), 37–58.

Ford, M.A., Simonsen, H.A., Viljoen, E.B., Janssen, M.S., Taylor, J.L., 1987. The first six years of the Chemwes uranium plant. J. South. Afr. Inst. Min. Metall. 87 (4), 113–124.

Gherghel, C., De Souza, E. 2008. Ventilation requirements for uranium mines. In: 12th U.S./North American Mine Ventilation Symposium 2008, Reno, NV, pp. 65–69.

Gomiero, L.A. 2008. Uranium Production in Caetité, Brazil. In: Proc. Regional Technical Meeting "Uranium Exploration, Mining, Production, Mine Remediation and Environmental Issues," Mendoza, Argentina, 2006. International Atomic Energy Agency <http://www.iaea.org/OurWork/ST/NE/NEFW/documents/RawMaterials/TC%20BRA/24%20Caetite%20Mining%20&%20Milling%20-%20Brazil.pdf>.

Gupta, C., Singh, H., 2003. Uranium Resource Processing: Secondary Resources. Springer-Verlag, Berlin, Heidelberg, New York.

IAEA, 1971. The recovery of uranium. In: Proceedings of a Symposium, Sao Paulo, 17–21 August 1970. International Atomic Energy Agency, Vienna, Austria.

IAEA, 1993. Uranium Extraction Technology, Tech. Rep. Series No. 359. International Atomic Energy Agency, Vienna, Austria.

IAEA, 2004. Occupational Radiation Protection in the Mining and Processing of Raw Materials, Safety Guide, RS-G-1.6. International Atomic Energy Agency, Vienna, Austria.

ICMM, 2008. Planning for Integrated Mine Closure: Toolkit. International Council on Mining and Metals, London, UK.

INAP, 2014. Global Acid Rock Drainage Guide. The International Network for Acid Prevention, <http://www.gardguide.com>.

Jakubick, A. 2009, Involvement of the private sector in the legacy tailings remediation in Central Asia: Reprocessing of tailings for recovery of valuable constituents. Unpublished report (pers. comm).

Jamieson, B.W. 2002. Mining the high grade McArthur River uranium deposit. In: Proc. Int. Symp. The Uranium Production Cycle and the Environment, Vienna, Austria, 2–6 Oct. 2000, IAEA C&S Papers Series No. 10/P, IAEA-SM-362/29, 272–286.

Jenk, U., Zimmermann, U., Uhlig, U., Schöpke, R., Paul, M., 2014. In situ mine water treatment: field experiment at the flooded Königstein Uranium Mine (Germany). Mine Water Environ. 33 (1), 39–47.

Kurkov, A., Shatalov, V. 2010. Development of technology for complex uranium ores processing on the flotation process basis. In: Proc. XXV International Mineral Processing Congress (IMPC) 2010, Brisbane, Australia, 6–10 September 2010, pp. 2097–2106.

Liefferink, M., Liefferink, S.L. 2015. Current reclamation of historical uraniferous tailings dams and sand dumps—exacerbating the mess or minimizing the mining footprint? Case studies within the Witwatersrand goldfields. In: Merkel, B., Arab, A. (Eds.), Uranium—Past and Future Challenges (Proc. Uranium Mining and Hydrogeology VII, Freiberg, Saxony, Germany, 21–25 September 2014), pp. 387–400.

Lunt, D., Boshoff, P., Boylett, M., El-Ansary, Z., 2007. Uranium extraction: the key process drivers. J. South. Afr. Inst. Min. Metall. 107, 419–426.

Marsh, D. 2014. Development and expansion of the Langer Heinrich Operation in Namibia. In: Proc. International Symposium on Uranium Raw Material for the Nuclear Fuel Cycle: Exploration, Mining, Production, Supply and Demand, Economics and Environmental Issues (URAM-2009), Vienna, Austria, 22–26 June 2009, IAEA-TECDOC-1739, pp. 160–170.

Marsh, D., Hladun, D. 2010. An overview of the Kayelekera uranium mine in Malawi. In: ALTA 2010 Uranium Conference, Perth, Western Australia, 27–28 May 2010.

McKay, A. D., Miezitis, Y. 2001. Australia's uranium resources, geology and development of deposits. AGSO-Geoscience Australia. Mineral Resource Report 1.

Michálek, B. 2011. History and present state of uranium mining in the Czech Republic. In: The Uranium Mining Remediation Exchange Group (UMREG) Selected Papers 1995–2007. International Atomic Energy Agency, Vienna, Austria, STI/PUB/1524, pp. 88–100.

Mishra, B., Gubel, N.R., Bhola, R., 2013. Uranium processing. In: Morrell, J.S., Jackson, M.J. (Eds.), Uranium Processing and Properties. Springer Science + Business Media, New York, pp. 123–172. Chapter 6.

Muthuswami, S.V., Vijayan, S., Woods, D.R., Banerjee, S., 1983. Flotation of uranium from uranium ores in Canada: Part I-flotation results with Elliot Lake uranium ores using chelating agents as collectors. Can. J. Chem. Eng. 61 (5), 728–744.

Newman, G., Newman, L., Chapman, D., Harbicht, T. 2011. Artificial ground freezing: an environmental best practice at Cameco's Uranium mining operations in Northern Saskatchewan, Canada. In: Rüde, T.R., Freund, A., Wolkersdorfer, C. (Eds.), 11th International Mine Water Association Congress—Mine Water—Managing the Challenges, Aachen, Germany, pp. 113–117.

Oddie, M.E., Pascoe, M.J. 2005. Stope performance at Olympic Dam mine. In: Proc. 9th Underground Operators' Conference, Perth, Western Australia, 7–3 March 2005. Australasian Institute of Mining and Metallurgy, Melbourne, Australia, pp. 265–272.

OECD-NEA (Organisation for Economic Cooperation and Development – Nuclear Energy Agency), 2006. Forty Years of Uranium Resources, Production and Demand in Perspective: "The Red Book Retrospective." Organization for Economic Cooperation and Development, Paris.

OECD-NEA (Organisation for Economic Cooperation and Development – Nuclear Energy Agency), 2014. Managing Environmental and Health Impacts in Uranium Mining, NEA No. 7062. Organization for Economic Cooperation and Development, Paris. <http://www.oecd-nea.org/ndd/pubs/2014/7062-mehium.pdf>.

OECD-NEA/IAEA (Organisation for Economic Cooperation and Development – Nuclear Energy Agency/International Atomic Energy Agency), 2012. Uranium 2011: Resources, Production and Demand. Organization for Economic Cooperation and Development, Paris.

OECD-NEA/IAEA (Organisation for Economic Cooperation and Development – Nuclear Energy Agency/International Atomic Energy Agency), 2014. Uranium 2014: Resources, Production and Demand. Organization for Economic Cooperation and Development, Paris.

Panigrahi, D.C, Gupta, R., Bhowmik, S.C 2005. Modelling of the ventilation system of a deep uranium mine. In: Eighth International Mine Ventilation Congress, Brisbane, Australia, 6–8 July 2005, pp. 439–445.

Pekkala, P., Hämäläinen, A., Gullen, K., Paatero, E. 2012. Uranium recovery as a by-product at Talvivaara mine. In: ALTA 2012 Uranium Conference, Perth, Western Australia, 31 May–1 June 2012, pp. 273–280.

Princep, D., Owers, M. 2012. How to sort the mineralisation from the background—processing of radiometric logging data. In: AusIMM International Uranium Conference, June 2012, Adelaide, South Australia. Australasian Institute of Mining and Metallurgy, Melbourne, Australia (Abstracts Vol. p59 and presentation).

Robertson, A., Grant, D., Liebezeit, V., Ehrig, K., Badenhorst, C., Durandt, G. 2013. Olympic Dam Mine, BHP Billiton, Australasian Mining and Metallurgical Operating Practices, third ed. AusIMM Monograph 28. The Australasian Institute of Mining and Metallurgy, Melbourne, Australia 1, pp. 793–799.

Robertson, S.W., Van Staden, P.J. 2009. The progression of metallurgical testwork during heap leach design. In: Hydrometallurgy Conference 2009. The Southern African Institute of Mining and Metallurgy, pp. 31–41.

Rössing Uranium, 2014. History and location of Rössing. <http://www.rossing.com/history.htm> (accessed February 2014).

Scheffel, R.E. 1981. Retreatment and stabilization of the Naturita Tailing Pile using heap leaching techniques. In: Interfacing Technologies in Solution Mining, Proceedings of the 2nd SME-SPE International Solution Mining Symposium, Denver, Colorado, 18–20 November 1981, Chapter 27, pp. 311–321.

Schmitke, B.W. 2004. Cigar Lake's jet boring mining Method. In: World Nuclear Association Annual Symposium, London, UK, 8–10 September 2004.

Schnell, H., 2014. Uranium processing practices, innovations, and trends. In: Anderson, C.G., Dunne, R.C., Uhrie, J.L. (Eds.), Mineral Processing and Extractive Metallurgy: 100 Years of Innovation. Society for Mining, Metallurgy & Exploration (SME), Englewood, CO, pp. 457–465.

Schnell, H.A., Corpus, E., 2000. Cluff lake mill operations: a historical review. CIM bull. 93 (1045), 81–88.

Scriven, D. 2014. Ablation—breakthrough technology to reduce uranium mining cost and increase resources. In: Abstracts, International Symposium on Uranium Raw Material for the Nuclear Fuel Cycle: Exploration, Mining, Production, Supply and Demand, Economics and Environmental Issues (URAM-2014), 23–27 June 2014. IAEA, Vienna, Austria, p. 171.

Seidel, D.C., 1980. Extracting uranium from its ores. IAEA Bull. 23 (2), 24–28.

Singh, R., Sengupta, P. L., Pathak, P. N., Chattopadhyay, A., Maulik, S. C. 2001. Floatation studies on processing of a low grade uranium ore. In: International Conference on Challenges in Coal and Mineral Beneficiation, Jamshedpur, India.

Shatalov, V.V., Nikonov, V.I., Tatarnikov, A.P., Zvonarev, V.N., Balakina, I.G., Asonova, N.I., 2001. Improvement of the technology of radiometric enrichment of uranium ores. At. Energy. 90 (3), 184–187 (translated from the Russian, Atomnaya Energiya, 90(3), 176–179, March 2001).

Sreenivas, T., Rajan, K.C., 2013. Studies on the separation of dissolved uranium from alkaline carbonate leach slurries by resin-in-pulp process. Sep. Purif. Technol. 112, 54–60.

Strahan, D., 2001. Uranium in glass, glazes and enamels: history, identification and handling. Stud. Conserv. 46 (3), 181–195.

Taylor, A. 2011. Review of uranium ore leaching systems. In: ALTA 2011 Uranium Conference, Perth, Western Australia, 26–27 May 2011.

Tietzel, A., Woodrow, G., Miller, R. 2013. Ranger Mine, Energy Resources of Australia Ltd, Australasian Mining and Metallurgical Operating Practices, third ed. AusIMM Monograph 28. The Australasian Institute of Mining and Metallurgy, Melbourne, Australia 2, pp. 1783–1798.

Toman, F., Ježova, V., 2007. Uranium ore mill at Dolni Rozinka: 40 years of operation. Uhli-Rudy-Geologicky Pruzkum. 14 (10), 12–14 (in Czech, English abstract).

Toro Energy Limited, 2015. Wiluna Uranium Project, Extension to the Wiluna Uranium Project (CMS 14025) Assessment No: 2002 Environmental Scoping Document, February 2015. <http://www.toroenergy.com.au/wp-content/uploads/2015/02/OEPA-Reviewed-ESD-December-2014-FINAL-to-EPA-Board-Feb15.pdf> (accessed March 2015).

Uggalla, S. 1999. Olympic Dam operation—an overview of underground design and planning. In: Tenth Australian Tunnelling Conference, Melbourne, Australia, 21–24 March 1999. Australasian Institute of Mining and Metallurgy, Melbourne, Australia, pp. 127–132.

UX Consulting, 2011. Uranium Suppliers Annual. Ux Consulting, Roswell, GA.

Van Tonder, D., 2012. Comparison of resin-in-pulp and direct precipitation for recovery of uranium in alkaline leach circuits. CIM J. 3 (2), 131–139.

Van Zyl, D., Caldwell, J. 2013. Proc. Heap Leach Solutions Conference, Vancouver, Canada, 22–25 September 2013 (46 papers). InfoMine Inc.

Webb, D. 1998. Olympic Dam operations review. In: Proc. Underground Operators' Conference, Townsville, Queensland, Australia, 30 June–3 July 1998. Australasian Institute of Mining and Metallurgy, Melbourne, Australia, pp. 87–94.

Westfield, J., Flinn, R.H., Look, A.D., Morgis, G.G. 1958. Engineering control of health and safety hazards in uranium mines. U.S. Dept. Interior, Bureau of Mines, Information Circular 7834.

Wines, D.R., Hulls, I., Woods, E., Creighton, A., 2013. The use of numerical modelling, slope monitoring and operational procedures to manage slope deformations at the Ranger 3 pit. In: Dight, P.M. (Ed.), Slope Stability 2013. Australian Centre for Geomechanics, Perth, Australia, pp. 611–624.

Wismut, 2014. The Lichtenberg open pit uranium mine. <http://www.wismut.de/www/web-root/en/ronneburg_lichtenberg-pit-mine.php> (accessed February 2014).

Wheatley, K.L. 2014. The successful application of modern exploration techniques to previously explored areas in the Athabasca Basin, Canada. In: Proc. International Symposium on Uranium Raw Material for the Nuclear Fuel Cycle: Exploration, Mining, Production, Supply and Demand, Economics and Environmental Issues (URAM-2014), 23–27 June 2014. IAEA, Vienna, Austria (Abstract published 2014).

Woods, P. 2015. Environmental and social aspects of feasibility studies, mining operations and closure; balancing realities and expectations from different angles. In: Proceedings ALTA 2015, Perth, Western Australia, 23–30 May 2015.

Woods, P., Edge, R., Fairclough, M., Fan, Z., Hanly, A., Miko Dit Angoula, I., et al. 2015. IAEA initiatives supporting good practice in uranium mining worldwide. In: Merkel, B., Arab, A., (Eds.), Uranium—Past and Future Challenges (Proc. Uranium Mining and Hydrogeology VII, Freiberg, Saxony, Germany, 21–25 September 2014), pp. 31–39.

Introduction to uranium in situ recovery technology

7

Mark S. Pelizza[a,1] *and Craig S. Bartels*[b,2]
[a]M.S. Pelizza & Associates LLC, Plano, TX, United States, [b]Heathgate Resources Pty Ltd, Adelaide, SA, Australia

7.1 General description

7.1.1 The ISR technique

In situ is a Latin word that translates literally to "on site" or "in position." Unlike conventional mining methods, where the uranium mineral and host rock are excavated together and the uranium is recovered on the surface, in situ recovery (ISR) technique removes the uranium while leaving the host rock in place. ISR utilizes wells to inject amended groundwater into the ore zone. This groundwater solution is commonly referred to as lixiviant. The lixiviant dissolves uranium as it is drawn through the uranium-bearing host rock by a pump in a nearby production well, which then sends the uranium-rich water to the processing plant where the uranium is recovered. The water is then refortified and sent back to the ore zone through the injection wells to recover more uranium. The cycle continues until the desired uranium extraction is complete.

ISR[3] involves the *entire* recovery process, in which a fortified groundwater leaching solution (lixiviant) is used to extract the uranium from the geologic formation in which it occurs and then the recovered uranium is concentrated and packaged into a form that

[1] Mr Pelizza's career in the uranium industry has spanned 36 years, with 34 years in various positions with Uranium Resources, Inc. where he most recently served as Sr. Vice President of Environment, Health, Safety and Public Affairs. Currently, Mr Pelizza is a consultant to the uranium industry through his firm M.S. Pelizza & Associates LLC. He serves as a Director with enCore Energy Corp. Mr Pelizza holds a BS degree from Fort Lewis College and a MS from Colorado Sc. of Mines. He is a registered Professional Geoscientist in Texas. He has served as President of the Uranium Producers of America, and Chairman of the Texas Mining and Reclamation Association.

[2] Mr Bartels has worked in ISR (uranium and copper) for 37 years with various companies, and has been involved in all aspects of ISR during that time (well field design and operation, resource estimation, process facilities design and operation, company management). He is currently President of Heathgate Resources Pty Ltd in Australia, which owns and operates the Beverley uranium ISR operation, and is operator of the Four Mile ISR project.

[3] In situ recovery (ISR) is commonly called in situ leaching (ISL). ISR is the commonly used term in the United States. In this chapter, the term ISR is appropriately used because the entire uranium recovery cycle is described from subsurface dissolution, through processing, drying, and packaging.

Uranium for Nuclear Power. DOI: http://dx.doi.org/10.1016/B978-0-08-100307-7.00007-7

can be sold for nuclear fuel. This is accomplished by injecting the lixiviant through injection wells completed in the zone of interest, leaching the target minerals, recovering the uranium-rich lixiviant by pumping production wells and then processing by ion exchange (IX), chemical treatment, drying, and packaging (Fig. 7.1, Photo 7.1).

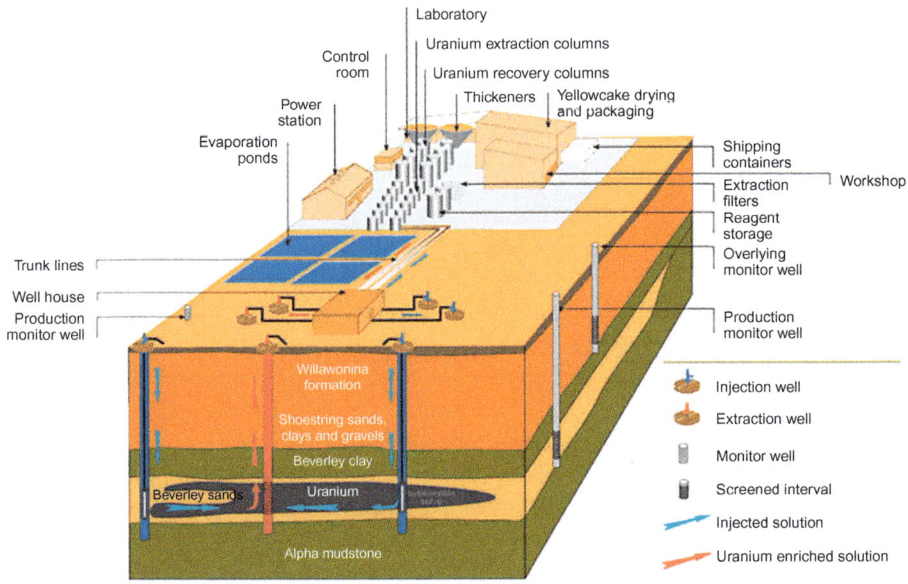

Figure 7.1 Conceptual block diagram of the ISR technique.
Source: World Nuclear Association/Heathgate Resources.

Photo 7.1 ISR well field at the Beverly ISR project, Australia.
Source: c.2004, Heathgate Resources Pty. Ltd.

Various well patterns are typically used for uranium ISR. Each well field area consists of groups of these patterns that are installed to correspond to the specific geometry of the orebodies.

To be amenable to ISR, the lixiviant must be able to contact the uranium mineral in the host rock, the uranium mineral must be a species that is soluble to the specific lixiviant, and there must be adequate groundwater and transmissivity to allow for hydrodynamic control. Generally, this means that uranium ISR is conducted in saturated permeable sandstone aquifer systems that are rich in redistributed uranium minerals. The natural groundwater in these aquifer systems is the fundamental component of the lixiviant and varies from to low total dissolved solid (TDS) "fresh" water to high TDS "saline" water that would be unsuitable for consumption (Hunter, 2001). However, it is important to note that even in freshwater aquifers, in the mineralized portion of the aquifer system that will be subjected to ISR, the groundwater is not potable because the concentrations of uranium and uranium progeny such as ^{226}Ra and ^{222}Rn exceed acceptable drinking water standards by a large margin (Pelizza, 2014).

Depending on regulatory requirements and natural geotechnical variables, an acidic or an alkaline lixiviant (leaching solution) may be used at a given location. Typical acidic lixiviants consist of native groundwater to which sulfuric acid and perhaps an oxidant is added. A typical alkaline lixiviant consists of native groundwater to which gaseous carbon dioxide (or some form of sodium bicarbonate) and oxygen (or other form of oxidant) are added. After injection of the barren lixiviant into injection wells, and recovery through production wells, it is piped to the IX facility, where the uranium is removed by circulating the uranium-rich lixiviant through IX resin. The barren lixiviant will then be returned to the well field. When loaded with uranium to capacity, the IX is washed of uranium and the uranium is further processed to its final form. IX resin is returned to service after it has been stripped of uranium.

The chief components of an ISR project include the leaching process and the plant in which the uranium is processed. In the leaching process, a lixiviant stream is continuously recirculated from the recovery plant into injection wells through the ore bearing strata. From this, a uranium-rich lixiviant is withdrawn (via production wells), pumped through the recovery plant [IX or, less often, solvent extraction (SX)],[4] refortified with reagents, and circulated to the injection wells. In the processing plant, the uranium is removed (eluted) from the IX resin, or SX circuit, and further treated (precipitated, dried, packaged). The IX process may be integral to the entire recovery process or remote at satellite plants and the resin transported to the remaining recovery steps. Acid consumption depends greatly on minerals other than uranium in the orebody.

7.1.2 Advantages of ISR

Uranium ISR is a proven technology that has been successfully demonstrated commercially worldwide. ISR of uranium has environmental advantages over conventional open pit uranium mining, as evidenced by the following.

[4] Ion exchange (IX), rather than solvent extraction (SX), is more generally considered here since it is, by far, the most prevalent method in ISR for removing uranium from lixiviant.

- ISR results in significantly less surface disturbance. Mine pits, waste dumps, haul roads, and tailings ponds are not needed.
- Compared to conventional mining, ISR reduces the short- and long-term exposure of the general population to the extremely low levels of radioactivity because almost all of the source term (notably the radioactive decay products of ^{238}U) remains underground in its natural location. Very little residual radioactive waste is produced and there are no tailings. Land and water are returned to their original pre-ISR uses and quality.
- ISR requires much less water than pit or underground mine dewatering, or conventional milling.
- Minimal use of heavy equipment, combined with the lack of haul roads, waste dumps, etc., results in virtually no air quality degradation at ISR sites.
- Following the initial construction activities, fewer employees are needed at ISR sites, thereby reducing transportation and socioeconomic concerns.
- Aquifers are not excavated, but remain intact during and after ISR so after any required rehabilitation they remain available for future uses. Avoiding the creation of large excavations preserves the surrounding land for grazing, raising crops, and other traditional uses.
- The technology of recirculating groundwater through the IX facility reduces the amount of solids to a negligible quantity, and tailings ponds are not used, thereby eliminating a major groundwater pollution concern.

7.1.3 Overview of global ISR operation

Uranium produced by ISR accounts for 46% of the uranium produced worldwide (World Nuclear Association, 2015a). Major historic and current production centers and operations are found in Australia, Asia, Eastern Europe, and the United States, as summarized next.

7.1.3.1 Australia

The Beverley in South Australia (520 km north of Adelaide, on the plains northwest of Lake Frome) was Australia's first ISR project, starting operation late in 2000. The project uses sulfuric acid leach chemistry. It was licensed to produce 1000 tU/year (2.24 mm lb U) and reached this level in 2004, though production has declined since. Mining of Beverley ceased at the end of 2013 (World Nuclear Association, 2015b).

The Four Mile leases are contiguous with Beverley, and ISR of the east orebody commenced in April 2014. Uranium recovery is through the neighboring Beverley Pannikan satellite IX plant with the loaded resin transported by truck and trailer to the main Beverley plant for stripping (elution) and precipitation, as is done at many US projects. Production is currently at about 1000 tU/year (2.24 mm lb U) (Fig. 7.2, Photo 7.2).

The Honeymoon ISR project in South Australia commenced operation in 2011, and, because of the very high sulfate levels in the native groundwater, has been one of the few ISR projects to use SX, rather than IX, to remove uranium from well field lixiviant. In November 2013, the project was closed and put on care-and-maintenance until uranium prices improved.

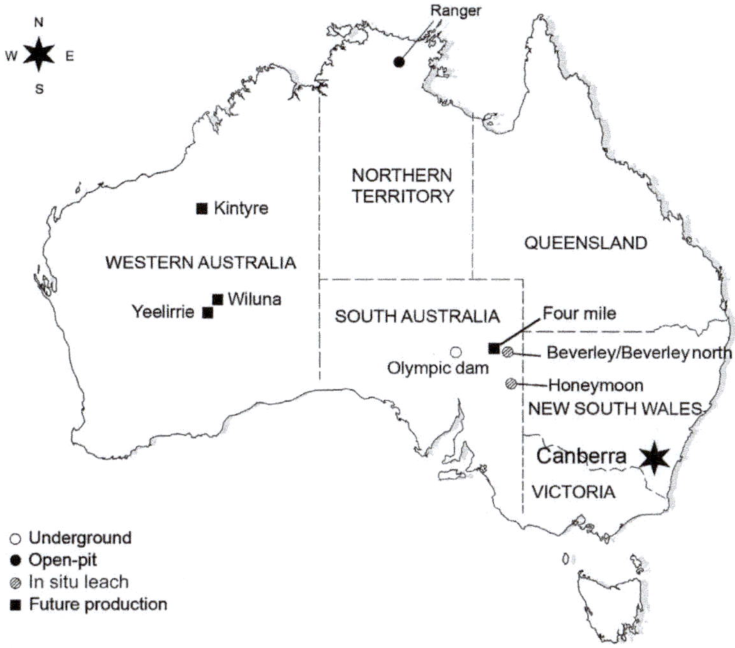

Figure 7.2 Australia uranium mines.
Source: OECD (2014).

Photo 7.2 The Beverly ISR process facility (2001), Australia.
Source: Heathgate Resources Pty. Ltd.

7.1.3.2 Asia

China. Prior to the 1990s, China's uranium resource development activities were mainly carried out on hydrothermal-related granite-type and volcanic-type uranium deposits in the Jiangxi, Hunan, and Guangdong provinces and the Guangxi Autonomous Region of Southern China. Facing the challenge of meeting demand for economic uranium resources for China's mid-term and long-term nuclear energy development plan, the direction was changed from "hard rock"-type deposits to ISR amenable deposits in Northern and Northwest China (OECD, 2014). ISR began in 1993 at the Yining production center Kujieertai deposit using sulfuric acid chemistry. Reported production in 2012 was 380 tU (.84 mm lb U) (OECD, 2014). The Shihongtan deposit, located in Tuha basin of Xinjiang, is a new uranium discovery. Preliminary ISR field testing began in 2000, including lab tests and field tests. During field testing with sulfuric acid, numerous problems were encountered because of the high content of CO_2 in the ore. Therefore, a NH_4HCO_3 ISR test was to be conducted (Xuebin et al., 2004) (Fig. 7.3).

Kazakhstan. Uranium exploration began in the 1940s in Kazakhstan, making it one of the first countries to develop uranium resources (Kim, 1997). In 2009, Kazakhstan became the world's leading source of mined uranium, producing almost 28% then, 33% in 2010, 36% in 2011, 36.5% in 2012, and 38% in 2013, almost all using ISR methods in recent decades. ISR uranium production in Kazakhstan

Figure 7.3 China uranium mines.
Source: OECD (2014).

Table 7.1 Kazakh uranium production by mines (tU) (World Nuclear Association, 2015a)

Province/ Group	Mine	2010	2012	2013
Chu-Sarysu, Eastern	Tortkuduk (Katco)	2439.3	2661	3558 both
	Moinkum (Northern, Katco)	889.1	1000	
	Southern Moinkum (Taukent/GRK)	442.5		1129 both
	Kanzhugan (Taukent/GRK)	561.9		
	Uvanas (Stepnoye-RU/GRK)	300.3		1192 both
	Eastern Mynkuduk (Stepnoye-RU/GRK)	1029.2		
	Central Mynkuduk (Ken Dala.kz)	1242.4		1800
Chu-Sarysu, Northern	Western Mynkuduk (Appak)	442.2		998
	Inkai-1, 2, 3 (Inkai)	1636.7	1701	2047
	Inkai-4 (South Inkai)	1701.4	1870	2030
	Akdala (Betpak Dala)	1027.1	1095	1020
	Budyonovskoye 1, 3 (Akbastau)	739.6	1203	1499
	Budyonovskoye 2 (Karatau)	1708.4	2135	2115
	North and South Karamurun (GRK)	1016.7		1000
Syrdarya, Western	Irkol (Semizbai-U)	750		750
	Kharasan 1 (Kyzylkum)	260.1	583	752
	Kharasan 2 (Baiken-U)	262.2	603	888
Syrdarya, Southern	Zarechnoye (Zarechnoye)	778.2	942	931
Northern, Akmola region	Semizbay (Semizbai-U)	224		411
	RU-1 (Vostok, Zvezdnoye)	352.1		331
Total		17,803.4	21,317	22,451

requires large quantities of sulfuric acid due to relatively high levels of carbonate in the orebodies (World Nuclear Association, 2015c) (Table 7.1).

Mongolia. From 1945–60, numerous uranium occurrences were discovered in Eastern Mongolia. Currently no uranium is being produced in Mongolia; however, a number of projects are in the planning stage of development, with start dates at the end of this decade. The following deposits are planned for development using sulfuric acid chemistry: the Kharaat, Khairkhan, Gurvansaikhan, and Ulziit deposits at the Gurvansaikhan Production Center; and the Dulaan Uul and Zoovch Ovoo deposits at the Coge-Gobi Production Center (OECD, 2014) (Fig. 7.4, Photo 7.3).

Russia. According Fazlullin et al. (2004), uranium ISR is performed in Russia on two deposits, the Dalmatovskoe/Khohlovskoe and Khiagdinskoye. The Dalmatovskoe/Klolovskoe deposit is located in the Trans-Ural region. The annual uranium production in 2012 was 578 tU (1.28 mm lbs) (World Nuclear Association, 2015d). The Khiagdinskoye deposit, the only ISR deposit in permafrost, is located within the Republic of Buryatia (World Nuclear Association, 2015e). Multiwell experiments began there in 1999 and in 2002 50 tU (110 m lbs) were produced at about

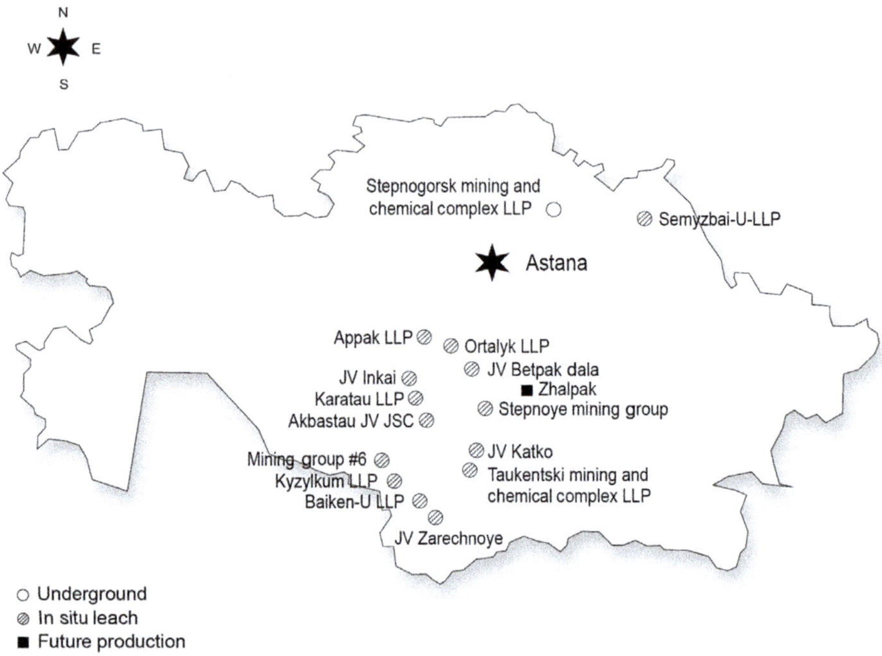

Figure 7.4 Kazakhstan uranium mines.
Source: OECD (2014).

Photo 7.3 Kharasan ISR processing facility, Kazakhstan.
Source: Uranium1.com.

50% recovery and average uranium concentrations in solutions of 118 mg/L (Boitsov and Razumov, 2004). Production in 2012 was 332 tU (.73 mm lbs). Sulfuric acid leach chemistry is used in both deposits. Uranium and associated mineralization occurs along redox interfaces (Boitsov and Razumov, 2004).

Uzbekistan. According to Grutsynov (2002), in the Republic of Uzbekistan, uranium is currently produced from deposits along redox boundaries located in the Central Kyzylkum Desert. ISR was first introduced in 1962, and since 1994, it has become the sole method for recovery of uranium in the Republic (Fig. 7.5). In the northern district, since 1965 at the Uchkuduk and 1995 at the Kendyk-Tyube, annual ISR production has been 700–750 tU (1.54–1.65 mm lbs). In the Zarafshan or eastern mining district, the Sugraly deposit was mined by ISR from 1977 until 1994, when it was closed. In the Bukhara province, the Bukinay group of uranium deposits being developed includes North and South Bukinay from 1970, the Beshkak from 1978, the Istiklol, Kukhnur, and Lyavlyakan from 1998, and the Tokhumbet and South Sugraly. Annual production from the eastern district is 1000–1200 tU (2.2–2.65 mm lbs) and rising to 2100 tU (4.63 mm lbs) from 2004. ISR began in the southern mining district in the Samarkand province, and specifically at the Sabirsay deposit in 1983. Other projects in the district are the Ketmenchi, including the Jaarkuduk, Yogdu, Shark, and Ulus. Annual production in the southern district is 600–650 tU (1.32–1.43 mm lbs) (World Nuclear

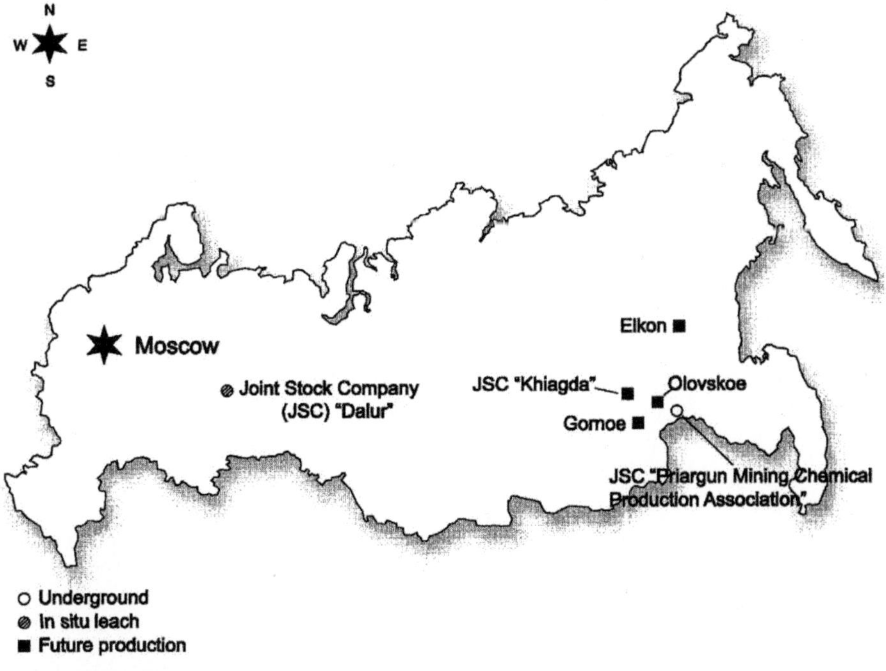

Figure 7.5 Russian uranium mines.
Source: OECD (2014).

Photo 7.4 Historic photo of the Zafarabad ISR project, Uzbekistan.
Source: unknown.

Association, 2015f). Pilot plants at Northern Kanimekh and Alendy were commissioned in 2008–09 and the two commercial ISR projects completed there at the end of 2013 are expected to achieve full capacity in 2015. Also, pilot plants for ISR were developed at the Yarkuduk and Tokhumbet deposits. In 2013, operations ramped up at the Aulbek, Kyzylkum, Meilysai, and Tutlinskaya. However, in August 2013, construction of Meilysai and Kyzylkum were terminated due to high carbonate content in the ore, rendering sulfuric acid ISR inefficient (World Nuclear Association, 2015f) (Photo 7.4).

7.1.3.3 Eastern Europe

In Bulgaria, in situ leaching (ISL) using sulfuric acid chemistry was used in many locations. Nedyalkov (1996) reported the first ISR activity at the Sclishte in 1961. From 1981, ISL of broken hard rock was also used to increase the yield from mined-out conventional underground mines. In 1990, 70% of the uranium produced was from ISL of ore deposits with very low grades of 0.02–0.07% of uranium (WISE Uranium Project). The Plovdiv ISR facility provided resin loaded with uranium solution to a mill at Elesnica Production Center for final processing. ISR was practiced at 21 sites by 4 companies. Production was reported at approximately 345 tU/year (0.761 mm lb) (Norman, 1993). In 1992, uranium development in Bulgaria was closed by decree. (Nedyalkov, 1996). Norman (1993) reported 14,000 wells in 15 well fields and 4 satellite recovery units.

Czech Republic. ISL using sulfuric acid chemistry began in Stráz pod Ralskem in North Bohemia in 1967 (Ekert and Muzak, 2010). Tomas (1997) reported the lesser use of nitric acid and hydrofluoric acid and ammonia in the leach solutions as well. The ore deposit is located in Cretaceous sandstones at depths of up to 488 m (1600 ft) with grades of about 0.12% uranium. The Stráz ISR project is adjacent to the Hamr underground uranium mine. Flowrate at the Stráz project was about 37,854 L/min (10,000 gal) at a production rate of about 575 tU/year (1.27 mm lb U) (Norman, 1993) (Fig. 7.6, Photo 7.5).

Figure 7.6 Czechoslovakia uranium mines.
Source: OECD (2014).

Photo 7.5 Injection wells at the Stráz ISR project, Czech Republic.
Source: Vlady Beneš.

Germany. Germany has been the most secretive about uranium production in Eastern Europe (Norman, 1993). In Eastern Germany, an underground mine converted to an ISL facility was in operation at Königstein near Dresden until the end of 1990, when all uranium production ceased in Germany. It produced a total of 18,000 tU (39.69 mm lbs), 30% of which were from ISR with sulfuric acid (http://www.wise-uranium.org).

Ukraine. In 1961, Ukraine began testing ISR. From 1966 to 1983, uranium was produced by ISR in the Devladovo of Sofiivela District, Drivipropetrovska Provence, and Bratske of Mikolaivska Provence using sulfuric acid chemistry. Futurc plans call for changing to alkaline chemistry (Sukhovarov-Jornoviy, 2005; Rudy, 1997; OECD, 2014).

7.1.3.4 United States

In the United States, several pilot tests using acid leach systems were conducted but there have been no commercial ISR operations in the United States using an acid leach system primarily because restoration of the orebody aquifer following ISR is the major problem for acid leach systems. In the United States, regulators require that the orebody aquifer quality be restored to premining quality (Underhill, 1992). Similarly, restoration of ammonia carbonate leach systems have been tested but the restoration on ammonia has been a difficult problem (Humenick et al., 1979). As a result, commercial operations in the United States have narrowed to a sodium carbonate–type alkaline leach chemistry. Numerous ISR pilot and/or commercial operations have been conducted in the states of Colorado, New Mexico, Nebraska, Texas, and Wyoming (Fig. 7.7).

According to the OECD (2014), the following US projects were in production or committed to production in 2012 (Fig. 7.8, Photo 7.6).

Figure 7.7 Ukraine uranium mines.
Source: OECD (2014).

Figure 7.8 United States uranium mines.
Source: OECD (2014).

Photo 7.6 The Crow Butte ISR project in Nebraska, USA.
Source: Cameco.com.

State	Project	Nominal tU/year
Nebraska	Crow Butte	385
Wyoming	Smith Ranch/Highland	2116
	Willow Creek	500
	Lost Creek	769
	Hank/Nichols	769
	Moore Ranch	192
Texas	Hobson/Palangana	385
	Alta Mesa	385

7.2 Geology and hydrogeologic conditions

7.2.1 Host rock amenability

Generally, deposits that are amenable to uranium ISR are associated with permeable sandstones. The uranium mineral is typically epigenetically redistributed and deposited on the surface of the sand grains through reduction-oxidation (REDOX) processes. Interstial pores in the sandstone provide the permeable conduit through which groundwater and hence the lixiviant can be circulated. The sandstone must be saturated or below the water table to allow for controlled circulation of the contained groundwater, which is fortified with leaching reagents during the ISR activity. Finally, it is preferred that the sandstone deposit be confined by strata of lesser permeability to facilitate containment of leaching solutions (Photos 7.7 and 7.8).

The uranium mineral must be amenable to dissolution in place using commercially available technology. In almost all cases, the formation of sandstone-hosted uranium deposits occurs when uranium is redistributed by transported in oxygen-rich groundwater until it interacts with a reduced host rock. During this interaction, the soluble hexavalent uranium (U^{6+}) ion is converted to the insoluble tetravalent (U^{4+}) ion that, in turn, forms uranium mineral species. The resulting mineralization

Photo 7.7 The Smith Ranch/Highland project in Wyoming, USA.
Source: Cameco.com.

Photo 7.8 The Nichols Ranch project in Wyoming, USA.
Source: Uranerz.com.

is fine-grained (often less than 20 μm) and comprises reduced uranium species (World Nuclear Association, 2015g). The ISR process uses the same groundwater to which an oxidant is added to reverse the redox process; that is, the insoluble tetravalent (U^{4+}) ion is converted to the soluble hexavalent uranium (U^{6+}) ion, which is flushed out.

7.2.1.1 Deposit types

According to the IAEA (2001), based on the genesis and ore formation, ISR amenable uranium deposits belong to two groups: roll-type (Fig. 7.9, Photo 7.9) and tabular (Fig. 7.10), with the roll front as the most common. Both are redistributed and occur along a redox interface. They differ in the orebody morphology and will lend themselves to different ISR well field patterns and perhaps approaches.

7.2.1.2 Uranium mineralogy

The most common uranium minerals in roll front deposits are the oxides, uraninite (UO_2), pitchblende, and coffinite [$U(SiO_4)(OH)_4$]. (NRC, 2009; IAEA, 2001). Uranium usually forms finely dispersed particles ranging in size from tens to one micron to visually discernable impregnated nest-like accumulations. Coffinite occurs in various forms. Uraninite, pitchblende, and coffinite all contribute to successful ISR performance.

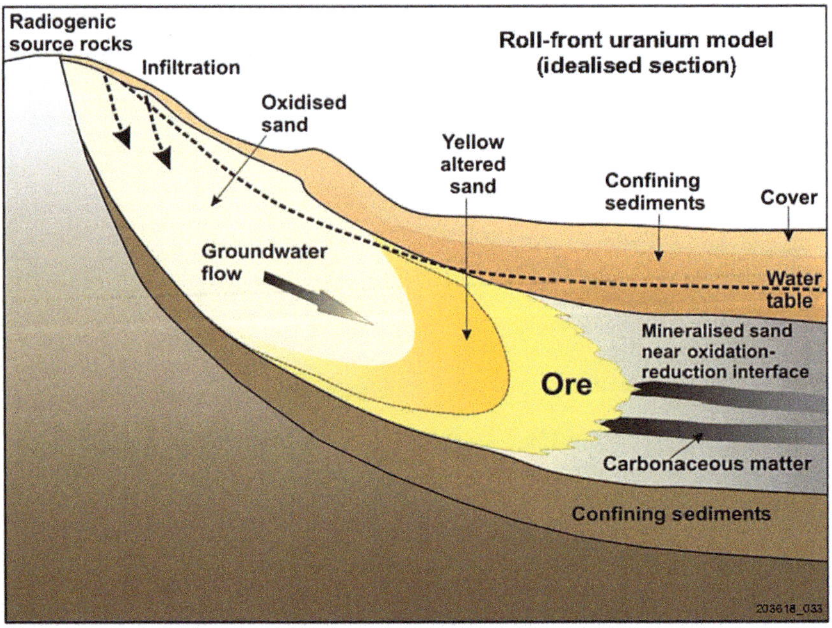

Figure 7.9 General roll front deposit diagram.
Source: DeVoto 1978 modified by Hou et al. (2007).

Photo 7.9 Side wall pit cut illustrating a classical roll front in cross section.
Source: Wyoming Mining Association.

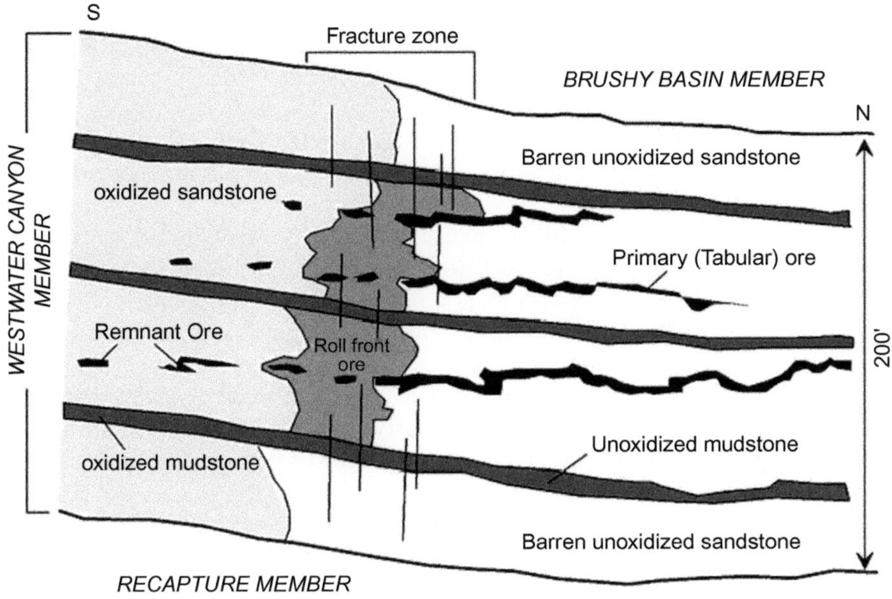

Figure 7.10 Tabular deposit schematic.
Source: NRC (2009).

7.2.2 Hydrogeologic conditions

7.2.2.1 Confinement

Hydrogeologic geometry must prevent uranium-bearing fluids lixiviant from vertically migrating. Typically, low permeability layers such as shale or clay confine the uranium-bearing sandstone both above and below. This isolates the uranium producing horizon from overlying and underlying aquifers (NRC, 2009). Confinement is routinely verified during mine unit development, as described in Section 7.3.3.

7.2.2.2 Groundwater conditions

Groundwater is fundamental to ISR development because it forms the lixiviant and the quality of the water that will affect lixiviant composition and other factors within the process. Section 7.3.1 guides the reader through a general groundwater-sampling scheme for ISR.

The general salinity of groundwater is an important measure of quality. Where the salinity in an aquifer is low (as expressed by TDS), water in the contiguous regional aquifer may be used for domestic purposes and groundwater restoration becomes a priority. This is generally the case in the United States, where water quality contains relatively low TDS and groundwater restoration is required. In Kazakhstan and Australia, TDS values are high and the groundwater in aquifers is not considered suitable for domestic purposes. As such, in those jurisdictions, groundwater restoration is not a priority. The wide variation in TDS in different locales is reported by Hunter (2001): at the Smith Ranch in Wyoming at 500 mg/L, at the Beverly in Australia as 3000–13,000 mg/L, and the Honeymoon in Australia as 17,000–20,000 mg/L.

General water quality is also an important parameter when evaluating the process design. Salinity values will affect IX efficiency and other components of the water treatment process.

In addition to the general salinity as a measure of the quality of water in and around uranium deposits, uranium and decay products[5] such as radium (^{226}Ra) and radon (^{222}Rn), which are uranium's natural decay products, are found in water near uranium deposits and significantly affect the toxicity and hence the suitability of water for domestic purposes. When the uranium mineralization is in sufficient concentrations in the aquifer rock, uranium and its decay products and the resulting radioactivity are generally found in anomalously high concentrations in the groundwater in the vicinity of the uranium. In the United States, extensive baseline sampling has been required at ISR sites by regulation since the 1970s. In the United States, it has been documented that the parameters that exceed federal drinking water limits are uranium, gross alpha radiation, ^{222}Rn, and ^{226}Ra. Pelizza (2014) reported documented testing results of 155 ISR mine units where over 80% exceeded EPA primary drinking standards for uranium, and essentially all mine units for ^{226}Ra. Pelizza also found that gross alpha radiation and ^{222}Rn exceeded primary drinking standards for each mine unit tested. The same data showed that radon may exceed 37,000 Bq/L (1,000,000 pCi/L) in groundwater in a mine unit.

7.2.3 Exploration and delineation methods

Initial green field uranium exploration is generally noninvasive. Methods may include aerial radiometric surveys, identification of regional mineral geochemical trends through existing databases such as oil and gas records, and water well driller reports. Also, groundwater analysis from existing wells for concentrations of uranium (Fig. 7.11), radium, or radon gas indicate the presence of natural uranium mineralization.

Regional exploration programs typically consist of widely spaced reconnaissance drilling in prospective areas where record review indicated uranium mineralization. Drilling is conducted by portable, rubber tire–mounted, water well drill rigs (Photo 7.10). The primary goal of regional drilling is to identify geochemistry relevant to redox conditions by observing cutting lithology. Specifically, areas of up-gradient oxidation are generally rust colored due to limonite staining and down-gradient reduction is grey with the presence of pyrite and perhaps other organic material (Photo 7.11). Intercepts of uranium mineralization are an obvious plus. During the regional drilling program, other important geologic features such as

[5] Uranium is a naturally occurring radioactive element. As such, radioactive decay of uranium into other elements is continuous. Some decay products make the water radioactive. The elements ^{226}Ra and ^{222}Rn decay products and gross alpha radiations all are constituents that impart toxicity to drinking water and are regulated in public drinking supplies. Their presence means that the element uranium must be close by in the host rock, so they are commonly used by geologists as a uranium exploration screening tool.

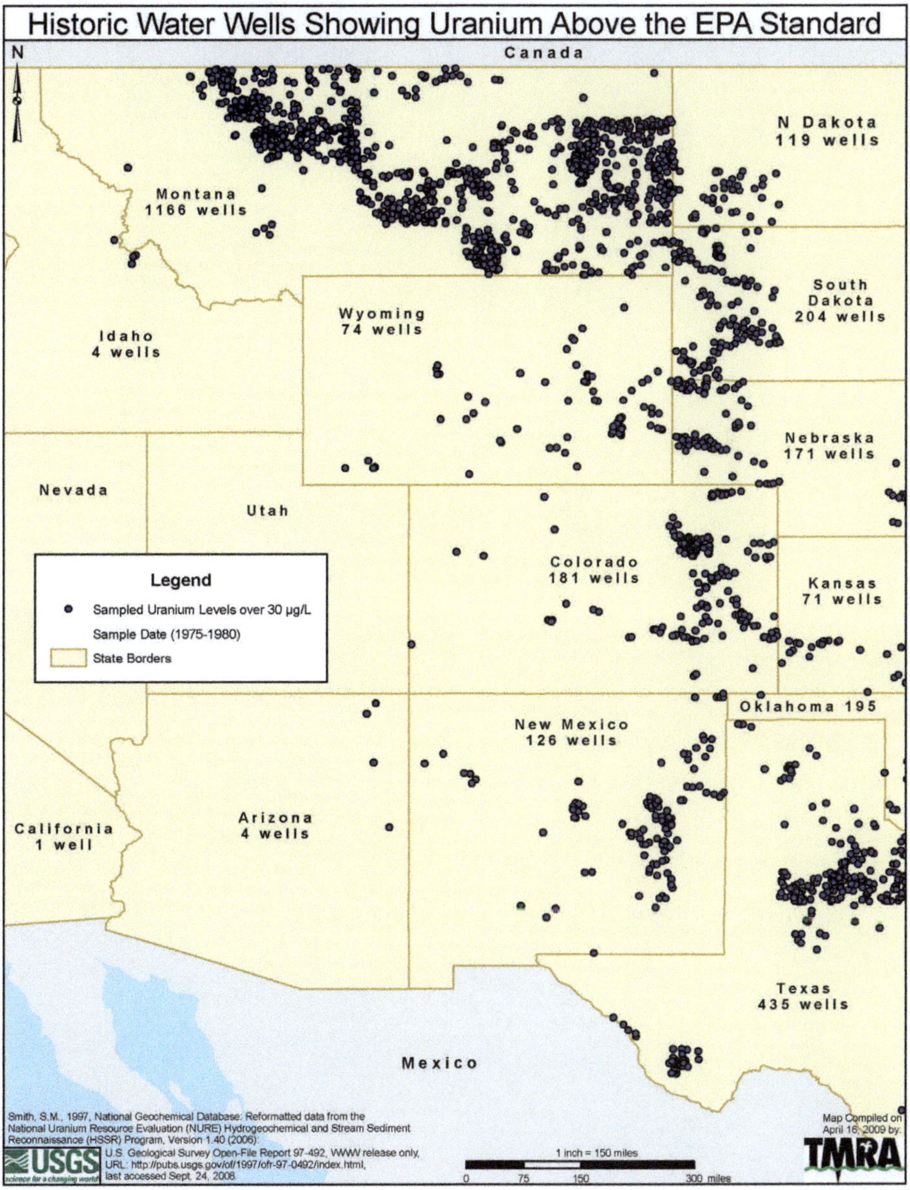

Figure 7.11 Map of the western United States that shows water wells that were screened for uranium concentrations that exceed 30 mg/L. Exploration geologists use anomalies such as these as evidence of the presence of uranium mineralization in the aquifer.
Source: Data from USGS, map courtesy of Texas Mining and Reclamation Association.

Photo 7.10 Rubber tire rotary drill rig and water truck.
Source: M.S. Pelizza.

Photo 7.11 Drill cuttings in 1.6 m (5 ft) intervals illustrating redox coloration (courtesy Uranium Energy Corp). Oxidized and reduced sand coloration example.
Source: Campbell et al. (2007).

sand quality, stratigraphic features such as channel morphology, and structural features such as faults will also be observed.

If the regional drilling program provides favorable results, a targeted drilling program may be conducted. Drill spacing density is decreased from hundreds of meters to tens of meters (thousands of feet to hundreds of feet) based on the success of the regional drilling. Additional dense drilling, that is, at less than 60 m (200 ft) spacing, is conducted only in the areas where high uranium grades are encountered. The purpose is to identify roll front morphology (Fig. 7.12) by adequate drill spacing that will lead to detailed mapping of the deposit by grade × thickness (GT) contour or equivalent (Photo 7.12).

7.2.3.1 Preliminary determination of amenability

If a uranium deposit is found, it must be evaluated by geological and engineering professionals for ore quality, ISR amenability, and project economics. Factors include quality of mineral (ie, grade, thickness, width, and continuity), depth, geology of the surrounding rock, and hydrology. Either historical project experience or detailed delineation drilling is required to properly conduct this evaluation. Laboratory analysis of core material and bench testing of core material for ISR amenability are commonly conducted.

7.2.3.2 Logging and subsurface evaluation

Spontaneous potential, resistivity, and gamma ray logs are generally obtained for all exploration holes. Gamma logs are used as an indirect measure of uranium mineralization. Where gamma logs indicate the presence of uranium mineralization, prompt fission neutron (PFN) logging may be conducted to measure actual uranium grades (Photo 7.13).

Uranium is a radioactive material. Natural uranium isotopes ^{235}U and ^{238}U have half-lives of 703.8 million years and 4.4 billion years, respectively. In nature, local uranium redistribution absent the radioactive daughter products occurs rapidly relative to radioactive decay, leading to uranium deposits that are not in secular equilibrium with decay products (Rosholt, 1959). Thus, in some instances, gamma log measurements alone will not be a reliable measurement of in-place uranium, and direct uranium measurements by coring or specialized logging tools, such as PFN, are required for proper resource evaluation.

Givens and Stromswold (1989) reported of a method using PFN as a direct logging technique for uranium. It utilizes pulsed source neutrons from a neutron accelerator (Photo 7.14) in a logging tool, which leads to neutron-induced fission of ^{235}U in the formation. Epithermal neutrons and thermal neutrons returning from the formation following fission are counted separately in detectors in the logging tool. The ratio of epithermal to thermal neutron counts provides a measure of uranium content. Logs obtained with the technique have shown good agreement with core measurements, and they have provided a major data source for delineation and exploitation of uranium mineralization. PFN is widely used in uranium exploration and delineation.

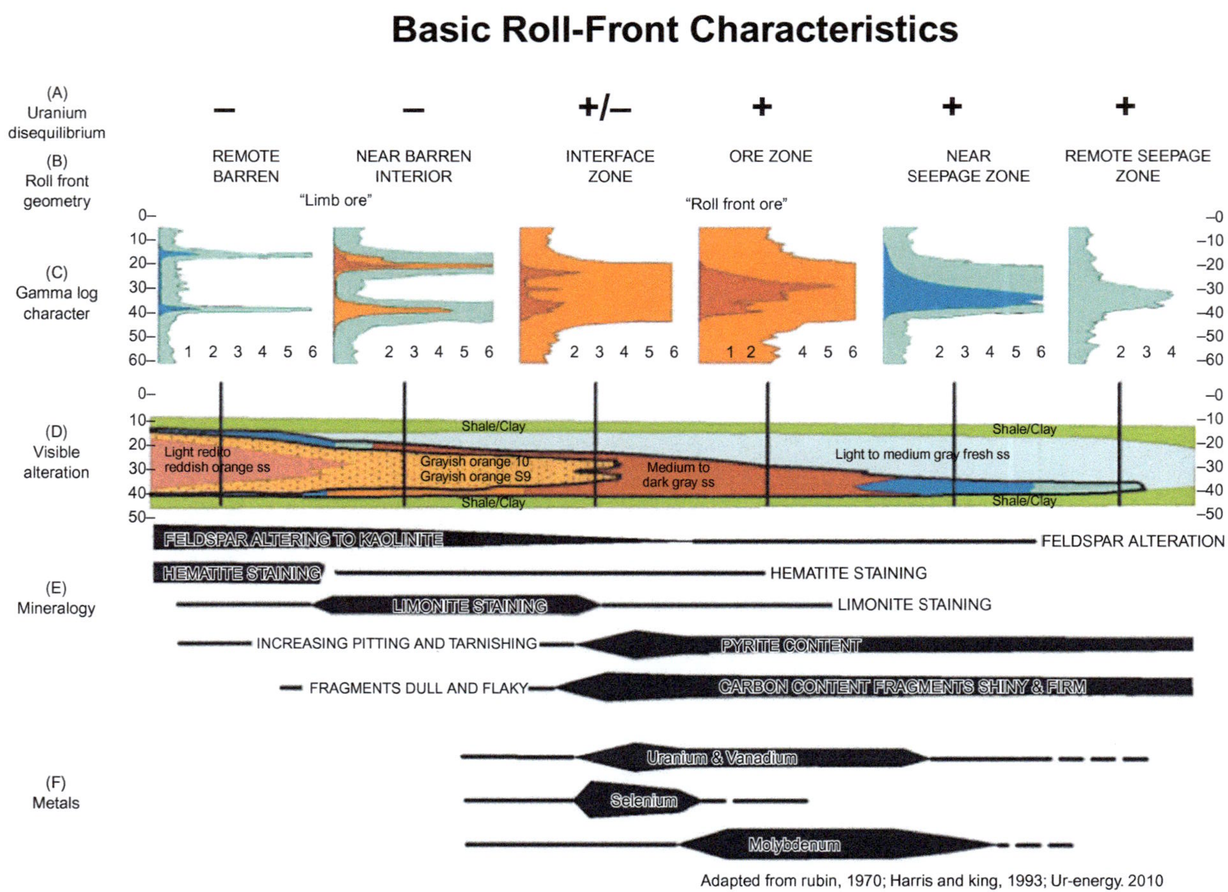

Figure 7.12 Cross section of logs illustrating roll front development.
Source: Boberg (2012).

Photo 7.12 Exploration drilling at the Palangana ISR project where drill "fences" are apparent from the brush clearing patterns.
Source: Uranium Energy Corp.

Photo 7.13 Logging truck in the field.
Source: Uranium Energy Corp.

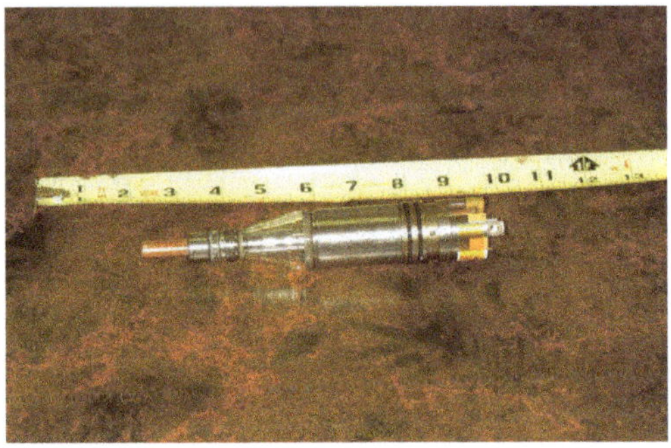

Photo 7.14 PFN generating tube.
Source: M. Pelizza.

More recently, Märten et al. (2014) described an advanced logging tool for use in uranium exploration and production, combining an improved PFN with gamma spectroscopy, which produces curves of uranium percent, disequilibrium factor, density, permeability, porosity, and abundance of main reactive minerals including sulfidic (eg, pyrite), calcareous (eg, calcite), dominating clay mineral (eg, kaolinite), and lignite (amongst others). They go on to describe the use of this tool with high-resolution shallow seismic and geochemical models at the Beverley deposit in South Australia.

7.2.4 Mine area site conditions

Site geology is continuously closely evaluated in conjunction with well field development to assure proper placement of monitor and production wells. The project geologists and hydrologists work together to perform hydrological tests and compile the geologic/hydrologic data into reports to include cross-sections, GT contour maps, isopach maps of the overlying and underlying confining units, well completion reports, models, and other information regarding the morphology of the uranium deposit.

7.3 Environmental evaluations

7.3.1 Baseline water quality

Water is fundamental to ISR development and an important natural resource. The collection of baseline water quality data, and determination of baseline water quality conditions is important as project geochemistry and leaching/processing data, the upper control limits for operational monitoring programs, and groundwater restoration objectives all require a good understanding of native water quality.

Depending on site hydrogeologic variables, baseline water quality may be determined from samples collected from wells installed in the various aquifer locations including production zone baseline wells, production area monitor wells, and overlying

Table 7.2 Common water quality parameters

Alkalinity	Lead
Ammonium	Magnesium
Arsenic	Manganese
Bicarbonate	Mercury
Boron	Molybdenum
Cadmium	Nickel
Calcium	Nitrate
Carbonate	pH (s.u.)
Chloride	Potassium
Chromium	Radium-226 (pCi/L)
Copper	Radon-222 (pCi/L)
Conductivity (μmho/cm)	Selenium
Fluoride	Sulfate
Gross alpha radiation (pCi/L)	Uranium
Iron	Vanadium

and underlying monitor wells. Sampling, preservation, and analytical quality control methods are according to standard methods, such as referenced in USEPA 1983.

Baseline water quality is generally determined from the data from baseline water well sample analysis collected for such parameters as those shown in Table 7.2.

A common practice is to use the arithmetic average of multiple samples from multiple wells to determine the well field baseline. However, uranium geologists generally accept that grades are log normally distributed in redistributed uranium deposits. If so, an infinite amount of samples covering every portion of the ore would yield a log normal distribution of uranium and uranium-related progeny in water samples. But in the real world, the industry is forced to deal with a limited sample well configuration. The distribution of each parameter may be validated by using a statistical software package such as ProUCL 4.00.04 (Singh et al., 2009). ProUCL first determines if the data for a given constituent follow a normal, gamma, or logarithmic distribution. If the data follow one of these statistical distributions, a relevant upper tolerance limit (UTL) may be calculated. If the data follow no discernible distribution, a nonparametric UTL may be calculated.

7.3.2 Operational groundwater monitoring

During production operations, a groundwater quality monitoring program is utilized to ensure that production fluids are contained within the defined production zone. If production fluids exit the production zone, increases in concentration of the upper control limit parameters at the affected monitoring wells will occur. Also, if production fluids exit the production zone into barren sandstone, the production process is less efficient.

Moreover, during production operations, quality in the lixiviant stream should be monitored to assure that uranium feed grades and other lixiviant chemical parameters stay within the prescribed limits for the operation.

7.3.3 Hydrological characterization

Prior to well field development, it is necessary to collect and assemble detailed information on geologic and hydrologic conditions so that ore zones can be defined, geologic and hydrologic parameters quantified, well fields planned, hydrologic monitoring programs developed, and baseline groundwater quality sufficiently determined. To accomplish this, an intensive pump test program is typically conducted.

The primary goal of pump testing in new mine units for ISR is to determine the degree of communication between the production zone and the overlying zones, and the production zone monitor wells. This will reflect the effects of hydraulic pathways, such as unplugged holes, and other pathways to the overlying zones, as well as ascertain the ability of production zone monitor wells to respond to changing flow conditions within the production area. The degree of communication at the production zone monitor wells surrounding the well fields will also directly indicate the magnitude of horizontal formation anisotropy. In addition, physical flow parameters (transmissivity, storage, permeability) are determined for the producing horizon for use in well field design, and lixiviant control during operations.

7.3.4 Environmental monitoring

Proper environmental effluent monitoring requires both a preoperational phase to document natural conditions absent any ISR development and then an operational phase program to document the impacts, if any, of ISR.

Shown in Table 7.3, a preoperational program typically includes at least 12 consecutive months of data, including complete sampling of water wells, soil, vegetation, direct radiation, and radon and radon flux data prior to any major site construction. During construction and operation these same media and parameters continue to be sampled (NRC, 1980).

7.4 Facilities

7.4.1 Well fields

An ISR mine consists of well fields that are progressively established as mine units over the orebody as uranium is depleted from sections of the orebody. A well field consists of injection wells, which are used to inject the barren lixiviant into the orebody, and production wells, which pump the uranium-rich solution to the surface. The well patterns are developed for a specific site, and installation for a given well field is based on the subsurface geometry of the ore deposit (Photo 7.15). Various pattern shapes are used, although five-spot and seven-spot patterns are common (NRC, 2009). A typical well arrangement using five- and seven-spot patterns is shown in Fig. 7.13. Because uranium deposits have irregular shapes, some of the well patterns in a given well field are also irregular, and the well patterns fit the size, shape, and boundaries of individual orebodies. Depending on orebody

Table 7.3 **Environmental effluent monitoring in the United States**

Type of sample	Number	Location	Method	Sample frequency	Analysis frequency	Type of analysis
Air radon gas	Five	One upwind and two downwind of the plant or satellite site, one at the nearest residence or occupied structure within 10 km of the plant or satellite site, one control	Track Etch	Continuous	Quarterly	^{222}Rn
Process fluids	Two	Lixiviant trunk lines in and out of process, one from lixiviant intake and one from lixiviant outlet	Grab	Quarterly	Quarterly	^{222}Rn
Groundwater	One from each well	Potable, livestock, and irrigation water supply wells within 2 km license area	Grab	Quarterly	Quarterly	Dissolved and suspended U- Nat., ^{226}Ra, ^{230}Th, ^{210}Pb, ^{210}Po, gross α, and β
Monitor wells	One from each well	As designated in UIC Permit	Grab	Bi-monthly	Bi-monthly	Ec, Cl, Dissolved U-Nat, HCO_3

(*Continued*)

Table 7.3 **(Continued)**

Type of sample	Number	Location	Method	Sample frequency	Analysis frequency	Type of analysis
Surface water	One from each impoundment and a minimum of two from each stream	Permanent impoundments and upstream and downstream in surface waters passing through the license area; also adjacent impoundments subject to drainage from the license area	Grab	Quarterly	Quarterly	Dissolves and Suspended U-Nat, total and soluble ^{226}Ra, ^{230}Th, ^{210}Po & ^{210}Pb
Soil & sediment	Same as surface water	At surface water sampling locations	Grab	Quarterly	Quarterly	U-Nat, ^{226}Ra & ^{210}Pb
Vegetation forage	Three	Grazing area near the facility in the direction of the highest predicted radionuclide values	Grab	Three times during grazing season	Each sample	^{226}Ra & ^{210}Pb
Direct radiation	Five	At radon gas sampling stations	TLD	Continuous	Quarterly	γ Exposure rate
Soil & sludge	One	Septic system drain field	Grab	Once	Prior to requesting termination of license	U-Nat, ^{226}Ra & ^{210}Pb

Photo 7.15 ISR well field in northern USA. All piping is buried below the surface (file photo).

geometry and surface topography, well spacing for common well patterns is typically between 12 and 50 m [40 and 150 ft] apart (NRC, 2009) (Photo 7.16).

ISR well fields are often complex, incorporating many different well patterns configured to the orebody. In addition, although many ISR operators initially plan to complete each of their injection/extraction wells into a single uranium horizon, economic dictates or reinterpretation of the orebody often result in a number of the individual wells being "multicompleted," meaning open to more than one production zone even if the open interval within a well is continuous. This can create a complex movement of lixiviant in the subsurface such that computer models are often used in well field planning (Fig. 7.14).

This complexity is compounded by the effect that flow rates from each individual well have in directing lixiviant movement throughout the producing aquifer. In the past, operators often paid little attention to lixiviant flow rates at individual wells, instead simply ensuring that injection and extraction flow rates across an area or multiple patterns were equal or "balanced." Excursions and poor recovery often resulted. Today, most ISR operators "balance" injection and extraction flow rates well by well, using computer models to compare actual lixiviant movement with predicted movement.

7.4.2 Processing plant equipment

Process pad. Process pads are generally made of concrete and provided with sumps, drains, and a curb at the periphery of sufficient size to assure secondary

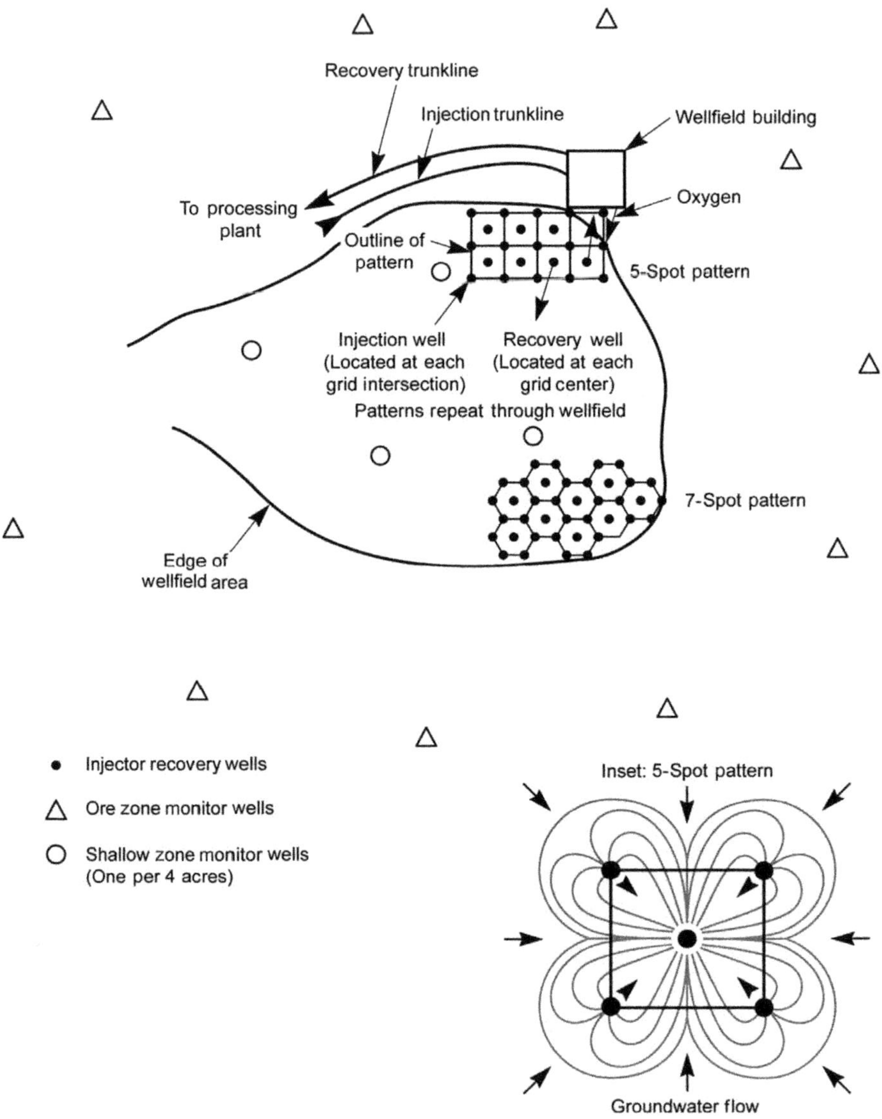

Figure 7.13 Generalized ISR well field layouts.
Source: NRC (2009).

containment. Often the pad is underlain by a synthetic liner that would capture potential leakage through cracks in the concrete. The curbed design is to confine and hold potential spills in the process, and potentially contaminated runoff from the processing equipment area.

Retention ponds. Where practical, retention ponds are constructed such that all retained fluid is below ground level. This will eliminate the potential for

Photo 7.16 ISR retention pond construction at the Beverly ISR facility, Australia.
Source: c.2000, Heathgate Resources Pty. Ltd.

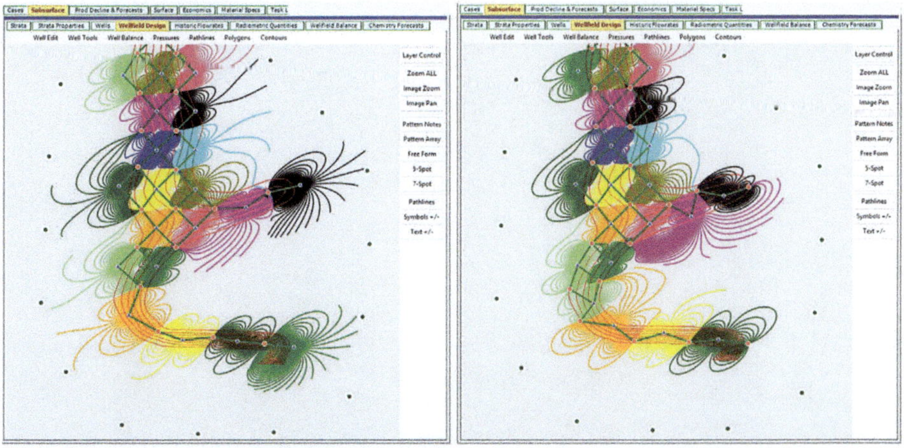

Figure 7.14 Well field planning using subsurface path lines from injectors (courtesy C. Bartels). Left: original well field plan and flow balance resulting in excursions. Right: new well field plan with well reversals and flow rebalance. Circles (monitor wells); Blue Triangles (injectors); Red Squares (extractors).
Source: PathCAD with 2:1 horizontal anisotropy at N60°E.

embankment failure. Standard provisions for the ponds are synthetic membrane liners and leak detection systems.

Tankage. Tankage at ISR facilities is generally fiberglass or steel, which are constructed to safety standards such as Voluntary Product Standard PS 15–69 or the equivalent for fiberglass (ASTM, 2010) or the American Society of Metallurgical Engineers guide for the design and fabrication of pressure vessels or the equivalent for steel (ASME, 2012).

Piping. The fluids handling system at ISR operations encompasses various pumps, meters, pipelines, fittings, and connections, and will generally consist of polyethylene, PVC, fiberglass, steel, and stainless steel materials, which are used universally in ISR. In materials technology, the ISR setting is considered both low pressure and low temperature, allowing use of "off the shelf" items and materials that will easily be available. In all cases, the components of this fluid handling system are rated to withstand ambient temperatures and pressures of their environment, and the pressures and temperatures of the fluids with which they are in contact, using published, generally accepted ratings. The materials are chemically resistant, over their useful life, to the fluids and solids with which they are in contact. Specifications are determined to maintain structural integrity throughout anticipated life of the component. All well field piping systems and equipment will either be housed in containment buildings, placed on the surface (Photos 7.17 and 7.18), or buried (Photo 7.19).

Weatherization. ISR operations are essentially water treatment systems. As such, the need for weatherization is driven by the freezing point of water. In most locations where temperatures fall below 0°C (32°F), well field piping systems will either be buried below freezing depths or be heated (Photo 7.19).

Photo 7.17 Well field piping for individual wells placed on the surface in Texas, USA. The blue hose delivers metered oxygen into individual injection wells.
Source: c.2007, Mestena Uranium LLC.

Photo 7.18 Header system placed on the surface in Texas, USA.
Source: Mestena Uranium LLC.

Photo 7.19 Preparation for pipe burial at the Nichols Ranch ISR project in Wyoming, USA.
Source: Uranerz Energy Corporation.

Individual wellheads need to be protected from the elements (Photo 7.20). Tankage and other process infrastructure are generally housed in temperature-controlled buildings (Photo 7.21).

In some locations, such Texas in the United States, freezing is not a major factor, and well fields, metering/gathering systems, and process facilities are exposed (Photos 7.22, 7.23, 7.17 and 7.18).

Photo 7.20 Weatherized wellhead enclosures in Wyoming, USA (file photo).

Photo 7.21 Indoor process facility at the Lost Creek ISR project in Wyoming, USA.
Source: Ur-Energy.

7.4.2.1 Yellowcake dryer systems

Wet yellowcake must be dried prior to final packaging. Historically, two kinds of yellowcake dryers have been used: multihearth dryers and vacuum dryers. Older uranium ISR facilities used gas-fired multihearth dryers (NRC, 2009).

Photo 7.22 The Alta Mesa ISR project in Texas, USA.
Source: c.2007, Mestena Uranium LLC.

In the United States, batch-type rotary vacuum dryer systems have become common. The drying and packaging occur in the same area. The systems include:

- A drying chamber, equipped with an internal mixing auger and a mechanism for directly discharging the dried product into 55-gallon drums
- A bag filter to capture and return to the drying chamber the entrained solid particles present in the exiting vapor stream
- A water-cooled condensing unit to cool and liquefy water evaporated from the yellowcake slurry
- A vacuum pump
- A recirculating closed-loop hot oil heating system that uses a propane, natural gas-fired, or electric boiler to heat the oil

With this type of vacuum dryer, the heating source is contained in a separate, isolated system so that no radioactive materials are entrained in the heating system or the exhaust it generates. The drying chamber containing yellowcake slurry is subjected to vacuum pressure. Moisture in the yellowcake is the only source of vapor remaining in the system. Any potential leak would result in outside air flowing into the drying chamber.

The bag filter is designed to recover all of the solids entrained in the water vapor, and any solids escaping this filter would be captured by the circulating sealant water within the vacuum pump. This water, which is kept cool by passage through a cooling tower, is periodically diverted to the production circuit to recover

Photo 7.23 The Palangana/Hobson satellite in Texas (2010), USA.
Source: Uranium Energy Corp.

collected yellowcake particles, or is diverted to the wastewater circuit. The vapor discharge line from the vacuum pump is vented to the atmosphere (Photo 7.24).

7.5 Processes

Uranium in the orebody becomes soluble in the oxidized phase, and once oxidized, it may be mobilized. Oxidization and mobilization of uranium into a lixiviant are basic to both acid and alkaline lixiviant chemistries. After the lixiviant is injected into wells and recovered from production wells, the uranium is removed by passing the pregnant (uranium-rich) lixiviant across IX resin. Uranyl anions are exchanged

Photo 7.24 Rotary vacuum dryer showing bag filter above drying chamber.
Source: Dennis Stover.

onto the surface of this IX resin. ISR proceed with the continuous recirculation of fortified groundwater solution through the uranium ore from the injection to the production wells. When the resin is fully charged, an eluant solution is used to release the uranyl ion and to regenerate the IX resins. The eluant is then chemically treated and the uranium precipitated. The wet uranium is then dried and the "yellowcake" product is packaged for transport (Fig. 7.15).

7.5.1 Lixiviant types

The Manual of Acid In Situ Leaching Uranium Mining Technology (IAEA, 2001) thoroughly reviews acid and carbonate- or alkaline-type lixiviants that can be used in the ISR of uranium. While a number of these lixiviant types have been tested, commercial operations across the globe have narrowed the selection to sulfuric acid or sodium bicarbonate lixiviant systems.

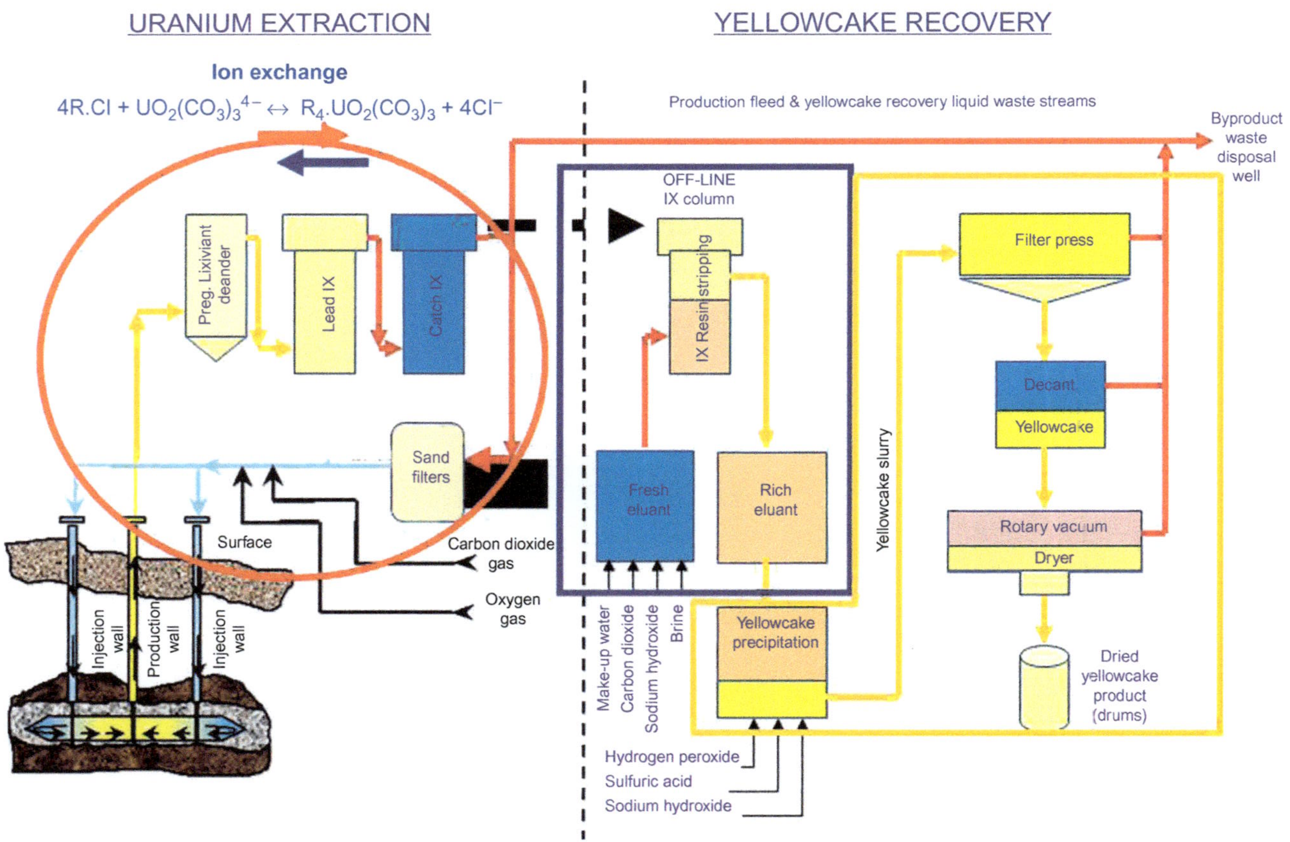

Figure 7.15 General ISR process schematic. The uranium extraction portion of the process may be conducted separate from the full process by RIX. *Source*: Mestena Uranium LLC.

Photo 7.25 Sulfuric acid storage at the Akdala ISR project in the South Kazakhstan Provence. *Source*: Uranium1.com.

Choosing between acid or alkaline is fundamentally dependent on the geochemistry of the host rock, in particular the concentration of calcium carbonate in the host rock matrix. According to IAEA (2001), if calcium carbonate content exceeds 2% in the host rock, a carbonate process is preferred because of excess acid consumption. With calcium carbonate below 2%, sulfuric acid chemistry is preferred because of accelerated leaching kinetics and higher uranium recoveries.

An alkaline lixiviant may also be chosen if groundwater restoration is required, such as in the United States. In the United States, carbonate-based leach solutions have become standard because ISR activity is conducted in aquifers with relatively low dissolved solids where domestic water supplies surrounding the ore deposit dictate groundwater restoration. Likewise, sodium bicarbonate type chemistry has become standard over ammonia carbonate because of the difficulty restoring ammonia that is sorbed onto the clay minerals (Humenick et al., 1979). In other countries, such as Kazakhstan and Australia, water in the extended aquifer is saline and not suitable for domestic use, so no restoration is warranted (Photo 7.25).

7.5.1.1 Acid

The IAEA (2001) has described the chemistry of leaching ores containing hexavalent uranium results in the formation of various uranyl sulfate complexes in solution as follows:

$$UO_3 + H_2SO_4 \rightarrow UO_2SO_4 + H_2O$$
$$UO_2SO_4 + SO_4 \leftrightarrow [UO_2(SO_4)_2]^{2-}$$
$$[UO_2(SO_4)_2]^{2-} + SO_4^{2-} \leftrightarrow [UO_2(SO_4)_3]^{4-}$$

The initial concentration of sulfuric acid is generally 15–25 g/dm^3 (g/L), which intensifies oxidation. However, the acid concentration drops off significantly en route to the production well. With sufficient trivalent iron ions, uranium is usually oxidized to tetravalent uranium that enters into solution. The rate at which uranium enters the solution is increased by raising the concentration of trivalent iron ions with the addition of iron oxide sulfate or some other oxidant.

In addition to the trivalent iron, oxidants such as oxygen, hydrogen peroxide, sodium chlorate, or nitrate ions may be added. Oxygen is most commonly used because of its availability, comparative low cost, and relatively strong oxidation potential. Oxygen is not effective in acid environments, so it may be introduced into the mineralized zone prior to injecting acid fed into wells through a special pipeline extending to the well screen depth.

The uranium-rich groundwater lixiviant solution containing the uranyl sulfate complex is received at the IX columns. Uranium is exchanged on the reacting sites of the strong base anionic resin for a sulfate ion (if the resin is in sulfate form) according to the following reaction (IAEA, 2001) (where R is a reacting site of the IX resin):

$$(R_4N^+)_2SO_4 + UO_2^{2+} + SO_4^{2-} = (R_4N^+)_2[UO_2(SO_4)_2]$$
$$2(R_4N^+)_2SO_4 + UO_2^{2+} + SO_4^{2-} = (R_4N^+)_4[UO_2(SO_4)_3] + SO_4^{2-}$$

or, if the anionic resin is in a nonsulfate form (eg, in nitrate):

$$2(R_4N^+)NO_3 + UO_2^{2+} + 2SO_4^{2-} = (R_4N^+)_2[UO_2(SO_4)_2] + 2NO_3^-$$
$$4(R_4N^+)NO_3 + UO_2^{2+} + 3SO_4^{2-} = (R_4N^+)_4[UO_2(SO_4)_3] + 4NO_3^-$$

7.5.1.2 Alkaline

Alkaline lixiviant consists of native groundwater to which gaseous oxygen (O_2) (or another oxidant, such as hydrogen peroxide) and gaseous carbon dioxide and/or sodium bicarbonate have been added. Uranium, present in the host ore in a reduced insoluble form, is oxidized by the lixiviant solution injected into the ore zone. Once uranium is oxidized, it complexes with bicarbonate anions in the groundwater and becomes mobile. Uranium in the ore reacts with the lixiviant to form a soluble uranyl di-carbonate complex.

$$2UO_2 + O_2 \rightarrow 2UO_3$$
$$UO_3 + 2NaHCO_3 \rightarrow Na_2UO_2(CO_3)_2 + H_2O$$

After passing through the ore zone, the uranium-rich lixiviant is pumped from production wells to then be extracted by IX onto resin. The resulting uranium-depleted (barren) water will then be refortified with O_2 and reinjected into the well field to repeat the uranium recovery cycle. O_2 may be introduced into each well through a drop tube and sparger or on the surface into the main injection trunk line.

Uranium is exchanged on the reacting sites of the resin for chloride ion (if the resin is in chloride form) according to the following reaction:

$$Na_2UO_2(CO_3)_2 + 2RCl \rightarrow R_2UO_2(CO_3)_2 + 2NaCl,$$

where R is a reacting site of the IX resin.

When the IX resin in a column has captured uranium to its optimum loading capacity, the uranium concentration in the barren water exiting the IX column begins to rise. At this point, the column is taken out of the operating circuit and another column with fresh IX resin is placed online (Photo 7.26).

After the uranium is removed by the IX columns, the process bleed is removed from the lixiviant stream. The bleed may be treated by reverse osmosis (R.O.), and if it is, the "product," or cleaned water, is returned to the lixiviant injection or to the formation outside the well field pattern, or disposed of by an approved method. The process bleed insures that more water is withdrawn than is injected, thereby keeping the lixiviant laterally within the production zone.

Photo 7.26 IX columns at the Nichols Ranch ISR project (courtesy of Uranerz Energy Corporation). Insert shows IX resin in hand.

7.5.2 Plant configurations

7.5.2.1 Central plant

A central processing plant (CPP) is capable of processing uranium through the finished form, dried yellowcake. As such, the CPP includes elution columns, yellowcake precipitation tanks, other tanks to hold barren eluant, and yellowcake slurry. The CPP also includes drying and packaging equipment. A CPP is often located in close proximity to the uranium deposit and lixiviant is treated as part of the central plant operation. If so, the CPP will include columns containing IX resin, piping, and pumps to circulate lixiviant to and from the well fields (Photo 7.27).

7.5.2.2 Remote ion exchange satellite plant

Remote IX (RIX) units are limited to columns with IX resin for recovering the uranium from the leachate stream and any required filtration. The resin, once loaded with uranium, is transferred out of the IX columns, drained of free water, and trucked to a CPP for further processing. The clean IX resin that has been stripped of uranium at the CPP is transferred back to the RIX in the same way.

The RIX system consists of three main components; pipeline to and from the well fields, IX columns, and a transfer system (Photo 7.28).

In addition to improved operating efficiencies, RIX has the potential to improve leaching characteristics by isolation production from individual mine units. With a large process facility servicing many mine units at long distances, it is usually necessary to comingle water across production areas through a common IX circuit, resulting in a progressive increase of ionic constituents. In an alkaline groundwater lixiviant chemistry example, this ionic build-up is illustrated, where iron oxidation

Photo 7.27 Central processing facility at the Inkai ISR project Kazakhstan.
Source: Cameco.com.

Photo 7.28 RIX satellite connected to the resin transport at the Nichols Ranch project in Wyoming.
Source: c.2013, Uranerz Energy Corporation.

(pyrite) results insoluble sulfate and a pH drop from the sulfuric acid that results from the sulfide oxidation, as follows:

$$\text{Oxidation of iron sulfides} - \text{Fe(ppt)SO}_4 \uparrow \text{pH} \downarrow$$

$$FeS_2 + 15/4O_2 + 7/2H_2O \rightarrow Fe(OH)_3[ppt] + 4H^+ + 2SO_4^=$$

Next, the host rock, with an abundance of calcium carbonate in its matrix, experiences dissolution of calcium carbonate from lowered pH. The resulting calcium, bicarbonate, and sulfate build-up is progressive, cumulative, and becomes more detrimental as one mine unit's water is comingled into the next. In time, saturation of calcium, carbonate, and sulfate is reached in the leach solution that potentially results in the precipitation of calcium carbonate (calcite) and calcium sulfate (gypsum) onto the uranium-coated sand grains in the subsurface, resulting in poor uranium recoveries.

Continuous IX also results in the build-up of sodium chloride in the leaching solution as follows:

$$Na_2UO_2(CO_3)_2 + 2RCl \rightarrow R2UO_2(CO_3)_2 + 2NaCl$$

Because of the increasing concentrations of sodium chloride, this "dirty water" will compete with uranium loading in the IX process, decrease efficiency, and result in high tails to the well field.

7.5.3 Elution and precipitation

According to IAEA (2001), the acid lixiviant system uranyl sulfate complexes are eluted from the anionic resin phase with concentrated solutions of sulfuric acid (often with additions of small quantities of nitric acid as an anticorrosion measure) or with solutions of nitrate, chloride, and ammonium carbonate salts or sodium chloride with addition of soda ash. When the elution is performed with sulfuric acid solutions, the resin can directly be contacted with recovery solutions. The use of other eluting reagents requires regeneration, that is, conversion of the resin into a form corresponding to the treated solution to prevent depressing anions (Cl^-, NO_3) entering into these solutions. Therefore, uranium stripped from the resin only with sulfuric acid solutions is preferred.

In the alkaline lixiviant system, once loaded with complexed uranyl di-carbonate, brine and soda ash solution are used to remove the uranium from the resin in a two-step process. The following chemical reaction occurs:

$$R_2UO_2(CO_3)_2 + 2NaCl + Na_2CO_3 \rightarrow Na_4UO_2(CO_3)_3 + 2RCl$$

Next eluant rich in uranyl d-, and tri-carbonate is acidified using hydrochloric acid (HCl) or sulfuric acid (H_2SO_4) to destroy the uranyl carbonate complex as shown.

Hydrochloric Acid:

$$Na_4UO_2(CO_3)_3 + 6HCl \rightarrow UO_2Cl_2 + 4NaCl + 3CO_2 + 3H_2O$$
$$Na_2UO_2(CO_3)_2 + 4HCl \rightarrow UO_2Cl_2 + 2NaCl + 2CO_2 + 2H_2O$$

Sulfuric Acid:

$$Na_4UO_2(CO_3)_3 + 3H_2SO_4 \rightarrow UO_2SO_4 + 2Na_2SO_4 + 3CO_2 + 3H_2O$$
$$Na_2UO_2(CO_3)_3 + 2H_2SO_4 \rightarrow UO_2SO_4 + Na_2SO_4 + 2CO_2 + 2H_2O$$

In the next step, hydrogen peroxide is added to the solution to oxidize the uranium even further and to cause it to precipitate according to one of the following reactions:

Hydrochloric Acid:

$$UO_2Cl_2 + H_2O_2 + xH_2O \rightarrow UO_4 \cdot xH_2O + 2HCl$$

Sulfuric Acid:

$$UO_2SO_4 + H_2O_2 + xH_2O \rightarrow UO_4 \cdot xH_2O + H_2SO_4$$

Peroxide may be used to precipitate uranium using either an alkaline or acid lixiviant system, although there are numerous other methods available to precipitate uranium in the acid system by decreasing pH (IAEA, 2001). The crystalline uranyl peroxide is adjusted for pH (if required) and then is allowed to settle. The yellowcake is further dewatered using a filter press. Finally, the yellowcake is washed with clean water to remove impurities such as sorbed chloride and then dried.

7.6 Waste management

As is commonplace for all ISR projects, the recovery of uranium from ores generates radioactive byproduct waste. All solid radioactive byproduct waste must be disposed of at properly licensed disposal facilities as specified by regulation or in the case of liquids using appropriate liquid disposal methods (eg, treatment and discharge, evaporation, deep well disposal).

7.6.1 Liquid radioactive byproduct waste

During ISR uranium production, the major source of liquid radioactive byproduct waste is the process bleed. Liquid radioactive byproduct waste also may be present as wastewater resulting from plant decontamination wash-downs and other water treatment activities.

If groundwater/aquifer restoration is necessary, the primary source of liquid radioactive byproduct waste is contaminated groundwater from restoration processes. Radioactive liquid byproduct waste may be disposed of in evaporative retention ponds and deep well disposal or a combination of these methods. In some instances, liquid radioactive waste can be treated to regulatory standards and either broadcast to land areas or surface discharged to waterways.

7.6.1.1 Reverse Osmosis or R.O.

R.O. is a water treatment process whereby the majority of dissolved "ions" are separated from the wastewater and concentrated into a smaller concentrated brine volume. The resulting purified water fraction typically exceeds background mine water quality, and during restoration activities, the purified water is reinjected back into the well field, further diluting the underground recovery solutions toward baseline quality. The concentrated brine fraction, representing 5–30% of the feed volume, still must be disposed of. However, because the volume is significantly reduced, the disposal capacity requirements are not as great (Photo 7.29).

7.6.1.2 Deep disposal well

A cost-effective and commonly used method for disposal of wastewater and brines from ISR operations is the use of a deep disposal well, where technically feasible. Preferred geologic formations are repositories containing TDS in excess of 10,000 ppm. Additionally, confinement from overlying and underlying freshwater aquifers must be demonstrated (Photo 7.30).

Photo 7.29 A R.O. unit at the Lost Creek ISR project in Wyoming, USA.
Source: Ur-Energy.

Photo 7.30 Liquid waste disposal well at the Alta Mesa ISR project in Texas, USA.
Source: Mestena Uranium LLC.

7.6.1.3 Evaporation ponds

In this system, liquid wastes are evaporated. It is common to install a spraying system in the ponds to enhance evaporation. The use of R.O. technology would greatly reduce the sizing of ponds, as only the brine fraction requires disposal.

7.6.1.4 Surface discharge and land application

To surface discharge radioactive byproduct wastewater, a company must first be able to demonstrate that treated wastewater quality comply with established standards. Land application is a disposal technique that uses agricultural irrigation equipment to broadcast wastewater on a relatively large area of land. Wastewater released by surface discharge or land application will require uranium and radium removal. Uranium may be removed using the same process described in Section 7.5. Radium can be removed from wastewater by barium chloride precipitation in retention ponds. The barium chloride will form barium sulfate that, in time, coprecipitates with soluble radium.

7.6.2 Solid radioactive byproduct waste

Solid radioactive byproduct waste generated at ISR projects may be spent or spilled resin as a result of normal operations and groundwater restoration, spilled yellowcake product, soils from well field or other facility spills, and site equipment and other materials resulting from well field and surface reclamation and from site decommissioning and decontamination. Solid waste must be characterized by scintillation probe and separated into radioactive and nonradioactive. All equipment or waste that does not meet the designation of nonradioactive may be cleaned and resurveyed, or disposed of in a regulated disposal facility. Materials that cannot be decontaminated to meet the nonradioactive standard must be disposed of at an appropriately regulated radioactive waste disposal facility.

7.7 Well drilling, installation, completion, and operation

7.7.1 Layout and patterns

Several types of wells are installed to facilitate the ISR process. Injection wells are installed to allow the injection of the lixiviant. Production wells are installed to allow the recovery (pumping) of the uranium-rich lixiviant (production fluid). Depending on regulatory requirements, wells are installed within the production zone to determine baseline water quality conditions, as well as monitor wells around the outside of the production zone (monitor well ring) to document the lateral control of the lixiviant. Monitor wells are also installed in the overlying aquifers above the production zone to ensure that the lixiviant does not migrate vertically from the production zone. All wells are constructed to assure that the well annulus is sufficiently cemented to prevent interformational transfer of water.

7.7.2 Construction and installation

Wells are constructed so as to perform for the life expectancy of the well. All holes are rotary-drilled with water well-type drill rigs that are capable of circulating drilling fluids to the surface.

Logging. Each hole is checked for deviation and logged. Spontaneous potential, resistivity, and gamma ray logs are generally obtained for all holes. Where gamma logs indicate the presence of uranium mineralization, PFN logging may be conducted. In the event that washouts are indicated, caliper logs are called for to determine the cement quantities required for subsequent casing activities.

Casing. Injection, production, and monitor wells are generally cased with threaded fiberglass or PVC, and perforated, under-reamed, or screened. When the casing is run into the hole, it will include centralizers with each being spaced between 45 and 60 m (150–200 ft) along the total casing length to assure uniform placement of the cement along the annular space.

Cementing. Once the casing is run into a well, the cement is pumped through the casing and up the annular volume between the casing and borehole to the surface.

Completion. The integral screen completion is typically used for shallower wells with very long completion intervals and satisfactory vertical isolation. The cement basket is set in confining shale above the completion interval and the screen is suspended below the basket. Perforated and under-reamed casing completion are both used to open wells with casing placed across the target interval. The perforated casing completion utilizes hollow charge shots to punch holes through the casing and cement, and into the formation. The under-reamed casing completion uses a mechanical down-hole tool to cut away the casing, cement, and the filter cake on the sand face. Perforated and under-reamed completions provide very good vertical isolation of the interval due to cement remaining above and below the opening to seal the annulus of the casing.

Mechanical integrity testing. After the interval has been completed, a mechanical integrity test (MIT) is performed to further test the casing for possible leaks. An inflatable packer is run into the well to a depth directly above the open interval. The packer is inflated and the casing is filled with water. The casing test pressure will vary with the maximum allowed wellhead injection pressure by a safety margin. After the test pressure is reached, the well is sealed to hold pressure, and allowed to stand for a prescribed period of time. The MIT is passed before the well is considered operational (Fig. 7.16).

7.7.3 Well operation

Each production well is operated at the maximum continuous flow rate achievable for that pattern area. The primary consideration in determining maximum continuous flow rate is to assure the well field is collectively balanced (Photo 7.31).

Generally, the overall injection flow rates into the well fields are less than the total extraction flow rate by an amount known as "process bleed," resulting in a hydraulic pressure sink that causes native groundwater outside of the ore zone to migrate into the well field. This process bleed is used to help protect the monitor

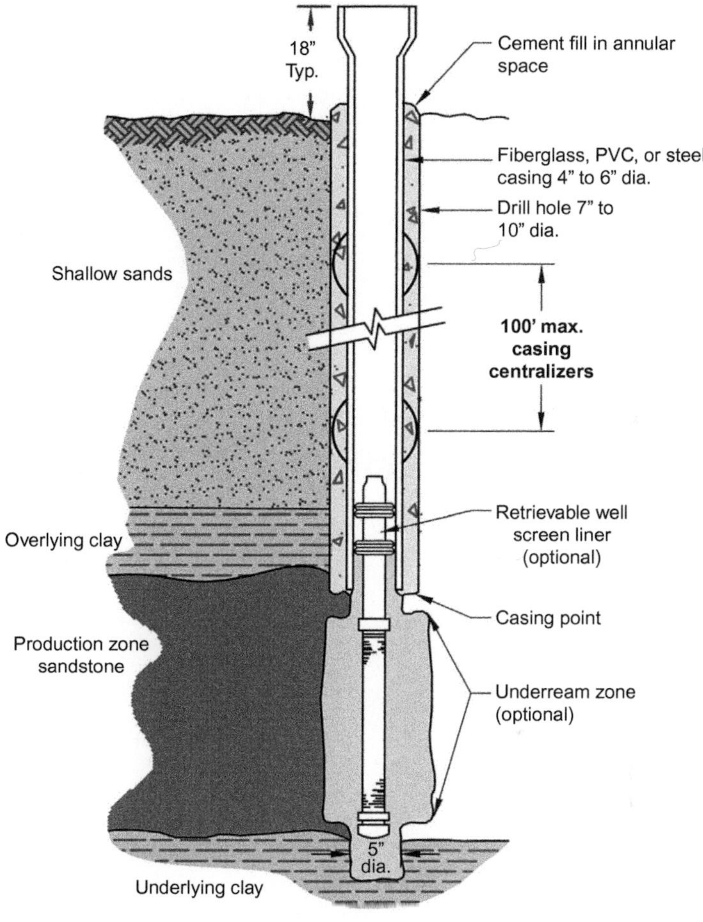

Figure 7.16 Typical ISR well construction using the under-ream method.
Source: NRC (2009).

wells against lixiviant excursion, and varies according ore geometry, well pattern and magnitude, and direction of the natural groundwater velocity (Fig. 7.17)

7.7.4 Well field instrumentation

Injection and production flow rates are monitored so injection can be balanced with production across the entire well field, with the injection flow smaller than the production flow by the amount of the bleed rate. A combination of meters is used in the well field and the plant (Photo 7.32).

To assure the maximum allowable injection pressure is not exceeded, continuous monitoring of pressure is generally located in the trunk lines. In addition, the pressure meters will detect changes in line pressure indicative of a pipeline leak or rupture.

Photo 7.31 A completed ISR well at the Beverly ISR Facility, Australia. The drip pan and auto shut off switch is designed to catch potential leaks and drips from the wellhead to prevent soil contamination.
Source: Heathgate Resources Pty. Ltd.

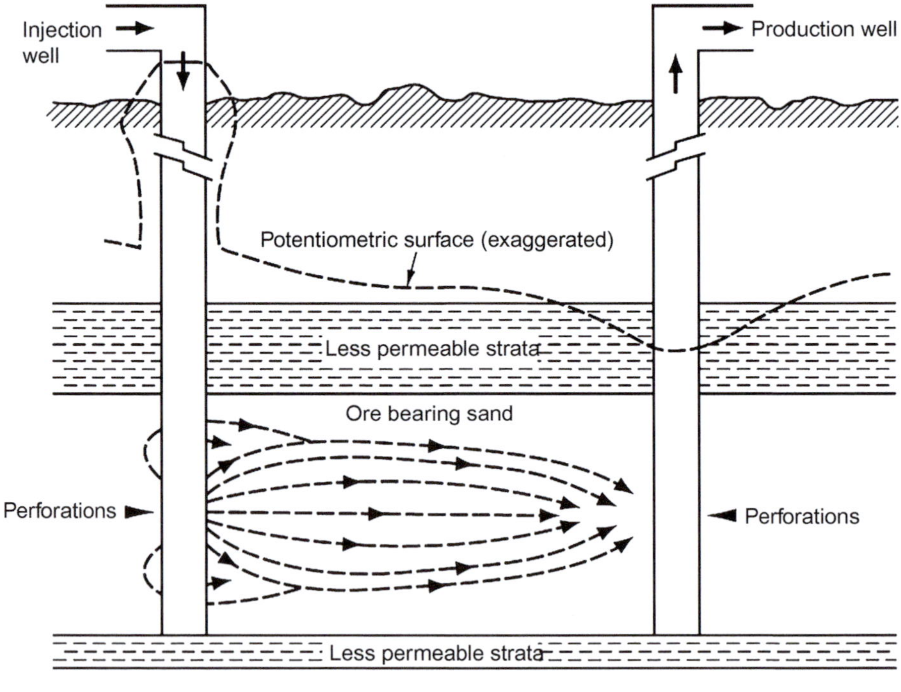

Figure 7.17 Schematic cross section with bleed response.
Source: NRC (2009).

Photo 7.32 Contained header house at the Nichols Ranch ISR project in Wyoming, USA. *Source*: Uranerz Energy Corporation.

7.7.5 Automation and telemetries

ISR facilities have different levels of sophistication for instrumentation and automation. In many industries (eg, chemical, oil refineries), increasing productivity, reducing labor requirements, and increasing operational awareness drive advances in and use of instrumentation and automation. However, honest reflection indicates that results attained in other industries have not been achieved for ISR: labor requirements for more highly qualified personnel often increase for calibration and maintenance, and for control room operators with no corresponding decrease in other positions (well field and processing plant operators), or corresponding increase in production levels. Productivity, therefore, deteriorates. In addition, examination suggests there has been little increase in operational awareness as a consequence of increased levels of instrumentation and automation; in fact, an overload of data and marginal alarms may cause just the opposite in some cases. The resulting enormous volume of recorded data, often considered a significant benefit when first considering additional instrumentation and automation in this era of "data mining" or "big data," requires IT staff, hardware, and other resources to manage and likely additional professional staff for analysis or that data simply goes unviewed and unanalyzed.

Photo 7.33 Employee at central process plant central control panel.
Source: Uranium Energy Corp.

The reasons for this outcome are likely threefold. First, analysis shows the economics of ISR is determined by the cost and quality of well fields, not by the surface facilities (satellite IX, processing plant). Processing is such a small percentage of total costs that even dramatic increases in efficiencies in surface facilities cannot offset high well field installation costs and/or poor well field recovery in the necessary timeframe. Second, the low cost typically associated with ISR results in large part from the much smaller workforce required compared to the combined mining/milling of a conventional operation covering the same scope (taking the uranium from rock to the final packaging). Third, production levels are typically somewhat smaller for ISR than conventional operations, that is, scale. The combined effect is that ISR operations, already under considerable pressure over a long period of time to minimize costs and labor force, have difficulty realizing meaningful efficiencies outside of the well fields, and will have a loss in productivity with even small, unintended increases in staffing levels.

Countering this, however, is that instrumentation and automation are often associated with sophistication, and as such, their use will likely increase to meet expectations of regulators and of the public (Photo 7.33).

7.8 Reclamation

Reclamation requirements at an ISR project site are dependent on the regulatory jurisdiction and generally are comprised of radiological decontamination of buildings, process vessels, and other structures or affected areas; surface reclamation and revegetation of process areas and restored well fields; and groundwater restoration within affected well fields, including production and monitor well plugging.

Decommissioning and reclamation of sites will generally take place after operations are complete. If required, groundwater restoration and well field decommissioning are accomplished as well fields are completely mined out.

Both surface reclamation and groundwater restoration are intended to return areas affected by ISR activities to a condition that supports the premining land use and water use (if any).

7.8.1 Radiological decontamination

All radiologically contaminated buildings, process vessels, other structures, and affected areas are decontaminated prior to final reclamation to unrestricted release standards in accordance requirements, or removed to the appropriate disposal facilities.

7.8.2 Reclamation and revegetation

The purpose of the reclamation program is to stabilize the site with self-sustaining vegetative cover, and to restore all land disturbed by ISR operations and related activities to a productive condition, consistent with the historical use of the area. (Photo 7.34).

7.8.3 Groundwater restoration and well plugging

Restoration of the production zone may be achieved by a combination of groundwater sweep and R.O. The restoration of groundwater at the ISR sites will have the benefit of a previously engineered array of injection and production wells that were initially installed in a configuration to maximize sweep efficiently throughout the uranium orebody, and to maximize uranium recovery.

With the R.O. techniques, injection and extraction operations continue at the facility, except produced water is processed through a R.O. unit that produces a deionized fluid for reinjection. The injection solution passes through the pores of the aquifer formation and replaces the affected solutions, which are pumped to the surface. The net effect is that the resulting interstitial groundwater quality becomes consistent with, and in many cases better than, premining quality. The primary benefit of R.O. treatment is that a large fraction of the total water extracted is purified

Photo 7.34 Surface reclamation of the Trevino ISR site in Texas. The top photo was taken during operations in 1982 and the bottom in 2010 after the facility was decommissioned and the surface reclaimed. One office building remains in view.
Source: H.L. Anthony.

and reinjected, resulting in less water consumption and less groundwater drawdown in the area.

All wells are permanently plugged and abandoned with an approved grout medium upon completion of groundwater restoration.

References

American Society of Mechanical Engineers, 2012. ASME Standard—BVPC-VIII-1-2013 Rules for the Construction of Pressure Vessels. American Society of Mechanical Engineers, New York, NY.

American Standard for Testing and Materials, 2010. ASTM D4097-01(2010) Standard Specification for Contact-Molded Glass-Fiber-Reinforced Thermoset Resin Corrosion-Resistant Tanks. American Society for Testing and Materials International, West Conshohocken, PA, http://dx.doi.org/10.1520/D4097-01R10. <www.astm.org>.

Boberg, W.W., 2012. Sandstone uranium deposits—roll-fronts, solution fronts, redox fronts. In: IAEA Technical Meeting on the Origen of Sandstone Uranium Deposits: A Global Perspective, Vienna, Austria.

Boitsov, A.V., Razumov, V.I., 2004. Uranium related activities in the Russian federation. In: Recent Developments in Uranium Resources and Production with Emphasis on In Situ Leach Mining, Proceedings of a Technical Meeting Organized by the IAEA in Co-operation with the OECD Nuclear Energy Agency, the Bureau of Geology, and China National Nuclear Corporation Held in Beijing, International Atomic Energy Agency, 18–23 September 2002. Report IAEA-TECDOC-1396.

Campbell, M.D., Henry, P.H., Wise, M., Rackley, R., 2007. Uranium in situ leach (recovery) development and associated environmental issues. In: 2007 Gulf Coast Geological Societies Convention, Corpus Christi, Texas, 21–23 October 2007.

Ekert, W., Muzak, J., 2010. Mining and remediation at the straz pod ralskem uranium deposit. Geosci. Eng. LVI (3) (special issue), 1–6.

Fazlullin, M.I., Shatalov, V.V., Smirnova, R.N., Boitsov, A.V., Razumov, V.I., 2004. In situ leach uranium mining in the Russian federation (status and prospects). In: Recent Developments in Uranium Resources and Production with Emphasis on In Situ Leach Mining, Proceedings of a Technical Meeting Organized by the IAEA in Co-operation with the OECD Nuclear Energy Agency, the Bureau of Geology, and China National Nuclear Corporation Held in Beijing, International Atomic Energy Agency, 18–23 September 2002. Report IAEA-TECDOC-1396.

Givens, W.W., Stromswold, D.C., 1989. Prompt Fission Neutron Logging for Uranium. In Nuclear Geophysics. 3 (4), 299–307.

Grutsynov, V.A., 2002. Environmental Protection at ISL Uranium Mining Sites in Uzbekistan. Navoi Mining and Metallurgical Combinat, Navoi, Uzbekistan. IAEA-SM-362/44.

Hou, B., Fabris, A.J., Keeling, J.L., Fairclough, M.C., 2007. Cainozoic palaeochannel-hosted uranium and current exploration methods. SA MESA J. 46, 34–39.

Humenick, M.J., Schechter, R.S., Breland, M., Garwacka, K., Johnson, D., Walsh, M., 1979. Investigation of the Fate of Ammonia from In-Situ Uranium Solution Mining. Technical Report – EHE- 79-01. The University of Texas at Austin, Austin, TX.

Hunter, T., 2001. Developments in uranium solution mining in Australia. In: ANA Conference Paper. IAEA-TECDOC-720.

International Atomic Energy Agency, 2001. Manual of Acid In Situ Leach Uranium Mining Technology. IAEA-TECDOC-1239.

Kim, A., 1997. Environmental restoration plans and activities in Kazakhstan. In: Planning for Environmental Restoration of Uranium Mining and Milling Sites in Central and Eastern Europe, Proceedings of a Workshop Held Under the Technical Co-operation Project RER/9/022 on Environmental Restoration in Central and Eastern Europe, Felix, Romania, International Atomic Energy Agency, 4–8 November 1997.

Märten, H., Marsland-Smith, A., Ross, J., Haschle, M., Kalka, H., Schubert, J., 2014. Optimization of uranium in-situ recovery based on advanced geophysical surveying and borehole logging technologies. In: Uranium—Past and Future Challenges, Merkel, B.J., Arab, A. (Eds.), Proceedings of the 7th International Conference on Uranium

Mining and Hydrogeology, Freiberg (Germany), Springer, 21–25 September 2014, pp. 105–112.

Nedyalkov, K., 1996. Plans for environmental restoration of uranium mining and milling sites in Bulgaria. In: Planning for Environmental Restoration of Uranium Mining and Milling Sites in Central and Eastern Europe, Proceedings of a Workshop Held Under the Technical Co-operation Project RER/9/022 on Environmental Restoration in Central and Eastern Europe, Felix, Romania, International Atomic Energy Agency, 4–8 November 1997.

Norman, R.E., 1993. Uranium Production in Eastern Europe and Its Environmental Impact: A Literature Survey. Oak Ridge National Laboratory, Oak Ridge, TN.

NRC, 1980. Regulatory Guide 4.14. Radiological Effluent and Environmental Monitoring at Uranium Mills. Washington, DC.

NRC, 2009. Generic Environmental Impact Statement for In Situ Leach Uranium Facilities. NUREG-1910.

OECD, 2014. Resources, Production and Demand. Nuclear Energy Agency Organization for Economic Cooperation and Development. NEA Report No. 7209. Paris, France.

Pelizza, M.S., 2014. Uranium and uranium progeny in Groundwater Associated with Uranium Ore Bearing Formations. In: U.S. Nuclear Regulatory Commission/National Mining Association Uranium Recovery Workshop, Denver, CO.

Rosholt, J.N., 1959. Natural Radioactive Disequilibrium of the Uranium Series. U.S. Geological Survey Bulletin 1084-a, Washington, DC.

Rudy, C., 1997. Environmental restoration in regions of Ukraine; progress, problems and perspectives. In: Planning for Environmental Restoration of Uranium Mining and Milling Sites in Central and Eastern Europe, Proceedings of a Workshop Held Under the Technical Co-operation Project RER/9/022 on Environmental Restoration in Central and Eastern Europe, Felix, Romania, International Atomic Energy Agency, 4–8 November 1997.

Singh, A., Maichle, R., Singh, A., Lee, S.E., Armbya, N., 2009. ProUCL Version 4.00.04 User Guide, Draft. Prepared for the U.S. Environmental Protection Agency, EPA/600/R-07/038.

Sukhovarov-Jornoviy, 2005, 2005. Uranium deposits of Ukraine for ISL mining: developments in uranium resources of Ukraine for in situ leach (ISL) uranium mining—historical analysis, operational, geological, environmental and economic aspects. In: Developments in Uranium Resources, Production, Demand and the Environment. Proceedings of a Technical Committee Meeting Held in Vienna, 15–18 June 1999. IAEA-TECDOC-1425.

Tomas, J., 1997. Planning environmental restoration in the Northern Bohemian Uranium District, Czech Republic: progress report. In: Planning for Environmental Restoration of Uranium Mining and Milling Sites in Central and Eastern Europe, Proceedings of a Workshop Held Under the Technical Co-operation Project RER/9/022 on Environmental Restoration in Central and Eastern Europe, Felix, Romania, International Atomic Energy Agency. 4–8 November 1997.

Underhill, D., 1992. In situ leach uranium mining in the United States of America: past, present and future. In: Proceedings of a Technical Committee Meeting held in Vienna, 5–8 October 1992.

USEPA, 1983. Methods for Chemical Analysis of Water and Wastes. U.S. Environmental Protection Agency. EPA 600/4-79-020. p. 552.

WISE Uranium Project, <http://www.wise-uranium.org>.

World Nuclear Association, 2015a. World Uranium Mining Production. <http://www.world-nuclear.org/info/Nuclear-Fuel-Cycle/Mining-of-Uranium/World-Uranium-Mining-Production/>.

World Nuclear Association, 2015b. Australia Uranium. <http://www.world-nuclear.org/info/Country-Profiles/Countries-A-F/Australia/>.

World Nuclear Association, 2015c. Uranium and Nuclear Power in Kazakhstan. <http://www.world-nuclear.org/info/Country-Profiles/Countries-G-N/Kazakhstan/>.

World Nuclear Association, 2015d. Russia's Nuclear Fuel Cycle. <http://www.world-nuclear.org/info/Country-Profiles/Countries-O-S/Russia--Nuclear-Fuel-Cycle/>.

World Nuclear Association, 2015e. Khiagdinskoye Uranium Output from End-2016. <http://www.world-nuclear-news.org/ENF-Khiagdinskoye-uranium-output-from-end-2016-02021501.html>.

World Nuclear Association, 2015f. Uzbekistan Uranium. <http://www.world-nuclear.org/info/Country-Profiles/Countries-T-Z/Uzbekistan/>.

World Nuclear Association, 2015g. In Situ Leach (ISL) Mining of Uranium. <http://www.world-nuclear.org/info/Nuclear-Fuel-Cycle/Mining-of-Uranium/In Situ-Leach-Mining-of-Uranium/>.

Xuebin, S., Haifeng, W., Yixuan, Y., 2004. Field test of ISL uranium mining at Shihongtan deposit in Xinjiang, China. In: Recent Developments in Uranium Resources and Production with Emphasis on In Situ Leach Mining, Proceedings of a Technical Meeting Organized by the IAEA in Cooperation with the OECD Nuclear Energy Agency, the Bureau of Geology, and China National Nuclear Corporation Held in Beijing, International Atomic Energy Agency, September 2002. Report IAEA-TECDOC-1396.

Nuclear fuel from secondary supplies of uranium and plutonium

8

Jeff Combs
The Ux Consulting Company, LLC, Roswell, GA, United States

8.1 Introduction

For the purposes of this chapter, secondary supplies refer to all uranium or uranium equivalents that are above ground, in various forms, that could be used to supply reactors. Secondary supplies, in all of their forms, are huge, as they include all uranium mined and processed in the past for both military and commercial purposes, which has not been consumed. In fact, in some years, the use of secondary supplies have met one-third or more of reactor uranium requirements.[1]

Determining the availability of secondary supplies to the market can be a rather complex process. Inventories held by different entities, especially government and financial institutions, must meet certain conditions to be released to the market. Other secondary supplies must be processed (or "deprocessed") further, or must be used in a certain types of reactor. Thus, in order for some sources of uranium or uranium equivalents to become available for consumption in reactors, they must be processed into an acceptable form and be approved for sale.

8.1.1 Secondary supply classification

If we define secondary supplies as uranium or uranium equivalents that have already been produced, it becomes evident that all of these supplies or potential supplies and supply sources are not equivalent when it comes to their availability to the market. This availability depends on the form of the inventory, whether it is commercial (civilian) or military, and the ability of reactors to use these supplies.

Fig. 8.1 represents one way of representing secondary supplies. This diagram has two columns and three rows. The columns are labeled "commercial" and "government," and the rows show the form of the inventory. Basically, the difference between the first and second rows is one of fungibility. Material in the second row has to be processed further to get it into a form in which it can be used in reactors. In this respect, highly enriched uranium (HEU) needs to be blended down to low-enriched uranium (LEU) suitable for use in reactors, and tails material needs to

[1] For example, in 2003, secondary supplies met over 55 million pounds of 147 million pounds of reactor demand, as estimated by Ux Consulting.

Uranium for Nuclear Power. DOI: http://dx.doi.org/10.1016/B978-0-08-100307-7.00008-9

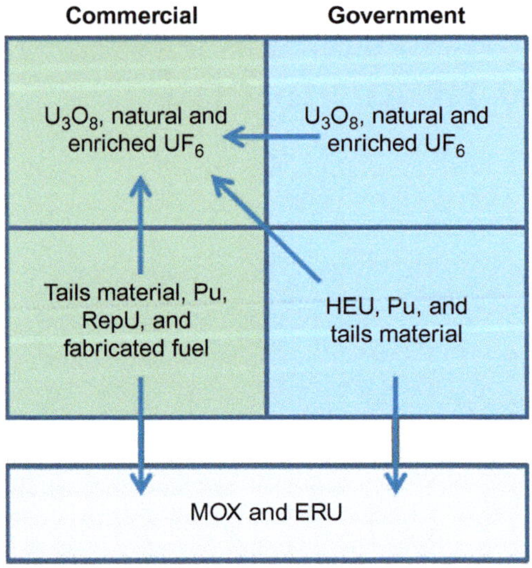

Figure 8.1 Secondary supplies.
Source: Original by author.

be enriched to either natural uranium in the form of UF_6 or LEU or blended with HEU to create LEU.[2] Plutonium and recycled uranium must also be recovered via reprocessing spent fuel and then processed into fuel for reactors.

The difference between the second and third rows is one of reactor usage. Only certain reactors can use mixed oxide (MOX) fuel that contains recycled plutonium and depleted uranium. The same is true for reprocessed uranium (RepU), which must be further enriched, yielding enriched reprocessed uranium (ERU), or blended with HEU or medium enriched uranium (MEU) to achieve a suitable reactivity. Another way of looking at this is that there is no broad commercial market for these types of fuel. Thus, the uranium or uranium derivative not only needs to be processed further (reprocessed), but it also needs a host reactor that can use it.

Note that it is possible to further subdivide the upper left-hand box, as the material in this box is not all equally available to the market for a variety of reasons:

- Pipeline (in-process) uranium held by utilities
- Strategic uranium inventory held by utilities
- Inventory being held by commercial banks as collateral
- Inventory being held by investment funds for exposure to uranium prices
- Inventory being held by fuel banks in case of supply disruptions.

The arrows show the steps that uranium or uranium equivalents must go through to be available to meet reactor needs. In this respect, governments might have some natural uranium they are looking to place in the commercial market. This uranium is not usually immediately available because the government must decide to release

[2] Normal uranium is uranium that has ^{235}U content of 0.7%, the fissile content of uranium in nature.

it, and in some cases this is subject to a public review. It is also the case that governments and international organizations have created fuel banks that hold inventories of LEU for potential use by utilities if certain conditions are met. These inventories properly belong in the upper right-hand box because they are not available to the general market and would only be available to specific utilities/countries under special circumstances.

As stated previously, secondary supplies in the second row require further processing. From a practical standpoint, much of these supplies fall in the right-hand column in that they are owned by governments. This is certainly true of all HEU and much of the tails material.[3] Thus, these supplies must be declared excess by governments and processed further to get them in a commercially usable form. This is not easy or straightforward, and when it comes to HEU, it can represent a herculean task. Even when it comes to tails material, there must be sufficient economic enrichment capacity to enrich the tails.

Thus, not all secondary supplies are equal, and their availability is conditional on their form and ownership. With this framework in mind, we will examine the different types of secondary supplies and some transformations in enrichment and reactor technology necessary to realize the full potential of these supplies.

8.1.2 The role of technology

In examining secondary supplies, it is important to understand the role of technology. Technology impacts the accessibility of secondary supplies in a variety of ways. Accessibility of supplies not only relates to any further processing that is necessary, but the ability of reactors to use different forms of supply; in both of these areas, technology can have a large impact on the role of secondary supplies for nuclear power.

For example, the necessary technology needed to be in place to convert and blend down HEU supplies so that they were available for reactor use. The same is true for the enrichment of tails material. Although this technology does exist (tails material is reenriched today), advances in enrichment technology can expand the amount of tails material and hence secondary supplies that are economically available. If laser enrichment can achieve both superior economics and can enhance the capability of enriching to lower tails assays, it will be able to transform more tails into usable material and reduce the need for newly mined uranium. Enrichment technology can also impact the amount of uranium available from underfeeding, the situation in which enrichment plants operate at a lower average tails assay than the one on which deliveries of uranium feed to the plants is predicated. However, the primary effect here is to reduce the demand for new uranium production, as lowering the tails assay reduces the amount of uranium needed in the production of nuclear fuel.

[3] The tails controlled by Russian and Chinese enrichers are considered government owned, due to the nature of these economies. Also, the DOE owns the tails inventories held in the US.

There is also the reactor dimension. Reactor fuel loadings must be designed (and licensed) to use mixed oxide fuel and recycled uranium. If reactors using this type of fuel are shut down and not replaced with others set up to utilize these fuel sources, then the contribution of MOX and RepU will decline. Importantly, advances in reactors in the future could allow reactors a greater use of recycled products or even enable them to create their own fuel, such as in the case of fast breeder reactors. Thus, the availability and usability of secondary supplies can change over time with advances in technology to either produce or use secondary supplies.

8.2 Commercial inventories of natural and enriched uranium

While commercial inventories of natural and enriched uranium are large, perhaps on the order of 300,000 tU or more, they are all not readily available to displace uranium production.[4] Some of these inventories are in process (in the pipeline), earmarked to meet reactors needs. Still others are held for strategic reasons, to be made available if there is a supply disruption or some other development. Suppliers and utilities hold inventories for this reason. In addition, uranium funds and fuel banks also hold inventories for very specific reasons.

Thus, there is a specific demand for inventories in addition to that associated with the ongoing fuel needs of reactors. This demand depends on the overall need for uranium and various reasons for holding it, and can be expected to vary over time as reactor requirements change, the market is deemed more or less secure, and prices fluctuate. This inventory demand itself can be quite large, and may be on the order of 200,000 tU based on current and expected reactor requirements and market conditions. Inventory supplies available to meet reactor requirements are those in excess of these various needs. However, even though entities may be holding inventories in excess of their needs, it does not mean these supplies are readily available to be sold into market, as there may be various factors restricting their immediate sale or use.

Next is information on inventories held in key regions of the world.

8.2.1 United States

Data on inventories held by US companies is collected and reported by the US Energy Information Administration (EIA). In its preliminary results for 2014, EIA reports that US companies held almost 52,000 tU (135 million pounds U_3O_8 equivalent) in the form of uranium concentrates (U_3O_8), natural UF_6, enriched UF_6, and fabricated fuel.[5] Of this amount, 44,600 tU (116 million pounds U_3O_8) were held by utilities and 7300 tU (19 million pounds U_3O_8) were held by suppliers. Only a small percentage of

[4] This figure does not include uranium contained in tails material and HEU scheduled to be blended down, reprocessed uranium, and the uranium equivalent of plutonium stocks.

[5] U.S. Energy Information Administration, 2014 *Uranium Marketing Annual Report*, May 2015.

these inventories are thought to be excess and become available for sale into the market, although some of these inventories may have been purchased to meet future utility needs and thus will be drawn down internally by utilities at a later date.

8.2.1.1 US government inventories

The US government also holds inventories earmarked for commercial use (or potential commercial use). Inventories of natural uranium acquired for past military programs have largely been disposed of. In addition, the US Department of Energy (DOE) purchased some Russian HEU feed as part of the Russia-US HEU deal which it has been disposing of. More recently, some DOE tails material has been enriched and sold to US utilities. Most of DOE's remaining inventories are currently in the form of HEU or tails material, and will be discussed later in the chapter.

DOE's sales and transfers of uranium inventories are subject to existing legislation, where the DOE Secretary must make a determination of "no adverse material impact" to the domestic (US) nuclear fuel industry.[6] For a number of years, the DOE Secretary made this determination despite falling uranium and enrichment prices, most notably in the wake of the Fukushima accident. The US nuclear fuel industry objected to these determinations, and pushed for a set limit on how much DOE could sell or transfer as a percentage of US uranium requirements.

In 2014, ConverDyn, a US uranium convertor, filed a lawsuit against DOE relating to DOE sales and transfers of uranium alleging, among other things, that the transfer of inventories was in violation of the USEC Privatization Act. Largely in reaction to this lawsuit, DOE slightly reduced the amount of its planned transfers, as shown in Table 8.1. Because it took this action, DOE

Table 8.1 Transfer volumes for Portsmouth cleanup & HEU down-blending in 2015 SD

	Concentrates (tU)	Conversion services (tU as UF_6)	Enrichment services (SWU)
2015	2500	2500	520,000
2016	2100	2100	520,000
2017	2100	2100	520,000
2018	2100	2100	520,000
2019	2100	2100	520,000
2020	992	992	520,000
2021	500	500	520,000
2022	500	500	520,000
2023	500	500	520,000
2024	500	500	520,000

Source: US Department of Energy (public information).

[6] The governing legislation is the USEC Privatization Act.

considered the matter settled and petitioned the court to end the case, but ConverDyn continued its challenge.

In May 2015, there was legislation introduced in the US Congress to restrict the ability of DOE to transfer uranium. This legislation, known as the Excess Uranium Transparency and Accountability Act, would restrict DOE transfers to 2100 tU in 2016–2023 and 2700 tU in 2024 and beyond. The lawsuit and the proposed legislation underscore the fact that DOE can face some constraints when making decisions to liquidate inventories.

8.2.2 Europe

The Euratom Supply Agency (ESA) reports inventories held by European Union (EU) nuclear utilities, both strategic and pipeline. In its report for 2014 activity, the ESA stated that EU utilities held 52,898 tU of uranium inventories, which it notes covers up to 3 years of EU gross reactor requirements.[7] The ESA advocates holding inventories for security of supply, so not much of this inventory is likely surplus to needs and thus would not available to the market. However, to the extent that some EU reactors are shut down early as countries move away from nuclear power, it is likely that some of this inventory would be drawn down to meet reactor needs or sold, as there would be no reason to hold strategic or pipeline inventories for reactors that will no longer be operating.

8.2.3 Japan

Typically, Japanese utilities have opted to hold rather large uranium inventories, as Japan has no indigenous supplies of uranium. Thus, they were holding very large inventories at the time of the Fukushima accident and these inventories have continued to grow as all Japanese reactors were shut down and remain offline, whereas fuel deliveries under some contracts continued.[8] UxC estimates that Japanese utilities held almost 75,000 tU in 2014 based on an analysis of information available in annual reports and other industry data.

A large share of these inventories are likely in excess of Japan's inventory needs as Japan will have fewer reactors operating in the future than prior to Fukushima. How much is excess depends on how many reactors eventually come back online, but most recently the Japanese government was targeting nuclear power to meet 20–22% of Japan's electricity needs by 2030, down from approximately 30% prior to the Fukushima accident.[9] Another reason that Japanese utilities may want to hold less inventory in the future is that its utility industry is undergoing deregulation and this could impact the economics of holding inventory. To the extent that Japanese

[7] Dariusz Kozak, Euratom Annual Report, 2014, PowerPoint presentation, ESA Advisory Committee meeting, April 28, 2015. http://ec.europa.eu/euratom/docs/ESA_%20Annual_report_2014_presentation_Kozak.pdf

[8] Only two reactors, Ohi 1 and 2, operated for a limited period following the routine post-accident shutdowns.

[9] "Japan Inc. not as keen as Abe government on nuclear power—Reuters poll," Reuters, May 24, 2015.

utilities use excess inventories internally, it is likely that they will not need to buy much uranium on the market for a relatively long time. In addition, it appears that some of these inventories are being sold on the market or at least sold back to suppliers which can deliver this material into other contracts.

8.2.4 China

Based on comparing China's uranium imports and domestic production versus its reactor needs, it is estimated that China has accumulated 105,000 tU of uranium inventory.[10] This is a huge amount, about twice the level of commercial inventory holding in the United States, which has the world's largest nuclear power program. China is in the process of rapidly expanding its overall nuclear power capacity and could overtake the US as the world's largest program by the next decade, and this prospective growth has likely motivated this large inventory buildup.

It is difficult to say how much, if any, of this inventory is excess to China's needs. Uranium is seen as having geopolitical importance in China, and because China has little in the way of domestic uranium resources and it will take time for its investments in foreign uranium mines to pay off, building large inventories relatively quickly is the best way to assure the availability of uranium supplies. It may be the case that some of these inventories are worked off internally as China's foreign uranium mines come into production and its domestic enrichment production capacity expands. Once China has more uranium and enrichment production in place, it may be more comfortable holding less material in inventory for strategic reasons.

A critical question here is China's nuclear power growth. If China greatly expands its nuclear power capacity, it is less likely to reduce its inventory holdings and may even add further to its inventory. However, if China's nuclear power growth stagnates, it may look to reduce its inventory holdings. Given the role of the Chinese government in China's nuclear power program and its economy in general, it makes little sense to try and differentiate inventories controlled by commercial entities inside of China and inventories controlled by the government.

8.2.5 Russia

Russia is also thought to hold large uranium inventories, as it does not have large, economic uranium resources. However, inventories held in Russia are a state secret, and thus there is no transparency with respect to inventory holding or disposition plans as is the case in the United States. In the past, the Soviet Union held large inventories of uranium based on deliveries from uranium producing regions like Kazakhstan and Uzbekistan and Eastern Europe (East Germany). These were used to fuel its growing nuclear power program which received a serious setback following the Chernobyl accident.

[10] This is an UxC estimate. Note that this amount would be in addition to any inventory that the Chinese government was holding for civil or military purposes prior to its importing of uranium.

It is thought that while Russia still holds notable stocks of natural uranium and UF_6, some of this from repatriated HEU feed, it has utilized much of this material over time. As a result, it may be the case that a larger portion of Russian inventories and thus potential secondary supplies are currently in the form of tails material and HEU, as well as being generated via underfeeding. As will be discussed later in this chapter, Russia plans to use these inventories to meet domestic demand and export commitments. Like the case with China, it makes little sense to differentiate inventory holding between government and commercial entities as the Russian government is heavily involved in all phases of Russian nuclear power and fuel cycle.

8.3 Other natural and enriched inventories

Inventories are held by other entities for financial or supply assurance and nonproliferation reasons.

8.3.1 Uranium participation corporation

In 2005, the Uranium Participation Corporation (UPC) was created as a publicly traded stock company listed on the Toronto Exchange as a way of giving investors the opportunity to participate in the then rapid appreciation of uranium prices. UPC purchases uranium (in the form of either U_3O_8 or UF_6) in the market and investors buy shares in UPC to underwrite these purchases and to own a share of UPC's uranium holdings. UPC currently holds a little more than 5500 tU (14.5 million pounds U_3O_8) equivalent in inventory in the form of U_3O_8 and UF_6.[11] About two-thirds of this inventory is held in the form of U_3O_8.

This type of inventory holding is solely driven by financial considerations to give investors exposure to the uranium price and market. So far, the fund has been operated in a way that there have only been purchases into the fund, with no sales from the fund. However, UPC has loaned out some of its inventory to defray some of its holding costs and has sold off some of the conversion component of the UF_6 it holds. UPC can sell any or all of its inventory at anytime (if it could not the inventory and the related stock would have no value). One potential exit strategy for UPC is to sell the entire inventory to one buyer.

8.3.2 Fuel banks

Another type of inventory that is theoretically available to meet reactor needs is that held in fuel banks. Fuel banks were established to provide a backup to the commercial market in case of supply disruptions. Access to supplies in fuel banks is generally conditioned on a supply disruption that cannot be remedied by the market and the receiving party having good nonproliferation credentials. While there is no

[11] "Uranium Participation Corporation Reports Net Asset Value at April 30, 2015," May 5, 2015.

requirement that countries forgo the right to develop their own fuel cycle facilities (including enrichment and reprocessing) to have access to fuel bank material, the aim of fuel banks is to provide fuel supply assurances such that development of these facilities may be deemed unnecessary.

There are currently two fuel banks, one in Russia and one in the United States. In addition, the International Atomic Energy Agency (IAEA) is in the process of developing a fuel bank, and the United States has instituted a MOX backup inventory program, which while different in function from the other fuel banks, holds inventories of LEU in case of disruptions/problems with supplies from commercial contracts signed for MOX supplies created in US facilities.[12] Table 8.2 next summarizes the various nuclear fuel banks.

The Russian fuel bank, which was set up under IAEA auspices, holds 120 t of LEU. The EUP for the IAEA fuel bank in Kazakhstan has not yet been purchased, although funds have been allocated. The irony is that the funds allocated for the fuel purchases can now be used to acquire considerably more EUP than when they were first committed because enrichment and uranium prices have fallen dramatically in recent years.[13] The US fuel bank, the American Assured Fuel Supply (AAFS) holds 230 t of LEU which is equivalent to about six reloads for a 1000 MWe reactor. Direct access to this fuel bank is limited to US companies, and non-US companies may only have access to this material via their US suppliers. The MOX Inventory Backup Program is estimated by UxC to contain about 168 t of LEU and continues to grow to the extent that material from the US highly enriched down-blending program is placed in this stockpile. It, too, would be limited to US companies, specifically utilities which would have contracts for MOX fuel produced in the United States, if such a program ever moves forward.

Fuel banks create their own demand separate from reactor requirements and normal inventory holding. The supplies held in fuel banks are rather low relative to overall reactor needs, representing less than 20 reloads for a 1000 MWe reactor in total, most of which is located in the US. Furthermore, it may be that uranium from fuel banks is used if certain situations arise, but in this case these inventories would be replenished. The exception to this would be the uranium contained in the MOX backup inventory program that would no longer be necessary if the United States does not pursue a MOX program. Thus, for the most part, fuel banks do not represent a net future supply to the market, as any outflows to the market would need to be replaced by purchases, unless it was decided that they were no longer necessary.

8.3.3 Future trends

Much of the current buildup and use of commercial uranium inventories is a vestige of the excess inventory supplies created by the Fukushima accident, along with the

[12] In June 2015, the IAEA Board of Governors approved agreements to establish a fuel bank in Kazakhstan. The IAEA has not yet procured the EUP. It has funds to purchase 70–74 t of EUP at today's prices.

[13] Funds were initially obtained in 2010, when enrichment prices were about twice what they are today.

Table 8.2 **Nuclear fuel banks**

	Russian fuel bank (under IAEA auspices)	**IAEA fuel bank**	**AAFS (US)**	**MOX backup inventory program (US)**
Status	Inaugurated in December 2010	Establishment approved in 2010; siting agreements signed in 2015	Availability announced in August 2011	Some material available, will be fully formed by the end of 2015
Location	Angarsk electrolysis chemical combine, (International Uranium Enrichment Center), Russia	Ulba Metallurgical Plant, Kazakhstan	United States	United States
Ownership while in storage	Russia	IAEA	United States	United States
Material available	120 t LEU	Financial funds: $125 million and €25 million	230 t LEU	~168 t LEU
Storage costs	Russia	IAEA	United States	United States
Supply criteria	Nonmarket-related supply disruption; good nonproliferation credentials	Nonmarket-related supply disruption; good nonproliferation credentials	Eligible recipients meeting certain nonproliferation criteria in the case of supply disruption	Backup to MOX commercial contracts
Price	Market prices	Market prices	Market prices	Market prices
Agreements	Russia and IAEA / IAEA and customer	Host country and IAEA / IAEA and customer	DOE will sell material to a US person who in turn will sell it to the foreign customer	N/A
Customer's commitment	Nonproliferation; nuclear safety and security; re-export, enrichment, reprocessing only with IAEA agreement	Nonproliferation; safety and security; enrichment, reprocessing, and re-export only with IAEA agreement	Good nonproliferation credentials	N/A

Source: The Ux Consulting Company, LLC.

desire of the Russian and US governments to dispose of certain inventories stemming from past military programs, or, in the case of Russia, utilize excess enrichment capacity from past military programs. In general, more inventories are being held than desired, and these inventories will be pared down over time as reactor requirements grow and utilities are better able to consume them. This may take some time, as witnessed by the difficulty that Japan is encountering in restarting its reactors. An exception to this could be China, which may want to continue to build inventories for strategic reasons, although this would be a function of its underlying nuclear capacity growth and development of its foreign uranium projects.

US government inventories available for commercial nuclear use are winding down, but their disposal may be even more protracted depending on how a current lawsuit and legislative initiatives play out. It also depends if DOE is able to enrich its tails material in the future, as its tails inventories need further processing. However, the enrichment of tails would occur over a number of years as there would be limits to how much could be processed in any one year.

In theory, inventories held by UPC and various fuel banks could be held indefinitely and thus may never be used. These inventories could build over time, depending on market and geopolitical events. For instance, if investors thought that uranium prices would appreciate notably, there may be more demand to gain exposure to uranium through UPC's holdings. There could be more fuel banks or fuel bank type arrangements as well. Recently, at the ATOMEXPO 2015 conference in Russia, it was announced that the BRICS states (Brazil, Russia, India, China, and South Africa) were considering setting up their own nuclear fuel reserve. In general, these types of arrangements create more demand for nuclear fuel; although they are potential sources of supply should certain developments occur in the future.

8.4 HEU supplies

8.4.1 Background

A large source of secondary uranium supplies has been HEU from dismantled nuclear warheads that have been blended down for reactor use. The enrichment level for nuclear weapons is typically over $90^{w}/_{o}$. HEU can also come from fuel for nuclear-powered submarines, although this is not as large a source of HEU as weapons material.[14]

The notion of blending down HEU for reactor fuel is relatively new, and the prospect for doing this was first mentioned in a 1989 article by Combs and Neff that explored unconventional possibilities for nuclear fuel trade between the United

[14] HEU is defined as uranium enriched to over $20^{w}/_{o}$ ^{235}U, and can include material from nuclear weapons, submarine fuel, and fuel for certain research reactors. Weapons material has enrichment levels in excess of $90^{w}/_{o}$.

States and the Soviet Union.[15] Neff went on to propose what later became the Russian–US HEU deal (discussed next) in an op-ed in *The New York Times*.[16]

For HEU from nuclear weapons to be transformed into reactor fuel, it has to undergo a number of steps, one it is removed from the warhead itself. It must be converted (in the case of Russia it was converted from a metal to highly enriched UF_6 gas) and be blended down to reactor grade (less than $5^{w}/_{o}$ ^{235}U). HEU can be blended down using natural uranium, tails material, or slightly enriched uranium. The ^{235}U content of the blendstock needs to be below the assay of the final product, and the closer the assay of the blendstock is to that of the final product, the more blendstock that needs to be utilized to get to the desired product assay.

8.4.2 The Russian–US HEU deal

One of the major sources of secondary supplies in recent years has been the HEU deal between Russia and the United States, which involved the delivery of 150,000 tU of uranium feed and 90 million SWU contained in about 14,400 t of LEU delivered over a 20-year period that ended in 2013.[17] As noted previously, the HEU deal was suggested by Neff and eventually led to an agreement between the Russian and US governments in 1993. The deal involved blending down 500 t of HEU, equivalent to the content of 20,000 nuclear warheads, into reactor-grade fuel. This government-to-government agreement was followed by an implementing contract between the United States Enrichment Corporation (USEC), acting as the executive agent for the United States, and Techsnabexport (TENEX) acting as the executive agent for Russia. USEC purchased the enrichment component of the down-blended LEU (about 90 million SWU), but not the feed component. The inability of the feed component to be sold at first threatened to scuttle the deal, but the US DOE stepped in to purchase some of the feed, and later a consortium of the uranium producers AREVA and Cameco, and the trader NUKEM, entered into an agreement to purchase much of the remaining feed, although a portion (38,000 tU) was retained by TENEX and returned to Russia.

The blendstock in the Russian–US HEU deal was slightly enriched tails material. The reason that this blendstock was used is because HEU has a relatively high concentration of ^{234}U and ^{236}U, while tails have a much lower proportion of these isotopes, both a function of the enrichment process. Blending HEU with a blendstock that has a lower concentration of these undesirable isotopes makes it possible to achieve a low enough concentration of these undesirable isotopes in the final product to meet American Society for Testing and Materials (ASTM) specifications.

[15] Combs, Jeff and Dr Thomas Neff, "The Soviets, SWU, and U—A Win-Win Solution?" *NYNCO Newsletter*, November 1989.

[16] Dr Thomas Neff, "A Grand Uranium Bargain," Op-ed, *The New York Times*, October 24, 1991.

[17] Some of this discussion comes from the WNA's "Military Warheads as a Source of Nuclear Fuel" updated August 2014.

Bukharin notes that in the US-Russia HEU deal 8555 t of depleted uranium (tails) at an assay of $0.25^{w}/_{o}$ were enriched using 5.34 million SWU to create 916.6 tU of LEU at a $1.5^{w}/_{o}$ product assay. This was blended with 30 t of HEU to produce 946.6 tU of LEU at a 4.4% product assay containing 5.52 million SWU and 9000 tU (about 24 million pounds U_3O_8).[18] It is noteworthy that almost as much enrichment was used to create the blendstock as was contained in the final product.

8.4.3 US HEU blend down

The United States has also blended down some of its HEU, but not nearly to the extent as Russia. The US initially declared 174 t of HEU excess to its needs and later added to this amount, although not all of this material has become available for down-blending. Some of it was utilized for naval propulsion and a portion was declared waste, reducing the total volumes available to DOE for down-blending into reactor-grade fuel. Note that US stocks of HEU are reported to be lower than those of Russia and some of these are contaminated, making them unsuitable to be easily converted to reactor fuel.

In 1997, DOE signed a memorandum of understanding with the Tennessee Valley Authority (TVA), to use off-spec LEU in TVA's reactors; after a demonstration was completed, DOE and TVA entered into an agreement in 2001. To create this fuel, DOE agreed to supply approximately 33 t of off-spec HEU to be blended down with natural uranium to reduce the ^{236}U content (along with other impurities) and make the resulting product more suitable for reactor use. This program is known as the blend low-enriched uranium (BLEU) program. In 2008, additional HEU was added to this program. To date, a total of 46 t of HEU has been processed as part of this program and it has been extended by down-blending an additional small quantity of off-spec HEU to be loaded between 2017 and 2023.

What has not gone into the BLEU program has largely gone into the AAFS fuel bank and the MOX backup inventory program discussed earlier. Less than 20 t of HEU remains to be blended down. Overall, DOE anticipates completing the down-blending of the available HEU by 2024 at an expected annual rate of 2−3 tU. A portion of the down-blended HEU has been used as payment to the companies providing the down-blending service (payment in kind) and has been subsequently sold into the market.

8.4.4 Future blend down of HEU

At the moment, there appears to be little prospect of another Russian−US deal, despite the great success of the deal that did take place and the fact that there are still large stocks of HEU. Importantly, Russia is not in the dire economic straits it was in during the 1990s and US and Russian relations have currently deteriorated quite a bit due to the situation in Ukraine.

[18] Oleg Bukharin, "Understanding Russia's Uranium Enrichment Complex," *Science and Global Security*, 12: 193−218, 2004.

At the same time, it is possible that Russia could choose to blend down additional quantities of HEU for its domestic use or perhaps for its export market. Overall, the USSR produced an estimated 1200 t of HEU, of which 500 t were used for the HEU Agreement, leaving ~700 t of HEU in Russia.[19] The Russian government has no formal process to declaring the material surplus and has given no indication at this time that it would consider doing so. It is interesting to note that, for internal use, Russia can blend its HEU with depleted uranium (tails) as it does not have to worry about meeting ASTM specifications for the final product. This approach is more economic as it preserves both Russian natural uranium supply and its enrichment capacity.

As mentioned previously, the United States is still planning to blend down some additional HEU, perhaps at the rate of 2–3 t per year, but this effort has been stymied recently by the decline in nuclear fuel market prices following the Fukushima accident, making the uranium and enrichment components of the HEU less valuable and less able to cover the costs of blending down the HEU. Because the United States no longer has enrichment capacity based on US technology, it cannot currently create more HEU and thus does not have SWU capacity based on US technology to enrich uranium that can be used in the production of tritium, which is used a booster in thermonuclear weapons.[20] This can be a motivation for the US to preserve its inventory of HEU, as it can be blended down and used for tritium production in the future.

8.5 Recycled uranium and plutonium[21]

Uranium and plutonium derived from reprocessing spent fuel can be used as fuel in light water reactors. The issue here is that when the original nuclear fuel is "burned" in a reactor, it is not completely consumed, but most of the uranium can be recovered along with some plutonium, which is created as part of the fission process. Uranium recovered from spent fuel after reprocessing (RepU) needs to be further enriched yielding ERU, but substitutes for freshly mined uranium.[22] Plutonium is mixed with depleted uranium (tails) to create a MOX fuel.

[19] The International Panel on Fissile Materials estimates that Russia had 695 ± 120 t of HEU at the end of 2012. International Panel on Fissile Materials, Global Fissile Material Report 2013.

[20] The only current enrichment technology in the United States that is commercially operable is that owned and operated by URENCO, a British/Dutch/German consortium. There are usually prohibitions against using non-U.S. enrichment technology and uranium for military purposes in the U.S., and in any case, the U.S. has limited itself to the use of U.S. enrichment technology and U.S.-origin uranium feed for these ends. The U.S. DOE has been funding the American Centrifuge Program in large part because it might need U.S. domestic technology in the future to satisfy some of its military programs.

[21] Much of this discussion is based on an essay entitled "The Estimated Impact of Recycling on the U_3O_8 Market," UxC *Uranium Market Outlook Q1 2015*, June 2015.

[22] RepU has about a $1^{w}/_{o}$ fissile content, which is greater than uranium found in nature ($0.7^{w}/_{o}$), but RepU must be enriched to higher levels than natural uranium to yield the same performance as freshly mined uranium due to the presence of undesirable isotopes. It also can be blended with HEU or MEU to achieve these same results.

Recycle can be done for economic reasons but is also done to reduce stocks of plutonium that have accumulated over time. It is a step that necessarily follows reprocessing of spent fuel, which is undertaken both to facilitate the management of spent fuel as well as to recover its fissile products and byproducts for further use as fuel. The economics of recycle, which a function of the costs of reprocessing and MOX fuel fabrication as well as market prices for uranium enrichment, were not very attractive prior to Fukushima, but after the accident and the subsequent decline in nuclear fuel demand and prices, recycle economics became even more problematic.[23] Whereas in 2007 some market participants were worried that the world was "running out" of uranium, there is now much less concern about the availability of future uranium supplies.

Current estimates of civil separated Pu place the Pu inventory at 260 t based on data reported to the IAEA. In addition, plutonium originally created for military purposes can be used as fuel. After the Cold War, more than 100 t of military Pu were declared excess by the US and Russia. In July 2011, the United States and Russia each agreed to dispose of 34 t of weapons plutonium by converting this material into MOX fuel for use in nuclear reactors.[24] While current inventories of RepU are not publicly available, UxC estimates these to be 75,000−90,000 t based on past reprocessing activity.

Looking at the future, UxC estimates that nearly 45,500 t will be reprocessed over the 2014−2030 period, yielding about 450 t of separated Pu and about 43,000 t of RepU. To the extent that the resulting RepU and plutonium are recycled in reactors, they will displace the need for newly mined uranium. For instance, the ESA notes that the use of MOX fuel resulted in a savings of 1156 tU in 2014.[25]

UxC estimates that MOX will displace about 16,500 tU (43 million pounds U_3O_8) over the 2014−2030 period, equaling about 1.2% of the almost 1.4 billion tU (3.566 billion pounds U_3O_8) projected to be required over the 17-year period. As far as ERU is concerned the reduction will be 20,270 tU (52.7 million pounds U3O8) over the same period equaling about 1.5% of the uranium required over this period. Thus, in total, MOX and ERU fuel are projected to displace about 36,800 tU (almost 100 million pounds U_3O_8) over the 2014−2030 period, accounting for about 2.7% of the uranium required to fuel reactors over this period, as shown in Fig. 8.2.

8.6 Future of recycled uranium and plutonium

As shown in Fig. 8.2, current plans are for the contribution of recycled uranium and plutonium to decline over time. A major reason for this is that some of the reactors that use MOX and RepU fuel will be shut down. The use of recycled

[23] When it comes to reenriching RepU, there is an added cost associated with dedicating a portion of an enrichment plant's capacity for this purpose to avoid contamination of centrifuges with undesirable isotopes.

[24] International Panel on Fissile Materials, *Global Fissile Material Report 2013*.

[25] EURATOM Supply Agency Annual Report 2014, p. 29.

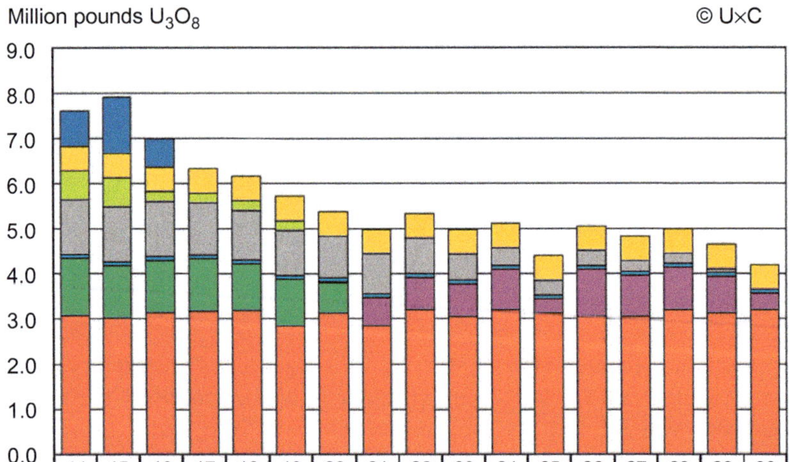

	14	15	16	17	18	19	20	21	22	23	24	25	26	27	28	29	30
United States	0.8	1.3	0.6	0.0	0.0	0.0	0.0	0.0	0.0	0.0	0.0	0.0	0.0	0.0	0.0	0.0	0.0
United Kingdom	0.5	0.5	0.5	0.5	0.5	0.5	0.5	0.5	0.5	0.5	0.5	0.5	0.5	0.5	0.5	0.5	0.5
Switzerland	0.6	0.6	0.2	0.2	0.2	0.2	0.0	0.0	0.0	0.0	0.0	0.0	0.0	0.0	0.0	0.0	0.0
Russia	1.2	1.2	1.2	1.1	1.1	1.0	0.9	0.9	0.8	0.6	0.4	0.3	0.3	0.2	0.2	0.1	0.0
Netherlands	0.1	0.1	0.1	0.1	0.1	0.1	0.1	0.1	0.1	0.1	0.1	0.1	0.1	0.1	0.1	0.1	0.1
Japan	0.0	0.0	0.0	0.0	0.0	0.0	0.0	0.6	0.7	0.7	0.9	0.3	1.1	0.9	1.0	0.8	0.4
Germany	1.3	1.2	1.2	1.2	1.0	1.0	0.7	0.0	0.0	0.0	0.0	0.0	0.0	0.0	0.0	0.0	0.0
France	3.1	3.0	3.1	3.2	3.2	2.8	3.1	2.8	3.2	3.0	3.2	3.1	3.0	3.0	3.2	3.1	3.2

Figure 8.2 Combined annual uranium displacement impact of MOX and ERU Fuel.
Source: The Ux Consulting Company, LLC.

products beyond 2030 will be a function of the future economics of uranium and enrichment production, future efforts to reduce plutonium stocks (both civilian and military), future operation of reactors that use these products, and future advances in reactors and reactor fuel design. This last point will be examined later in this chapter.

8.7 Enrichment of tails material

8.7.1 Overview

As discussed in the introduction, the fissile content of uranium that is mined is not completely consumed when it is processed or burned in a reactor. When uranium is enriched, some of the fissile content remains in the waste stream, or tails. The assay of the tails depends on whether more or less uranium (and less or more enrichment) was used to create the enriched product at the time of enrichment. Tails assays can generally range from $0.10^{w}/_{o}$ to $0.40^{w}/_{o}$ and can fall outside of this range under extreme situations. Because the ^{235}U assay of uranium found in nature is $0.711^{w}/_{o}$, tails with an assay of $0.40^{w}/_{o}$ have much of the fissile content remaining, while tails as low as $0.10^{w}/_{o}$ have little fissile content remaining, and thus are not economic to reenrich. The economics of enriching tails material depends both on the assay of the tails and the cost of enrichment, as well as the value of uranium feed. In the case of centrifuge technology, the cost of enrichment can be quite low as variable costs (chiefly the consumption of electricity) are low and enrichers generally prefer to keep centrifuges spinning because shutting them down and restarting can lead to failures and loss of capacity.

All enrichers can reenrich tails, but to date, the vast bulk of tails reenrichment has been done in Russia, where Russia has enriched not only domestic tails material, but also tails imported from Europe. The United States has had some well-publicized tails reenrichment activity, but its scope has been far less than that of Russia, although it could be expanded with advances in enrichment technology.

8.7.2 Russian tails reenrichment activities

Russia has represented by far the largest source of tails reenrichment, due to its large and relatively inexpensive enrichment capacity (developed in large part to meet earlier military needs), early transition to centrifuge technology, its historically large stock of tails, and its relatively low level of natural uranium resources.

8.7.2.1 Domestic consumption and export

The Chernobyl accident and the end of the Cold War left Russia with a considerable amount of excess enrichment production capacity. Furthermore, following the breakup of the Soviet Union, the major uranium producing regions of Kazakhstan and Uzbekistan were no longer captive to Russia. Because Russia does not have large economic uranium resources, it made economic sense for it to enrich tails material using its excess enrichment capacity and its abundant supply of tails to fuel its domestic reactors and for export. Bukharin notes that in 1992, Minatom (the then Russian Ministry of Atomic Energy) planned to increase the enrichment capacity devoted to tails enrichment from 1.29 million SWU in 1993 to 6.44 million SWU in 2000 through 2010.[26] Also, the Russian enrichment complex was originally supposed to enrich low assay tails and move to higher assay tails, but it appears that it did not follow this program as it was enriching $0.36^{w}/_{o}$ (high assay) tails as early as 1992.[27]

It is thought that by now Russia has enriched most or all of its high assay tails, and currently enriches tails material with average assays of around $0.20^{w}/_{o}$. Russia does have considerable enrichment capacity in excess of the demand for its enrichment services, and is thought to operate this capacity at a tails assay of $0.10^{w}/_{o}$ to further strip its available tails material. It is estimated that Russia currently uses one-third of its excess SWU capacity to enrich tails. This could produce on the

[26] Bukharin, op. cit.
[27] Bukharin, op. cit.

order of 1153 tU (3 million pounds U_3O_8) per year to the extent that this tails feedstock and SWU capacity are available in the future.[28]

8.7.2.2 HEU blendstock

As noted previously, uranium tails enriched to an assay of 1.5$^{w}/_{o}$ were used as the blendstock for HEU in the Russian–US HEU deal. This process consumed a little over 5 million SWU of Russia's enrichment capacity on an annual basis. While the enrichment component of the LEU derived from the HEU was sold to USEC, a portion of the uranium feed was returned to Russia and the remainder was purchased by a Western consortium for sale in the market. The material returned to Russia has become part of the state-owned inventory and is not readily available to the Russian nuclear industry at this time.

8.7.2.3 AREVA and URENCO tails

In the late 1990s, Russia entered into agreements to enrich tails material from AREVA and URENCO. The tails shipped to Russia were relatively high assay (0.30$^{w}/_{o}$ and above). While some of the tails were likely used for the HEU blendstock, Russia returned both natural (0.711$^{w}/_{o}$) uranium and enriched uranium to the European enrichers. Overall, these deals represented a sizeable amount of natural uranium supply, over 2300 tU (6 million pounds U_3O_8) per year that were returned to the European enrichers plus another 923 tU (2.4 million pounds U_3O_8) contained in EUP. With the contracts expiring in 2009 and 2010, Russia decided it would not extend the contracts as they were a public relations nightmare and also they were economically unjustified in light of the then new conditions in the enrichment market.

Russia continues to enrich tails material with its excess SWU capacity, as it looks to utilize its excess enrichment capacity to generate additional uranium supplies. The Fukushima accident left Russia with considerable excess SWU capacity and the enrichment market oversupplied in general. As a consequence of these developments, Russia is likely to have ample SWU capacity to devote to the enrichment of tails for the foreseeable future. One issue that may confront Russia is the lack of relatively high assay tails material to enrich.

8.7.3 US tails enrichment

Tails material has also been enriched to natural uranium levels in the United States. There have been two tails deals in the United States, both involving Energy Northwest (ENW) with the enrichment done at USEC's then operating gaseous diffusion plant in Paducah, Kentucky. The first, done in 2005, involved tails enriched by USEC and supplied to ENW and involved about 1923 tU (5 million pounds

[28] Given the poor quality of Russian tails, UxC estimates that about one-third of Russia's excess enrichment capacity is devoted to tails enrichment, while the remaining two-thirds of capacity is utilized for underfeeding. This analysis assumes stripping form 0.20$^{w}/_{o}$ to 0.10$^{w}/_{o}$, the assumed operational tails for Russia.

U_3O_8). The second, from 2012, involved 9075 t of tails enriched by USEC to 4.4$^{w}/_{o}$ with the product being delivered to ENW and TVA. This deal resulted in the production and sale of 482 t of EUP and another 1600 tU of UF_6.

Enrichment of tails in the United States was performed using gaseous diffusion enrichment technology, which is more expensive than centrifuge but does not incur the same operational penalty for being shut down or having its output cut back. Thus, it is more sensitive to market price pressures but can react more definitively toward them. In this regard, there was to have been a third tails enrichment campaign in the US following the 1-year tails enrichment program of ENW/TVA/USEC, but it became uneconomic when uranium prices fell following the Fukushima accident. Given the relatively high cost of enrichment using the antiquated gaseous diffusion technology employed by USEC, the enrichment of tails became uneconomic once the uranium price dropped below $60.

8.7.4 Future of enriching tails material

As previously noted, Russia is likely to have excess enrichment capacity for a long time in the future, so it will clearly have the ability to reenrich tails; the question is what assays of tails material it will have at its disposal to reenrich. In this respect, if Russia is enriching at a 0.10$^{w}/_{o}$ tails assay now, it will not be economic or even technically feasible to further strip the resulting tails due to their very low fissile content.

As mentioned previously, DOE is currently considering enriching tails at Paducah using an enrichment plant based on the Silex laser technology that would be built by GE-Hitachi Global Laser Enrichment (GLE). This technology promises lower costs and an ability to strip tails to lower levels, making further tails enrichment more economic. As of now, plans are to enrich tails to create 2000 tU (5.2 million pounds U_3O_8) of natural uranium per year for a period of 15 years, resulting in 30,000 tU (78 million pounds U_3O_8) of additional secondary supplies via tails enrichment.[29] In addition to the economics of future tails enrichment in the US, the ability of DOE to sell uranium into the market has been challenged and some restrictions may be placed on the annual quantities of uranium that can be sold by DOE, including those generated by the enrichment of tails, based on legislation that has recently been proposed.

Further in the future, the enrichment of tails could represent an important source of uranium feed to the extent that enrichment technology improves. There are two facets of this improvement. First is the economics of enrichment, which depends on advances in enrichment technology. The lower the cost of enrichment, the greater the quantity of tails that can be economically reenriched. Second, also associated with technology advances is the ability to physically enrich at lower tails assays, which expands the recovery of uranium. Thus, if advances in technology occur which both reduce costs and lower the assay at which the enrichment plant can physically strip tails, then more tails material can be economically processed.

[29] The RFP for this work has a provision for two optional extensions of five years each.

Ultimately, there will be diminishing returns even with tremendous advances in technology as the assays of the tails feedstock (the tails material being enriched) would continuously decline as more of the fissile content of uranium was recovered through the more efficient processes. However, this would likely take some time to occur, as there is an abundance of tails material, especially at lower tails assays.

8.8 Underfeeding

Another type of secondary supply is created when enrichers operate at lower tails assays than those at which they transact. The result of this activity, which is commonly referred to as underfeeding, is that enrichers end up with extra uranium feed they sell into the market. This material ends up competing with newly mined uranium and inventories in meeting the demand for uranium.

The amount of underfeeding that will take place in the future depends on both the tails assay at which enrichers sell their services to utilities and the tails assay at which they operate their plants. This, in turn, depends on the competitiveness of the enrichment market and the overall need for enrichment. In the recent past, the average transaction tails assay has been notably above the average operating tails assay and even well above the optimal tails assay, which has dipped below $0.20^{w}/_{o}$.[30] As a result, enrichers have accumulated a lot of uranium and have been big sellers on the spot, mid-term, and even long-term contract markets.

8.8.1 Underfeeding vs tails enrichment

Tails enrichment and underfeeding are sometimes thought of as essentially the same thing—utilizing excess enrichment capacity to create uranium, but they are different in important respects. The enrichment of tails creates new uranium supply by enriching tails material (depleted uranium) up to the fissile level at which uranium is found in nature. Underfeeding, on the other hand, reduces the amount of uranium that is needed in the future by using less uranium in the enrichment process to produce the same amount of enriched product. Uranium "created" as the result of underfeeding can be thought of as a secondary supply (and perhaps a major source of future secondary supply) in that it does not come *directly* from uranium producers. However, it only becomes a supply if more uranium is delivered to enrichers by utilities than is actually needed in the enrichment process; no new uranium is created.

While there are inventories that exist today that have been created by underfeeding, future supplies of uranium from underfeeding depend on the contracting prerogatives of enrichers and utilities, and thus do not relate to any current physical supplies of uranium. That is, if all enrichers transacted enrichment services at the

[30] The optimal tails assay is the assay that provides the lowest-cost mix of uranium feed and enrichment in the creation of enriched product. This valuation is usually based on the prevailing spot prices of enrichment, uranium, and conversion.

same tails assays at which they operated their plants, there no uranium would come from underfeeding in the future.

The difference between underfeeding and tails enrichment can be looked at another way as well. Underfeeding relates to the reduction of the future need for uranium, while tails enrichment, in essence, reduces the past requirements for uranium to the extent that tails from previous enrichment activities are still available to be reenriched.

Finally, another difference is that enrichment plants can be built specifically to enrich tails, and in this sense they can be thought of as being akin to uranium mines. Indeed, the proposed enrichment plant to be built at Paducah by GLE with laser enrichment technology developed originally by Silex Systems in Australia would only be used for the enrichment of tails, and not for providing conventional toll enrichment services.[31] The plant would be configured a little differently for this purpose.[32] Other plants or cascades within existing centrifuge plants can be similarly configured to enrich tails in the future.

8.8.2 Future underfeeding trends

Most recently, as competition has intensified in the enrichment market, it appears that enrichers are giving their customers more flexibility in selecting their tails assay, allowing them to choose lower tails assays, perhaps below 0.20w/o. To the extent this flexibility exists and utilities opt to transact at lower tails, the amount of uranium they need to deliver to enrichers will decline and thus the secondary supplies generated from underfeeding will also decline. There will likely always be some degree of underfeeding even if utilities are given complete flexibility in selecting their tails because the optimal tails for enrichers to operate their plants will be lower than the optimal tails at which utilities transact.[33] Finally, declining secondary supplies from underfeeding should not be taken as a sign that more uranium needs to be produced. To the extent that secondary supplies from underfeeding decline, so does the amount of uranium that utilities need to supply to enrichers to meet their enrichment contract obligations. These two developments offset each other. All that is occurring is that enriches are giving their utility customers a greater ability to select tails and thus to optimize their fuel costs.

8.9 Implications of generation IV nuclear reactors

Advances in reactors can bring with them new fuel designs that need less, or perhaps no, natural uranium at all to fuel them. In this case, some of the secondary supplies discussed here, including plutonium and depleted uranium, can meet more of future demand. However, this is more of a demand-side development than

[31] Silex Systems announcements: www.silex.com.au
[32] Private communication: Michael Goldsworthy, Silex Systems.
[33] This is because enrichers' costs are typically lower than market prices of enrichment.

a supply-side one, because it depends mostly on what happens with respect to reactor technology.

As an example of this type of consumption, Russia is planning to use weapons plutonium manufactured into MOX fuel in its BN-600 and BN-800 fast neutron, breeder reactors.[34] Fast reactors can consume "nonstandard" secondary supplies including plutonium and depleted uranium. In general, generation IV reactors can use a wide variety of fuel, including thorium, and thus are less dependent on traditional uranium supplies.[35]

Advanced reactors and different fuel cycles are most likely to be sought by countries that have low uranium resources. The chief candidates here are China, India, and Russia, all of which possess relatively low levels of uranium resources and production, but all of which have relatively large and expanding nuclear power programs. China and Russia have active breeder programs. India, which has abundant thorium reserves, is pursuing the goal of moving to a thorium cycle. However, the development of these reactors is progressing slowly, and so far they have had minimal impact on the consumption of secondary supplies and overall uranium requirements.

8.9.1 Future trends

To the extent that the movement to advanced reactors is driven by the desire to economize on the future use of uranium, recent market developments have likely reduced this incentive. In this respect, the need to economize on uranium and nuclear fuel in general has declined following Fukushima and the lower demands and prices it brought. Also, advances in enrichment technology (later generation centrifuges and the potential of laser enrichment) should continue to reduce the need for uranium. One question is whether advances in enrichment technology can outpace those in reactor technology and which will have the greatest relative impact on uranium demand. So far, it appears that enrichment technology is winning this battle.

8.10 Summary: The declining but continuing role of secondary supplies

Fig. 8.3 shows the estimated historical use (from 2008) and projected future use (through 2030) of secondary supplies relative to both uranium production from existing and new facilities and a forecast range of future uranium requirements.[36] The sources of secondary supplies include drawdown of excess commercial inventories by utilities and suppliers, US government inventory sales and transfers, tails

[34] Pavel Podvig, "Disposition of Excess Military Nuclear Material, February 2012."

[35] Six types of generation IV reactors along with their fuel supplies, are discussed in the World Nuclear Association's "Generation IV Reactors" (updated June 2015).

[36] This is based on UxC information, including its midcase uranium production forecast, which includes projected production from existing and new mines.

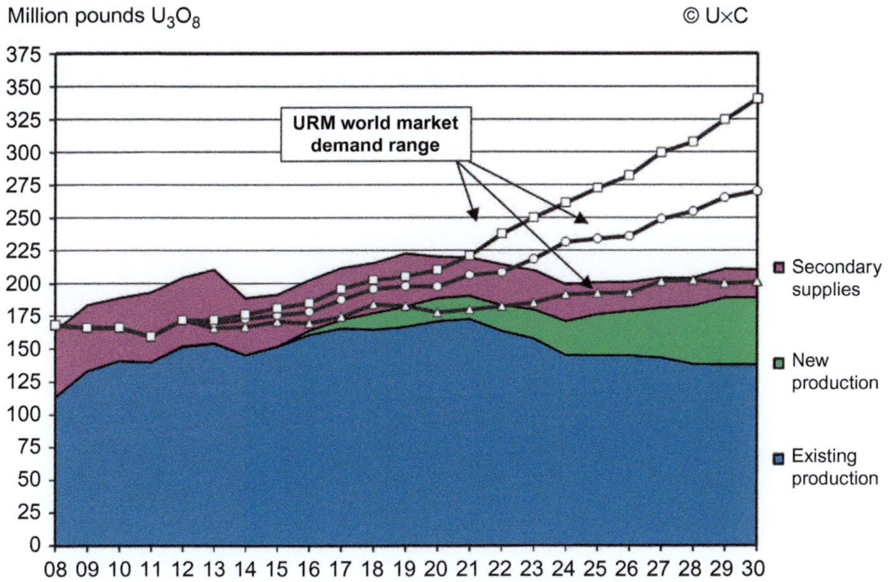

Figure 8.3 Uranium supplies by source versus demand, 2008–2030.
Source: UxC Market Outlook, Q2 2015, The Ux Consulting Company, LLC.

enrichment and underfeeding (particularly in Russia), ERU, and uranium equivalents contained in MOX fuel. It also includes uranium deliveries from the Russian–US HEU deal for the period prior to 2014.

On inspection of Fig. 8.3, it is clear that the contribution of secondary supplies has declined recently, and is projected to decline further over time. The large drop-off in secondary supplies after 2013 relates to the end of the HEU deal. From 2013 to 2030, the annual level of secondary supplies is essentially halved. Pretty much all categories of secondary supplies fall off during this period, and supplies from the drawdown/sale of currently excess utility inventories are projected to disappear entirely by 2030. During this time, the excess inventories built up as a result of the Fukushima accident are expected to be worked off. Of course, these projections are based on current forecasts of uranium requirements, contracting practices of enrichers, and other factors, which are all subject to change.

Despite their levels dropping off, Fig. 8.3 shows that secondary supplies are projected to contribute 200,000 tU (over 500 million pounds U_3O_8) toward meeting reactor requirements over the 2015–2030 period. Further, these supplies will continue to meet a portion of reactor needs well beyond 2030. It is likely that underfeeding, tails enrichment, and recycle of MOX and ERU fuel will continue, at least to some degree, with Russia continuing to be the primary source of tails enrichment and underfeeding. Developments in enrichment and reactor technology will affect the availability and usability of secondary supplies to some extent, but any advancement is more likely to impact secondary supplies in the post-2030 period.

There is a large amount of uranium contained in HEU stocks, but there is no indication that much of this material will come to the market anytime soon, if ever. It is more likely that there will be a low level of HEU blend down in the US and Russia for some time, but this will not likely have a large impact on meeting future reactor requirements.

Production of byproduct uranium and uranium from unconventional resources

9

Ian Hore-Lacy
World Nuclear Association, London, United Kingdom

9.1 Introduction

The "Red Books" from the International Atomic Energy Agency (IAEA) and the OECD Nuclear Energy Agency point out that conventional resources are defined as resources from which uranium is recoverable as a primary product, a coproduct, or an important byproduct, whereas unconventional resources are those from which uranium is only recoverable as a minor byproduct, such as uranium associated with phosphate rocks, nonferrous ores, carbonatite, rare earths, black shale, and lignite. Most of the unconventional uranium resources reported are associated with uranium in phosphate rocks, and most of the interest in recovery is focused there and in one major rare earths deposit. For the purpose of this chapter, the definition is extended to all subeconomic uranium resources, even seawater, on the basis that economics can and do change.

The figures presented in Table 9.1 from the 2014 Red Book (OECD NEA & IAEA, 2014) are expected to evolve, and are clearly incomplete because large uranium resources associated with the Chattanooga (United States) and Ronneburg (Germany) black shales that, combined are estimated to contain a total of 4.2 MtU, and those in Uzbekistan, are not listed. Neither are large uranium resources associated with monazite-bearing coastal sands in Brazil, India, Egypt, Malaysia, Sri Lanka, and the United States. Rare earths deposits also do not feature. Unconventional resources in former Soviet countries are not listed here, apart from Kazakh phosphates.

9.2 Uranium from phosphate deposits

The principal source of uranium in unconventional resources is rock phosphate, or phosphorite. Estimates of the amount available range from 9 to 22 MtU. The IAEA's World Distribution of Uranium Deposits (UDEPO, IAEA, 2009) database tabulates 14 Mt, though the 2014 Red Book tabulates only 7.6 Mt, while suggesting that the 22 Mt may be realistic.

Uranium for Nuclear Power. DOI: http://dx.doi.org/10.1016/B978-0-08-100307-7.00009-0

Table 9.1 Unconventional uranium resources (1000 tU) reported in Red Books (most recent figures)

Country	Phosphate rocks	Nonferrous ores	Carbonatite	Black schist/ shales, lignite
Brazil[a]	84.4	2	13	
Chile	7.2	4.5–5.2		
Columbia	20.0–60.0			
Egypt[b]	35.0–100.0			
Finland	1		2.5	22
Greece	0.5			
India	17–2.5	6.6–22.9		4
Jordan	60			
Kazakhstan	58			
Mexico	240	1		
Morocco	6526			
Peru	21.6	0.14–1.41		
South Africa	No estimate			77
Sweden	42.3			967.6
Syria	60.0–80.0			
Thailand	0.5–1.5			
United States	140–330	1.8		
Venezuela	42			
Vietnam				0.5

[a]Considered a conventional resource in Brazil and is thus included in conventional resource figures.
[b]Includes an unknown quantity of uranium contained in monazite.
Source: OECD 2014 Red Book.

With uranium as a minor byproduct of phosphates, the potential supply is tied to the economics of phosphate production, coupled with the environmental benefits of removing uranium from the waste stream and/or product. World phosphorous pentoxide (P_2O_5) production capacity is about 50 Mt/year according to PhosEnergy, 9.5 Mt in North America, 9.4 Mt in Africa, and 19.2 Mt in Asia.

About 20% of US uranium came from central Florida's phosphate deposits as a byproduct in the mid-1990s, but recovery then became uneconomic. From 1981 to 1992, US production from this source averaged just over 1000 tU/year, then fell away sharply and finished in 1998. The IAEA Red Book (OECD NEA & IAEA, 2009) also reports significant US production of uranium from phosphates from 1954–1962. With higher uranium prices today, the US resource is being examined again, as is a lower-grade resource in Morocco. Plans for Florida extend only to 400 tU/year at this stage.

In Brazil, where uranium is essentially a coproduct with phosphate, the Santa Quitéria joint venture between the government company, Indústrias Nucleares do Brasil, and Galvani phosphates has a prime customer in the form of Eletrobras, owner of the national nuclear power operator Eletronuclear. This project based on the Santa Quiteria and Itataia mines will produce both uranium concentrate and diammonium phosphate in a single integrated process. The mine was expected to produce 970 tU/year from 2015, and ramp up to 1270 tU/year in 2017 as byproduct or coproduct of phosphate. Reserves are 76,000 tU at 0.08% U, though resources are reported as 140,000 tU at Santa Quiteria and 80,000 tU at Itataia, grading 0.054% U in P_2O_5.

In the United States, Cameco and Uranium Equities Ltd have run a demonstration plant using a refined process—PhosEnergy—and estimate that some 7700 tU could be recovered annually as byproduct from phosphate production, including 2300 tU in the US. The prefeasibility study on the PhosEnergy process was completed early in 2015 and confirmed its potential as a low-cost process.

9.3 Processing phosphates

Phosphate rock (phosphorite) is a marine sedimentary rock that contains 18–40% P_2O_5, as well as some uranium with all its decay products, often 70–200 ppm U, and sometimes up to 800 ppm. The main mineral in the phosphate rock is apatite, and most commonly, fluorapatite—$Ca_5(PO_4)_3F$ or $Ca_{10}(PO_4)_6(F,OH)_2$. This is insoluble, so cannot directly be used as a fertilizer (unless in very acid soils) so it must first be processed. This is normally completed in a wet process phosphoric acid (WPA) plant, where it is first dissolved in sulfuric acid. About 2–4% fluorine is usually present. There are about 400 plants using this wet process worldwide, with capacity of some 50 Mt P_2O_5 per year.

Some phosphate deposits—about 4% of total known—are igneous, created by magmatic extrusion as an alkaline chimney. The main mineral is apatite, with some fluorapatite.

When phosphate rock is treated with sulfuric acid in substoichiometric quantity, normal superphosphate is formed. If more sulfuric acid is added, a mixture of phosphoric acid and gypsum (calcium sulfate) is obtained. After the gypsum is filtered out, the resultant phosphoric acid can be treated to recover uranium.

The basic reaction is:

$$Ca_3(PO_4)_2 + 3H_2SO_4 + 6H_2O \rightarrow 2H_3PO_4 + 3CaSO_4 \cdot 2H_2O \quad - \text{exothermic}$$

An improved higher-temperature process produces hemihydrate: $CaSO_4.1/2H_2O$.

Fluorides need to be controlled as gases and in-process effluents (HF, fluorosilicic acid), and about half the fluorine reports with the gypsum. In the process a lot of crud—an emulsion stabilized by fine solids—is generated, and this has been disposed of with gypsum tailings, despite its low-level radioactivity.

After the gypsum precipitation stage, triple superphosphate is obtained by reacting the phosphoric acid with further phosphate rock. Otherwise, various ammonium phosphate fertilizers can be produced by reacting the phosphoric acid with ammonia.

The uranium is normally recovered from the phosphoric acid (H_3PO_4—bearing about 28% P_2O_5) by some form of solvent extraction (SX).

PhosEnergy, an ion exchange (IX) process, represents a step-change refinement of the old processes. It was announced in 2009, offering uranium recovery costs of \$25–30/lb U_3O_8, compared with historical costs of double this.[1] A Demonstration Plant was built in Adelaide, South Australia, and shipped to the USA to operate at a US fertilizer producer, where it was commissioned in May 2012. The test campaign with four trials on two feed sources was successful, showing recoveries of over 92% at \$20–25/lb operating cost, and with Cameco very fully involved. Full evaluation of the project operation with an engineering study was reported in March 2013, and \$18/lb operating cost with capital cost of \$156 million for a base case 400 t/year U_3O_8 plant were quoted. Cameco reaffirmed its commitment in subscribing a further \$4 million. PhosEnergy Ltd (PEL, 27%) and Cameco (73%) are progressing the project through Urtek LLC.

The process involves taking 27% acid, redox reduction to ferrous iron giving oxidation of uranium, primary IX, then recycle of acid with improved quality and a secondary IX uranium recovery being much the same as that in Cameco's US in situ leach (ISL) operations, so potential synergy. The process achieves 95% U recovery, leaving no radwaste, at \$18/lb U_3O_8 cost, with improved acid quality for the main plant, for \$120 million bolt-on plant cost.

Continuous on-site PhosEnergy demonstration plant operation at the site of an existing US phosphate producer took place over 10 weeks to May 2014 with consistent recoveries of over 92% from the filter-grade acid stream. The product was shipped to a licensed uranium mill in Wyoming where it was converted into saleable product. This phase will underpin a Definitive Feasibility Study (DFS) and be the basis for a full-scale commercial facility, which might be constructed and commissioned within 3 years of the commencement of a DFS.

9.4 Production potential

In the USA eight plants for the recovery of uranium from phosphoric acid have been built and operated since the 1970s (six in Florida, two in Louisiana). Plants have also been built in Canada, Spain, Belgium (for Moroccan phosphate), Israel, and Taiwan.

[1] The PhosEnergy process was being developed by Uranium Equities Limited (UEQ) through a US registered company, Urtek LLC. Cameco secured rights to earn up to a 73% interest in the technology, and initially paid \$12.5 of the \$16.5 million required for this to UEQ and a further \$4.5 million for the founder's 10%. UEQ subsequently agreed to pay Cameco a share of that 10%, to hold 27% of the rights for the process. Cameco paid a further \$4 million in March 2013 to reach 73% share. On the basis of its earlier 70% interest, Cameco agreed to provide funding for a minimum of 50% of UEQ's portion of capital expenditure for the construction of the first commercial plant, repayable out of earnings. Cameco and UEQ are seeking commercial arrangements with phosphate producers where the process would provide a technical solution for the recovery of uranium from phosphates. The capital required to install the process would be in exchange for off-take from the facility.

The economic benefit of recovering uranium from WPA phosphoric acid streams will be both in the value of the uranium and in reduced regulatory demands on disposal of low-level radioactive wastes arising from the WPA process. Estimated uranium production costs will put the new process in the lowest quartile of new uranium production.

The USA has reserves of 1400 Mt phosphates containing about 170,000 tU. At 9.6 Mt/year P_2O_5 production, this would yield 2300–2680 tU/year byproduct.

Jordan has reserves of 1500 Mt phosphates containing up to 140,000 tU. At 676,000 t/year P_2O_5 production the uranium potential is 135 tU/year. The government is putting out the Qatrana phosphorites for tender to develop, containing 52 Mt phosphate and 22,000 tU with vanadium.

Morocco has reserves of 50 billion tonnes of phosphates containing about 6.9 MtU. At 4.8 Mt/year P_2O_5 production, this would yield 960 tU/year byproduct. There was some expectation of 1900 tU/year production from 2017.

Egypt has reserves of 100 Mt phosphates containing about 42,000 tU at 50–200 ppm U.

Tunisia has reserves of 100 Mt phosphates containing about 50,000 tU. At 1.6 Mt/year P_2O_5 production, this would yield 265 tU/year byproduct.

India is setting up a commercial-scale uranium extraction plant for recovering uranium from fertilizer-grade phosphoric acid at Paradeep in Odisha, at Indian Farmers Fertiliser Cooperative Limited's phosphatic fertilizers complex. A second project for extracting uranium and other rare earth elements (REE) from WPA is planned for Paradeep Phosphates Limited, a privatized fertilizer enterprise. The proposed Rare Material Recovery project has been declared as "strategic" by the government.

9.5 Uranium from rare earths deposits

A significant unconventional resource is rare earths, where China has a preeminent position as supplier. This gives rise to commercial pressure for development of other rare earths deposits outside China. Some of these contain significant uranium mineralization.

REE are a set of 17 chemical elements in the periodic table, specifically the 15 contiguous lanthanides plus the lighter scandium and yttrium. Scandium and yttrium are considered REE because they tend to occur in the same ore deposits as the lanthanides and exhibit similar physical and chemical properties. The REEs have unique catalytic, metallurgical, nuclear, electrical, magnetic and luminescent properties. Most REEs are in fact not rare. However, because of their geochemical properties, REE are typically dispersed and not often found in concentrated and economically exploitable forms. REEs are often found together, and are difficult to separate. One of the minerals involved in the progressive discovery and identification of REEs was samarskite—containing uranium. Many more REE minerals contain thorium.

Economically and geologically there is a distinction between light REE, or cerium earths (scandium, lanthanum, cerium, praseodymium, neodymium, promethium, samarium, europium, gadolinium—elements 21, 57–64) and heavy REE, or yttrium earths (yttrium, terbium, dysprosium, holmium, erbium, thulium, ytterbium, lutetium—elements 39, 65–71). Samarium, europium and gadolinium can fall into either category, or with terbium and dysprosium form a middle REE category. (Geoscience Australia, 2013) In 2015 Greenland Minerals said that neodymium, dysprosium, terbium, europium and yttrium were critically important for clean energy uses and had high supply risk. Resources are usually reported as rare earth oxides (REO).

Large orebodies of the cerium earths are known around the world, and are being exploited. Monazite includes cerium, and associated light REE. Corresponding orebodies for heavy REE yttrium earths tend to be rarer, smaller, and less concentrated. Most of the current supply of yttrium originates in the "ion adsorption clay" ores of Southern China. Some of these provide concentrates containing about 65% yttrium oxide, with the heavy lanthanides. Uranium ores from Ontario have occasionally yielded yttrium as a byproduct. Xenotime incorporates yttrium and heavy REEs.

Since 1998 China has produced over 80% of the world's REE, mostly from Inner Mongolia, and the proportion is now higher. All of the world's heavy REE (such as dysprosium) come from China.

China cut back exports of all REE from the 35,000 t/year level in 2010, and banned exports of terbium and dysprosium. This led to recommissioning of Molycorp's Mountain Pass mine in California, and enhanced interest in bringing new orebodies into production, including Lynas' Mount Weld in Western Australia, Nolan's Bore in Central Australia and Kvanefjeld in Greenland, the latter two having significant uranium content with the REE.

Considering rare earths alone, China has 55 Mt (47.7%) of the world's 115 Mt of economic resources for REO including yttrium, followed by the USA with 13 Mt (11.3%), according to Geoscience Australia 2012 data, which excludes Greenland (now 11 Mt including inferred resources). The largest REE deposit in the world is the Bayan Obo/ Baiyuebo deposit in China, with at least 48 Mt of REE. Some of the REE deposits have significant uranium as potential byproduct.

Molycorp Inc's Mountain Pass deposit in California was discovered by a uranium prospector in 1949. It dominated worldwide REE production from the 1960s to the 1990s, and has been recommissioned. Molycorp was spending $781 million on modernising and expanding its plant, which reached 19,000 t/year capacity in 2012—slightly more than US demand, but in June 2015 it filed for bankruptcy. Remaining resources in carbonatite are about 1.6 Mt REE averaging 8.24%, with 5% cutoff. The company makes no mention of uranium as byproduct, but thorium is present in monazite.

Kvanefjeld in southern Greenland is the main REE deposit with major potential for uranium production, along with Sorensen and Zone 3 orebodies in the same Ilimaussac intrusive complex. It was investigated intensively over 1955–86, then dropped for commercial reasons. Greenland Minerals & Energy acquired the project in 2007 then with 6.5 Mt of REO at 1.07% known in Lujavrite. About 3.6% of its REOs are terbium, dysprosium and yttrium, that is, "heavy."

The full Kvanefjeld project uranium resource is now estimated at 228,000 tU in four deposits: Kvanefjeld, Sorensen, Zone 3 and Steenstrupfjeld. This includes 36,600 U measured resources at 0.0257% U, 66,000 tU indicated resources at 0.0215% U (all in the Kvanefjeld deposit), and 125,000 tU inferred resources (in the first three deposits), all with 0.015% U_3O_8 cutoff, and 11.14 Mt of REO including 0.40 Mt of heavy REO. Also 2.42 Mt zinc as sphalerite. Relative to other world REO deposits (especially China), Kvanefjeld has a high proportion of terbium through to lutetium (elements 65–71).

The maiden ore reserves (May 2015) are 43 Mt proven reserves with 352 ppm U_3O_8 (12,834 tU) and 500 ppm heavy REO, and 64 Mt probable reserves with 368 ppm U_3O_8 (19,970 tU) and 490 ppm heavy REO. These are in the upper part of the Kvanefjeld deposit. Light REO are not included in reserves.

The Kvanefjeld mine and concentrator with flotation circuit will produce a REE-U concentrate plus zinc concentrate and fluorspar byproducts. Then the refinery will employ atmospheric sulfuric acid leach to produce uranium oxide byproduct, then stronger acid leach followed by SX recovery, and precipitation to separate REEs. Primarily it will produce a critical rare earth concentrate (Nd, Pr, Eu, Dy, Tb, Y), with lanthanum and cerium byproducts. Production is envisaged as 7900 t of critical mixed REO per year, 434 tU, 15,100 t lanthanum and cerium oxides, 15,000 t zinc concentrate and 16,000 t fluorspar per year. Construction could commence in 2016 with first production in 2018.

Start-up costs for a 3 Mtpa plant are estimated at $1121 million for mine, concentrator and refinery, plus $240 million for infrastructure, in the May 2015 feasibility study. Unit production costs are low—$8.56/kg for the exported concentrate after byproduct credits, compared with an anticipated market price of 10 times this, and an incremental cost of $6/lb U_3O_8 for the uranium.

Greenland Minerals & Energy had been seeking partners for the refinery, then envisaged offshore, and in March 2014 signed an agreement with China Non-ferrous Metal Industry Engineering & Construction (NFC), a subsidiary of China Non-Ferrous Metal Mining Group to provide concentrates of heavy REOs to NFC's new 7000 tpa rare earth separation facility in China, now under construction. A further agreement with NFC was signed in April 2015. NFC's participation in the rare earth industry comes through its subsidiary Guangdong Zhujiang Rare Earths Company, which was the first group to carry out full separation of 15 REE in China, and is recognized globally as a leader in rare earth separation technology. Uranium is not included in the NFC agreement.

In November 2012 the Greenland government voted unanimously to support the project, including uranium, and in October 2013 it repealed the long-standing policy banning uranium development. Denmark has responsibility to ensure that international conventions, such as nonproliferation, are respected. Both states expect to have a cooperation agreement for the mining and export of uranium finalized in 2015.

The company is involved with the EU-funded EURARE program, designed to support the development of REO deposits in Europe. The Kvanefjeld project was selected by EURARE as a major one for demonstration of REO production, and a 26 t ore sample (about 1.3% REO) was processed over 100 h at the beneficiation pilot plant in Finland by simple froth flotation to yield 2 t of concentrate. The next test work stage will produce 15% REO, then pilot plant hydrometallurgy at Aachen in Germany to give a mixed carbonate of 60% REO. At Aachen a separation plant will sort the Nd + Pr from La & Ce and from other materials. The products will be evaluated.

In Ontario, Canada, Pele Mountain Resources Inc has received mining leases for its Eco Ridge uranium and REE orebody, 11 km east of Elliot Lake. In the decade to closure of the Stanleigh mine at Elliot Lake in 1996, the area produced REE as a commercial byproduct of uranium production. The Eco Ridge deposit contains a full range of REE at 0.164% grade, including 9.5% heavy REE (including scandium, europium and gadolinium in this category, 7.4% if excluding them). Uranium grade is 0.041% U with cutoff grade 0.024%. Some 65% of the heavy REE report in the uranium leach solution without any additional milling cost. The deposit includes indicated resources of 36,600 t REO and 8690 tU, plus 56,900 t REO with 14,500 tU inferred resources. An underground mine is envisaged, using underground bio-leaching and surface heap leaching, the ore being moved to the surface in haul trucks. The leachate from underground and surface will be piped to the processing facility where it will be clarified and processed in a SX circuit. SX raffinate will be recycled to the two leach operations.

A preliminary economic assessment in 2012 suggested uranium production of 10,500 tU over 11 years. However weakening REE prices coupled with the need for hardrock underground mining meant that the Eco Ridge project was deferred in favor of processing monazite at Elliott Lake, focused on neodymium and praseodymium. The monazite would come from mineral sands tailings in other countries. In March 2015 Pele announced an agreement with Sheng Kang Ning (Shanghai) Mining Investment Co. Ltd, a subsidiary of Shenghe Resources, to develop monazite processing facilities to recover REE at Elliott Lake.

Appia Energy reports 3080 tU indicated resources and 7750 tU inferred resources in connection with rare earths at Teasdale Lake, Elliot Lake, Ontario.

Texas Rare Earth Resources Corp earlier signed an agreement with AREVA to take up to 116 tU/year byproduct uranium from its Round Top heavy rare earth project in Hudspeth County, Texas. The 2013 preliminary economic assessment includes no uranium product and only a recommendation to perform a preliminary review of uranium potential, though 37,000 tU is identified in all classes of REE resources in rhyolite (with the company saying that 70% is heavy REE). A 2015 agreement establishes a joint venture with K-Technologies Inc to develop processing technology for REE.

Nolans Bore, Northern Territory, in Australia is a deposit of light REE with 7500 tU content at 0.016% U with 1.2 Mt REO grading 2.6%, about 135 km north of Alice Springs. It also contains 5 Mt of 12.9% P_2O_5 and 0.27% thorium (Dec 2014). Arafura Resources intended to develop it as a REE mine and plans to rail its concentrate to Whyalla, South Australia where it would be processed to produce

20,000 t/year REO, 130 tU/year, plus phosphoric acid and gypsum coproducts. The combined neodymium/praseodymium (Nd + Pr) content of the ore means that Nd + Pr oxide would generate 77% of projected revenue, though these are both light REE. The thorium would be separated as an iron-thorium precipitate and transported back to the Nolans Bore mine site in NT for long-term storage as a possible future energy source.

Alkane Resources' Dubbo Zirconia Project, New South Wales, in Australia has REE as potential byproduct of zirconium production—700,000 t zirconia as measured resources in 2011. These have 0.75% REO and 0.014% U—about 4200 tU, and the same in inferred resources. The 2012 reserve figures also have 0.75% REO at same tonnage but do not include uranium.

Mary Kathleen, Queensland, Australia was a uranium mine in a rare earths deposit, and from which no rare earths were recovered commercially. It operated 1958–63 and 1975–82. The uranium grade was 0.10–0.15%, the REE grade about 4–5%, and the company initially had great difficulty keeping the REE assay in product below specified levels of 0.1%. The REOs ended up in tailings, and in 2015 the Queensland government was seeking expressions of interest in recovering them.

The Khiagda ISL uranium mine in Siberia is planning to recover REOs as byproduct.

9.6 Uranium from black shales

There is considerable uranium in black shale deposits in Scandinavia and Central Asia, Chattanooga (United States) and Ronneburg (Germany). The latter two combined are estimated to contain a total of 4.2 MtU, according to the "Red Book." Development of the projects described next is faltering due at least partly to low uranium prices.

Early in 2010, Finland's Talvivaara Mining Company Plc announced that it planned to recover 350 tU/year as a byproduct of nickel and zinc production from suphidic black shales (schists) using bacterial heap leaching at Sotkamo in northeastern Finland, over 46 years. The uranium resource is 26,000 tU. In April 2010, Talvivaara applied to the Ministry of Employment and Economy for a license to extract uranium as a byproduct, in accordance with the Nuclear Energy Act. The company signed an agreement with Cameco early in 2011 to build a €45 million plant for uranium recovery, to operate from 2013 using SX. Cameco would take all Sotkamo uranium production to 2027, this agreement being approved by the European Commission in November 2011 and the process licensed by Finland in March 2012. The government approved the company's application for the uranium recovery early in 2012, and in April 2014 the Regional State Administrative Agency licensed uranium recovery. In January 2012 STUK said that uranium recovery could be done with minimal radiation release, and the European Commission expressed a positive view of the uranium recovery project.

The heap leach pads at Sotkamo cover 210 ha and after 18 months the ore (0.002% U) is moved to a secondary pad for 36 months. The pregnant liquor (20 ppm U) is first stripped of copper and zinc, then the uranium will be recovered by bacterial leaching. Hitherto a lot of uranium has remained in the waste stream and has accumulated at the bottoms of nearby lakes, giving rise to concern. However, as announced in 2015, the company "faced various difficulties since the commissioning of the mine in 2008 and 2009," these being both operational and environmental.

In October 2013, Talvivaara announced that its liquidity position was poor and constraining development. This created some uncertainty for Cameco regarding timing of production start-up and length of the ramp-up period. It had then invested a total of US$ 70 million in the uranium extraction process, and construction of that plant was about 98% complete. In November 2014 Talvivaara Sotkamo Ltd was declared bankrupt and in mid August 2015 a new state-owned company Terrafame Oy took over Sotkamo from Talvivaara to revive nickel mining and secure long-term investment in the mine.

In Central Sweden, Australia's Aura Energy has inferred resources for its Haggan project of 308,000 tU at 0.02% U in the alum black shales (schists) near Storasen and Vasterasen. Molybdenum, nickel, zinc and vanadium are present and are potential coproducts. The Haggan deposit is flat, with low mining costs, and though amenable to acid leach it has high carbonate levels, so bacterial heap leaching is being investigated, which shows 85% U recovery. A scoping study confirmed the economic potential of bio-heap leach producing 3000 tU/year plus nickel and molybdenum, but at the end of 2013 a smaller initial operation was being proposed. Expected production costs even for the smallest scale operation (385 tU/year) are in the lower half of 2010 industry cost curve, under \$25/lb U_3O_8 after byproduct credits. Production costs for the large operation are put at \$13.50/lb U_3O_8.

Also in Sweden, some 200 tU was produced from a black shale deposit in Ranstad in the 1960s. Another deposit is Pleutajokk, near the Arctic Circle.

In Uzbekistan a 2006 Japan-Uzbek intergovernmental agreement was aimed at financing Uzbek uranium development, and in October 2007 Itochu Corporation agreed with Navoi Mining Co to develop technology to mine and mill the black shales, particularly the Rudnoye deposit, and to take about 300 tU/year from 2007. A 50−50 joint venture was envisaged, but no more was heard of this. Then in mid 2008 Mitsui & Co. signed a basic agreement with the Uzbek government's Goscomgeo to establish a joint venture for geological investigations regarding the development of black shale uranium resources at the Zapadno-Kokpatasskaya mine, 300 km northwest of Navoi.

In August 2009 Goscomgeo and China Guangdong Nuclear Uranium Corp. (CGN-URC) set up a 50−50 U exploration joint venture: Uz-China Uran, to focus on the black shale deposits in the Boztau-skaya area in the Central Kyzylkum Desert of the Navoi region. Some 5500 tU resources are reported. Over 2011−13 CGN-URC was to develop technology for the separate production of uranium and vanadium from these black shale deposits with a view to commencing production from them. In May 2014 Goskomgeo said CGN-URC planned to start mining in 2014, with production being sold to China.

9.7 Uranium from seawater

Seawater has long been regarded as a possible source of uranium due to the large amount contained there (over 4 billion tU). However, its low concentration (about three parts per billion) means that cost-effective extraction remains problematical.

Research on uranium recovery from seawater was carried out in several countries through to the 1980s, and more recently in Japan. Japanese (JAERI) research in 2002 using a polymeric absorbent in a nonwoven fabric containing an amidoxime group that was capable of forming a complex with uranyl tricarbonate ions, suggested about $300/kgU as cost of recovery.

In 2012, researchers at the US Department of Energy's Oak Ridge National Laboratory and Pacific Northwest National Laboratory reported encouraging results through the use of plastic fibers with a surface area ten times greater than the Japanese design, and production costs about $660/kgU. Further cost reducing improvements were being tested.

Chinese research groups in universities and institutions have also shown interest in uranium extraction from seawater.

9.8 Uranium from gold tailings

In South Africa, there are about 400 tailings dams and dumps arising from gold mining in the Witwatersrand area of Gauteng province, and much of the available uranium today is in these. There are further tailings near Klerksdorp, close to the Vaal River. There is some radionuclide and heavy metal pollution arising from some of these, and acid mine drainage. Many of the tailings dams and dumps are being retreated to recover gold and sometimes uranium.

AngloGold Ashanti quoted byproduct resources (mostly indicated resources) of 110,700 tU including byproduct reserves (mostly probable resources) of 49,094 tU at the end of 2013 for its Vaal River operations. Some 60% of the resources were in its Mine Waste Solutions project or other surface tailings. Production from Vaal River in 2014 was 504 tU.

Sibanye Gold produces some byproduct uranium from its Ezulwini-Cooke operations—69 tU in 2014. Its measured resources/ probable reserves at its West Rand Tailings Retreatment project (Driefontein and Kloof) adjacent to Cooke were 16,600 tU at the end of 2013.

9.9 Uranium from coal ash

Most coal contains uranium and thorium, as well as their decay products and potassium-40. The total levels of individual radionuclides typically are not great and are generally about the same as in other rocks near the coal, which varies according to region and geology.

In the USA, 858 Mt of coal was used in 2013 for electricity production. With an average content of 1.3 ppm U and 3.2 ppm thorium, US coal-fired electricity generation in that year gave rise to 1100 tU and 2700 t of thorium in coal ash, none of which was recovered. In Victoria, Australia, some 65 Mt of brown coal is burned annually for electricity production. This contains about 1.6 ppm U and 3.0–3.5 ppm thorium, hence about 100 tU and 200 t of thorium in fly ash and bottom ash is buried in landfill each year in the Latrobe Valley.

In the 1960s and 1970s, some 1100 tU was recovered from coal ash in the USA. The feasibility depends on grade and the composition of the ash—high acid consumption makes recovery uneconomic.

In 2007, China National Nuclear Corp commissioned Sparton Resources of Canada with the Beijing No. 5 Testing Institute to undertake advanced trials on leaching uranium from coal ash in Central Yunnan. In early 2007, Sparton signed an agreement with the Xiaolongtang Guodian Power Company of Yunnan for a program to test and possibly commercialize the extraction of uranium from waste coal ash. Some 250 km southwest of Kunming, the Xiaolongtang, Dalongtang and the Kaiyuan power stations, all located within 20 km of each other burn coal from a centrally located open pit lignite mine with high ash content (20–30%) and very high uranium content. The coal uranium content varies from about 20 to 315 ppm and averages about 65 ppm. The ash averages about 210 ppm U (0.021% U)—above the cutoff level for some uranium mines. The power station ash heap contains over 1000 tU, with annual arisings of 190 tU (Recovery of this by acid leaching is reported to be about 70%).

A joint venture company, Yunnan Sparton New Environ-Tech Consulting Co was set up to operate "the secondary uranium recovery programs in Yunnan," notably at Lincang, but no commercial recovery of uranium has been reported. Sparton also had an agreement to extract uranium from coal ash following germanium recovery in the Bangmai and Mengwang basins in Yunnan. This ash ranges from 150 to over 4000 ppm U (0.40% U), averaging 250 ppm U (0.025%). Sparton had an 85% interest in the Huajun germanium and coal mine, but its interest in uranium is not mentioned here. However, Sparton's website at the end of 2014 had no mention of these projects.

In South Africa HolGoun's Uranium and Power Project was investigating uranium recovery from the Springbok Flats coal field, estimated to contain 84,000 tU at grades of 0.06–0.10% U. The project is investigating the feasibility of mining the low-grade coal, using it to fire a conventional electricity generation plant, and extracting the uranium from the residual ash.

Sources and references

The majority of this chapter is taken from the 2014 Red Book and World Nuclear Association information papers on *Uranium from Phosphates* and *Uranium from Rare Earths deposits*. The author is responsible for those two papers, but worked as editor and collator on this work, not as an authority.

Arafura Resources Ltd web site <http://www.arultd.com>

Aura Energy web site <www.auraenergy.com.au>

Bruneton P., Cuney M., Dahlkamp F., Zaluski G., 2014. IAEA geological classification of uranium deposits, URAM 2014.

Geoscience Australia, 2013. Australian Mineral Atlas, Rare Earths <http://www.australianminesatlas.gov.au/aimr/commodity/rare_earths.html>

Greenland Minerals & Energy web site <www.ggg.gl>

Guzman, E.T.R. et al., 1995. *Uranium in Phosphate Rock and Derivatives.*

IAEA, 2009. World Distribution of Uranium Deposits (UDEPO) with Uranium Deposit Classification, IAEA-TECDOC-1629.

IAEA, 2013. Radiation Protection And Management of NORM Residues in the Phosphate Industry, Safety Reports series #78.

Molycorp Inc web site <http://www.molycorp.com>

OECD NEA & IAEA, 2009. Uranium 2009: Resources, Production and Demand ("Red Book").

OECD NEA & IAEA, 2014. *Uranium 2014: Resources, Production and Demand* ("Red Book").

Pele Mountain Resources web site <http://www.pelemountain.com>

PhosEnergy Ltd web site <www.phosenergy.com> and Fact Sheet.

Texas Rare Earth Resources Corp web site <www.trer.com/projects/uranium>

Urtek LLC web site <http://www.urtekllc.com/>

Wikipedia: Rare Earth Elements.

World Nuclear Association, August 2015, Uranium from Phosphates.

World Nuclear Association, 24 August 2015, Uranium from Rare Earths deposits.

Thorium as a nuclear fuel

10

Bradley S. Van Gosen[1] *and Harikrishnan Tulsidas*[2]
[1]US Geological Survey, Denver, CO, United States,
[2]International Atomic Energy Agency, Vienna, Austria

10.1 Introduction

Interest in using thorium as a nuclear fuel remerged during the first part of the 21st century. Thorium-based nuclear power has yet to be fully commercialized despite numerous projects and pilot test reactors designed to evaluate thorium as a viable reactor fuel. This research has been conducted over several decades in several countries. In fact, the history of research devoted to thorium-based nuclear reactor design is about as long as the history of uranium-plutonium-fueled reactors (Martin, 2012). Currently, research and development is being carried out on several concepts for advanced reactors including high-temperature gas-cooled reactors (HTGR); molten salt reactors (MSR); Canada Deuterium Uranium (CANDU)-type reactors; advanced heavy water reactors (AHWR); fast breeder reactors (FBR); and pressurized heavy water reactors (PHWR). Federal government-supported projects, particularly in India and China, as well as France, Norway, Canada, Brazil, and Russia, have recently devoted research toward the development of thorium-based nuclear power (NEA-IAEA, 2014, 2015). These projects build upon past pilot-scale reactor projects in Canada, Germany, India, Japan, the Netherlands, Norway, the Russian Federation, Sweden, Switzerland, the United Kingdom, and the United States (discussed next). Based on the current activities, utilization of thorium as fuel is expected after 2020 (NEA-IAEA, 2015).

Because of the low demand for thorium, it has not been a primary target of exploration in the past. Ongoing thorium fuel research and reactor development efforts may increase the demands for thorium and likewise increase national and global evaluations of thorium deposits. If India, China, or another country or entity can prove successful in generating electricity safely and efficiently from a thorium-based nuclear power plant, then considerable interest and activity may be drawn to thorium exploration across the globe and to recovering thorium as a byproduct of other mining operations.

Some of the stated advantages of thorium as a nuclear fuel source (NEA-IAEA, 2014) are:

- Thorium is about 3–4 times more abundant than uranium, based on average concentrations in the Earth's crust.
- Energy potential may be maximized by utilizing easily available thorium resources, which can be readily obtained as a coproduct or byproduct from deposits mined for other commodities.

Uranium for Nuclear Power. DOI: http://dx.doi.org/10.1016/B978-0-08-100307-7.00010-7

This is because most significant thorium deposits occur in deposits containing other mineral resources of value, such as rare earth elements (REEs) and (or) titanium, as examples.

- The spent fuel waste products of thorium fission are not the types used in nuclear weaponry (such as plutonium, a byproduct of uranium power generation).
- Thorium-rich spent fuels contain fewer radioactive elements and are smaller in volume and mass than conventional uranium-based nuclear wastes.
- The thorium fuel cycle provides an efficient way to reduce existing plutonium stocks by using plutonium to initiate the thorium fission chain reaction.

This chapter: (1) summarizes past and present research on thorium as a nuclear fuel that could generate electricity from small and (or) large reactors; (2) describes the geochemistry, mineralogy, and geology of thorium deposits; (3) explains the potential for byproduct recovery of thorium from deposits of other valuable mineral resources; and (4) describes current research and future goals for utilizing thorium as a nuclear fuel source.

10.2 Thorium fuel cycle

Even though thorium has been considered as a sustainable fuel cycle option, due to the abundance of uranium and its relative ease of handling, serious attention was not paid in the past toward developing a commercial thorium fuel cycle. Recently, focus has renewed toward thorium utilization because of the favorable aspects of thorium fuel explained previously.

Some countries, such as China and India, have ambitious programs in nuclear power, but have limited resources of uranium and considerably larger resources of thorium. In terms of energy security, utilization of thorium is favored in these countries (Anantharaman et al., 2008). Other countries, including Norway and the United States, seek to maximize energy potential by utilization of abundantly available, domestic thorium resources.

The thorium fuel cycle differs from the uranium fuel cycle in a number of ways. Natural thorium contains only trace amounts of fissile material (such as ^{231}Th), which are insufficient to initiate a nuclear chain reaction. In a Th-fueled reactor, ^{232}Th absorbs neutrons to eventually produce ^{233}U. The ^{233}U either fissions in situ or is chemically separated from the used nuclear fuel and formed into new nuclear fuel. The sustained fission chain reaction could be started with existing ^{233}U or some other fissile material such as ^{235}U or ^{239}Pu. Subsequently, a breeding cycle similar to, but more efficient than, that with ^{238}U$-^{239}$Pu can be created (IAEA, 2005; WNA, 2015a).

10.2.1 Advantages of the thorium fuel cycle

Thorium-based fuels exhibit several attractive nuclear properties relative to uranium-based fuels, such as:

- In a thermal reactor, fertile conversion of thorium is more efficient than that of ^{238}U.
- Fewer nonfissile neutron absorptions take place and neutron economy is improved.

- Th fuels can be the basis for a thermal breeder reactor.
- Thorium dioxide (ThO_2) has a higher melting point, higher thermal conductivity, lower coefficient of thermal expansion, and greater chemical stability than U-Pu fuel.

In the thermal spectrum, the reproduction factor (η) (number of neutrons produced per neutron absorbed in the fissile nuclide) is larger for ^{233}U than for either ^{235}U or ^{239}Pu. Of these, ^{239}Pu has the lowest η in the thermal spectrum (US Atomic Energy Commission, 1969). Although thermal breeders have no limitation in the possibility for using any available fissile fuel, the use of ^{233}U may be preferable because of its superior neutronic characteristics (IAEA, 1979). The possibility to breed fissile material in slow neutron systems is a unique feature for thorium-based fuels and is not possible with uranium fuels. The relatively large η for ^{233}U in the thermal spectrum is the most important factor contributing to the potentially larger conversion ratios achievable with the thorium cycle. But for realizing thermal breeding with thorium, the neutron economy in the reactor has to be very good.

Nuclear reactor designs can be of two basic types, employing either homogenous or heterogeneous core configurations. In a homogenous reactor core, nuclear properties such as neutron flux and average cross sections are spatially uniform, whereas nuclear properties change from one region to another in the heterogeneous cores (Jevremovic, 2009). In a homogenous reactor, the active core consists of a homogenous mixture of nuclear fuel and a moderator (Lane et al., 1958). On the other hand, in a heterogeneous reactor, there is a sharp boundary between nuclear fuel, the moderator, and the coolant. The thorium cycle tends to be more economical than the uranium cycle in high-temperature reactors in which the fuels are homogenous to neutrons. In relatively homogenous thermal spectrum reactors, such as the HTGR and the MSR, which operate at higher temperatures, the reproduction factor (η) remains high and the conversion ratio larger with the thorium cycle than when using the uranium cycle. Thus, the thorium cycle tends to lower the fuel cycle costs, even though the fuel manufacturing costs are relatively on the higher side.

The uranium cycle tends to be economical in soft-spectrum reactors heterogeneous to neutrons, such as the light water reactor (LWR) and heavy water reactor (HWR), in which the fissile enrichment requirement is low. ^{238}U has larger resonance absorption (absorption by epi-thermal neutron) than ^{232}Th, and to reduce the absorption by ^{238}U, heterogeneous core configuration is required when the fuel region is small. On the other hand, in homogenous configuration, resonance absorption occurs over a wide fuel region.

The problem of uneven power distributions can be less severe in heterogeneous reactors utilizing the thorium cycle. The higher conversion ratio of this cycle leads to smaller changes in power due to depletion effects. In addition, the thermal cross section characteristics of ^{233}U are closer to those of ^{235}U than are those of ^{239}Pu to ^{235}U. When ^{239}Pu and ^{235}U are present together, the ^{239}Pu will burn out proportionately faster because of its relatively larger cross section (Van Den Durpel et al., 2011).

The thorium offers an advantage from the waste perspective. Because the mass number of ^{232}Th is 6 units less than that of ^{238}U, the production of plutonium (Pu) and minor actinides (MA) (neptunium, americium, curium), which are the major contributors to the radiotoxicity of the wastes in the uranium-plutonium cycle, is drastically reduced if actinides are recycled (David et al., 2007). Long-term radiotoxicity of spent fuel is mainly from Pu and MA, which is significantly lower in Th-^{233}U fuel cycle; this effect becomes observable only when the fuel involves Pu or ^{238}U. Elimination of at least the transuranic portion of the nuclear waste problem is possible in the thorium fuel cycle, but there are some actinides that constitute a somewhat "long-term" radiological impact, such as ^{229}Th (half-life 7340 years). ^{229}Th is a daughter of ^{233}U (half-life 159,200 years), which can be easily eliminated if actinide recycle is the option.

In uranium fuels, decay (residual) heat is due to transuranium radionuclides after 100 years, in particular the Pu isotopes. In both Th-Pu and Th-U fuels, the influence of fission products in decay heat is dominant for 100 years, and after that, actinides are mainly responsible for the decay heat (IAEA, 2005). Decay heat is similar for both U and Th fuels, if they are not recycled.

The ^{233}U produced in thorium fuels is inevitably contaminated with ^{232}U, a hard gamma emitter; therefore, heavily shielded facilities are required for handling it. Due to this issue, thorium-based used nuclear fuels possess inherent proliferation resistance. However, the chemical separation of ^{233}Pa can be possible, which can avoid the ^{232}U problem (Ashley et al., 2012). But chemical separation of ^{233}Pa from Th has not been established as yet, and even if reprocessing of highly radioactive fuel is possible within 6 months, ^{233}Pa decays with half-life of 27 days, and decreases to less than 1%. Therefore, acquiring enough ^{233}Pa would, in practice, be a difficult proposition. Thorium blankets irradiated in FBR can potentially have very low levels of ^{232}U, if the blankets are located away from the core (Keshavamurthy and Mohanakrishnan, 2011). This also has not been yet demonstrated on a large scale.

10.2.2 Disadvantages of the thorium fuel cycle

Some of the unique features of the thorium fuel cycle often prove to be the major challenges in its application. Initial fissile requirements for ^{235}U, ^{233}U, or Pu make thorium unsuitable for rapid expansion of nuclear energy. Thorium introduction could be preferred only after a good stock of fissile material (in form of either Pu or ^{233}U) has been built up (IAEA, 2012).

^{232}Th breeds into ^{233}U over a comparatively long interval because ^{233}Pa, the intermediate isotope, has a half-life of about 27 days. As a result, some amount of ^{233}Pa builds in thorium-based fuels. ^{233}Pa is a significant neutron absorber, and although it eventually breeds into fissile ^{235}U, this requires two more neutron absorptions, which degrades neutron economy and increases the likelihood of transuranic production. Hence, it is often difficult to design thorium fuels that produce more ^{233}U in thermal reactors than the fissile material they consume, even though the use of ^{233}U in a thermal reactor makes it possible to achieve higher fuel conversion ratios and longer fuel burnups than is practical with either ^{235}U or ^{239}Pu. But the effect of ^{233}Pa is not a major problem in some reactor designs, as it quantitatively estimated to be only 1/1000 by weight in Th fuel. For example, in MSR, it was demonstrated that the Breeding Ratio (BR) was 1.07 with ^{233}Pa removal by online reprocessing, which was reduced only to 1.05 without the ^{233}Pa removal (Briggs, 1967).

If thorium is used in an open fuel cycle (ie, utilizing ^{233}U in situ), higher burnup is necessary to achieve a favorable neutron economy. Although ThO_2 has performed well at burnups of 170,000 MWd/t and 150,000 MWd/t, there are challenges associated with achieving this burnup in LWRs. These high burnup levels are beyond those currently attainable in LWRs with zirconium cladding. High concentration of ^{235}U or Pu is required to increase burnup. While ^{235}U enrichment of more than 19.9% is not desirable from a proliferation perspective, high Pu is not desirable from a waste point of view, as this negates the advantages of having lesser concentrations of MA, the main contributor for long-term radiotoxicity in spent fuel. In general, the merit of incorporating thorium in open fuel cycle is often questioned (eg, see Ashley et al., 2014).

If thorium is used in a closed fuel cycle in which ^{233}U is recycled, remote handling is necessary because of the high radiation dose resulting from the decay products of ^{232}U. This is also true of recycled thorium because of the presence of ^{228}Th, which is generated from decay of ^{232}U, but is only 1% of ^{232}U. Although there is substantial worldwide experience recycling uranium fuels (eg, Plutonium and Uranium Recovery by EXtraction (PUREX)), similar technology for thorium (eg, Thorium Recovery by EXtraction (THOREX)) is still immature. Remote handling for the fuel fabrication will also make the fuel cycle more expensive.

10.2.3 *Fast reactor thorium utilization*

Fast fission of ^{232}Th is possible, but it is only a 10th as efficient as ^{238}U. Fissile plutonium produced from ^{238}U offers the potential of high breeding gains in fast reactors with the production of 40–50% more fissile fuel than is consumed. BRs of 1.009–1.115 can be obtained with ^{233}U in a very high-energy, fast neutron spectrum reactor, but it is less than that of ^{238}U (1.284–1.582). On the other hand, in a degraded (10–100 keV) fast spectrum, ^{233}U would probably be as good as, or better than, ^{239}Pu.

The variation of ^{233}U and ^{239}Pu cross sections with energy is such that improved reactivity coefficients would be obtained with the use of ^{233}U in a sodium-cooled FBR. This leads to improved nuclear safety characteristics. However, a fast reactor using Th/^{233}U cycle is not considered to have a significant advantage over thermal reactors, even if the safety aspects are better. To breed ^{233}U efficiently, a Pu-fueled fast reactor with thorium as blanket would be preferable.

While thorium appears to have the nuclear disadvantages in fast spectrum, its physical behavior as a metallic fuel material is far better than that of uranium. Consequently, the use of thorium as a fertile material in fast spectrum reactors may offer some advantages if it is used in the metallic form.

10.2.4 Accelerator-driven systems

A subcritical reactor is a nuclear fission reactor that produces fission without achieving criticality. Instead of a sustaining chain reaction, a subcritical reactor uses additional neutrons from an outside source. The neutron source can be a particle accelerator producing neutrons by spallation. Such a device with a reactor coupled to an accelerator is called an accelerator-driven system (ADS). Different fuel cycles that include thorium have been proposed with these systems. The concept of using an ADS that is based on the thorium-^{233}U fuel cycle has been proposed by several studies (Carminati et al., 1993; Rubbia, 1995; Furukawa, 1997).

In an ADS, the long-lived transuranic elements (such as Np and Am) in nuclear waste can in principle fission, releasing energy in the process and leaving behind the fission products that are shorter lived. This would shorten considerably the time for disposal of radioactive waste (IAEA, 2003).

Various ADS designs propose a high-intensity proton accelerator with energy of about 1 GeV, directed toward a spallation target made of thorium (either thorium containing molten salt or solid thorium target that is cooled by liquid lead-bismuth in the core of the reactor). In that way, for each proton interacting in the target, 20–40 neutrons on average are created to irradiate the surrounding fuel. Thus, the neutron balance can be regulated so that the reactor would be below criticality if the additional neutrons by the accelerator were not provided. Whenever the neutron source is turned off, the reaction ceases.

There are technical difficulties to overcome before ADS can become economical and eventually integrated into nuclear energy systems. The accelerator must provide a high-intensity proton beam and should be highly reliable. There are concerns about the accelerator-reactor coupling, especially the beam window and the spallation target, which are subject to complex degradation phenomena. Beam window is an integral part of the target containment structure. Due to combined thermos-mechanical loads, high-energy particle irradiation, and contact with liquid heavy metals, these components are subjected to complex stress, corrosion, and irradiation conditions and may require frequent replacement. An emergency situation of losing containment can be caused by window rupture.

10.3 Previous work on the thorium fuel cycle

Research toward utilizing thorium as a nuclear fuel has occurred for over 50 years. Basic research and development, as well as operation of test reactors with thorium fuel, has been conducted in Canada, Germany, India, Japan, the Netherlands, Norway, the Russian Federation, Sweden, Switzerland, the United Kingdom, and the United States (also see Martin, 2012, for a detailed history of past thorium-reactor research in the United States). Historical thorium utilization in various reactors is shown in Table 10.1 and discussed next.

Table 10.1 Thorium utilization in experimental and nuclear power reactors

No	Name and country	Type	Power	Fuel	Operation period
1	Indian Point 1, United States	LWBR PWR, (Pin Assemblies)	285 MWe	Th + ^{233}U Driver Fuel, Oxide Pellets	1962–80
2	Shippingport, United States	LWBR PWR, (Pin Assemblies)	100 MWe	Th + ^{233}U Driver Fuel, Oxide Pellets	1977–82
3	Elk River, United States	BWR	24 MWe	Th + ^{235}U-^{238}U Fuel, Oxide Pellets	1963–68
4	Lingen, Germany	BWR Irradiation-testing	60 MWe	Test Fuel $(Th,Pu)O_2$ pellets	Terminated in 1973
5	KAPS 1 & 2; KGS 1 & 2; RAPS 2, 3 & 4, India	PHWR, (Pin Assemblies)	220 MWe	ThO_2 Pellets (For neutron flux flattening of initial core after start-up)	Continuing in all new PHWRs
6	Peach Bottom, United States	HTGR Experimental (Prismatic Block)	40 MWe	Th + ^{235}U Driver Fuel, Coated fuel particles, Oxide & dicarbides	1966–72
7	Fort St Vrain, United States	HTGR, Power (Prismatic Block)	330 MWe	Th + ^{235}U Driver Fuel, Coated fuel particles, Dicarbide	1976–89
8	AVR, Germany	HTGR, Experimental (Pebble-bed reactor)	15 MWe	Th + ^{235}U Driver Fuel, Coated fuel particles, Oxide & dicarbides	1967–88
9	THTR-300, Germany	HTGR, Power (Pebble Type)	300 MWe	Th + ^{235}U, Driver Fuel, Coated fuel particles, Oxide & dicarbides	1985–89
10	Dragon, UK OECD-Euratom also Sweden, Norway, and Switzerland	HTGR, Experimental (Pin-in-Block Design)	20 MWt	Th + ^{235}U Driver Fuel, Coated fuel particles, Oxide & Dicarbides	1966–73

(*Continued*)

Table 10.1 **(Continued)**

No	Name and country	Type	Power	Fuel	Operation period
11	MSRE ORNL, United States	MSR	7.5 MWt	^{233}U Molten Fluorides	1964–69
12	FBTR, India	LMFBR, (Pin Assemblies)	40 MWt	ThO_2 blanket	1985–present
13	SUSPOP/KSTR KEMA, Netherlands	Aqueous Homogenous Suspension (Pin Assemblies)	1 MWt	Th + HEU, Oxide Pellets	1974–77
14	NRU & NRX, Canada	MTR (Pin Assemblies)	–	Th + ^{235}U, Test Fuel	Irradiation-testing of a few fuel elements
15	KAMINI, India	MTR Thermal	30 kWt	Al + ^{233}U Driver Fuel,	Research reactor in operation 1996–present
16	CIRUS, India	MTR Thermal	40 MWt	"J" rod of Th & ThO_2	Research reactor 1960–2010
17	DHRUVA, India	MTR Thermal	100 MWt	"J" rod of ThO2	Research reactor in operation 1985–present

MWe, Megawatt electric. Reactor types: LWBR, light water breeder reactor; PWR, pressurized water reactor; BWR, boiling water reactor; PHWR, pressurized heavy water reactor; HTGR, high-temperature gas-cooled reactors; MSR, molten salt breeder reactor; LMFBR, liquid metal fast breeder reactor; MTR, materials testing reactor.

10.3.1 Light water reactors

LWRs can be operated with thorium fuel. In the United States, thorium fuel was tested in LWBR pressurized water reactors (PWRs) at the Indian Point plant in New York initially, but it was later converted to uranium fuel cycle. Shippingport reactor in Pennsylvania, a "seed-blanket" LWBR PWR, operated with thorium fuel between August 1977 and October 1982. At the end of core life, 1.39% more fissile material was present, proving that LWR breeding was possible with thorium.

Boiling water reactors (BWR) offer a design flexibility that can be optimized for thorium fuels. Thorium was also used in the BWR at Elk River Reactor in Minnesota, United States. Thorium fuel was also tested at the 60 MWe BWR in Lingen, Germany.

Germany and Brazil collaborated on various theoretical and some experimental studies in thorium utilization in PWRs during 1979–88; they concluded that for once-through operation the U/Th cycle was as competitive as the U/U cycle (Pinheiro et al., 1989). Studies by EDF in France also arrived at similar conclusions (IAEA, 1985). Studies were also carried out in Russia for thorium utilization in water–water energetic reactors (Morozov et al., 1999).

10.3.2 Heavy water reactors

HWRs could offer excellent neutron economy and faster neutron energy, so they are considered better for breeding ^{233}U. Conceptual design studies have indicated that thorium and uranium fuel concepts have many common design characteristics and that the thorium cycle could be used in a plant designed for the uranium cycle without large performance penalties. In Canada, Atomic Energy Canada Limited has more than 50 years of experience with thorium-based fuels, including burnup to 47 GWd/t. So far, about 25 tests have been performed in three research reactors and one precommercial reactor. India is continuing the use of ThO_2 pellets in PHWRs, for neutron flux flattening of initial core after start-up.

10.3.3 Fast breeder reactors

There is no relative advantage in using thorium instead of depleted uranium as a fertile fuel matrix in FBR systems due to a higher fast fission rate for ^{238}U and the fission contribution from residual ^{235}U in this material. The fast fission cross section of ^{238}U is significantly larger than that of ^{232}Th by a factor of 4–5. Because large stocks of depleted uranium are available as tails in uranium enrichment, there is little incentive to develop the use of thorium in these reactors.

^{232}Th in the blanket can be advantageous in a mixed reactor scenario. India has a three-stage nuclear energy scenario in which FBR play an important role; the use of thorium in the blanket to breed ^{233}U was tested in a 40 MWt fast breeder test reactor (FBTR), Kalpakkam, which has been operational since 1985. The Kamini 30 kWt experimental neutron-source research reactor, adjacent to the FBTR, uses ^{233}U as fuel. Thorium blanket is planned to be used in a 500 MWe prototype fast breeder reactor (PFBR), which is expected to be operational by 2015 (WNA, 2015b).

While the use of the ^{232}Th- ^{233}U cycle in a FBR does not, in general, appear to be as attractive as the ^{238}U- ^{239}Pu cycle, the use of ^{233}U in the core may provide advantages in reactor safety and control. The use of either thorium or ^{233}U in a fast spectrum can mitigate the issue of positive sodium void coefficient that is obtained with the uranium cycle under corresponding design conditions.

Studies were done in Russia on thorium utilization in BN-type fast reactors. Utilization of Pu/Th fuel has been proposed for the BN-800 type reactor (expected to be commercially operational in 2015).

10.3.4 High-temperature gas-cooled reactors

HTGRs are thermal spectrum reactors moderated with graphite and cooled by helium. Tristructural-isotropic (TRISO) fuel, composed of particles of thorium mixed with plutonium or enriched uranium, is the fuel in pebble-bed or prismatic arrangement. Deep burn of the fissile content and flexibility in fuel management makes high-temperature gas-cooled reactors (HTGR) a good option for thorium utilization. The possible use of beryllium oxide (BeO) in the fuel element, either as a matrix for the fuel particles or in place of some of the bulk graphite in the element, would greatly enhance the conversion ratio of the system.

In Germany, a 15 MWe AVR (*Arbeitsgemeinschaft Versuchsreaktor*) experimental pebble-bed nuclear reactor at Jülich operated from 1967 to 1988, partly as a test bed for various fuel pebbles, including thorium. The 300 MWe thorium high-temperature reactor (THTR), developed from the AVR, operated between 1983 and 1989 with 674,000 pebbles, over half containing Th/highly enriched uranium (HEU) fuel.

In the United States, General Atomics' Peach Bottom high-temperature, graphite-moderated, helium-cooled reactor operated between 1967 and 1974 at 110 MWth, using high-enriched uranium with thorium. The Fort St Vrain reactor, the only commercial thorium-fueled nuclear plant in the United States, was a high-temperature (700°C), graphite-moderated, helium-cooled reactor with a Th/HEU fuel designed to operate at 842 MWth (330 MWe). The fuel in the Fort St Vrain reactor was arranged in hexagonal columns ("prisms") rather than as pebbles. Almost 25 tTh were used as fuel for the reactor, and this achieved 170 GWd/t burn-up; the reactor operated from 1979 to 1989.

In the United Kingdom, thorium fuel elements with a 10:1 Th/U (HEU) ratio were irradiated in the 20 MWth Dragon reactor at Winfrith for 741 full power days. The Dragon reactor ran between 1964 and 1973 as an OECD/Euratom cooperation project, involving Austria, Denmark, Sweden, Norway, and Switzerland, in addition to the United Kingdom.

Because there was no established thorium ceramic fuel reprocessing route, commercial attention to the THTR has been directed to a once-through mode of operation. Even though this requires nearly double the amount of ^{235}U over a reactor lifetime compared with the ^{233}U recycle case, the ^{235}U requirement is much less than that of the current LWRs operating in the once-through mode.

10.3.5 Molten salt reactors

MSRs offer attractive concepts for thorium utilization. The primary coolant, and even the fuel itself, is a molten salt mixture, with graphite usually used as the moderator (Serp et al., 2014). In theory, this design can operate as a full recycle system, with continuous "online" pyroprocessing for separation of fission products. Some designs of MSRs are capable to "burn" problematic MA. MSRs may operate with epithermal or fast neutron spectrums, and with a variety of fuels, but good neutron economy makes the MSR attractive for the thorium fuel cycle.

The molten salts coolant, which is mostly lithium-beryllium fluoride (called FLiBe, $2LiF\text{-}BeF_2$), remain liquid without pressurization from about 500°C up to about 1400°C, in marked contrast to a PWR which operates at about 315°C under 150 atmospheres pressure. The advanced MSR concept is to have the fuel dissolved in the coolant as fuel salt. Intermediate designs have fuel particles in solid graphite and have less potential for thorium use (WNA, 2015c). Safety is achieved with a freeze plug, which if power is cut, allows the fuel to drain into a drain tank, which has a subcritical geometry. There is also a negative temperature coefficient of reactivity due to expansion of the fuel.

In 1954, Oak Ridge National Laboratory (ORNL), in the United States, conducted an aircraft reactor experiment (ARE), which was a 2.5 MWt nuclear reactor experiment designed to attain a high power density for use as an engine in a nuclear-powered aircraft (Rosenthal, 2009). The ARE used molten fluoride salt $NaF\text{-}ZrF_4\text{-}UF_4$ (53-41-6 mol%) as fuel, moderated by BeO, with helium gas as a secondary coolant; the reaction had a peak temperature of 880°C. The experiments ran for only several days. The Pratt and Whitney Aircraft Reactor No. 1 (PWAR-1) was a zero power MSR that was tested in ORNL in 1957, which used $NaF\text{-}ZrF_4\text{-}UF_4$ as the primary fuel and coolant (Scott et al., 1958).

ORNL continued this work with the molten salt reactor experiment (MSRE), a 7.4 MWt test reactor (Briggs, 1964). It went critical in 1965 and ran for 4 years. The fuel for the MSRE was $LiF\text{-}BeF_2\text{-}ZrF_4\text{-}UF_4$ (65-29-5-1), the graphite core moderated it, and its secondary coolant was fertile blanket of lithium-beryllium fluoride (FLiBe) ($2LiF\text{-}BeF_2$). It primarily used two uranium fuels: first ^{235}U and later ^{233}U. The latter $^{233}UF_4$ was the result of breeding from thorium in other reactors. Because this was an engineering test, the large, expensive breeding blanket of thorium salt was omitted in favor of neutron measurements.

During 1970–76, ORNL proposed a design for a 1000 MWe molten salt breeder reactor (MSBR) that would use $LiF\text{-}BeF_2\text{-}ThF_4\text{-}UF_4$ (72-16-12-0.4) as fuel and be moderated by graphite (MacPherson, 1985). The MSBR offered the potential of a BR of 1.07, a specific inventory on the order of 1.0 kg fissile/MWe or less, a power doubling time of less than 15 years. Even though the BR was smaller than that of the fast breeder designs proposed at that time, the doubling time was almost the same (Yoshioka, 2013). However, this program in the United States was closed in favor of the liquid metal fast breeder reactor.

The UK conducted theoretical and experimental research on a lead-cooled, 2.5 GWe molten salt fast reactor (MSFR) using chloride salt and fueled by plutonium, and either helium gas or molten lead as the secondary coolant (Endicott, 2014). Theoretical work on the concept was conducted between 1964 and 1966, and experimental work was done between 1968 and 1973. Russia conducted some theoretical and experimental research on MSRs during the 1970s at the Kurchatov Institute (Novikov, 1995).

As molten salts are present as eutectic mixture of LiF-BeF2, with fertile Th and fissile U/Pu dissolved in the medium, it serves as fuel element, heat transfer, and a fuel-processing medium. For reprocessing the solid spent fuel, PUREX is commercially available. In this process, spent fuel elements are dissolved as aqueous solution from which Pu and U are recovered by organic solvent extraction. However, molten salt processing can be possible through the "dry" pyroprocessing route developed in recent years, which could be simpler and compact than the PUREX process (Yoshioka, 2013). Dry processing technology was proposed and developed at Oak Ridge National Laboratory (ORNL) in Oak Ridge, Tennessee (US) in the 1970s and later in European countries; this technology was tried for the solid fuels. The online processing of molten fuel salts in MSR designs usually involves removal of U, Pa and FP (Fission Products) and reintroduction of U, Pu, or Th. MA remain in the reactor until they fission.

10.3.6 Resource requirements

Because ^{232}Th is not fissile, a certain quantity of ^{235}U, ^{233}U, or Pu is required in the fuel cycle. The amount of fissile material required depends on the reactor system and the fuel cycle options (once-through, limited recycle, or continuous recycle). Natural uranium as a starting point for ^{235}U and Pu will be required for thorium-based nuclear energy systems. Depending on the system, natural uranium will be required either in the driver fuel or as a continuous top-up.

A thorium fuel cycle can be self-sustaining if enough ^{233}U is created to replace the quantities that are consumed. It will remain necessary to provide fissile material from elsewhere (^{235}U or plutonium) to provide inventories for new construction. To achieve this self-sustaining cycle, fuel burnup has to be limited to about 13,000 MWd/t HM in the case of a PHWR. This is called the self-sustaining equilibrium thorium cycle. Depending on the various costs involved, however, it will be more economical to aim for higher fuel burnup and accept the need for a continuing supply of fissile material "driver" for topping up.

Taking a "high burnup" case of 40,000 MWd/t HM and assuming that its plutonium requirements are to be supplied by natural uranium CANDUs, the thorium-using reactors would use only about 30 tTh, plus 140 Kg Pu/year (generated by 45 t of natural uranium), in comparison to 140 t of natural uranium per GW-year for a once-through natural uranium cycle. A mixed fuel cycle of this type could be competitive in resource utilization (Lung, 1997).

Wigeland et al. (2014) evaluated various options in nuclear fuel cycles in terms of different criteria, including fuel resources utilization. Of the 40 evaluation groups considered in the study, 15 have thorium as a fertile material in the fuel cycle. Natural thorium and uranium resources requirements for the various evaluation groups in the study are summarized in Table 10.2.

Table 10.2 **Natural thorium and uranium resources requirement for various utilization scenarios**

Evaluation group	Short description	System considered	Natural Th use (t/GWe)	Natural U use (t/GWe)
Once-through				
EG05	Once-through using enriched-U/Th fuel in thermal or fast critical reactors	High-Conversion HTGR with LEU and Th fuel	4.65	289.20
EG06	Once-through using Th fuel to very high burnup in thermal EDS	Breed and Burn ^{233}U/Th in Thermal spectrum FFH	9.88	0.00
EG08	Once-through using Th fuel to very high burnup in fast EDS	Breed and Burn ^{233}U/Th in Fast spectrum Fusion-Fission Hybrid (FFH)	1.62	0.00
Limited recycle				
EG10	Limited recycle of ^{233}U/Th with new Th fuel in fast and/or thermal critical reactors	Limited recycle of ^{233}U/Th in MSR	10.86	0.00
EG11	Limited recycle of ^{233}U/Th with new enriched-U/Th fuel in fast or thermal critical reactors	Breed and Burn ^{233}U/Th with LEU Support in sodium-cooled fast reactor (SFR) with Partial Separation	2.55	106.80
EG17	Limited recycle of Pu/Th with new enriched-U/Th fuel in thermal critical reactors	Limited recycle of Pu from PWR in an PWR burner fueled with Thorium	1.88	172.41
EG18	Limited recycle of ^{233}U/Th with new enriched-U/Th fuel in thermal critical reactors	Limited recycle of ^{233}U/Th from PWR in a PWR burner	3.42	152.16
Continuous recycle				
EG25	Continuous recycle of ^{233}U/Th with new enriched-U/Th fuel in thermal critical reactors	Continuous recycle of ^{233}U/Th in PWR with LEU Support	0.85	113.54

(Continued)

Table 10.2 (Continued)

Evaluation group	Short description	System considered	Natural Th use (t/GWe)	Natural U use (t/GWe)
EG26	Continuous recycle of $^{233}U/Th$ with new Th fuel in thermal critical reactors	Continuous recycle of $^{233}U/Th$ in MSR	1.25	0.00
EG27	Continuous recycle of $^{233}U/Th$ with new enriched-U/Th fuel in fast critical reactors	Continuous recycle of $^{233}U/Th$ in SFR with LEU Support	0.45	186.62
EG28	Continuous recycle of $^{233}U/Th$ with new Th fuel in fast critical reactors	Continuous recycle of $^{233}U/Th$ in SFR	1.68	0.00
EG37	Continuous recycle of $^{233}U/Th$ with new enriched-U/Th fuel in both fast and thermal critical reactors	Continuous Recycle of TRU/U from PWR in SFR and produce $^{233}U/Th$ for recycle in Advanced PWR	0.43	24.36
EG38	Continuous recycle of $^{233}U/Th$ with new Th fuel in both fast and thermal critical reactors	Continuous Recycle of $^{233}U/Th$ produced in SFR in PWR	1.93	0.00
EG39	Continuous recycle of $^{233}U/Th$ with new enriched-U fuel in both thermal critical reactors and fast EDS	Continuous Recycle of ^{233}U in PWRs and burn TRU in ADS	0.75	114.85
EG40	Continuous recycle of $^{233}U/Th$ with new Th fuel in fast EDS and thermal critical reactors	Produce $^{233}U/Th$ in ADS and continuously recycle in PWR	1.51	0.00

The maximum natural thorium requirement of 11 t/GWe is estimated in order for a limited recycle of ^{233}U/Th in a MSR system. Other scenarios with limited recycle indicate thorium requirements of 1.88–3.42 t/GWe, but all scenarios have higher natural uranium use of 107–172 t/GWe. Some of the continuous recycle scenarios also have a very low thorium requirement, as well as no or limited natural uranium requirement, such as the continuous recycle in a MSR.

10.4 Current research and future possibilities

Despite many projects and pilot test reactors in the second half of the 20th century, the use of thorium as a reactor fuel has yet to be commercialized in a modern power reactor. However, today ongoing research and development on advanced reactor designs may employ thorium as a nuclear fuel, described next. These projects include HTGR; MSR; CANDU-type reactors; AHWR; FBR, and PHWR.

10.4.1 China

China in collaboration with the USA has very active ongoing research on thorium utilization in MSR designs (Li et al., 2015). This is a dual program involving an early solid fuel stream and advanced liquid fuel stream (WNA, 2015c). In January 2011, the China Academy of Sciences launched a research and development program on a liquid-fluoride thorium reactor (LFTR), called the thorium-breeding molten salt reactor (TMSR). In 2013, the National Energy Administration included the TMSR project among the 25 "National Energy Major Application-Technology Research and Demonstration Projects" in its "Plan of Energy Development Strategy." In 2014, the local government of Shanghai launched a major TMSR project to support the TMSR technology development.

The TMSR program is divided into three stages. In the early stage, a 10 MWt solid-fueled molten salt test reactor (TMSR- SF1) and a 2 MWt liquid-fueled molten salt experimental reactor (TMSR-LF1) are planned for construction and operation by 2016. In the engineering experimental stage, a 100 MWt solid-fueled TMSR demonstration system (TMSR-SF2) is planned by 2025 and a 10 MWt liquid-fueled molten salt experimental reactor (TMSR-LF2) is planned by 2018. The third industrial promotion stage will aim for the commercialization of a 1 GW TMSR-SF3 by 2030. A fast spectrum TMSFR-LF fast reactor optimized for burning of MA is also envisaged.

Solid fuel MSR technology was preferred in the early stage, due to the technical difficulty associated with high radioactivity of the molten salt when they contain dissolved fuels and wastes. After the accumulation of experience is gained with component design, operation, and maintenance of clean salts, use of liquid salt will be applied. Molten salt fuel is considered superior to the TRISO fuel in effectively unlimited burnup, less waste, and lower fabricating cost (WNA, 2015c).

Solid fuel envisages only partial utilization of thorium with an open fuel cycle, whereas liquid fuel designs will have a fully closed Th-U breeding cycle. Solid fuel TRISO particles will be with both low-enriched uranium and thorium, separately. The first step will be to develop solid fuel, bypassing the difficult reprocessing and refabrication options, and subsequently mastering the complex fluid fuel technology. The US cooperation with this project is primarily on the solid fuel technology, which is considered as the realistic first step.

China is developing HTR-PM, which is a graphite-moderated, helium-cooled high-temperature reactor. It is possible to use thorium in this type of reactor. Construction of a twin HTR-PM unit started in 2014, and is expected to be operational by 2017.

10.4.2 Canada

Since 2008, CANDU Energy of Canada and the China National Nuclear Corporation (CNNC) are cooperating in the development of thorium and recycled uranium as alternative fuels for new CANDU reactors. CANDU Energy is now part of SNC-Lavalin, and works on Advanced Fuel CANDU Reactor (AFCR) technology, which aims at thorium utilization. AFCR will be designed to use recycled uranium or thorium as fuel, thus reducing spent fuel inventories and significantly reducing fresh uranium requirements. Spent fuel from four conventional PWR reactors can fully supply one AFCR unit (as well as providing recycled plutonium for mixed oxide fuel (MOX)). CNNC is currently preparing an AFCR feasibility study and proposal to China's National Energy Administration, to be submitted by early 2015.

The integral molten salt reactor is proposed by the Terrestrial Energy in Canada (LeBlanc, 2013). It is based on the MSBR, but is a kind of tank-type reactor, where heat exchangers are enclosed in the reactor vessel.

10.4.3 India

In India, research on thorium utilization has been carried out since the 1950s. A three-stage nuclear energy program with uranium-fueled PHWRs, plutonium-fueled FBRs, and thorium-^{233}U-based AHWRs has been envisaged. A 500 MWe PFBR is in the final stages of construction, and is expected to be completed in 2015. More 500 MWe FBRs are planned for immediate deployment and beyond 2025; a series of 1000 MWe FBRs with metallic fuel, capable of high breeding potential are proposed (Chetal et al., 2011). The large-scale deployment of thorium is expected to be about three to four decades after the commercial operation of FBR with short doubling time, when thorium can be introduced to generate ^{233}U.

Demonstration of the use of thorium on an industrial scale is planned in an AHWR. AHWR is a 300 MWe boiling light water cooled and heavy water moderated vertical pressure tube type reactor (Sinha and Kakodkar, 2006). The reactor is designed with the dual objective of utilization of abundant thorium resources and to meet the future demands to nuclear power, which includes enhanced safety and reliability, improved economics, and a high level of proliferation resistance. It has many passive and inherently safe features so that the reactor can be located close to population centers.

During mid-2010, a prelicensing safety appraisal of the planned experimental thorium-fueled 300 MW(e) AHWR was completed by the (India) Atomic Energy Regulatory Board. The site-selection process started in 2011; the reactor is expected to become operational by 2020. An experimental assembly with AHWR type (Th-Pu) MOX fuel pins completed its test irradiation, and another with (Th-LEU) MOX fuel pins has been loaded in the Dhruva research reactor. Several test facilities have been setup for the AHWR design validation.

India is also considering the use of thorium in a compact high-temperature reactor (CHTR) and innovative high-temperature reactor for hydrogen production (IHTR). Both of the designs use ^{233}U–Th-based TRISO coated particle fuel. CHTR uses lead-bismuth as coolant and IHTR molten salt.

India's nuclear energy program, in general, and thorium utilization in particular, are heavily linked to solid fuels, its fabrication, reprocessing, and refabrication. There are technical challenges in Th-based fuel fabrication that require higher sintering temperature. Thorium-based fuels are reprocessed using the THOREX process, which is similar to the PUREX process. The THOREX process is still in the developmental stage and need extensive modifications prior to large-scale deployment. There are challenges in dissolving the very inert thorium fuel in aqueous solutions. The presence of ^{232}U requires automated reprocessing and fuel fabrication in shielded facilities. Breeding of ^{233}U from ^{232}Th goes through an intermediate ^{233}Pa with about 27 days half-life. Therefore, a cooling time of at least 1 year is required for maximizing the recovery of ^{233}U. Remaining ^{233}Pa in the fuel will go into the waste stream, which could have a long-term radiological impact. To avoid this, it is essential to separate ^{233}Pa from the spent fuel prior to extraction of ^{233}U and Th (Vijayan et al., 2013).

India currently has only small uranium spent fuel reprocessing plants at three sites with a total of 330 t/year capacity, with mostly the PUREX process in use (except for reprocessing fast reactor carbide fuel). To close the FBR fuel cycle, a fast reactor fuel cycle facility is under construction. Power reactor thorium reprocessing facility has been constructed to reprocess ThO_2 fuel bundles irradiated in PHWRs; this facility is under commissioning. India considers that premature deployment of thorium would lead to suboptimal use of indigenous energy resources, and that it would be necessary to build up a significant amount of fissile material prior to launching the thorium cycle in a big way. Therefore, thorium-based reactors are expected to be deployed only beyond 2070.

Recently, India started to also consider MSRs as one of the promising options for thorium utilization. Conceptual design of the Indian molten salt breeder reactors (IMSBRs) is currently under research and design. India considers IMSBRs to provide significant advantages over metallic fueled FBRs, such as reduced fissile inventory and better BR. The major difference in practical terms will be the bypassing of the comparatively difficult fuel fabrication, reprocessing, and fuel refabrication requirements.

India also has an active research program for thorium utilization in ADS systems (Nema, 2011; Sinha, 2011).

10.4.4 Norway

In April 2013, Thor Energy of Norway commenced a thorium MOX testing program in the Halden research reactor in Norway. Fuel irradiation is being tested to determine if thorium-plutonium (Th-Pu) MOX can be used in commercial nuclear power plants. Thor has commenced discussions with several utilities about the use of these thorium-mixture fuels in commercial LWRs and is conducting feasibility studies with one utility. Thor has commenced discussions with several regulators about the licensing of thorium fuels for use in these LWRs.

10.4.5 Europe

Safety Assessment of the Molten Salt Fast Reactor (SAMOFAR) is a consortium that consists of 11 participants, which include universities and research laboratories, such as CNRS (France), JRC (European Commonwealth), CIRTEN (Italy), TU Delft (Netherlands), CINVESTAV (Mexico), and PSI (Switzerland). Industrial partners include IRSN (France), AREVA (France), KIT (Germany), EDF (France), and CEA (France). The objective of SAMOFAR is to prove the innovative safety concepts of the MSFR by advanced experimental and numerical techniques.

10.4.6 South Africa

Steenkampskraal Thorium Ltd (STL), South Africa is undertaking several activities related to the thorium fuel cycle. STL owns the rights to the thorium that will be produced at the Steenkampskraal rare earths and thorium mine in South Africa. STL has designed a thorium refinery for the production of reactor-grade thorium. STL is designing a generation IV high-temperature, gas-cooled, pebble-bed reactor, the HTMR100 (100 MWth high-temperature modular reactor). The HTMR100 will be cooled with helium gas and operate at temperatures around 750°C.

10.4.7 United States

Several companies are developing innovative thorium fuel for LWRs, and HTGR and MSR concepts for thorium utilization.

Lightbridge is developing advanced metallic nuclear fuel for maximum power levels and operating cycle length extension in LWRs. The metallic fuel rod also forms the central seed region of the seed-and-blanket thorium-based LWR fuels. Current development efforts are focused on demonstrating the performance of the metallic fuel in a prototypic LWR environment and developing data and methodologies that will enable evaluation of the fuel for nuclear regulatory licensing.

X-energy is designing the Xe-100 reactor, a high-temperature gas-cooled pebble-bed nuclear reactor. Xe-100's small footprint and safe design allow operation close to population centers, and it is suited for installation at geographically constrained sites. The Xe-100 is being designed specifically to accommodate a variety of fuel types, including thorium, without design changes. X-energy's current plans leverage the US Department of Energy's investment in fuels through the Next Generation Nuclear Plant program.

Transatomic Power Corp has finalized the preliminary design of an advanced MSR, and began experimental testing of key materials and components. Corrosion, radiation, and high-temperature materials testing are being conducted under a 3-year sponsored research agreement with the Department of Nuclear Science and Engineering at the Massachusetts Institute of Technology.

Martingale is developing a simple thorium MSR called ThorCon. Preliminary detailed design has been completed.

Flibe Energy is planning to develop and commercialize a LFTR. LFTR is a heterogeneous MSR design that breeds its ^{233}U fuel from a FLiBe salts with thorium fluoride. A study of the LFTR design to understand the potential and challenges has commenced.

10.4.8 United Kingdom

Moltex Energy's stable salt reactor (SSR) is a conceptual design with no pumps (only impellers in the secondary salt bath) and relies on convection from vertical fuel tubes in the core to convey heat to the integral steam generators. The SSR can be run with thermal or fast neutron spectrums and thorium can be potentially used.

10.4.9 Generation IV international forum

The Generation IV International Forum (GIF) has Canada, China, the European Atomic Energy Community (Euratom), France, Japan, Russia, South Africa, South Korea, Switzerland, and the United States as active members. The GIF explore areas of mutual interest and make recommendations regarding research and development areas and processes for development of generation IV nuclear energy systems. The GIF program for the MSR includes the concepts of MSFR where Th fuel can be used. The GIF 2014 annual report (The Generation IV International Forum, 2015) said that a lot of work must be done on salts before demonstration reactors were operational, and they suggested 2025 as the end of the viability research and development phase.

10.4.10 Russia

Russia's molten salt actinide recycler and transmuter is a fast reactor fueled only by transuranic fluorides from uranium and MOX LWR used fuel. It is part of the MARS (minor actinide recycling in molten salt) project involving Scientific Research Institute of Atomic Reactors (Russia) (RIAR), Kurchatov, and other research organizations. The 2400 MWt design has a homogenous core of Li-Na-Be or Li-Be fluorides without graphite moderator and has reduced reprocessing compared with the original US design. Thorium may also be used, though it is described as a burner-converter rather than a breeder.

10.4.11 Germany

The Institut für Festkörper-Kernphysik gGmbH (Institute for Solid-State Nuclear Physics—IFK) is designing a reactor concept called the dual fluid reactor (DFR). The DFR can use versatile nuclear fuels, for example thorium, natural uranium, or even depleted uranium and spent nuclear fuel. It has a hard neutron spectrum that favors the poor neutron economy of the Th-U cycle, yielding better doubling times. It allows for the production of custom nuclides for medical use, including the scarce Molybdenum-99/Technetium-99m.

10.4.12 Denmark

Copenhagen Atomics plans to design a MSR-based system, preferably to fit in a shipping container, called the Copenhagen Atomics Waste Burner (CAWB). The CAWB will use thorium to burn out actinides from spent nuclear fuel to convert long-lived radioactive waste into short-lived radioactive waste, while producing energy. The aim is to optimize the CAWB so that it can start on fissile material from spent nuclear fuel alone, or with added external fissile material. The first version of CAWB will have a 50 MWt capacity.

10.4.13 Japan

International Thorium Molten Salt Forum (ITMSF) and Thorium Tech Solution Inc. (TTS) are developing the FUJI MSR. The proposed design is rated at 200 MWe output. The consortium plans to first build a much smaller MiniFUJI, a 7 MWe reactor of the same design (Yoshioka, 2013). A 1000 MWe capacity Super-FUJI is also under preliminary design phase. Continuous chemical processing of fuel salt is not performed in FUJI, as the system proposed to use ADS for production of fissile ^{233}U. But radioactive Xe, Kr, and tritium are removed from the fuel salt continuously. The total fissile ^{233}U required for 30 years operation at 70% load factor is 1.48 t, and the remaining fissile at the end of the reactor life is 1.50 t. A total of 3.9 tTh will be consumed in 30 years. Estimated production of Pu will be 0.3 kg after 30 years of operation, which is 0.03% of that of a BWR with same capacity. The production of MA is 5.3 kg after 30 years, which is 5% of that of a same sized BWR.

Kyoto Neutronics is designing a small thorium MSR integrated with an accelerator neutron source, which they refer to as UNOMI (Universally Operable Molten salt reactor Integrated).

10.4.14 Thorium production for the foreseeable future

In current market conditions, primary production of thorium is not likely to be as economic as an independent private enterprise. Recovery of thorium as a byproduct, extracted during REE recovery from monazite [$(REEs,Th)PO_4)$], seems to be the most feasible source of thorium production at this time.

The recovery of monazite from raw sand or crushed ore is possible by gravity separation techniques (utilizing the high density of monazite) and electrostatic methods (monazite displays weak magnetism). The monazite is then dissolved in either sodium hydroxide or sulfuric acid. The resulting solutions contain REEs, uranium and thorium. This is followed by a multistage process using organic phases to achieve separation with a final product of Th oxide.

Processing of monazite to recover rare earths and thorium has been effectively done in the past in many countries. Today, recovery of monazite (and possibly thorite) as a coproduct of REE mining and processing, made efficient and relatively inexpensive by modern heavy-mineral separation techniques, may provide an economical means of acquiring thorium for the foreseeable future. This thorium then may become a source of fuel stock for a new generation of nuclear power reactors.

10.5 Thorium geology and resources

10.5.1 Thorium geochemistry and mineralogy

Thorium is among a group of elements sometimes referred to as the high field strength elements (HFSE). These are elements having a valence state greater than two (high charge) and small-to-medium-size ionic radii, thus producing a high electric field (high field strength). Besides thorium, these elements also include the REEs (including yttrium), niobium, zirconium, hafnium, tungsten, phosphorus, titanium, tantalum, scandium, and uranium.

These HFSE are associated, as incompatible ions, in alkaline igneous melts that form from low degrees of partial melting in the upper mantle. Alkaline igneous rocks—sometimes referred to as alkalic rocks—are a series of igneous rocks that formed from magmas and fluids so enriched in alkali elements that sodium- and potassium-bearing minerals form constituents of the rock in much greater proportion than "normal" (siliceous) igneous rocks (Fitton and Upton, 1987; Sørensen, 1974). Neodymium isotopic data derived from carbonatite and alkaline igneous complexes are consistent with the thorium and REEs being derived from the parental magma (Verplanck et al., 2014). For the most part, alkaline intrusions—carbonatite, lamprophyre, syenite, feldspathic rocks, kimberlite, and related rock compositions—have silica undersaturated compositions, often with unusual mineralogy. Thorium-REE deposits and associated rock types tend to occur within stable continental tectonic units, in areas defined as shields, cratons, and crystalline blocks; they are generally associated with intracontinental rift and fault systems. For detailed discussions about alkaline igneous rocks and their geographic distribution, refer to Sørensen (1974), Fitton and Upton (1987), and Woolley (1987, 1989, 2001).

In magmatic systems, the high charge of the HFSE ions impedes the ability of these elements to achieve charge balance and thus fit into the structure of the most common igneous minerals, which have coordination sites best suited for elements of larger ionic radii and valence states of 2 or 1. As a result, when common silicate minerals crystallize—such as amphiboles, feldspars, pyroxene, olivine, quartz, and micas—most of the REEs, thorium, and HFSE tend to preferentially remain in the coexisting melt. Successive generations of this process, referred to as crystal fractionation, increase the concentrations of REEs and thorium in the remaining melt until individual mineral-rich phases crystallize that can accommodate the HFSE. Commonly, the REEs and Th fractionate and crystallize from the magma together, as exemplified by the coexistence of REEs and Th in monazite. In addition to ionic radius and charge balance, partitioning of Th, the REEs, and other HFSE between a mineral and coexisting magma can also be dependent on temperature, pressure, fluid composition, magmatic convection, and variation in mineral settling velocities in the magma chamber (Henderson, 1984; McKay, 1989; Kogarko et al., 2006). Subsequent hydrothermal metasomatic mineralization and (or) supergene processes can enhance the concentrations of Th, REEs, and other HFSE (Mariano, 1989; Chakhmouradian and Zaitsev, 2012).

Thorium, in contrast to uranium, has low mobility in most surface waters and groundwaters, and thorium minerals are typically insoluble in circa-neutral waters (Langmuir and Herman, 1980). The low chemical solubility and high mechanical strength of thorium minerals, particularly monazite, is the primary reason that these minerals are able to be transported intact in placer deposits (heavy-mineral sands). The low mobility of thorium in waters in dissolved form is an important factor to consider in developing exploration plans and environmental assessments.

Natural radioactivity is the principal exploration guide in locating thorium deposits. Radioactivity is the environmental factor of most concern associated with the development or disturbance of these deposits (Verplanck et al., 2014).

The REE-Th-phosphate mineral ***monazite*** is the principal thorium mineral (Fig. 10.1). Monazite, recovered primarily as a source of the REEs, can contain as much as 20% thorium oxide (Long et al., 2010), although 5–10% is more typical. It is a common accessory mineral in carbonatites and alkaline intrusions and is a minor constituent in some other types of igneous and some metamorphic rocks

Figure 10.1 Scanning electron microscope image of a zoned monazite crystal.
Source: Image courtesy of John Aleinikoff, US Geological Survey.

Figure 10.2 An example of monazite grains separated from a heavy-mineral sand; such deposits have historically been important sources of the REEs and thorium. Monazite has been separated from some ancient and modern beach sands as a coproduct to the recovery of economic titanium minerals (ilmenite and rutile), zircon, and other industrial minerals. *Source:* Photograph used with permission of www.geology.com.

(Force, 1976; Van Gosen et al., 2014). Monazite can also be an accessory mineral in some coastal deposits of heavy-mineral sands (Fig. 10.2); monazite produced from heavy-mineral sands has been an important historic source of REEs and thorium (Van Gosen et al., 2014).

Xenotime [YPO_4] is another REE-phosphate mineral that can likewise contain thorium. Similar to monazite, xenotime can be an accessory mineral in alkaline intrusions, in some acidic igneous rocks, and in some mica- and quartz-rich metamorphic rocks of high metamorphic grade. Xenotime also occurs as a detrital mineral that can deposit with other durable, dense minerals (including monazite) in the heavy-mineral assemblage of fluvial and coastal sediments. Xenotime has been recovered from fluvial placer (alluvium) deposits as a source of yttrium and other REEs (Van Gosen et al., 2014)

Thorite, a thorium silicate [$(Th,U)SiO_4$], is less common than monazite, but is the primary thorium mineral in some vein deposits. Examples of thorite-rich veins are those in the Lemhi Pass district in the United States (Fig. 10.3); the Lemhi Pass veins average about 0.4% thorium oxide content, mainly due to thorite (Staatz, 1972, 1979).

10.5.2 Thorium-bearing deposits

Due to the geochemistry of Th and other "incompatible" HFSE, these elements cooccur as anomalous concentrations in uncommon rock types, such as:

- Carbonatites
- Alkaline igneous intrusive complexes and associated veins and (or) dikes
- Massive magnetite-apatite bodies.

Additionally, some moderate- to high-grade metamorphic rocks (amphibolite facies and higher) can contain monazite as an accessory mineral (Force, 1976; Van Gosen et al., 2014).

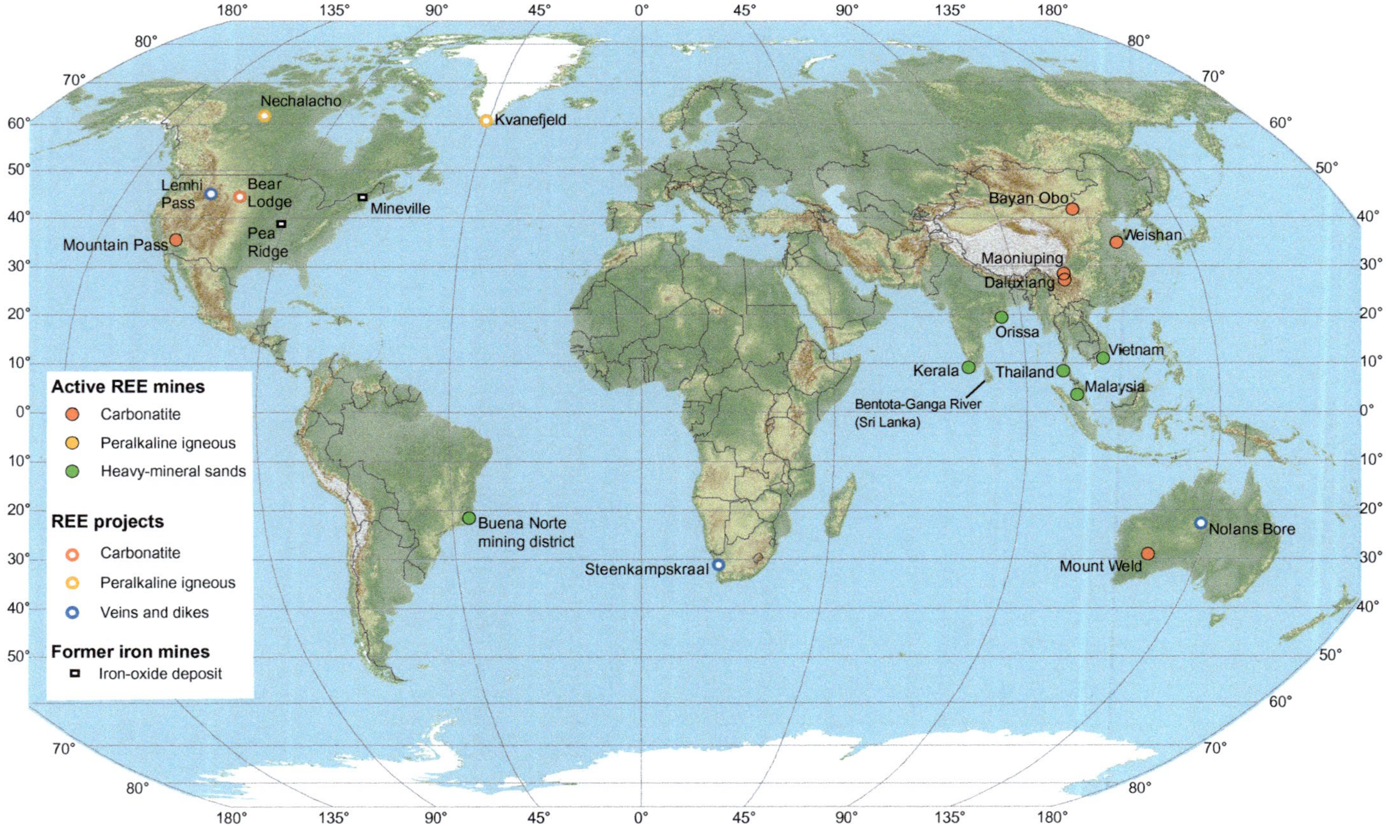

Figure 10.3 Index map showing active REE mines, selected REE exploration-development projects, and two former iron mines, used as examples of deposits containing REEs and thorium. An extensive dataset of reported thorium deposits worldwide is provided in the database **ThDEPO**, which is provided online by the International Atomic Energy Agency (IAEA) (available at https://infcis.iaea.org/THDEPO/About.cshtml).

Weathering and decomposition of these rock types can release monazite and/or xenotime as detrital grains into streams and rivers, which transport the monazite and/or xenotime to the coastal region. Here these mineral grains are sorted by waves, wind, and tides, ultimately to settle out in thin laminae with other heavy (dense) minerals within extensive deposits of heavy-mineral sands.

Carbonatites host the largest tonnage REE deposits and consistently have associated enrichments in thorium. Carbonatites are a rare, carbonate igneous rock formed by magmatic or metasomatic processes, which consist of 50% or more primary carbonate minerals, such as calcite, dolomite, and/or ankerite. They are genetically associated with, and therefore often occur near, alkaline igneous rocks. Thorough treatises on carbonatites are provided by Tuttle and Gittins (1966), Heinrich (1980), Bell (1989), and Verplanck et al. (2014).

Carbonatites occur throughout the world, but only about 400 carbonatites are known (Orris and Grauch, 2002; Berger et al., 2009; Verplanck et al., 2014). Of these, currently only five carbonatite complexes are being mined for REEs: Bayan Obo, Daluxiang, Maoniuping, and Weishan deposits in China, and the Mountain Pass deposit in California, United States (Fig. 10.3) (Verplanck et al., 2014). The Mount Weld mine in Western Australia, Australia, develops the deeply weathered (laterite) REE-rich zone that overlies a carbonatite (Lynas Corporation Ltd., 2014; Verplanck et al., 2014).

The thorium content in an actively mined carbonatite deposit is exemplified by the REE orebody mined at Mountain Pass in southeast California of the United States (Figs. 10.3 and 10.4). The Mountain Pass operation of Molycorp, Inc. is the only active REE mine and processing facility in the United States

Figure 10.4 The Mountain Pass mine of Molycorp, Inc., located in southeast California, is the only active producer of REEs in the United States (as of 2015). The orebody is a carbonatite intrusion, thought to host the largest REE resource in the United States.
Source: Photograph by Bradley Van Gosen, US Geological Survey.

(http://www.molycorp.com). The mine reopened in late 2010 after an 8-year hiatus. The orebody is a massive carbonatite intrusion—the Sulfide Queen Carbonatite—that reportedly contains 16.7 million tons of proven and probable reserves averaging 7.98% total REE oxides (Molycorp, Inc., 2012). This carbonatite is thought to be the largest REE resource in the United States. The Sulfide Queen is a calcite–dolomite–barite–strontianite carbonatite (Castor and Nason, 2004; Castor, 2008). The primary ore mineral is bastnäsite [(REEs)CO_3F], forming 10–15% of the carbonatite orebody, plus lesser REE contributions from the mineral parisite [$Ca[(REEs)_2(CO_3)_3F_2]$] (Castor and Nason, 2004; Castor, 2008). The Sulfide Queen is the only known carbonatite where the REE ore minerals—bastnäsite and parisite—are interpreted to have crystallized directly from the magma, thus representing primary magmatic mineralization (Mariano, 1989). In the other carbonatites, the REE-bearing minerals are interpreted to have formed by secondary processes, such as hydrothermal events and/or metasomatism.

Monazite is an accessory mineral in the Sulfide Queen Carbonatite, varying in amounts from a trace mineral to locally abundant within the orebody; the monazite is the principal Th-bearing mineral in the ore. In 2011, a bulk sample (about 1 t) of high-grade REE carbonatite ore was collected in the mine by the US Geological Survey (USGS); this sample revealed an average Th content of about 0.025% (elemental weight percent) (unpublished author data). This estimate of Th content is nearly identical to the thorium concentrations found in an earlier geochemical study of this carbonatite (Staatz et al., 1979).

For discussion purposes, using an approximate thorium concentration of 0.025% throughout the Sulfide Queen Carbonatite, this suggests that each ton of ore mined and processed at Mountain Pass contains, on average, about 0.25 kg of thorium. By then applying 16.7 million tons of proven and probable REE-bearing ore material in the Sulfide Queen, this equates to an estimated in-place resource of about 4175 tTh. This estimate of thorium content and resource in the Sulfide Queen orebody is an approximation based on limited sampling; monazite concentrations may prove to vary considerably across the carbonatite as mining progresses.

Presently (2015), when the carbonatite of Mountain Pass is mined, processed, and the REEs are separated, the Th moves with other residues into the tailings impoundment. Thus, it would require modifications to the ore processing system or processing of the tailings to recover the Th in the future.

The Bear Lodge project of Rare Element Resources Ltd. plans to mine a highly altered carbonatite-alkaline complex located in the Bear Lodge Mountains near Sundance in northeastern Wyoming, United States (Fig. 10.3). This project is currently in an advanced stage of permitting for their proposed REE mine and processing plant. Based on considerable exploratory drilling over the last several years, the company reports total measured and indicated resources of 18 million tons of ore averaging 3.05% total REE oxides in the district, applying a cutoff grade of 1.5% total REE oxides (Rare Element Resources, 2014).

The Bear Lodge REE deposits occur in hydrothermally altered zones of an Eocene carbonatite-alkaline intrusive complex. The REE deposits are hosted by diatremes composed of fragmented, highly altered trachyte and phonolite intrusive rocks, which are cross cut by carbonatite and siliceous carbonatite dikes, veins, and stockwork veinlets. The primary REE ore minerals are ancylite [$SrCe(CO_3)_2(OH) \cdot H_2O$] and REE-fluorocarbonates of the bastnäsite group, and important Th-bearing accessory minerals are the phosphates monazite, xenotime, and brockite [$(Ca,Th,Ce)PO_4 \cdot H_2O$] (Anderson et al., 2013). In 2011, Rare Element Resources permitted the US Geological Survey to collect grab samples in an exploration trench excavated into ore at the site of the proposed open pit mine. The sampling found Th concentrations that average about 0.11–0.12% (elemental weight percent).

In a press release of January 21, 2014, Rare Element Resources announced that pilot testing of their proposed ore processing technology identified a process to "selectively isolate and economically remove thorium." Their proposed processing procedures were filed as a utility patent application with the US Patent and Trademark Office; the patent currently (April 2015) awaits evaluation and approval. Separation and capture of thorium from their waste stream was designed into their ore processing plan due to concerns of radioactivity generated by Th-bearing waste materials. The company states the Th will be disposed of by a third party, in an unspecified manner. Thus, the segregated Th may become either unavailable or difficult to obtain in the future and be a cost liability in the meantime.

Another type of REE-thorium enrichment in carbonatite systems—supergene processes—involves deep weathering of a carbonatite in tropical environments, forming a REE-enriched laterite deposit. This type of deposit is exemplified at the Mount Weld mine in Western Australia (Fig. 10.3). The thickness of the laterite zone overlying the Proterozoic carbonatite is variable, but as much as 69 m thick in places (Verplanck et al., 2014). According to the study of Lottermoser (1990) the REE enrichment at Mount Weld was formed by the adsorption of REEs in the laterite zone onto plumbogummite-group minerals [$PbAl_3(PO_4)_2(OH)_5 \cdot H_2O$], rhabdophane [$(Ce,La)(PO_4) \cdot H_2O$] and secondary monazite in upper parts of the weathered zone (laterite), adsorption of REEs onto churchite [$Y(PO_4) \cdot 2H_2O$] in lower zones. Microcrystalline, secondary monazite accounts for the thorium content in the laterite. Gangue minerals include goethite [$Fe^{3+}O(OH)$] and hematite [Fe_2O_3] with minor calcite [$CaCO_3$], cryptomelane [$KMn^{4+}{}_6Mn^{2+}{}_2O_{16}$], dolomite [$CaMg(CO_3)_2$], hollandite [$Ba(Mn_6{}^{4+}Mn_2{}^{3+})O_{16}$], kaolinite [$Al_2(Si_2O_5)(OH)_4$], maghemite [$Fe_2{}^{3+}O_3$], montmorillinite [$(Na,Ca)_{0.33}(Al,Mg)_2(Si_4O_{10})(OH)_2 \cdot nH_2O$], magnetite [$Fe^{2+}Fe_2{}^{3+}O_4$], pyrochlore [$(Na,Ca)_2Nb_2O_6(OH,F)$], ilmenite [$FeTiO_3$], and rutile [$TiO_2$], and traces of quartz [SiO_2], barite [$BaSO_4$], anatase [TiO_2], and vermiculite [$(Mg,Fe^{+2},Fe^{+3})_3(Al,Si)_4O_{10}(OH)_2 \cdot 4H_2O$] (Lottermoser, 1990).

Alkaline igneous rocks crystallize from silicate magmas and fluids enriched in alkali elements, precipitating sodium- and potassium-bearing minerals, such as sodium- or potassium-rich amphiboles and pyroxenes (Sørensen, 1974; Fitton and Upton, 1987). As described earlier, thorium and the REEs have a strong genetic association with alkaline igneous processes, particularly peralkaline magmatism. Alkaline rocks typically have higher enrichments in REEs than most other igneous rocks (Verplanck et al., 2014). Peralkaline igneous rocks are defined as those in which the molecular proportion of aluminum oxide is less than that of sodium and potassium oxides combined ($Al_2O_3 < Na_2O + K_2O$).

Due to the geochemistry of thorium and alkaline magmatism, some of highest concentrations of monazite (and thus thorium concentrations) occur in alkaline igneous intrusions. Examples occur in alkaline plutons, magmatic layered complexes, and associated epigenetic veins and (or) dikes. Thorium content is rarely reported for these intrusions, because the emphasis is usually placed on their REE content. However, due to the mineralogical cooccurrence of thorium and REEs, alkaline igneous rocks that are enriched in REEs are likewise anomalous in thorium.

An example of a REE-thorium-rich, layered peralkaline igneous complex is the Nechalacho deposit at Thor Lake in the Northwest Territories, Canada (Fig. 10.3) (Avalon Rare Metals Inc, 2014). The primary phase REE ore minerals are eudialyte (complex zirconium-silicate with typically about 7 wt% REE oxides) and zircon (with about 3 wt% REE oxides); these minerals are interpreted to have crystallized and settled by gravity in the alkaline magma emplaced near Thor Lake, accumulating in layers (Sheard et al., 2012). Multiple subsequent episodes of hydrothermal metasomatism dissolved these primary phases, depositing REEs in fergusonite [(Ce, La,Y)NbO_4], allanite [(CaCe)(Al_2Fe^{2+})(Si_2O_7)(SiO_4)O(OH)], and bastnäsite, along with precipitation of REEs and thorium in secondary (late-stage) zircon and monazite (Sheard et al., 2012).

Another well-developed example of a layered, alkaline magmatic complex is the Ilímaussaq alkaline complex, South Greenland (Sørensen et al., 2006). The Kvanefjeld deposit within the Ilímaussaq complex (Fig. 10.3) is being evaluated for the development of its substantial REE, Zn, and U resources (Greenland Minerals and Energy Ltd, 2014). The rocks hosting REE-Zn-U-Th mineralization are lujavrites, which rare nepheline syenites that represent the most evolved igneous sequence of peralkaline intrusive complexes, containing numerous rare and obscure minerals. Monazite is among the more than 200 minerals identified in this complex (Sørensen et al., 2006).

Thorium-rich veins of uncertain origin also exist. Most of these types of vein deposits are interpreted to be related to concealed alkaline magmatism. Examples of vein-hosted thorium mineralization occur in the Lemhi Pass district along the Montana-Idaho border in the United States (Fig. 10.3). Thorite-bearing veins in this district appear to host the largest thorium resource in the United States (Van Gosen et al., 2009). Within the Lemhi Pass district, Staatz (1972, 1979) mapped 219 veins enriched in thorium and REEs. Most of these veins are quartz−hematite−thorite veins, which fill fractures, shears, and brecciated zones in Mesoproterozoic quartzite and siltite host rocks (Fig. 10.5). Thorium also occurs in monazite-thorite-apatite shears and replacements with specularite, biotite, and alkali feldspar. The thorium veins of the district range from 1 m to at least 1325 m in length and from a few centimeters to as much as 12 m in width (Staatz, 1979; Staatz et al., 1979). Fifteen thorium veins in the district exceed 300 m in length. Some of the veins contain carbonate minerals, such as calcite, siderite [($FeCO_3$)], and ankerite [Ca(Fe,Mg,Mn) $(CO_3)_2$], and local fluorite [CaF_2]. Rare earth- and thorium-bearing allanite and monazite are locally abundant. Other reported ore minerals include brockite [(Ca, Th,Ce)$PO_4 \cdot H_2O$] and xenotime. The primary gangue minerals include quartz,

Figure 10.5 Example of the quartz–iron oxide–thorite veins typical of those in the Lemhi Pass district, located along the border of eastern Idaho with southwestern Montana (United States). Thorite-rich veins of this type in the district cut Precambrian quartzite and siltite, but have an undetermined origin. Individual veins can crop out for at least 1325 m in length and as much as 12 m in width (Staatz, 1979). The veins of the Lemhi Pass district may form the largest thorium resource in the United States (Van Gosen et al., 2009).
Source: Photograph courtesy of Virginia Gillerman, Idaho Geological Survey.

hematite, limonite [$FeO(OH) \cdot nH_2O$], fluorapatite [$Ca_2Ca_3(PO_4)_3F$], potassium feldspar [$\mathbf{K}AlSi_3O_8$], biotite [$KFe_3^{2+}(AlSi_3O_{10})(OH)_2$], albite [$NaAlSi_3O_8$], and barite [$BaSO_4$]. The thorite veins of the Lemhi Pass district contain approximately equal concentrations of thorium and REEs. Staatz (1972) reported the REE analyses of 31 vein samples, which showed total REE oxide contents with an average value of 0.428% in the 10 largest veins in the district, very similar to their average thorium oxide content of 0.43%.

The Nolans Bore REE-P-U-Th-F deposit in the Northern Territory, Australia (Fig. 10.3), consists of multiple veins and breccia zones filled with abundant fluorapatite, with associated REE and thorium mineralization (Hoatson et al., 2011). These complex, multiphase deposits are hosted by fault zones and are spatially, perhaps genetically, associated with nearby carbonatite/alkaline complexes and tin- and tantalum-rich pegmatites. Features of the vein and breccia deposits indicate several overprinting events of hydrothermal mineralization. The primary ore minerals are cheralite, thorite, allanite, bastnäsite, monazite, and several REE-bearing fluorcarbonates (Hoatson et al., 2011). As of June 2012, Arafura Resources Limited reported total probable reserves in the Nolans Bore deposits, using a cutoff grade of 1.0% REE oxides, totaling 24 million tons of ore containing 2.8% total REE oxides, 12% P_2O_5, and 0.45 lb/t of U_3O_8 (Arafura Resources Limited, 2014); an earlier estimate reported a Th content of 0.27% (Hoatson et al., 2011). Veins of massive fluorapatite host most of the REE and thorium resources. When the REE concentrate is processed at Whyalla, South Australia, the thorium will be separated as an iron-thorium precipitate and transported back to the Nolans Bore mine site in Northern Territory, Australia, for long-term storage as a possible future energy source.

Massive iron oxide deposits of magmatic-hydrothermal origin can contain elevated concentrations of thorium and REEs, albeit in relatively small amounts. Two examples of massive iron oxide deposits in the United States with enrichments in thorium and REEs are the Pea Ridge deposit in southeast Missouri and the Mineville district in northern New York (Fig. 10.3).

The Pea Ridge magnetite-hematite orebody is hosted by Mesoproterozoic volcanic rocks of the St Francois terrane of southeastern Missouri. The magnetite-rich body is interpreted as a high-temperature, magmatic-hydrothermal deposit (Sidder et al., 1993) in ash-flow tuffs and lavas, which may have formed in the root of a volcanic caldera (Nuelle et al., 1991). Four REE-bearing breccia pipes (exposed in underground mine workings) steeply crosscut the magnetite-hematite orebody and the altered rhyolite host rock. Exposed portions of the breccia pipes are as much as 60 m in horizontal length and as much as 15 m in width; the pipes extend below the mined levels to an undetermined depth (Seeger et al., 2001). Identified REE-bearing minerals in the breccia pipes are monazite, xenotime, and minor amounts of bastnäsite and britholite. According to Seeger et al. (2001, p. 2), "total REE oxide concentrations of grab samples [of breccia] range from about 2.5 to 19 wt%." Mine tailings on the surface contain additional lanthanide resources, primarily in fine-grained REE-bearing minerals, chiefly monazite and xenotime, which form inclusions within apatite (Vierrether and Cornell, 1993). The monazite in the orebody, breccia pipes, and tailings comprises a thorium resource that has not been estimated. However, limited analyses of the breccia pipes found thorium concentrations from 0.23 to 0.41 wt% (Long et al., 2010, p. 59).

Thorium and REEs reside within apatite in iron bodies once mined in the Mineville iron district, New York, which is located in the northeastern part of the Adirondack Mountains. The orebodies are magnetite deposits intricately folded and faulted within a complex suite of Precambrian metamorphic and igneous rocks. The host rocks have both mafic and felsic compositions that include gabbro, diorite, augite syenites, and granite (Staatz et al., 1980). The iron deposits are mainly magnetite, martite, and apatite, with accessory augite, hornblende, albite, quartz, pyrite, and tourmaline (McKeown and Klemic, 1956). Iron deposits in the Mineville–Port Henry area contain apatite and are likewise rich in phosphorous, thorium, and REEs because these elements are concentrated within the apatite grains. In addition, the magnetite (the primary iron ore mineral) is intergrown with 1- to 3-mm-long, rice-shaped grains of apatite.

Currently, large tailings piles and unmined parts of magnetite orebodies in the Mineville district contain REE-bearing apatite-rich rock. This apatite contains accessory REEs and thorium. Staatz et al. (1980) estimated that about two-thirds of the tailings piles were derived from apatite-rich ores, which would represent about 9 million tons of the tailings. Using an average grade of about 8% apatite, approximately 720,000 t of apatite could be present in the tailings dumps in the district. McKeown and Klemic (1956) reported an average REE oxide content of 11.14% in

14 samples of apatite separated from the Old Bed, Joker, and Smith orebodies. Thus, the tailings dump piles could contain approximately 80,200 t of REE oxides. Uranium and thorium contents average 0.032% and 0.15%, respectively, suggesting approximate in-place resources of about 230 tU and 1080 tTh in the tailings. Spectrographic analyses reveal that yttrium is one of the main REE contained in this apatite, likely occurring as inclusions of xenotime in the apatite grains. This makes the apatite-rich tailings a potential source for yttrium and other heavy REEs (McKeown and Klemic, 1956).

Metamorphic rocks host some thorium-rich deposits, exemplified by the Steenkampskraal monazite deposit in South Africa (Fig. 10.3). The deposit consists of dikes of monazite-apatite-chalcopyrite-magnetite within the Proterozoic Namaqualand Metamorphic Complex (Andreoli et al., 1994, 2006). A study by Andreoli et al. (1994, p. 994) interpreted that the Steenkampskraal veins "formed under water-undersaturated conditions in granulite facies (T[empratures] approximately 800–860°C, P[ressure] approximately 5–6 kbars." Monazite in these veins has thorium contents of 8–8.8 wt% (Andreoli et al., 1994), thereby forming a vein-type deposit of high thorium grade along with the REE resource. Great Western Minerals Group (2014) reported a NI 43–101 estimate for the Steenkampskraal deposit of 559,000 t of measured plus indicated resources containing 15% total REE oxides (including Y oxide) and 2.14% Th oxide.

Metamorphic rocks of regional extent can contain monazite as an accessory mineral. Examples are: (1) sillimanite-grade schists, amphibolite, and gneisses in the southern Blue Ridge and inner Piedmont regions of the southeastern United States (Overstreet, 1967; Force, 1976); and (2) khondalites, charnockites, and leptynolites of granulite facies in the Eastern Ghats group of rocks in Odisha state (formerly known as Orissa) of northeastern India (Mohanty et al., 2003). These two examples are thought to be the bedrock sources of the detrital monazite in the extensive deposits of heavy-mineral sands in the coastal regions of the southeastern United States and eastern India, respectively.

Heavy-mineral sands are sedimentary deposits of dense (heavy) minerals that accumulate with sand, silt, and clay in coastal and alluvial environments, locally forming economic concentrations of heavy minerals (Van Gosen et al., 2014). Expansive coastal deposits of heavy-mineral sands are the main source of titanium feedstock for the titanium dioxide (TiO_2) pigments industry, through recovery of the minerals ilmenite ($Fe^{2+}TiO_3$), rutile (TiO_2), and leucoxene (an alteration product of ilmenite). Heavy-mineral sands are also the principal source of zircon ($ZrSiO_4$); it is often recovered as a coproduct. Other heavy minerals produced as coproducts from some deposits are sillimanite/kyanite, staurolite, garnet, and, for the recovery of REEs and thorium, monazite.

Monazite, one the "heaviest" minerals in the heavy-mineral suite, is the dominant REE- and Th-mineral in heavy-mineral sands (Fig. 10.2). This monazite is mainly derived from the erosion of igneous and high-grade metamorphic rocks that lie inland of the coastal plain. Monazite is resistant to chemical and physical weathering. For these reasons, monazite survives the arduous trip from a distant bedrock source to its deposition in a stream, river, or coastal plain environment.

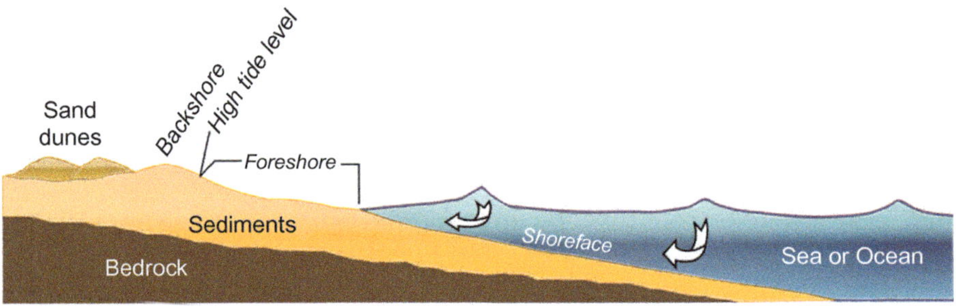

Figure 10.6 Features commonly used to describe shoreline (strandline) depositional environments associated with deposits of heavy-mineral sands. Not to scale.

Byproduct monazite has been recovered in Australia, Brazil, China, Indonesia, Korea, Malaysia, New Zealand, Sri Lanka, Thailand, Zaire, and the United States, derived mainly from the mining of heavy-mineral sands deposits in ancient and modern coastlines, as well as from some inland stream and river (alluvial) deposits. Recently, monazite has been recovered from modern coasts in India, Sri Lanka, and Brazil, and alluvial placers in Malaysia and Thailand (Fig. 10.3) (described in more detail in (Van Gosen et al., 2014).

The processes that form coastal deposits of heavy-mineral sands begin inland. High-grade metamorphic and igneous rocks that contain heavy minerals weather and erode, contributing detritus composed of sand, silt, clay, and heavy minerals to fluvial systems. Streams and rivers carry the detritus to the coast, where they are deposited in a variety of coastal environments (Fig. 10.6), such as deltas, the beach face (foreshore), the near-shore, barrier islands or dunes, and tidal lagoons, as well as the channels and floodplains of streams and rivers in the coastal plain. The sediments are reworked by waves, tides, longshore currents, and wind, which are effective mechanisms for sorting the mineral grains on the basis of differences in their size and density. The finest-grained, most dense heavy minerals are the most effectively sorted. The result is that heavy minerals accumulate together, forming laminated or lens-shaped, heavy-mineral-rich sedimentary packages that can be several meters and to as much as tens of meters thick. Most economic deposits of heavy-mineral sands are Paleogene, Neogene, and Quaternary in age; some are modern coastal deposits.

In some heavy-mineral-sands deposits, monazite is a coproduct mineral purposely recovered for its content of REEs and thorium. In particular, India oversees a state-administered program designed specifically to recover monazite, along with other heavy minerals (Anantharaman et al., 2008). These minerals are derived from mining and processing multiple, Quaternary heavy-mineral sands deposits along India's southern coast, principally in the states of Kerala and Orissa (Fig. 10.3). Governmental entities and partners, such as the Indian Rare Earths Limited (2014) and Kerala Minerals and Metals Ltd. (2014), are involved in the production and marketing of the coastal heavy minerals, as well as the separation of monazite. India has a particular interest in stockpiling monazite as a

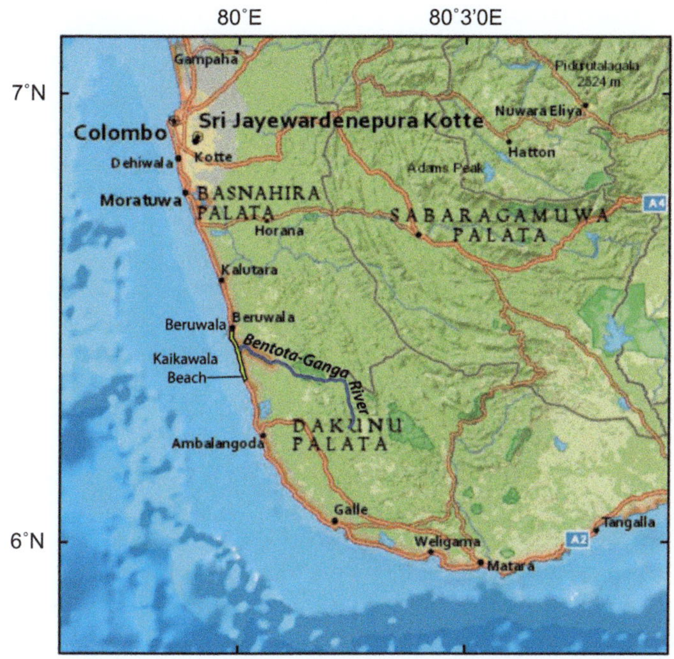

Figure 10.7 Southwestern Sri Lanka showing the Bentota-Ganga River. Monazite and other heavy minerals are transported by this river, which deposits them in seasonal beaches that extend from near Beruwala on the north to Kaikawala Beach on the south.

source of thorium, for use in nuclear fuel needed for their nationalized effort to develop and operate thorium-based nuclear power plants (Anantharaman et al., 2008; Martin, 2012).

The coastal sand deposits of Brazil have some of the highest monazite concentrations known in the world—as much as 8% average monazite in some sand bodies (Overstreet, 1967). The Guarapari coastline of Espiroto Santo is a popular tourist destination known for its white sand beaches, but this shoreline is also known for its very high level of background radioactivity due to abundant monazite. The Buena Norte coastal deposit near Campos in Espirito Santo (Fig. 10.3) has a reported monazite content of 0.83% (Jackson and Christiansen, 1993). Analyses of Brazilian monazites suggest that their average REE oxide content is typically 57–60% and that they are preferentially enriched in light REEs (Overstreet, 1967; Orris and Grauch, 2002).

Monazite-bearing alluvium in southwestern Sri Lanka, specifically stream sediments of the Bentota-Ganga River (Figs. 10.3 and 10.7), have been described as "one of the world's most thorium-rich sediments" (Rupasinghe et al., 1983, p. 1). Monazite is carried by this river system into seasonal beach sand deposits at Kaikawala and Beruwala (Fig. 10.7). Monazite was once mined on a small scale at Kaikawala beach.

10.5.3 Thorium resources

The coproduct occurrence of thorium and a general lack of economic interest in thorium have meant that thorium quantities were rarely, if ever, accurately defined in most countries. Information on estimated quantities of thorium was published between 1965 and 1981 in the biennial publication the "Red Book," published jointly by the Nuclear Energy Agency (NEA) Organization for Economic Co-operation and Development (OECD) and the International Atomic Energy Agency (IAEA); the thorium estimates reported in these publications applied the same terminology as was used for uranium resources at that time. The "Red Book" provides the most recent overview of the world uranium market fundamentals and industry, including uranium exploration, resources, production, and projected reactor-related requirements.

Information on estimated thorium resources by country was once again included in the 2014 edition of the "Red Book" (NEA-IAEA, 2014) (Table 10.3). In 2011 and 2013, the IAEA conducted technical meetings on thorium resources. Based on the inputs given in the meetings and details available in other open sources, total thorium resources, regardless of resource category or cost category, could be updated for 16 of the 35 countries that have known thorium resources. The largest identified thorium resources (from largest to smallest) occur in India, Brazil, the United States, Egypt, Turkey, South Africa, China, and the Russian Federation; total global thorium resources (in situ) are estimated at about 6.4 million tons (NEA-IAEA, 2014) (Table 10.3).

Globally, the principal thorium resources occur in a variety of REE deposits and some heavy-mineral sands (Fig. 10.3), which are mineral deposits in which the thorium minerals are accessory to other ore minerals. Significant REE deposits of all deposit types, as well as monazite-bearing deposits of heavy-mineral sands, are not only the largest sources of REEs but also represent the largest thorium deposits. Actively mined REE ore deposits are economic on the basis of their REE production. Heavy-mineral sands operations are typically economic based on their production of titanium minerals (ilmenite and rutile), zircon, and sometimes additional

Table 10.3 Identified thorium resources

Region	Country	Total thorium resources, tTh (*in situ*)
Europe	Turkey[a]	374,000
	Norway	87,000
	Greenland (Denmark)	86,000–93,000
	Finland[a]	60,000
	Russian Federation	55,000
	Sweden	50,000
	France	1000
	Total	*713,000–720,000*

(*Continued*)

Table 10.3 (Continued)

Region	Country	Total thorium resources, tTh (*in situ*)
Americas	United States	598,000
	Brazil	632,000
	Venezuela[a]	300,000
	Canada	172,000
	Peru	20,000
	Uruguay[a]	3000
	Argentina	1300
	Total	*1,726,300*
Africa	Egypt[a]	380,000
	South Africa	148,000
	Morocco[a]	30,000
	Nigeria[a]	29,000
	Madagascar[a]	22,000
	Angola[a]	10,000
	Mozambique	10,000
	Malawi[a]	9000
	Kenya[a]	8000
	Democratic Republic of the Congo[a]	2500
	Others[a]	1000
	Total	*649,500*
Asia	CIS[a] (excluding Russian Federation)	1,500,000
	–Includes Kazakhstan, estimated	(>50,000)
	–Includes Russian Federation, Asian part, estimated	(>100,000)
	–Uzbekistan, estimated	(5000–10,000)
	–Others	Unknown
	India	846,500
	China, estimated	>100,000 (including 9000[a] Chinese Taipei)
	Iran, the Islamic Republic of[a]	30,000
	Malaysia	18,000
	Thailand[a], estimated	10,000
	Viet Nam[a], estimated	5000–10,000
	Korea, Republic of[a]	6000
	Sri Lanka[a], estimated	*4000*
	Total	>2,647,500–2,684,500
Australia		595,000
World total		6,358,300–6,375,300

CIS, Commonwealth of independent states.
[a]Data not updated in 2013.
Source: Modified from NEA-IAEA (2014).

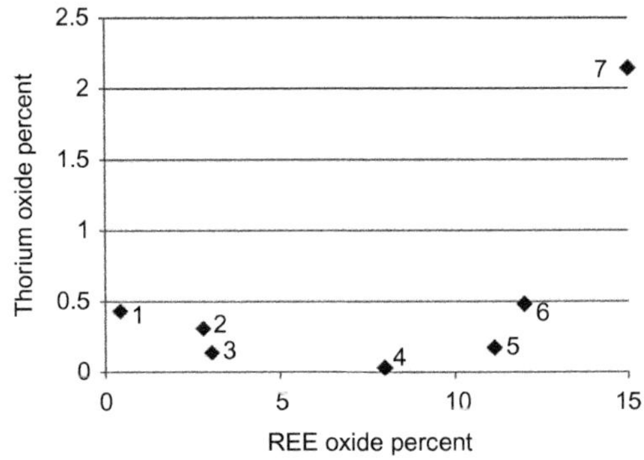

Figure 10.8 Comparison of REE and thorium content in a variety of deposit types, which are described in the text. For comparison, monazites separated from beach placers (heavy-mineral sands) in Kerala, southwest coast of India (Figure 10.3), reportedly contain an average REE oxide content of 57.5% and a thorium oxide content of 2.14% (Kerala Minerals and Metals Limited, 2014). Deposits shown are:

1. Lemhi Pass district, Idaho, United States; quartz–iron oxide–thorite veins. Data from Staatz (1972)
2. Nolans Bore deposit, Northern Territory, Australia; fluorapatite veins. Data from Arafura Resources Limited (2014) and Hoatson et al. (2011)
3. Bear Lodge deposit, Wyoming, United States; altered carbonatite-alkaline intrusive complex. Data from Rare Element Resources Ltd. (2014) and author data
4. Mountain Pass deposit, California, United States; carbonatite. Data from Molycorp, Inc. (2012), Staatz et al. (1979), and author data
5. Mineville iron mining district, New York, United States; apatite separated from magnetite deposits. Data from McKeown and Klemic (1956)
6. Pea Ridge iron deposit, Missouri, United States; four breccia pipes that cut the massive magnetite deposit. Data from Long et al. (2010)
7. Steenkampskraal deposit, South Africa; monazite-apatite veins. Data from Great Western Minerals Group (2014)

industrial minerals. Economic mineral deposits with coexisting thorium-rich minerals could be evaluated as sources of byproduct or potentially coproduct thorium (Fig. 10.8) if a market develops for thorium in the future.

The Indian Bureau of Mines describes India's coordinated effort to recover monazite from heavy-mineral sands. The monazite separate is then chemically treated to separate rare earths in composite chloride form and thorium as hydroxide upgrade. India reports large resources of REEs and thorium, mostly in monazite within coastal sediments along its southern coastlines. According to India's Atomic Mineral Directorate for Exploration & Research, "the reserves of monazite in India is about 10.70 million tonnes which translates to approx. 5 million tonnes of rare earth oxide" (Government of India, Department of Atomic Energy, 2011, p. 2).

The recovery of thorium from the monazite-apatite veins that form the Steenkampskraal deposit in South Africa is being considered (Fig. 10.3), which is a REE mining and production project under development in 2014 (Gambogi, 2014b;

Great Western Minerals Group, 2014). STL has the rights to this thorium and is planning to refine it, in collaboration with Thor Energy AS in Norway.

For now, the production of thorium as the primary product is generally regarded as uneconomic (NEA-IAEA, 2014). In the foreseeable future, the economic production of thorium is most likely to occur through the byproduct or coproduct of production of another mineral commodity of value, such as REEs or titanium minerals in heavy-mineral sands. In the context of production, a byproduct can be defined as the output from a production process that is minor in quantity and/or net realizable value relative compared to the primary products (Wouters et al., 2012). By convention, byproducts also are not inventoried, but the net realizable value from byproducts is typically recognized as "other income" or as a reduction of joint production processing costs when the byproduct is produced. Coproduct on the other hand is a major output from a joint production process that is significant in quantity and/or net realizable value. Coproducts play an important role in the economic analysis of a mineral project.

10.6 Current and recent thorium recovery

Monazite concentrate production has occurred in recent years in India, Malaysia, Thailand, Vietnam, and Brazil (Fig. 10.3), in decreasing order of production (Gambogi, 2014a).

Indian beach placers are the principal present-day source for the production of monazite, mainly obtained from shorelines in the states of Kerala and Orissa (Fig. 10.3). The monazite is stockpiled as part of a government-industry cooperative designed to provide source material for thorium-based nuclear power under development by the Department of Atomic Energy of the Indian government. The typical composition of monazite from the Kerala deposits is reportedly 57.5% REE oxide and 7.96% Th oxide (Kerala Minerals and Metals Ltd., 2014). Monazite is currently processed by the Rare Earths Division of Indian Rare Earths Limited (Indian Rare Earths Limited, 2014).

Prior to the domination of global yttrium (Y) production by China in the late 1980s, xenotime-bearing alluvial placer deposits in Malaysia were the largest sources of yttrium in the world (Castor and Hedrick, 2006). The sources within Malaysia are alluvial tin placer deposits, which carry considerable cassiterite accompanied by ilmenite, monazite, and xenotime. Recently, tailings produced from past tin placer mining have been reprocessed to recover monazite and xenotime, thereby providing another potential source of thorium. In this manner of recovery, 350 t of rare earth oxides were produced from Malaysia in 2012 (Gambogi, 2013).

Tin (in cassiterite) has been mined from alluvial placers in Thailand for over 1000 years, but much more recently Thailand initiated byproduct recovery of monazite (starting in 1969), ilmenite (in 1976), and xenotime (in 1977) (Economic and Social Commission for Asia and the Pacific, 2001, pp. 72–74). In Thailand the monazite and xenotime are recovered as byproducts from the retreatment of processing plant tailings earlier produced from tin placers. The host tin placers, those enriched in monazite and xenotime, are alluvial deposits that were mined by gravel pump; these deposits occur mainly in southern Thailand (shown in Economic and Social Commission for Asia and the Pacific, 2001, pp. 72–74).

Deposits of heavy-mineral sands are known at many places along the Vietnam coastline for a distance of more than 1500 km. The largest heavy-mineral sands deposits are in the Binh Thuan province of southern Vietnam, parallel to the coast. In this coastal area, GPM Asia (2014) produces ilmenite, zircon, rutile, and monazite from coastal heavy-mineral sands.

In contrast to most heavy-mineral sands operations, many of the Brazilian deposits were mined in the past primarily for their monazite, sought foremost as a source of thorium with titanium minerals and zircon treated as coproduct commodities. Overstreet (1967) reported that from 1900 to 1947 Brazil exported 56,350 t of monazite concentrate that was recovered from beach placers. Since the early 1990s, the Buena placer district, which includes the Buena Norte and Buena Sol deposits, has been the only active Brazilian producer of monazite; it participates in a state-administered program (Indústrias Nucleares do Brasil SA (INB), 2014).

10.7 Disclaimer

Any use of trade, product, or firm names is for descriptive purposes only and does not imply endorsement by the US government.

References

Anantharaman, K., Shivakumar, V., Saha, D., 2008. Utilisation of thorium in reactors. J. Nucl. Mater. 383 (1), 119–121.

Anderson, A.K., Larson, P.B., Clark, J.G., 2013. Rare earth element zonation and fractionation at the Bear Lodge REE deposit, Wyoming. Geological Society of America Abstracts with Programs 45 (5), 42.

Andreoli, M.A.G., Smith, C.B., Watkeys, M., Moore, J.M., Ashwal, L.D., Hart, R.J., 1994. The geology of the Steenkampskraal monazite deposit, South Africa—implications for REE-Th-Cu mineralization in charnockite-granulite terranes. Econ. Geol. 89 (5), 994–1016.

Andreoli, M.A.G., Hart, R.J., Ashwal, L.D., Coetzee, H., 2006. Correlations between U, Th content and metamorphic grade in the Western Namaqualand Belt, South Africa, with implications for radioactive heating of the crust. J. Petrol. 47 (6), 1095–1118.

Arafura Resources Limited, 2014. Resources & Reserves: Arafura Resources Limited. Available from: <http://www.arultd.com/our-projects/nolans/resources-reserves.html> (accessed 19.12.14.).

Ashley, S.F., Parks, G.T., Nuttall, W.J., Boxall, Colin, Grimes, R.W., 2012. Nuclear energy—thorium fuel has risks. Nature. 492 (7427), 31–33.

Ashley, S.F., Lindley, B.A., Parks, G.T., Nuttall, W.J., Gregg, R., Hesketh, K.W., et al., 2014. Fuel cycle modelling of open cycle thorium-fuelled nuclear energy systems. Ann. Nucl. Energy. 69, 314–330.

Avalon Rare Metals Inc, 2014. Advanced REE Project. Avalon Rare Metals Inc. <http://avalonraremetals.com/> (accessed 19.12.14.).

Bell, K., 1989. Carbonatites—Genesis and Evolution. Unwin Hyman Ltd, Boston, MA.

Berger, V.I., Singer, D.A., Orris, G.J., 2009. Carbonatites of the world, explored deposits of Nb and REE—database and grade and tonnage models. U.S. Geological Survey Open-File Report 2009-1139, 17 p. Available from: <http://pubs.usgs.gov/of/2009/1139/>.

Briggs, R.B., 1964. MSR Program Semiannual Progress Report for the period ending July 31, 1964. Oak Ridge National Laboratory, U.S. Atomic Energy Commission, Report ORNL-3708 (66.3 MB PDF document), published November 1964. Available from: <http://www.energyfromthorium.com/pdf/ORNL-3708.pdf>.

Briggs, R.B., 1967. Summary of the objectives, the design, and a program of development of molten-salt breeder reactors. Oak Ridge National Laboratory, U.S. Atomic Energy Commission, Report ORNL-TM-1851.

Carminati, F., Klapisch, R., Revol, J.P., Roche, Ch., Rubio, J.A., Rubbia, C., 1993. An energy amplifier for cleaner and inexhaustible nuclear energy production driven by a particle beam accelerator. European Organization for Nuclear Research, CERN/AT/93-47 (ET). Available from: <http://cds.cern.ch/record/256520/files/at-93-047.pdf>.

Castor, S.B., 2008. The Mountain Pass rare-earth carbonatite and associated ultrapotassic rocks. California. Can. Mineral. 46 (4), 779−806.

Castor, S.B., Hedrick, J.B., 2006. Rare earth elements. In: Kogel, J.E., Trivedi, N.C., Barker, J.M., Krukowski, S.T. (Eds.), Industrial Minerals and Rocks—Commodities, Markets, and Uses, seventh ed. Society for Mining, Metallurgy, and Exploration, Inc. (SME), Littleton, CO, pp. 769−792.

Castor, S.B., Nason, G.W., 2004. Mountain Pass rare earth deposit, California. In: Castor, S. B., Papke, K.G., Meeuwig, R.O. (Eds.), Betting on Industrial Minerals, Forum on the Geology of Industrial Minerals, 39, Reno/Sparks, Nev., 18−24 May, 2003, 33. Proceedings: Nevada Bureau of Mines and Geology Special Publication, pp. 68−81.

Chakhmouradian, A.R., Zaitsev, A.N., 2012. Rare earth mineralization in igneous rocks—sources and processes. Elements. 8 (5), 361−367.

Chetal, S.C., Chellapandi, P., Puthiyavinayagam, P., Raghupathy, S., Balasubramaniyan, V., Selvaraj, P., et al., 2011. Current status of fast reactors and future plans in India. Energy Procedia. 7, 64−73.

David, S., Huffer, E., Nifenecker, H., 2007. Revisiting the thorium-uranium nuclear fuel cycle. Europhys. News. 38 (2), 24−27.

Economic and Social Commission for Asia and the Pacific (ESCAP), 2001. Mineral resources of Thailand. United Nations publication, Atlas of Mineral Resources of the ESCAP Region 16, 239 p.

Endicott, N., 2014. The UK's forgotten molten salt reactor programme (Weblog post). The Alvin Weinberg Foundation. Available from: <http://www.the-weinberg-foundation.org/2014/11/13/the-uks-forgotten-molten-salt-reactor-programme/> (accessed 01.04.15.).

Fitton, J.G., Upton, B.G.L. (Eds.), 1987. Alkaline Igneous Rocks. Blackwell Scientific, Oxford, Geological Society Special Publication 30, 568 p.

Force, E.R., 1976. Metamorphic source rocks of titanium placer deposits—a geochemical cycle. U.S. Geological Survey Professional Paper 959−B, 16 p.

Furukawa, K., 1997. Thorium cycle implementation through plutonium incineration by thorium molten-salt nuclear energy synergetics. Vienna, Austria, International Atomic Energy Agency Technical Document IAEA-TECDOC-1319.

Gambogi, J., 2013. Rare earths. U.S. Geological Survey Mineral Commodity Summaries [2013]. Available from: <http://minerals.usgs.gov/minerals/pubs/commodity/rare_earths/>.

Gambogi, J., 2014a. Rare earths. U.S. Geological Survey Mineral Commodity Summaries [2014]. Available from: <http://minerals.usgs.gov/minerals/pubs/commodity/rare_earths/>.

Gambogi, J., 2014b. Thorium. U.S. Geological Survey Mineral Commodity Summaries [2014]. Available from: <http://minerals.usgs.gov/minerals/pubs/commodity/thorium/>.

Government of India, Department of Atomic Energy, 2011, Lok Sabha unstarred question no. 300—Deposits of rare earths: Government of India, Department of Atomic Energy, 2 p. Available from: <http://www.dae.nic.in/writereaddata/lsus300_231111.pdf> (accessed 23.11.15.).

GPM Asia, 2014. More about GPM Asia. GPM Asia Company. Available from: <http://www.gpm.asia/> (accessed 23.12.14).

Great Western Minerals Group, 2014, Great Western Minerals Group Company. Available from: <http://www.gwmg.ca/mining-operations/overview> (accessed 19.12.14.).

Greenland Minerals and Energy Ltd, 2014. Greenland Minerals and Energy Ltd. Available from: <http://www.ggg.gl/> (accessed 19.12.14.).

Heinrich, E.W., 1980. The Geology of Carbonatites. Robert E. Krieger Publishing Company, Huntington, NY, 585 p.

Henderson, P., 1984. General geochemical properties and abundances of the rare earth elements, chap. 1. In: Henderson, P. (Ed.), Rare Earth Element Geochemistry, v. 2. Elsevier, Amsterdam, pp. 1–32. , Developments in Geochemistry series.

Hoatson, D.M., Jaireth, S., Miezitis, Y., 2011. The major rare-earth-element deposits of Australia—geological setting, exploration, and resources. Geoscience Australia, 204 p.

IAEA, 1979. Status and prospects of thermal breeders and their effect on fuel utilization. Vienna, Austria, International Atomic Energy Agency Technical Report Series No. 195.

IAEA, 1985. Thorium-based nuclear fuels—current status and perspectives. Vienna, Austria, International Atomic Energy Agency Technical Document IAEA TECDOC 412.

IAEA, 2003. Review of national accelerator driven system programmes for partitioning and transmutation. Vienna, Austria, International Atomic Energy Agency Technical Document IAEA TECDOC 1365.

IAEA, 2005, Thorium fuel cycle—potential benefits and challenges. Vienna, Austria, International Atomic Energy Agency Technical Document IAEA TECDOC 1450.

IAEA, 2012, Role of thorium to supplement fuel cycles of future nuclear energy systems. Vienna, Austria, International Atomic Energy Agency, IAEA Nuclear Energy Series NF-T-2.4, 157 p.

Indian Rare Earths Limited, 2014. Unit profile. Indian Rare Earths Limited Company. Available from: <http://www.irel.gov.in/scripts/unit.asp> (accessed 24.11.14.).

Indústrias Nucleares do Brasil SA (INB), 2014. INB Buena—mineral treatment unit. Indústrias Nucleares do Brasil SA. Available from: <http://www.inb.gov.br/pt-br/WebForms/default.aspx> (accessed 23.12.14.).

Jackson, W.D., Christiansen, G., 1993. International strategic minerals inventory summary report—rare-earth oxides. U.S. Geological Survey Circular 930–N, 68 p.

Jevremovic, T., 2009. Nuclear Principles in Engineering. second ed. Springer US, New York, 546 p.

Kerala Minerals & Metals Ltd, 2014. KMML. The Kerala Minerals & Metals Ltd. Company. Available from: <http://www.kmml.com/> (accessed 24.11.14.).

Keshavamurthy, R.S., Mohanakrishnan, P., 2011, Thorium utilization in fast reactors. In: Presentation given at the IAEA Technical Meeting on World Thorium Resources, Thiruvananthapuram, India, October 2011.

Kogarko, L.N., Williams, C.T., Woolley, A.R., 2006. Compositional evolution and cryptic variation in pyroxenes of the peralkaline Lovozero intrusion, Kola Peninsula. Russ. Mineral. Mag. 70 (4), 347–359.

Lane, J.A., MacPherson, H.G., Maslan, F. (Eds.), 1958. Fluid Fuel Reactors. Addison-Wesley Publishing Company, Reading, Massachusetts, 979 p.

Langmuir, D., Herman, J.S., 1980. The mobility of thorium in natural waters at low temperatures. Geochim. Cosmochim. Acta. 44 (11), 1753–1766.

LeBlanc, D., 2013. The Integral Molten Salt Reactor. In: Thorium Energy Alliance Conference 5, Chicago, Illinois, 30–31 May 2013. Available from: <http://www.thoriumenergyalliance.com/ThoriumSite/TEAC5.html> (accessed 14.07.15.).

Li, X.X., Cai, X.Z., Jiang, D.Z., Ma, Y.W., Huang, J.F., Zou, C.Y., et al., 2015. Analysis of thorium and uranium based nuclear fuel options in fluoride salt-cooled high-temperature reactor. Prog. Nucl. Energy. 78, 285–290.

Long, K.R., Van Gosen, B.S., Foley, N.K., Cordier, D., 2010. The principal rare earth elements deposits of the United States—a summary of domestic deposits and a global perspective. U.S. Geological Survey Scientific Investigations Report 2010–5220, 96 p. Available from: <http://pubs.usgs.gov/sir/2010/5220/>.

Lottermoser, B.G., 1990. Rare-earth element mineralization within the Mt. Weld carbonatite laterite, Western Australia. Lithos. 24, 151–167.

Lung, M., 1997. A present review of the thorium nuclear fuel cycles. Luxembourg, European Commission, EUR 17771, Office for Official Publications of the European Communities.

Lynas Corporation Ltd, 2014. Mount Weld. Lynas Corporation Ltd. Available from: <https://www.lynascorp.com/Pages/Mt-Weld-Concentration-Plant.aspx> (accessed 14.12.15.).

MacPherson, H.G., 1985. The molten salt reactor adventure. Nucl. Sci. Eng. 90, 374–380.

Mariano, A.N., 1989. Economic geology of rare earth elements, Chapter 11. In: Lipin, B.R., et al., (Eds.), Geochemistry and mineralogy of rare earth elements, v. 21. The Mineralogical Society of America, Reviews in Mineralogy, Washington, D.C., pp. 309–337.

Martin, R., 2012. Superfuel—Thorium, the Green Energy Source for the Future. Palgrave Macmillian, New York, NY, 262 p.

McKay, G.A., 1989. Partitioning of rare earth elements between major silicate minerals and basaltic melts, chap. 3. In: Lipin, B.R., McKay, G.A. (Eds.), Geochemistry and Mineralogy of Rare Earth Elements, vol. 21. The Mineralogical Society of America, Washington, DC, pp. 45–77. , Reviews in Mineralogy series.

McKeown, F.A., Klemic, H., 1956. Rare-earth-bearing apatite at Mineville, Essex County. New York: U.S. Geological Survey Bulletin 1046–B, pp. 9–23.

Mohanty, A.K., Vijayan, V., Sengupta, D., Das, S.K., Saha, S.K., 2003. Geochemical characteristics of ilmenite sands of Chhatrapur beach placer deposit of Orissa, India—a PIXE study. Int. J. PIXE. 13, 121–131.

Molycorp, Inc, 2012. Molycorp's rare earth reserves at Mountain Pass increase by 36%. Molycorp, Inc., Press Release, 9 April 2012.

Morozov, A.G., Galperin, A.L.E.X., Todosow, M., 1999. A thorium-based fuel cycle for VVERs and PWRs—a nonproliferative solution to renew nuclear power. Nucl. Eng. Int. 44 (534), 13–14.

NEA-IAEA, 2014. Nuclear Energy Agency (NEA) Organization for Economic Co-operation and Development (OECD) and the International Atomic Energy Agency (IAEA), Uranium 2014—Resources, Production and Demand. OECD Report No. 7209, 504 p. [This is the most recent volume of the series informally referred to as the "Red Book"]. Available from: <http://www.oecd-nea.org/pub/>.

NEA-IAEA, 2015. Nuclear Energy Agency (NEA) Organization for Economic Co-operation and Development (OECD) and the International Atomic Energy Agency (IAEA), Introduction of thorium in the nuclear fuel cycle—short- to long-term considerations. NEA No. 7224, 136 p. Available from: <http://www.oecd-nea.org/pub/>.

Nema, P.K., 2011. Application of accelerators for nuclear systems—Accelerator Driven System (ADS). Energy Procedia. 7, 597–608.

Novikov, V.M. 1995. The results of the investigations of Russian Research Center—"Kurchatov Institute" on molten salt applications to problems of nuclear energy systems. In: The International Conference on Accelerator-Driven Transmutation Technologies and Applications: AIP Conference Proceedings 346, p. 138.

Nuelle, L.M., Kisvarsanyi, E.B., Seeger, C.M., Day, W.C., Sidder, G.B., 1991. Structural setting and control of the Pea Ridge magnetite deposit, Middle Proterozoic St. Francois terrane, Missouri [abs.]. Geological Society of America Abstracts with Programs, 23, 5, A292.

Orris, G.J., Grauch, R.I., 2002. Rare earth element mines, deposits, and occurrences. U.S. Geological Survey Open-File Report 02–189, 174 p.

Overstreet, W.C., 1967. The geologic occurrence of monazite. U.S. Geological Survey Professional Paper 530, 327 p., 2 plates.

Pinheiro, R.B., Lameiras, F., Peehs, M., Maly, V., 1989. Program of research and development on the thorium utilization in PWRs—Final Report (1979-1988)—German-Brazilien cooperation in scientific research and technological development: Kernforschungsanlage (Jülich), Internationales Büro, 226 p.

Rare Element Resources, 2014. Bear Lodge project. Rare Element Resources. Available from: <http://www.rareelementresources.com/> (accessed 19.12.14.).

Rosenthal, M., 2009. An account of Oak Ridge National Laboratory's thirteen nuclear reactors. Oak Ridge National Laboratory Report ORNL/TM-2009/181. Available from: <http://info.ornl.gov/sites/publications/Files/Pub20808.pdf>.

Rubbia, C., 1995. A high gain energy amplifier operated with fast neutrons, in International conference on accelerator-driven transmutation technologies and application. In: AIP Conference Proceedings, 346, pp. 44–53.

Rupasinghe, M.S., Gocht, W., Dissanayake, C.B., 1983. The genesis of thorium-rich monazite placer deposits in Sri Lanka. J. Natl. Sci. Counc. Sri Lanka. 11 (1), 99–110.

Scott, D., Alwang, G.W., Demski, E.F., Fader, W.J., Sandin, E.V., Malenfant, R.E., 1958. A zero power reflector-moderated reactor experiment at elevated temperature. Oak Ridge National Laboratory Report ORNL-2536. Available from: <http://energyfromthorium.com/pdf/ORNL-2536.pdf>.

Seeger, C.M., Nuelle, L.M., Day, W.C., Sidder, G.B., Marikos, M.A., Smith, D.C., 2001. Geologic maps and cross sections of mine levels at the Pea Ridge iron mine, Washington County, Missouri (version 1.0). U.S. Geological Survey Miscellaneous Field Studies Map 2353, 6 p. pamphlet, 5 sheets.

Serp, J., Allibert, M., Beneš, O., Delpech, S., Feynberg, O., Ghetta, V., et al., 2014. The molten salt reactor (MSR) in generation IV—overview and perspectives. Prog. Nucl. Energy. 77, 308–319.

Sheard, E.R., Williams-Jones, A.E., Heiligmann, M., Pederson, C., Trueman, D.L., 2012. Controls on the concentration of zirconium, niobium and rare earth elements in the Thor Lake rare metal deposit, Northwest Territories, Canada. Econ. Geol. 107, 81–104.

Sidder, G.B., Day, W.C., Nuelle, L.M., Seeger, C.M., Kisvarsanyi, E.B., 1993. Mineralogic and fluid-inclusion studies of the Pea Ridge iron-rare-earth-element deposit, southeast Missouri, chap. U. In: Scott, R.W., Jr, Detra, P.S., Berger, B.R. (Eds.), Advances

Related to United States and International Mineral Resources—Developing Frameworks and Exploration Technologies. U.S. Geological Survey Bulletin 2039, pp. 205–216.

Sinha, R.K., 2011. Advanced nuclear reactor systems—an Indian perspective. Energy Procedia. 7, 34–50.

Sinha, R.K., Kakodkar, A., 2006. Design and development of the AHWR—the Indian thorium fuelled innovative nuclear reactor. Nucl. Eng. Des. 236 (7), 683–700.

Sørensen, H. (Ed.), 1974. The Alkaline Rocks. John Wiley, London, 622 p.

Sørensen, H., Bohse, H., Bailey, J.C., 2006. The origin and mode of emplacement of lujavrites in the Ilímaussaq alkaline complex, South Greenland. Lithos. 91 (1–4), 286–300.

Staatz, M.H., 1972, Geology and description of the thorium-bearing veins, Lemhi Pass quadrangle, Idaho and Montana. U.S. Geological Survey Bulletin 1351, 94 p.

Staatz, M.H., 1979. Geology and mineral resources of the Lemhi Pass thorium district, Idaho and Montana. U.S. Geological Survey Professional Paper 1049–A, 90 p.

Staatz, M.H., Armbrustmacher, T.J., Olson, J.C., Brownfield, I.K., Brock, M.R., Lemons, J.F., et al., 1979. Principal thorium resources in the United States. U.S. Geol. Surv. Circ. 805, 42.

Staatz, M.H., Hall, R.B., Macke, D.L., Armbrustmacher, T.J., Brownfield, I.K., 1980. Thorium resources of selected regions in the United States. U.S. Geol. Surv. Circ. 824, 32.

The Generation IV International Forum, 2015. Annual Report 2014, 125 p. Available from: <https://www.gen-4.org/gif/upload/docs/application/pdf/2015-05/gif_2014_annual_report_final.pdf> (accessed 30.07.15.).

Tuttle, O.F., Gittins, J., 1966. Carbonatites. Interscience Publishers, New York, 591 p.

U.S. Atomic Energy Commission, 1969. The use of thorium in nuclear power reactors. Division of Reactor Development and Technology, U.S. Atomic Energy Commission, Report WASH 1097, 147 p.

Van Den Durpel, L., Guesdon, B., Lecomte, M., 2011. Industrial prospects for the optimized use of U, Pu and Th for sustainable nuclear energy deployment. Energy Procedia. 7, 21–33.

Van Gosen, B.S., Gillerman, V.S., Armbrustmacher, T.J., 2009. Thorium deposits of the United States—energy resources for the future? US Geol. Surv. Circ. 1336, 21. Available from: <http://pubs.usgs.gov/circ/1336>.

Van Gosen, B.S., Fey, D.L., Shah, A.K., Verplanck, P.L., Hoefen, T.M., 2014. Deposit model for heavy-mineral sands in coastal environments. U.S. Geological Survey Scientific Investigations Report 2010–5070–L, 51 p. Available from: <http://dx.doi.org/10.3133/sir20105070L>.

Verplanck, P.L., Van Gosen, B.S., Seal, R.R., McCafferty, A.E. 2014. A deposit model for carbonatite and peralkaline intrusion-related rare earth element deposits. U.S. Geological Survey Scientific Investigations Report 2010–5070–J, 58 p. Available from: <http://dx.doi.org/10.3133/sir20105070J>.

Vierrether, C.W., Cornell, W.L., 1993. Rare-earth occurrences in the Pea Ridge tailings. U.S. Bureau of Mines Report of Investigations, RI 9453, 10 p.

Vijayan, P.K., Dulera, I.V., Krishnani, P.D., Vaze, K.K., Basu, S., Sinha, R.K., 2013. Overview of the thorium programme in India. In: Presentation given at iThEC13 International Thorium Energy Conference, Geneva, Switzerland, 27–31 October 2013. Available from: <http://thoriumenergyconference.org/sites/default/files/pdf/Overview%20of%20the%20Thorium%20Programme%20in%20India%20-%20PK%20Vijayan.pdf>.

Wigeland, R., Taiwo, T., Ludewig, H., Todosow, M., Halsey, W., Gehim, J., et al., 2014. Nuclear fuel cycle evaluation and screening—final report. U.S. Department of Energy Report FCRD-FCO-2014-000106, 51 p.

WNA, 2015a. Thorium: World Nuclear Association. Available from: <http://www.world-nuclear.org/info/Current-and-Future-Generation/Thorium/> (accessed 16.04.15.).

WNA, 2015b. Nuclear Power in India: World Nuclear Association. Available from: <http://www.world-nuclear.org/info/Country-Profiles/Countries-G-N/India/> (accessed 16.04.15.).

WNA, 2015c. Molten Salt Reactors: World Nuclear Association. Available from: <http://www.world-nuclear.org/info/Current-and-Future-Generation/Molten-Salt-Reactors/> (accessed 16.04.15.).

Woolley, A.R., 1987. Alkaline Rocks and Carbonatites of the World—Part I, North and South America. University of Texas Press, Austin, TX, 216 p.

Woolley, A.R., 1989. The spatial and temporal distribution of carbonatites. In: Bell, K. (Ed.), Carbonatites—Genesis and Evolution. Unwin Hyman Ltd, Boston, MA, pp. 15−37.

Woolley, A.R., 2001. The Alkaline Rocks and Carbonatites of the World—Part 3—Africa. The Geological Society of London, London, UK, 372 p.

Wouters, M., Selto, F.H., Hilton, R.W., Maher, M.W., 2012. Cost Management—Strategies for Business Decisions. International edition, McGraw-Hill Education Europe, London.

Yoshioka, R., 2013. Nuclear energy based on thorium molten salt. In: Lantelme, F., Groult, H. (Eds.), Molten Salts Chemistry—From Lab to Applications. Elsevier Inc, Amsterdam, Netherlands, pp. 471−496.

Part III

Conversion, Enrichment and Fuel Fabrication

Conversion of natural uranium

11

Andrew J. Oliver[1] *and Engin Özberk*[2]
[1]Retired Cameco Corporation, Port Perry, ON, Canada, [2]Retired Cameco Corporation, Saskatoon, SK, Canada

11.1 Introduction

Processing of uranium beyond the milling stage is somewhat generic. While milling tends to be ore specific, mill concentrates are essentially fungible, and subsequent processing is more easily transferred from one location to another. Thus the subsequent operations, of which refining and conversion are the first step, are subject to much more scrutiny to prevent nuclear proliferation, as well as commercial interests in a very competitive business. These restrictions put many constraints on the level of technical details that can be published, including in this current chapter, which is limited to summarizing what is appropriate for general distribution in the public domain. Nevertheless the refining and conversion of uranium processing is described, including some history of technology development, how this history affected past technical and economic choices, and how these are likely to affect further developments and activities in this area.

11.2 Conversion and needs

The extraction of uranium containing minerals occurs through leaching where the initial purification also occurs (milling operations) at or near mine sites provides uranium ore concentrates (UOCs or "yellowcake"). Chemically, these are most commonly triuranium octoxide (U_3O_8) or uranyl peroxide (UO_4), but have also included sodium diuranate and magnesium diuranate.

The further processing of the concentrates has the two primary goals of providing:

- Refining to achieve impurity contents in the converted uranium products, which are sufficiently low that there are no serious interferences with the enrichment, fuel manufacturing, and reactor processes, hence contributing to stability and efficiency in those operations
- Conversion to a form suitable for feed to the enrichment process, on a commercial basis that form is exclusively uranium hexafluoride, UF_6: although other feed options have been suggested for some enrichment processes, none of these are currently in commercial operation, or even in late-stage development

The commercial purchase of the natural UF_6 is made by electrical utilities, which need a material that is deliverable to multiple enrichers, thereby encouraging competition and expected better pricing for this next step in nuclear fuel production. Conversion to UF_6 is only a small part of the fuel cycle, and is often not considered

Uranium for Nuclear Power. DOI: http://dx.doi.org/10.1016/B978-0-08-100307-7.00011-9

of strategic importance. Little attention is paid to this segment yet it is clearly essential for enrichment to take place.

Refining and conversion to natural UO_2 is also commercially important although only smaller in scale compared to conversion to natural UF_6. Natural UO_2 is used for manufacture of CANDU fuel and for blanket fuel in some light water reactors. For convenient matching with fuel manufacturing, fuel blanket material is often made via natural UF_6, but is more economically produced by a direct process described near the end of this chapter.

11.3 Conversion technologies

11.3.1 Process outline

The refining and conversion of UOC to UF_6 are discussed together as the two steps may be in different order. The most common process involves refining and subsequent conversion to UF_6, and this is often called the "wet process" as this refining route requires water solution. The alternative route to UF_6 involves conversion and then refining by distillation of the UF_6. Because no water is involved in the main process steps, this is called the "dry process."

UF_6 can be generated from the reaction of virtually any uranium compound with elemental fluorine at elevated temperature. However, fluorine is both expensive and hazardous to store and react, so cost-effective processing minimizes its use. Nevertheless, production of UF_6 requires more commercial F_2 production than any other use of F_2.

The commercial routes to UF_6 minimize F_2 usage by using three steps instead of just one. These steps are reduction (reaction with H_2), hydrofluorination (reaction with hydrogen fluoride (HF)), and fluorination (reaction with F_2). Thus, the steps may be represented by the equations:

$$UO_3 \text{ (or other oxide)} + H_2 \rightarrow UO_2 + H_2O$$

$$UO_2 + 4HF \rightarrow UF_4 + 2H_2O$$

$$UF_4 + F_2 \rightarrow UF_6$$

11.3.2 History

While downstream uranium processing today is focused on fuel supply for the peaceful production of electrical power, the initial focus for technology development was for nuclear weapons and a great deal of investment was made in those early days to develop and apply processing methods as rapidly as possible. Choices made in those early days still influence current processing routes and locations. Even the industry operations terminology, still in use today, comes from an earlier time when the word uranium was to be avoided for security reasons. For example, we often hear of "yellowcake" for mill concentrate (even UOC) and "green salt" as the precursor to UF_6.

The initial challenge of producing large quantities of high purity uranium in the 1940s was addressed by turning to the laboratory method of purifying uranium, which involved the extraction of uranium from aqueous nitrate solution into ether, as had been published by Peligot in 1842. Hence, this was the first process introduced at Mallinkrodt, near St. Louis, Missouri and at Springfields in the UK, operated by ICI (later British Nuclear Fuels Limited (BNFL) and then Springfields Fuels Limited (SFL)). The process required nitric acid dissolution, ether extraction, washing the ether extraction to remove impurities, reextracting with larger volumes of water, and conversion of the purified uranyl nitrate to uranium trioxide. With wartime urgency, production of purified uranium at more than 1 t/day was achieved within 2 months of starting the project (Mallinkrodt, 1962).

With further research, it was determined that tributyl phosphate (TBP) in kerosene was a superior extractant, principally for safety reasons. This TBP system was established in a large production facility, called the Feed Materials Production Centre operated by National Lead at Fernald, just outside Cincinnati, Ohio. In 1956, essentially a duplicate of the Fernald solvent extraction plant was built by Eldorado in Port Hope, Ontario. Likewise, similar commercial TBP solvent extraction processes were established in the UK and France.

The US Atomic Energy Commission (AEC) had installed UF_6 conversion capability at its US enrichment sites, but later decided this activity could be handled by the private sector. So, in late 1955, the AEC solicited proposals for the production of refined uranium products by privately owned and operated facilities. This led to support of a proposal from Allied Chemical Corporation as this was expected to provide the lowest overall cost. Hence, a plant was constructed in Metropolis, Illinois for production of refined UF_6, through conversion of concentrates and purification by distillation (Ruch et al., 1960).

With increasing demands at that time, the Kerr-McGee Corporation (later called Sequoyah Fuels) completed construction of a "wet process" conversion facility in Gore, Oklahoma, with capacity of 5000 tU/a in 1970, and later increased that capacity, but this facility is no longer operational.

The wet process is now exclusively used on a large scale in Canada, France, Russia, and China, and in smaller plants in various other countries, such as Brazil, Pakistan, and Iran. In Russia, the government enrichment and conversion company TVEL, a subsidiary of Atomenergoprom, is consolidating its UF_6 production from Angarsk to Seversk. Russian UF_6 production has historically received UF_4 feedstock from the JSC Chepetsky Mechanical Plant and concentrates from domestic and foreign mines.

The second commercial process (dry process) for nuclear-grade UF_6 is achieved by production of impure UF_6 from concentrates, again using the steps shown by the three preceding chemical equations, but then achieving the purification by distillation of the impure UF_6. This process is currently only used for natural uranium by Honeywell in Metropolis, Illinois, and the product is marketed through Converdyn, a 50:50 partnership of Honeywell and General Atomics. This plant was previously owned by Allied Chemical and then Allied Signal, prior to Honeywell.

11.3.3 Uranium oxide concentrate refining/conversion to UF_6 by wet process

The essential steps in refining by solvent extraction (wet process) are dissolution in nitric acid, solvent extraction, concentration and thermal decomposition to nuclear-pure UO_3 powder. This process is well documented (Harrrington and Ruele, 1959; Ashbrook and Smart, 1980; Ashbrook, 1982; Page, 1986).

The wet process is used in the current operations at Cameco Corp. in Canada, (formerly Eldorado Nuclear) in Blind River (Schisler et al., 1986; Astles and Green, 1998) and Port Hope, Canada, and by Comurhex/AREVA in France Comurhex/AREVA in Malvesi and Pierrelatte, France (Delannoy and Faron, 1982a,b). These are discussed in further detail next. Both operations receive concentrates from around the world. The concentrates are expected to meet the specifications that are based on the generic ASTM specification for such concentrates. A sliding scale of penalties may be applied in treatment pricing for deviations from the specifications.

11.3.3.1 Cameco uranium oxide concentrate refining to UO_3

The flowsheet outlining the Cameco refining process is shown in Fig. 11.1, and in its current form, was described by Cameco in 2011 in regulatory documentation (Cameco, 2012).

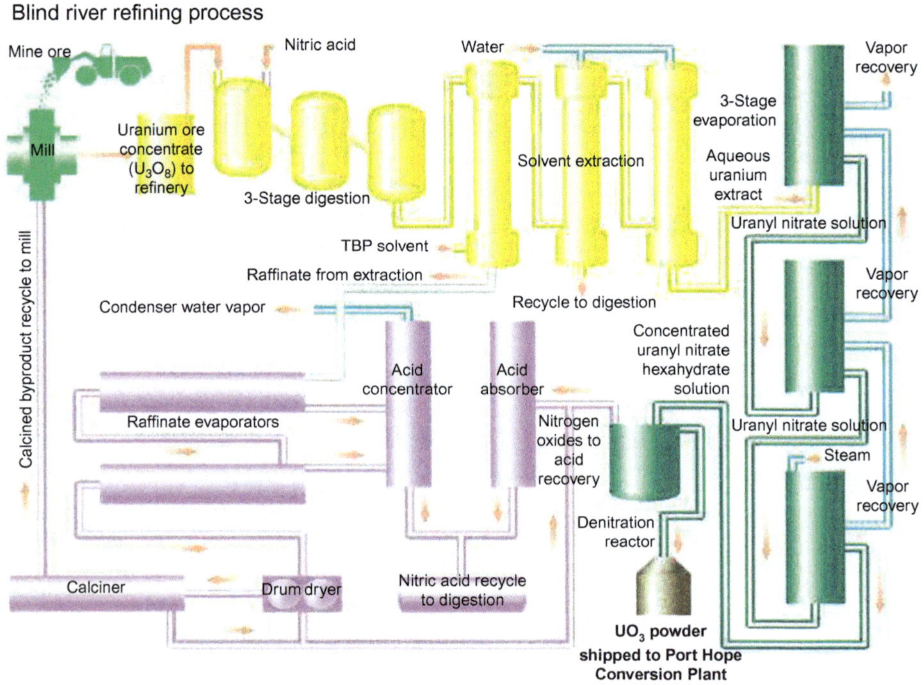

Figure 11.1 Outline of Cameco's uranium oxide concentrate refining process.

UOCs are sampled and split to provide required samples for the producer, the original mill, and for possible umpire analyses should any dispute arise. The ore concentrates are dissolved in nitric acid to produce a uranyl nitrate solution. Small amounts of scrap uranium-bearing materials are recycled to this digestion step. Impurities are removed from this solution using solvent extraction by TBP in a kerosene-type diluent. This has three principal steps: extraction, scrubbing (impurity washing), and reextraction (also called stripping). The extraction chemistry is described by the equation:

$$UO_2^{2+}(aq) + 2NO_3^-(aq) + 2TBP\,(org) \rightarrow UO_2(NO_3)_2.2TBP\,(org)$$

From the equation, it can be easily understood that higher nitrate concentrations push the equilibrium to the right. Hence, this encourages the uranium extraction in the first stage. Reextraction (also known as stripping) is achieved by promoting the reverse reaction by using little or no nitrate in the reextraction feed water. Intermediate scrubbing uses a relatively small amount of feed water to remove less well-extracted impurities, with minimum removal of uranium. The resulting scrub solution is recycled to extraction for recovery of the uranium content. The stream of impurities in aqueous solution from the extraction step is known as raffinate. This is evaporated to dryness and because the solids still contain uranium values, they are recycled as a feed to uranium mills.

The purified uranyl nitrate solution from reextraction is often given the historical name of "OK liquor." This solution is heated and concentrated, producing a nuclear-grade uranyl nitrate hexahydrate (UNH). The UNH is a yellow crystalline solid but, on warming, readily dissolves in its own water of hydration, and hence moves as a liquid between unit operations. This UNH is thermally decomposed in heated and stirred pots to form UO_3 powder. The UO_3 is stored in specially designed bulk containers and these are shipped to Cameco's conversion facility in Port Hope, but there is also ability to fill drums for customers who may require nuclear-grade UO_3.

Water is condensed from the UNH and raffinate evaporation steps, and this can be recycled to provide water for the solvent extraction steps. Nitrogen oxide fumes are generated from various steps in the process, and these are absorbed and concentrated to provide nitric acid that is recycled to digestion. (Özberk, 2011).

11.3.3.2 Comurhex refining and conversion to UF_4

The flowsheet published for the Comurhex refining process is shown in Fig. 11.2.

The process involves TBP solvent extraction with column contactors similar to the Cameco operation.

For Comurhex, a more reactive UO_3 is required for the downstream gas/solid reactions, so their published process involves evaporation and then precipitation with ammonia to make an ammonium diuranate (ADU), which is dried and calcined to UO_3. A byproduct ammonium nitrate solution is also generated.

$$(NH_4)_2U_2O_7 \rightarrow 2UO_3 + 2NH_3 + H_2O$$

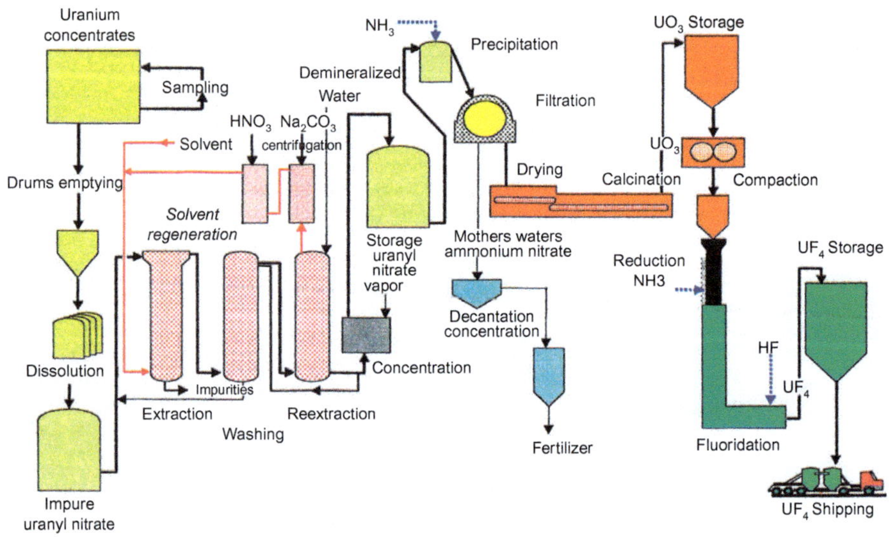

Figure 11.2 Outline of Comurhex refining process.

Smaller amounts of nitrogen oxide fumes from this process are collected and neutralized for disposal.

The direct route has the benefit of avoiding a byproduct ammonium nitrate stream requiring disposition.

11.3.3.3 Process byproducts and environmental aspects

Generally byproducts of processing result in some materials with low residual uranium concentrations, but still of some value. Typically, the solids are dissolved or leached in nitric acid to recover the uranium into solution. For these types of dilute solutions, the uranium is concentrated by precipitation with ammonia or sodium hydroxide. The impure concentrate can then be collected and recycled with mill concentrates to re-enter the main process stream. An alternative strategy is to use a weak liquor solvent extraction column. The uranium-rich organic stream can be fed into the main extraction column at a point equivalent to the same uranium concentration.

In the solvent extraction process, the organics are slowly oxidized and the resulting generally acidic degradation products are removed in a solvent treatment step. This process is generally known as solvent regeneration. This process results in a regeneration product high in the oxidation products as well as some residual uranium and this can be recycled as a mill feed for uranium recovery.

11.3.4 Uranium oxide conversion to UF_6 after refining

11.3.4.1 Reduction to UO_2

Reduction with hydrogen at high temperature is carried out in fluid beds at Cameco's Port Hope plant. Some increase in surface area is generally preferred prior to reduction. At Cameco, this is achieved by pulverizing. In the reduction, sufficient fluidizing gas is needed which may be achieved through recycling of hydrogen or use of inert gas with the hydrogen.

Gas/solid contact in a moving bed was used at National Lead and in the early days of Port Hope plant. This method remains in use with Comurhex at Malvesi. UO_3 is compacted and broken into suitable size pieces for entry into the top of the moving bed or lit courant (LC) reactor. The UO_3 moves by gravity into the high-temperature reduction zone of the reactor. Reduction is achieved by introduction of ammonia at the base of the reactor, where it dissociates into hydrogen and nitrogen, catalyzed by uranium oxide produced in the reduction. The solids then pass through a cooling zone and on to hydrofluorination.

11.3.4.2 Hydrofluorination to UF_4

At Comurhex, conversion to UF_4 is in the same type of LC reactors, where liquid anhydrous HF is fed through preheaters for vaporization, and into the UF_4 discharge end of the reactor. The gas moves counter-current to the uranium, so that most reaction to UF_4 takes place in the vertical section of the reactor where the temperature is relatively high. The upper portion, which mostly contains UO_2, acts as a clean-up to maximize use of HF (Delannoy and Faron, 1982a,b). The UF_4 product is transported from the Malvesi site to the Pierrelatte site for subsequent conversion to UF_6.

Both Comurhex and SFL in the past have historically diverted some best quality UF_4 to make metal, but both sites have ceased making metal. However, for Comurhex, it was the initial focus on UF_4 for metal that led to geographical separation of the UF_4 production at Malvesi and UF_6 production at Pierrelatte (Fig. 11.3).

Cameco chose to use a wet hydrofluorination process, which they call "wet way" (Lenahan, 1982). In this process, pulverized UO_2 powder is fed into stirred slurry reactors with water and hydrofluoric acid. On drying the UF_4, most of the vaporized hydrofluoric acid is recycled from scrubbers and this results in very effective use of the HF. The downside of this wet process is that subsequent drying inevitably results in some reversion to oxide and purity that would be unsuitable for metal production. However, this is no longer a significant market consideration; rather the focus is on processes with least environmental impact.

11.3.4.3 Fluorination to UF_6

Commercial conversion of UF_4 to UF_6 is carried out by reaction with fluorine gas. Most commonly, this is achieved in vertical shaft reactors where the fluorine and UF_4 powders are mixed together at the top of the shaft and the reaction is so exothermic and fast that the reaction is largely complete during the time of fall of the powder down the shaft. Due to the highly exothermic nature of the reaction, the reactor temperature is naturally very high. This type of reactor is used at Angarsk, Port Hope, and Pierrelatte, but with some differences in dimensions and number of reactors.

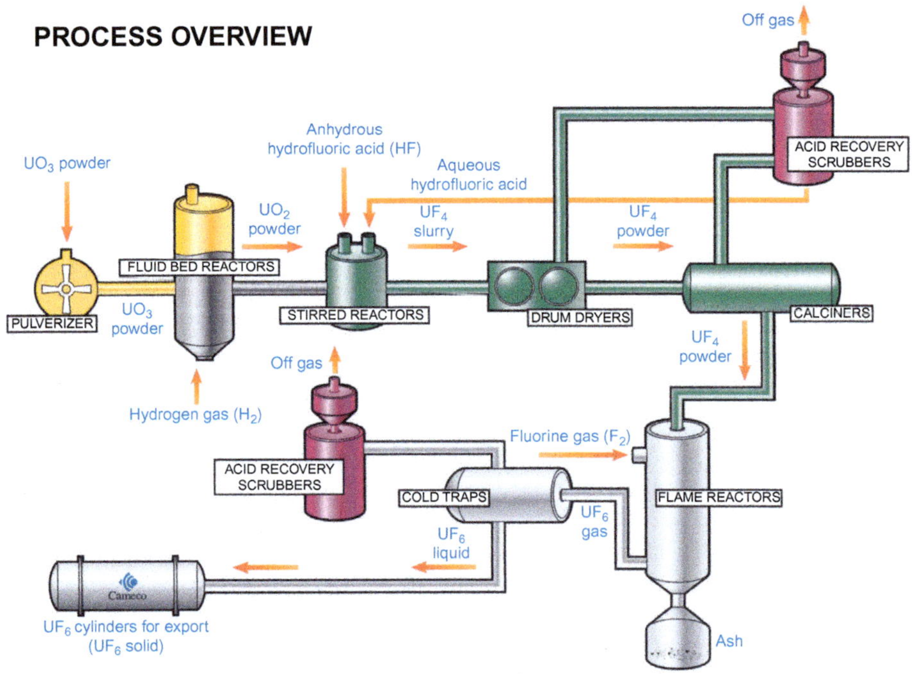

Figure 11.3 Outline of Cameco's UF_6 conversion process.

Both Cameco and Comurhex use two reactors in series; in the first reactor, there is excess fluorine encouraging complete conversion of all UF_4 and in the second reactor there is an excess of UF_4 thereby ensuring complete use of the expensive fluorine gas ("clean-up" reactors). Several reactors of each type operate in parallel. The remaining UF_4 from the clean-up reactors is recycled as part of the feed to the primary reactors. Solids collected at the bottom of the primary reactors, often called primary ash, are recycled to the primary reactor feed, while solids subsequently collected from the UF_6 gas stream are a secondary ash and treated by appropriate uranium recovery methods, possibly as mill or refinery feeds, but only after ageing for some months to reduce radiation fields associated with uranium daughters that are no longer shielded by uranium. Treatment to avoid HF fumes, possibly using basic compounds additions, is also necessary for safe shipping of this ash.

Gaseous UF_6 is condensed in cold traps with internal heat exchange pipes that contain a refrigerant. Typically when the first condenser is full, the second condenser is moved to the primary position and an empty condenser added in the secondary position. After a trap is filled in the primary position, it is heated to about 95°C and under the small overpressure, the UF_6 liquefies and can be drained off into storage/transport cylinders via remotely operated filling systems.

11.3.5 Uranium oxide concentrate conversion to UF_6 prior to refining

The process now uniquely used at Honeywell in Metropolis, Illinois, is the conversion of UOC to impure UF_6 and then purification by distillation (Lawroski et al., 1958; Ruch et al., 1960; Bishop and Hanson, 1982). Ore concentrates are sampled through a falling steam system, where a 1% sample is further split to give about 1500 g/batch. This is typically divided into six 150-g samples, two for Honeywell's analysis, two for the originating mill or designee and two are retained for umpire use as required.

In the Honeywell process, concentrates can be passed through a calciner to remove carbonates, water, organics, and other volatiles. The calcined material is blended and agglomerated to optimize particle size for fluid-bed operation. Feeds high in sodium are dissolved in sulfuric acid, ammonia added, the slurry settled, filtered, resuspended in water, again settled, and filtered. The resulting solids are returned to the raw concentrates calciner. The operation of the calciner is considered critical to good operation of the fluid-bed reactors downstream.

The chemistry for UF_6 production is essentially the same as described previously, and this is followed by distillation for purification. The gas/solid reactors used in each stage are fluidized bed reactors. For the conversion of UF_4 to UF_6, an inert bed of CaF_2 is used in the fluid bed because a bed of UF_4 can easily lead to fusion of bed material due to the high-energy release by this fluorination reaction and relatively low melting temperatures of UF_4 and intermediate fluorides.

The process flowsheet for the Honeywell plant, Metropolis is presented in Fig. 11.4 Uranium 2000 short course notes, Chapter 10, presented by Tom Rice, METSOC/CIM.

After the UF_6 gas leaves the fluorination step, it is filtered and then condensed in specially designed heat exchangers. When full, the collection units go through a cycle of heating and melting the UF_6, which is sent as liquid under a small overpressure to distillation. There are two distillation units in series, first the low boiling impurities are removed from the top of the first distillation column, and then high boiling impurities are removed at the bottom of the second column. Heels from UF_6 process have high radiation fields and are stored temporarily in lead shielded drums. Liquid UF_6 from the distillation product condenser is drained into transport cylinders. The UF_6 is sampled during filling and the cylinders weighed and removed to the storage area where they are allowed to sit for 5 days to solidify prior to further handling.

Byproduct high uranium materials are sent to a recovery unit and returned to the UOC calciner. Potassium hydroxide (KOH) scrub liquors are regenerated with lime. Weak HF solutions are also neutralized with lime. Other solutions mostly water potentially contaminated with insoluble uranium are passed through a series of settling ponds prior to discharge.

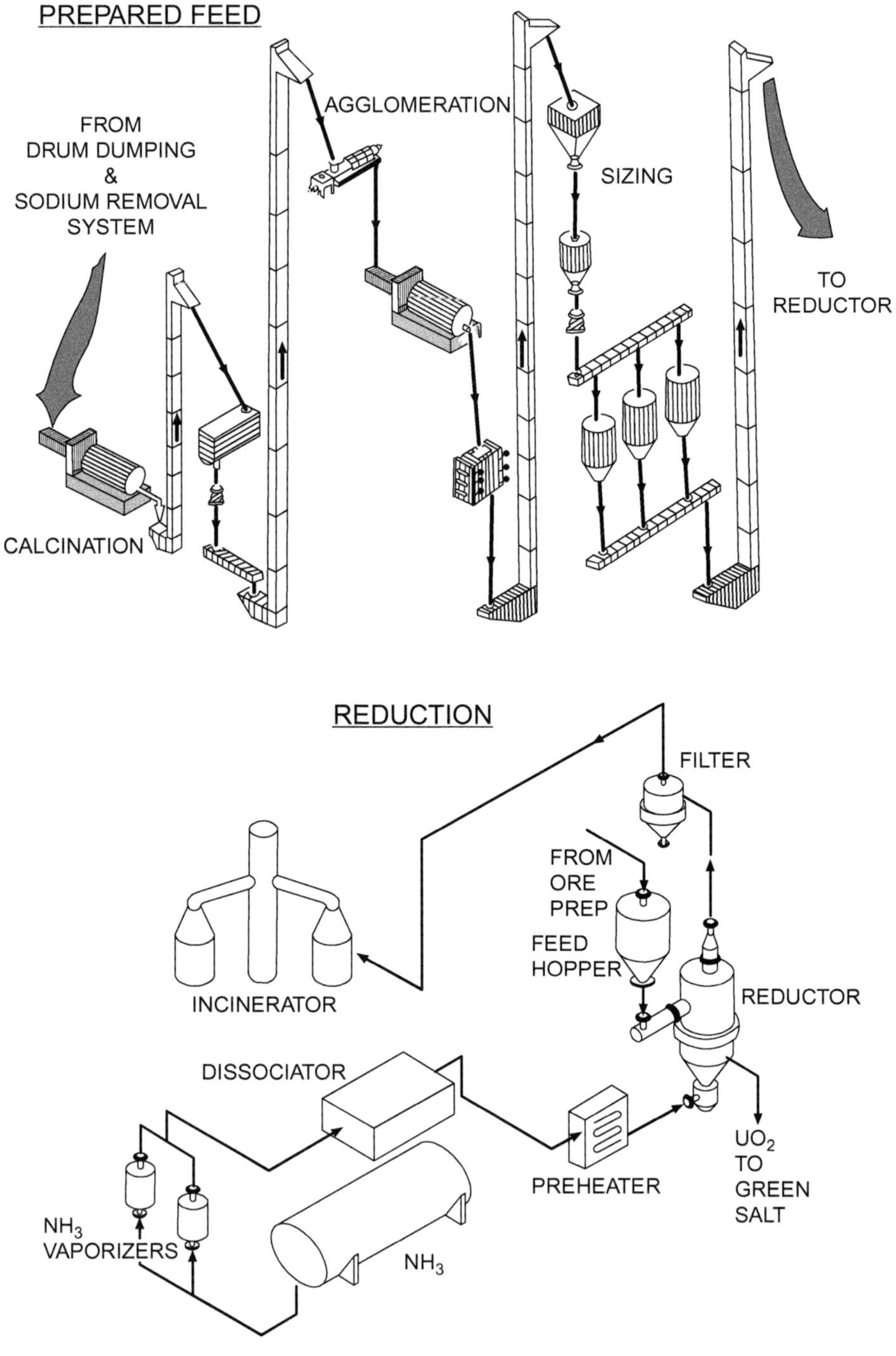

Figure 11.4 Outline of Honeywell conversion and refining process.

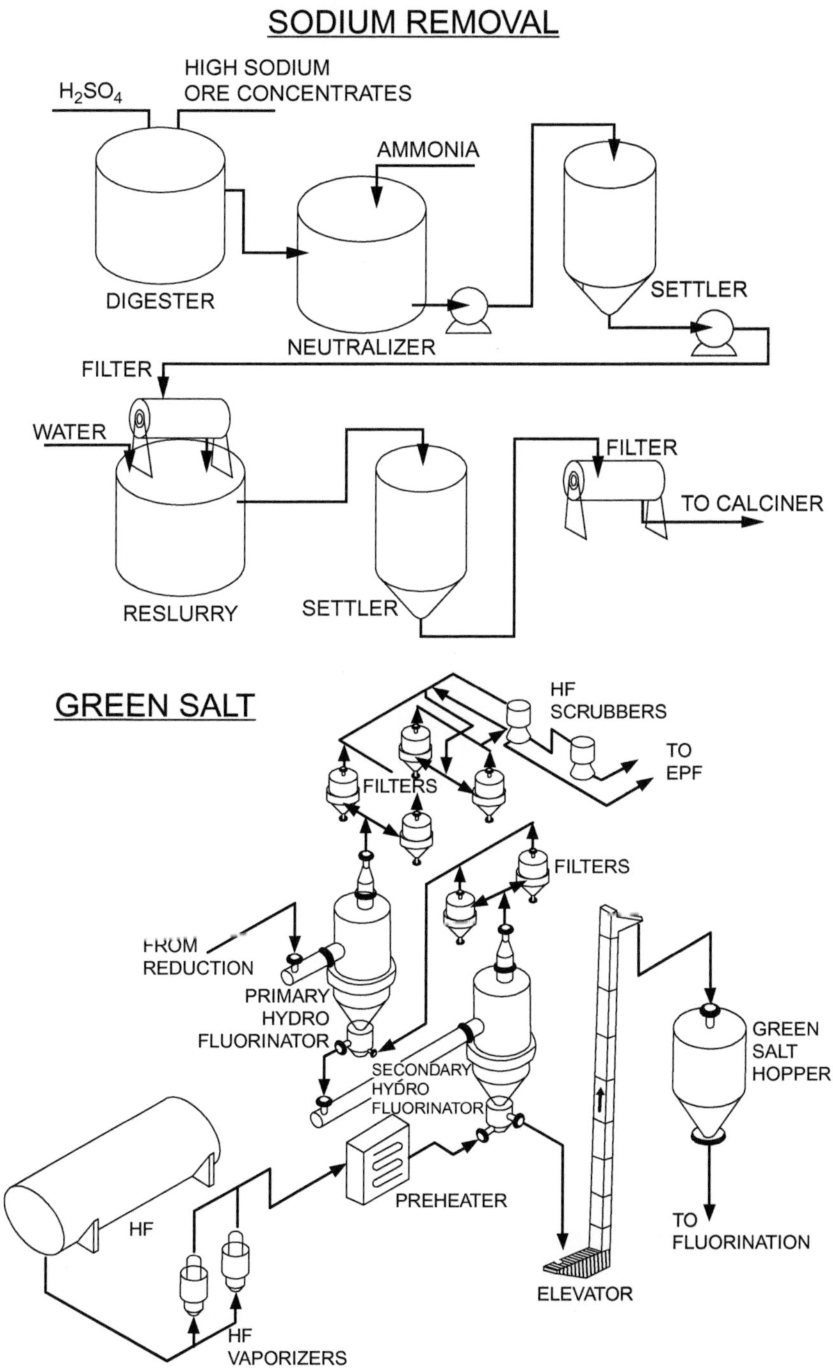

Figure 11.4 (Continued)

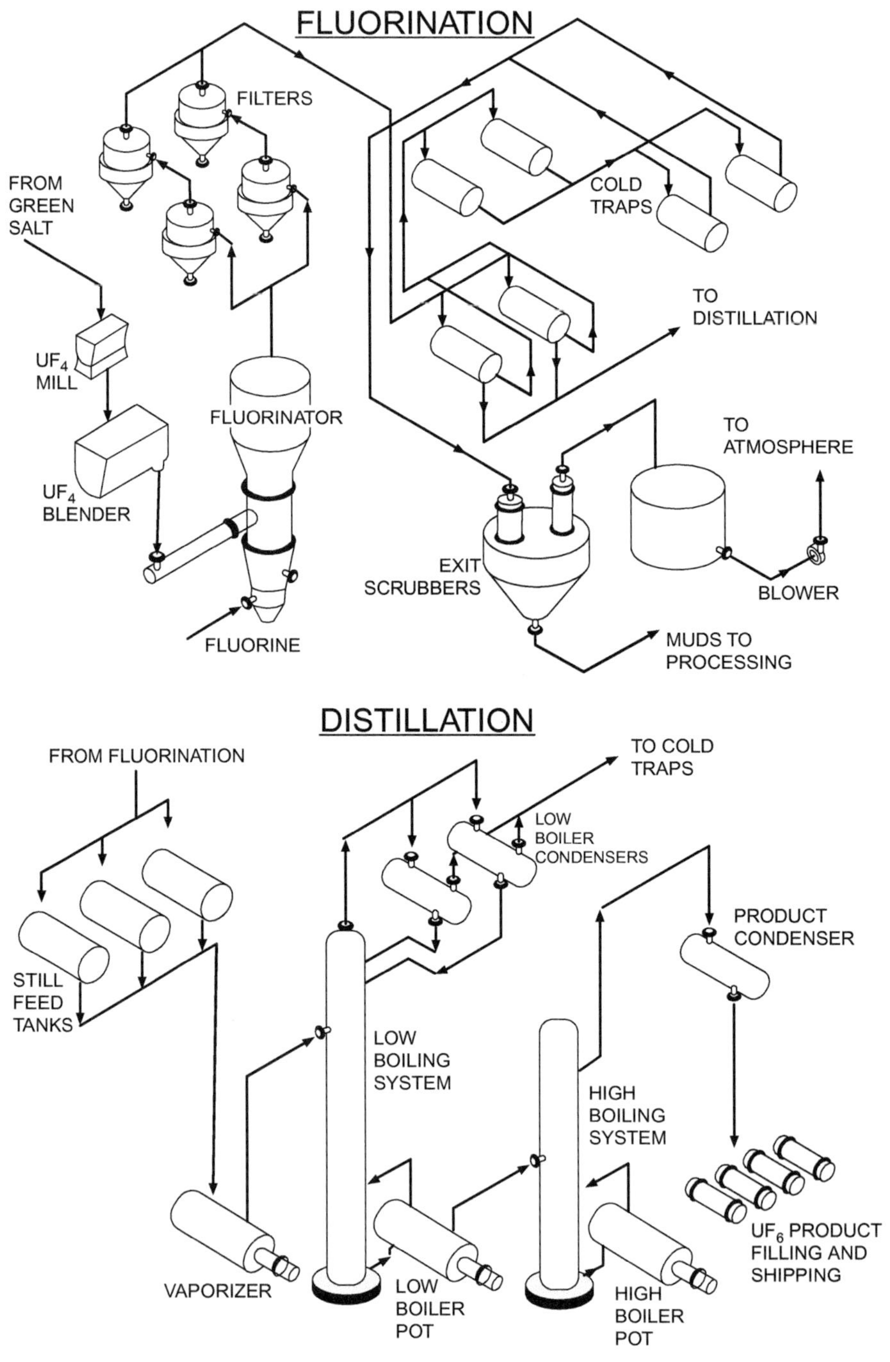

Figure 11.4 (Continued)

11.3.6 UF_6 product

Verifying of the quality of the UF_6 product is an important step. For Cameco, this is achieved by remelting of UF_6 in representative cylinders, and withdrawing samples for analysis on specification elements. Comurhex prefers to have an intermediate liquid UF_6 hold tank that is large enough to fill one cylinder completely and allow for homogeneity and settling of fine solids. The liquid is then decanted to the transport cylinders. Product quality assurance procedures must be rigorously maintained, although this does not mean every cylinder must be sampled, only that the sampling procedures are such that customers are assured of product quality.

The containment of UF_6 requires particular care, primarily due to three factors:

1. UF_6 expands by about 36% when solid is melted to liquid so it is essential that cylinders not be overfilled. If a mistake is made, it can result in bursting of a transportation cylinder.
2. UF_6 reacts immediately with any moisture to yield HF and a very visible cloud of UO_2F_2. This acts as a warning of leaking and it is now common that cylinder filling areas have the ability to freeze out any possible liquid UF_6 with solid carbon dioxide sprays.
3. UF_6 is a very strong oxidizer and should any hydrocarbon oils or greases mix with the UF_6, there is an explosion risk.

All producers take great care in the handling of UF_6, particularly as liquid; in some cases, UF_6 is only moved as a gas or as a solid, including collection of the solid in tote bins for transfer to an enrichment plant. The valving of containers with larger openings for solid and gas transfer adds some risk as well. All fluoride-containing gas streams are typically scrubbed with aqueous KOH.

11.3.6.1 Fluorine production

Fluorine is generated in greater amounts for UF_6 production than any other product, and is a very expensive and hazardous part of this production. Hence, fluorine production technology is very important to economic UF_6 production. Fluorine is produced by electrolysis of the fused salt, KF.2HF, at 85–90°C in mild steel cells, using amorphous carbon anodes and mild steel cathodes. The cells operate continuously with HF added appropriately by monitoring electrolytes levels. Water cooling is also needed. More in-depth information of this subject is available in references (Ellis and May, 1986; Shia, 1995; Slesser and Schram, 1951).

Fluorine leaving the cells contains about 7% HF, which may be used as is (Cameco) or the HF removed by reaction with sodium fluoride. With no HF removal, the HF passes through UF_6 collection and is usually trapped in very cold secondary traps, allowing it to recycle into the hydrofluorination process.

The byproduct hydrogen stream also contains about 7% HF. Typically the HF from the hydrogen stream is removed by cold traps and scrubbers before the hydrogen gas is vented to atmosphere.

11.3.6.2 Process comparisons

AEC chose to promote the distillation process based on cost, however the rest of converters have chosen the solvent extraction process. One published report (Fryer, 1968) suggests the economics are not much different between the two processes and other operating parameters drive the choice between the two options. These factors include:

1. The solvent extraction process is believed by most to be more tolerant of higher levels of impurities in the feed concentrate.
2. Sodium and potassium present in the feed concentrate tend to make nonvolatile salts such as Na_2UF_8, reducing uranium recovery and further lowering the sintering temperature of the bed in the fluid bed, causing lumping. These alkalis metals can be pretreated for removal but the solvent extraction process has the advantage of rejecting these elements without an additional step and increased wastes, if uranium oxide concentrates are expected to be routinely high in concentrations of the alkali metals.
3. Some organics and other impurities can promote emulsions in solvent extraction, whereas these impurities would not be as problematic in the fluorination/distillation process.
4. The presence of impurities during fluorination increases consumption of expensive HF and F_2 in the distillation process, so higher levels of most impurities support use of the solvent extraction process.
5. In the distillation process, the presence of impurities, such as S and As, in the initial reduction, produce higher concentrations of noxious gases (H_2S and AsH_3), which must be scrubbed at extra cost. The distillation process also produces small amounts of impurity streams rich in MoF_6 and VOF_3, which are difficult to dispose.
6. Distillation does have the advantage that if corrosion products from process equipment add to product impurity levels, then impurity removal near the end of processing has advantage.
7. The nuclear-pure UO_3 from the solvent extraction process may have other commercial value. For example, Cameco takes some of this stream to make sinterable natural UO_2 powder for fuel manufacturing mainly for CANDU reactors and making blanket fuels for light water reactors.
8. The nuclear-pure UF_4 from the solvent extraction process may have other commercial value. For example, SFL took some of this stream for uranium metal production for Magnox reactor fuel, and Comurhex also took advantage of this feature in the distant past.
9. The distillation process may be preferable for enriched uranium as the absence of water reduces the risk of criticality.

11.3.7 Safety

A major safety concern for the production of UF_6 is the safe handling of anhydrous HF. Different approaches can be taken to achieve acceptable levels of safety. At Malvesi, HF storage and handling of anhydrous hydrogen fluoride (AHF) is entirely outside. At Honeywell, the HF is unloaded from outside transport containers and significant HF releases addressed by a curtain of water cannon, as has become standard practice at HF production facilities in North America. At Cameco, the approach is to ensure all unloading is within the production facility, where collection of any fumes is ensured and discharge only after passage through the plant scrubbers. The system is designed so that even if there were significant leakage the HF would be contained within the building. (Clark and Kennedy, 2010).

Within plants, both HF and UF_6 present significant gaseous hazards. Where possible, systems operate at below atmospheric pressure so that any leak would be inward, not outward. However, this is not always possible particularly where liquid UF_6 is present because this requires a slight overpressure. Also, the more progressive plant designs have ability to segregate operating areas so fume collection capabilities can be focused in specific areas when considered necessary. After multiple dust collection and scrubbing systems, ideally there is one final discharge point at high level where there is online monitoring of HF and U emissions.

There is typically wide use of sensors for uranium dust, fluorine, hydrogen fluoride and closed circuit TV, with outputs to both a central control room and localized control rooms where necessary, for example, HF unloading, cylinder filling. Fluorine detection is usually through UV analyzers and stack fluoride emissions by absorption and ion-selective electrodes. Fine uranium dust levels are usually measured by filters that are taken off line as frequently as appropriate and are therefore not truly continuous for control room information. Even very small leaks of UF_6 immediately form very visible clouds of UO_2F_2, which may be detected by smoke detectors or by visual appearance on camera. Thus, although unlikely hazardous releases are quickly recognized and addressed with minimum exposures to operating staff.

Large parts of the systems are automated and use is made of remote control systems and closed control rooms with closed circuit television systems to minimize any chance of gas exposure to operators.

11.3.8 Environmental

Process byproducts, even with low uranium contamination are increasingly difficult to dispose of, mainly due to negative public perceptions. One would think, because very low uranium levels are already widely present in the environment it would not be a big issue. However, this is a reality of the industry and in recent years has probably been the greatest driver for process changes to ensure minimizing uranium emissions.

In general, sites scrub final off-gases from hydrofluorination and fluorination with KOH solution. This is preferred over NaOH solution due to the relatively poor solubility of NaF, over more soluble KF, produced from the scrubbing of HF.

Potassium diuranate may be separated off and recycled to refining or milling operation. The exhausted scrub solution may be evaporated to dryness to yield a uranium contaminated KF/K_2CO_3 byproduct, or treated with lime to regenerate the KOH and produce uranium containing $CaF_2/CaCO_3/Ca(OH)_2$ byproduct. Either product may be recycled to a mill for uranium recovery, although some have chosen to simply send this for disposal.

11.4 Current status of the conversion industry

As noted previously, the commercial UF_6 conversion plants, with best capacity estimates from published data, are as follows (WNA, 2014; IAEA, 2012, 2009):

Operator	Country	Location	Nominal capacity (tU/a)
Cameco	Canada	Port Hope, ON	12,500
Comurhex (AREVA)	France	Malvesi and Pierrelatte	15,000
Rosatom	Russia	Seversk	12,500
CNNC	China	Lanzhou	5,000
Honeywell (Converdyn)	United States	Metropolis, IL	15,000

Although nominal capacities for UF_6 production are given, the current low demands for conversion means that most plants are operating well below, and some near 50% of these nominal capacities. Also, nominal capacities for refined UO_3 may be substantially higher as some of this intermediate product may be diverted for production of other products such as uranium metal and ceramic-grade UO_2. Most significantly for Cameco, Blind River refinery has almost twice the capacity (24,000 tU as UO_3) relative to UF_6 production capacity due to requirements of producing other products during its history.

Major changes are expected in France with a new plant to be built in Pierrelatte to accommodate all refining and conversion operations with some eventual increase in overall capacity.

It is of note that Russia has historically produced UF_6 at two sites but has enrichment plants at four sites. Currently, UF_6 is produced at one site at Seversk and the Angarsk plant was shut down in 2014. It is believed partially enriched UF_6 is shipped as feed to the sites without UF_6 production capacity.

The China National Nuclear Corporation (CNNC) production is estimated to be ~5000 tU/a. It is expected to continue increasing to meet domestic enrichment requirements.

There was commercial capacity of about 6000 tU/a at Springfields in the UK, (Özberk, 2011) but this was closed in 2014. Construction of a future commercial capacity of about 6000 tU/a was publicized by Kazakhstan from a joint venture of Kazatomprom and Cameco, but timing remains uncertain. Kazakhstan appears committed to become an UF_6 producer adding value to its large uranium production.

There are also a number of small-scale or "pilot" plants for internal national supplies of UF_6, sometimes associated with nuclear weapons programs and therefore known with less certainty, notably:

Operator	Country	Location	Capacity (tU/a)
AEOI	Iran	Isfahan	~200
BARC	India	Trombay	Unknown
CNEA	Argentina	Rio Negro	Unknown
IPEN	Brazil	Sao Paulo	~100
	North Korea		Unknown
CPC	Pakistan	Dera Ghazi Khan	Unknown

These operations are all believed to use the wet process refining; certainly, the deliveries of TBP have been used as indicators of probable uranium refining activity. However, there are few published details on these operations.

AEOI of Iran started the plant in Isfahan in 2004 and this plant is under IAEA safeguards. Production is tracked by various organizations, such as IAEA.

India has an unsafeguarded plant called the Uranium Metal Plant at the Bhabha Atomic Research Center (BARC), which converts yellowcake to UF_4 and then either to metal or UF_6. There is also a larger refining capacity as some refined product is diverted to make natural UO_2 for their heavy water moderated reactors.

The National Atomic Energy Commission of Argentina (CNEA) has facilities at the Complejo Tecnologico Pilcaniyeu site, in the Rio Negro district, which includes a UF_6 conversion plant.

The Brazilian pilot-scale process is largely based on French technology and uses the ADU route to UO_3.

The Pakistan Chemical Plants Complex (CPC) converts yellowcake to UF_6, and also enriched UF_6 to metal and depleted UF_6 to metal. This undeclared and unsafeguarded nuclear site was built in the 1970s and early 1980s, and continues to be upgraded.

11.5 Factors that impact converters

A very significant issue is that the two North American converters are economically separate from enrichment, unlike all other such operations around the world. Thus, they work to a different business model, to justify their activity independent of enrichment revenue in this step of the fuel cycle. This is a challenge as historically refining/conversion to UF_6 has generated relatively little revenue and, on average, little profit. The cost structure is improved somewhat by increased capacity, as long as that capacity is well used, that is over 80%. Of course, in recent years this has not been achieved and is a significant part of the reasons why the smaller converter in the UK, Springfields Fuels Ltd ceased production recently.

Government operations tend to act to ensure conversion is available for their national enrichment programs, independent of the economics of this separate step. Probably in recent years, this type of operation is most evident in China, for as the national enrichment capacity has increased, so has the national capacity for conversion to UF_6.

Other risks must be considered along with the politics and economics of the business. Of note are:

Chemical contact risk: Several hazardous chemicals are essential parts of refining conversion. These include hydrogen fluoride, fluorine, UF_6, nitric acid, solvents, and TBP. Unintended contacts with employees or releases outside plant boundaries can have very negative impacts. Equipment and operating conditions are designed to minimize risks but some small risk remains. When upsets occur, there are many barriers to maintain containment, but some unplanned releases to the environment still occur, although are becoming less through proactive more stringent safety programs.

Chemical supply risk: Hydrogen fluoride is a strategically important chemical its major use is as refrigerants, and as world affluence grows so does the demand for air conditioning, and hence HF. World supplies are dominated by China, and China does not hesitate to use supply dominance to protect home industries. Hence, demand/supply risks for HF can be expected to increase. The extremely toxic nature of HF has also caused some jurisdictions to consider banning the chemical, and acceptance of its transport through communities can be expected to become more challenging. Organic phosphates, such as TBP, are considered a cancer risk, although there is no evidence of such a problem for TBP in uranium refining. Still supply disruptions must be considered a risk.

Radiological risk: Radiation doses are generally well below regulatory standards but still some risk is associated with the uranium daughters when procedures, shielding, and containment are not properly maintained. It is a feature of the industry that when releases might have impact outside a facility, the public express primary concern over radiation, although chemical impacts are likely to be greater. In turn, regulators usually target the focus of public concern.

Recycle risk: Refining and conversion result in byproducts, containing impurities but usually some residual levels of uranium that still have value. Recycle to mix such material with uranium mill feeds is usually possible, but the transport of such materials, which have at times wrongly been labelled as waste, usually attracts public and regulatory interest, resulting is some uncertainties in availability of these routes.

Waste risk: Contaminated equipment that is beyond useful lifetimes needs to be cleaned and residues disposed. Some other materials such as insulation, plastics, wood, and paper can be contaminated and not easily cleaned. Incineration may be useful in allowing some uranium recovery but such incinerators are no longer easily licensed. Although very small there is inevitably some waste, and disposal can be difficult. Some historic practices of simply storing are now generally a regulatory issue, demanding action.

Inventory risk: Refining/conversion sites have historically held the uranium inventories for customers and these have been as large as the feed/production for 1 year. The value of such an inventory can approach 1 billion dollars. Fortunately the weight is very substantial so theft is not likely, concerns are greater that in some way the inventory could be stranded by strikes, plant upset, transport or regulatory issues. For these reasons, there is a tendency in recent years for lower inventories to be held at refining/conversion sites and more to be held at enrichment sites.

11.6 Conversion to UO_2

Natural uranium is also converted to ceramic-grade UO_2 for production of reactor fuels. The demand is mainly for CANDU reactors, but some use also exists for pressurized water reactor (PWR) and boiling water reactor (BWR) blanket fuels.

Production may be through natural UF_6 just as enriched UO_2 is produced through enriched UF_6, as discussed further in Chapter 13 "Nuclear fuel fabrication." A more cost-effective process is through conversion of natural nuclear-grade UO_3. This process involves precipitation of ADU from ammonium nitrate solution using ammonia, separation and drying of the ADU and then thermal reduction with hydrogen at high temperature to yield the UO_2, as it is illustrated in the flowsheet published by Cameco Corp., shown in Figure 11.5.

In Cameco's current operation (Itzkovitch and Zawidski, 1985; Kwong and Kuchurean, 1997; Özberk, 2007), UO_3 from the refining at Blind River as described previously, is dissolved by nitric acid to give the feed uranyl nitrate solution. Fume collection is provided at the tank to prevent nitric acid vapors from escaping into the plant. In earlier days of Cameco operations, when refining and conversion were on the same site, the uranyl nitrate solution could be taken after solvent extraction/partial evaporation and prior to denitration, but now because UO_3 is the form used for uranium transport between distant sites the dissolution step is needed.

Prior to precipitation, the concentrated uranyl nitrate is diluted and then ammonia added to yield an ADU precipitate. These solids are collected, and reduced to UO_2 powder in externally heated reactors with counter-current flow of hydrogen. Conditions are set to produce a powder with an O:U atomic ratio of 2.05–2.18,

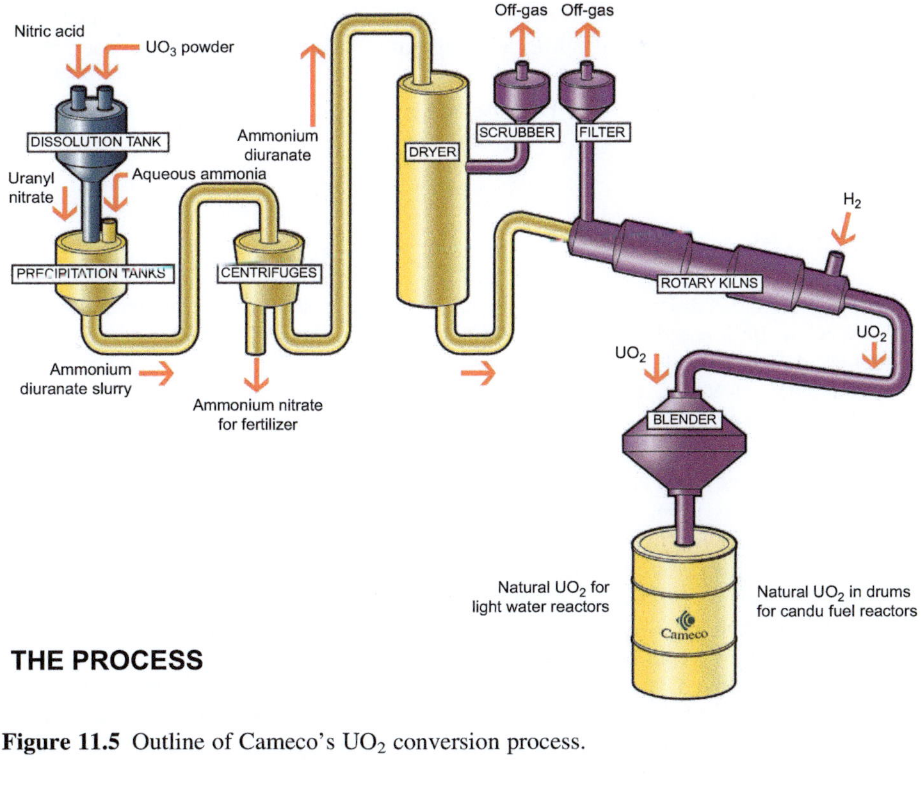

Figure 11.5 Outline of Cameco's UO_2 conversion process.

which can be handled without difficulty in air. The powder is then blended in batches, packaged and shipped to fuel manufacturers.

While Canada is the primary commercial producer of such natural UO_2, there are other small plants in Argentina, China, and Romania where national self-sufficiency is a priority.

11.7 Potential future developments

Technology, processing, and conversion are not likely to change significantly in the near future. No product other than UF_6 is likely to be in demand for enrichment and the overall demand will only increase slowly. Indeed as enrichers increase enriching efficiency and more high-^{235}U tails are used as feed, then there is reduced demand for conversion.

Potential increasing demand for conversion, resulting from the expansion of nuclear programs in Asia, particularly China and India, will be addressed by more rapid expansion of capacities as national control takes priority over economic considerations. This will result in little chance of increasing demand for conversion in the Western world and no substantial increases in price due to the competition inevitable from demands that do not cause plants to operate near capacity. The small profit margin results in limited research on process improvements, and that which does happen will address the emission and waste issues of public concern, as required to ensure continued public and regulatory acceptance of these operations, and security of supply for customers.

In short, limited changes can be expected in processing technologies, except for some production growth in Asia in the near future.

References

Ashbrook, A.W., 1982. Uranium refining and conversion practice in the western world: an overview. In: Proceedings of Uranium '82, Toronto, Canada, CIM, 1982.

Ashbrook, A.W., Smart, B.C., 1980. A review and update of refining practice in Canada. In: Production of Yellow Cake and Uranium Fluorides, 5–9 June 1979, Paris, Published IAEA proceedings, Vienna, 1980, pp. 261–287.

Astles, C.R., Green, C.A., 1998. New process for natural uranium by-product preparation in uranium refining operations. In: Proceedings of Pacific Nuclear Conference, Toronto, 1998, pp. 1501–1506.

Bishop, J.C., Hanson, B.J., 1982. Fluoride volatility process operating experience. In: Proceedings of Uranium '82, 12th Annual Hydrometallurgical Meeting of CIM, Toronto, Canada, August 29–September 1, 1982.

Cameco Corporation, Fuel Services Division, 2012. Licence renewal application for the blind river refinery, 15 April 2011.

Clark, D.R., Kennedy, T.W., 2010. Minimizing the risk and impact of uranium hexafluoride production. In: Lam, E.K., Rowson, J.W., Özberk, E. (Eds.), Proceedings Uranium 2010, 'the Future Is U', the 3rd International Conference on Uranium, 40th Annual

Hydrometallurgy Meeting, METSOC/CIM, Proceedings Volume 2, Saskatoon, Canada, pp. 83–96.

Delannoy, J., Faron, R., 1982a. Conversion of concentrates containing uranium into uranium hexafluoride. In: Proceedings of Uranium '82, 12th Annual Hydrometallurgical Meeting of CIM, Toronto, Canada, August 29–September 1, 1982.

Delannoy, J., Faron, R., 1982b, Conversion of concentrates containing uranium to uranium hexafluoride. In: Proceedings of Uranium '82, Toronto, Canada, CIM, 1982.

Ellis, J.F., May, G.F., 1986. Modern fluorine production. J. Fluorine Chem. 33, 133–147.

Fryer, R.M., 1968. Comments on processing natural uranium. In: Presentation at the Joint ANs-CAN Meeting, 11 June 1968, Toronto, Canada.

Harrrington, C.D., Ruele, A.E., 1959. Uranium Production Technology. D. Van Nostrand Company, Princeton, N.J.

IAEA, 2009, A Directory of Nuclear Fuel Cycle Facilities (NFCIS), IAEA-TECDOC-1613, IAEA, April 2009.

IAEA, 2012, List of Nuclear Fuel Cycle Facilities (NFCIS), IAEA, 2012.

Itzkovitch, I.J., Zawidski, T.W., 1985. Yellowcake to uranium dioxide. In: Presented at the 3rd International Conference Nuclear Technology, 14–16 October 1985, Madrid, Spain.

Kwong, A.K., Kuchurean, S.M., 1997. Ceramic UO2 powder production at Cameco corporation. In: Proceedings of CNS Fifth International Conference on CANDU Fuel, Toronto, 21–25 September 1997, pp. 188–194.

Lawroski, S., et al., 1958. Production of refined UF6 from ore concentrates by fluidization and fractional distillation techniques. In: Proceedings of 2nd International Conference Peaceful Uses of Atomic Energy, Paper P1552, United Nations, Geneva, Switzerland, pp. 44–57.

Lenahan, K.J. (1982), Eldorado wet way process. In: Proceedings of Uranium '82, Toronto, Canada, CIM.

Mallinkrodt Uranium Division News, 1962. Vol. 7, Nos 3 & 4.

Özberk, E., 2007. Uranium processing and sustainable development. In: European Metallurgical Conference 2007, EMC2007, Symposium Proceedings, Vol. 1, 11–14 June 2007, Dusseldorf, Germany.

Özberk, E., 2011. Innovation and uranium production at Cameco. In: Roderick Guthrie Honorary Symposium on Process Metallurgy, Published in Symposium Proceedings, 6–9 June 2011, Montreal, Quebec.

Page, H., 1986, Conversion of uranium ore concentrates and reprocessed uranium to nuclear fuel intermediates at BNFL. In: Paper T1-TC-576, Technical Committee Meeting on Advances in Uranium Refining and Conversion, IAEA, April 1986, Vienna, pp. 7–11.

Rice, T., 2000. Chapter ten, refining and conversion. In: Uranium 2000 Short Course Unit Operations of Uranium Process Metallurgy, 30th Annual Hydrometallurgy Meeting of CIM, 9–15 September 2000, Saskatoon, Saskatchewan, Canada.

Ruch, W.C., Peterson, D.A., Gaskill, E.A., Tepp, H.G., 1960. Production of pure uranium hexafluoride from ore concentrates. Nucl. Eng. Chem. Eng. Prog. Symp. Ser. 56 (28), 35–41.

Schisler, J.M., et al., 1986. Commissioning and startup of the blind river uranium refinery. In: Minerals and Metallurgical Processing, February 1986, pp. 14–19.

Shia, G., 1995. Fluorine. In: Howe-Grant, M. (Ed.), Fluorine Chemistry: A Comprehensive Treatment. John Wiley & Sons, New York, pp. 1–27.

Slesser, C., Schram, S.R. (Eds.), 1951. Preparation, Properties and Technology of Fluorine and Organic Fluorine Compounds. McGraw-Hill Book Company Inc., New York.

World Nuclear Association, Conversion and Deconversion <www.world-nuclear.org/info/Nuclear-Fuel-Cycle/Conversion-Enrichment-and-Fabrication/Conversion-and-Deconversion/> (updated September 2015).

Uranium enrichment

12

Paul Harding
URENCO Limited, Stoke Poges, United Kingdom

12.1 Introduction

12.1.1 What is uranium enrichment?

Naturally occurring uranium comprises about 99.29% of the isotope ^{238}U (146 neutrons, 92 protons) and 0.71% of the isotope ^{235}U (143 neutrons, 92 protons).

^{238}U is mildly radioactive, undergoing alpha decay to ^{234}Th, but it is not fissile, that is, it cannot be used on its own to sustain a nuclear chain reaction.

^{235}U, on the other hand, is a less stable isotope that undergoes fission when a thermal neutron collides with the nucleus. More neutrons are emitted and, in the right quantity and concentration, ^{235}U is capable of sustaining a nuclear chain reaction; for example, as is required for a power generating nuclear reactor.

Uranium enrichment is the name given to any isotope separation process applied to natural uranium resulting in a product in which the concentration of ^{235}U is increased above the natural concentration of 0.71%.

The capacity of any equipment or installation to enrich uranium in measured in separative work units (SWUs).

12.1.2 What is a SWU?

Separative work performed to enrich uranium is defined by a mathematical formula (see next), but in essence it is a measure of work performed by a process to take a quantity of feed material at a certain ^{235}U concentration and convert it into a quantity of enriched product with a higher ^{235}U concentration and a balancing quantity of depleted "tails" with a lower ^{235}U concentration.

An example is shown in Fig. 12.1 where the feed is 0.71% ^{235}U, enriched product is 4% ^{235}U, and depleted tails are 0.3% ^{235}U.

Precise calculation of SWUs involves differential equations. In ordinary practice, however, the following equation can be used:

$$S = P(2xp - 1)\ln\frac{xp}{1 - xp} + T(2xt - 1)\ln\frac{xt}{1 - xt} - F(2xf - 1)\ln\frac{xf}{1 - xf}$$

where:

S is the effort in SWUs,
P is the mass of the product, the enriched uranium,

Uranium for Nuclear Power. DOI: http://dx.doi.org/10.1016/B978-0-08-100307-7.00012-0

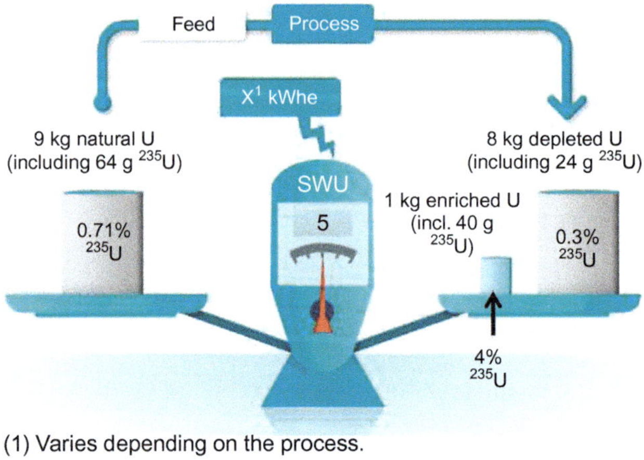

Figure 12.1 Enrichment calculation.
Source: AREVA.

T is the mass of the tails,
F is the mass of the feedstock,
xp is the assay of the product,
xt is the assay of the tails,
xf is the assay of the feedstock.

For a detailed theoretical explanation, see Cohen (1951).

The dimension of SWU is mass so:

1 SWU = 1 kg SWU

1000 SWU = 1000 kg SWU = 1 tSWU

SWU calculators are available on the Internet and as smartphone apps (search: SWU calculator).

A typical large nuclear power plant with an electrical output of 1300 MW requires about 25 t pa of enriched uranium with ^{235}U concentration of 4%. This is produced from about 210 t of natural uranium feed using about 120 tSWU. An enrichment facility with a capacity of 1000 tSWU pa is therefore able to enrich uranium sufficient to fuel about eight large nuclear power plants.

12.1.3 Who needs uranium enrichment services?

Uranium enrichment services are needed by entities that operate nuclear power plants. An up-to-date tabulation of World Nuclear Power Reactors can be found on the World Nuclear Association (WNA) website (www.world-nuclear.org/Info/Facts-and-Figures/World-Nuclear-Power-Reactors-and-Uranium-Requirements). As of February 2015, 438 civil power reactors were listed as operable.

Not all power reactors historically or in operation today require enriched uranium. For example, the first generation of gas-cooled power reactors commissioned in the United Kingdom 1956–71 (the magnox fleet) used natural assay uranium metal fuel. Pressurized heavy water reactors currently operated in a number of countries (Canada, China, South Korea, Romania, Argentina, Pakistan, and India) can utilize natural assay uranium fuel.

Enriched uranium is used by some governments for naval propulsion. Highly enriched uranium (HEU) has, and can be, used for weapons. Both these topics are outside the scope of this chapter.

The boundary of civil enrichment for nuclear power generation is generally considered to be 20% ^{235}U. Typically, to date, most reactors require enrichments up to 5% ^{235}U.

Research reactors can also use small quantities of enriched uranium. A database of research reactors around the world can be found at the International Atomic Energy Agency (IAEA) website (www.iaea.org, or Google nucleus.iaea.org/RRDB/RR/ReactorSearch.aspx). Historic and some current research reactors have used HEU. Globally there is an ongoing drive to convert research reactors to use low-enriched uranium (LEU) (<20% ^{235}U). More on this topic can be found at www.world-nuclear.org/Info/Non-Power-Nuclear-Applications/Radioisotopes/Research-Reactors/ and https://www.iaea.org/OurWork/NE/NEFW/Technical-Areas/RRS/conversion.html.

Using projections of the number of civil power reactors operating globally including considerations of new reactor construction and commissioning, reactor shutdowns, power upgrades, maintenance, and refueling outages, various scenarios for forward demand for civil nuclear enrichment services can be forecast. Fig. 12.2 shows the WNA low, reference, and high case forward forecasts. Fig. 12.3 shows a comparison of a number of published reference case forward forecasts.

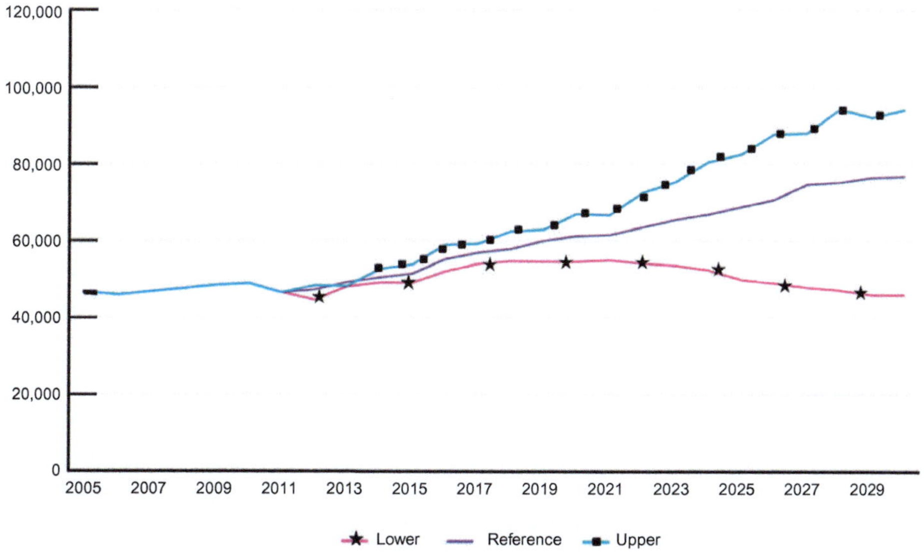

Figure 12.2 WNA enrichment demand scenarios.
Source: WNA.

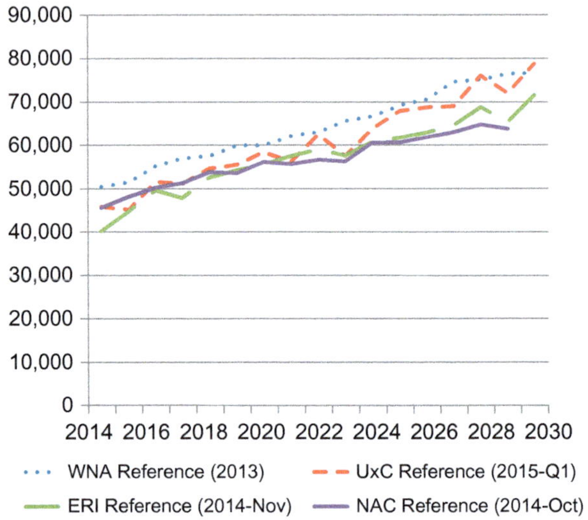

Figure 12.3 Comparison of demand scenarios.

In the reference cases, demand is forecast to rise from approximately 50,000 tSWU pa in 2015 to 60,000 tSWU pa by 2025.

Other nuclear materials can be used as fuel for power reactors, specifically plutonium mixed with uranium known as mixed oxide (MOX) fuel and thorium. Use of MOX is factored into the uranium enrichment demand forecasts. Thorium is not yet used routinely.

12.2 How is uranium enriched?

12.2.1 Current technology and civil facilities

All major commercial civil enrichment facilities today employ some form of gas centrifuge to achieve isotope separation (ie, enrichment) of ^{235}U from ^{238}U. To be suitable for enrichment in a gas centrifuge, uranium that has been mined is concentrated, purified, and chemically converted into uranium hexafluoride (UF_6), a compound of uranium that can easily be turned into a gas when heated for feeding to centrifuges.

Since the inception of uranium enrichment on an industrial scale in the 1950s, uranium isotope separation has predominantly utilized the "uranium hexafluoride" route and global industrial infrastructure to support uranium enrichment, historic and current, is built around conversion of mined uranium to uranium hexafluoride.

Disruptive technology could emerge to challenge this current norm, but it would require very large investment and changes to currently established processes and logistics.

12.2.2 Principles of a gas centrifuge

A gas centrifuge is a cylindrical device that spins at very high speed in a vacuum. When, for example, natural uranium hexafluoride is introduced into a gas centrifuge, the heavier ^{238}U containing molecules tend to migrate to the outer volume of the cylinder and the molecules incorporating the lighter ^{235}U tend to concentrate in the central volume (Fig. 12.4).

The degree of separation achieved by a single machine is small, so the process needs to be repeated hundreds of times to achieve ^{235}U concentrations in final product up to 5% assay required for typical nuclear power plants.

In industrial facilities, many machines are connected in series and in parallel trains to achieve the required enrichment of uranium hexafluoride feedstock in tonne quantities.

The configuration of machines to perform repeated isotope separations is known as a cascade. A representation of a cascade is shown (Fig. 12.5), together with a picture of an industrial installation (Fig. 12.6).

Feedstock is delivered to enrichment facilities from the facilities where it has been converted into uranium hexafluoride in International Standard "48Y" transport cylinders (Fig. 12.7). Enriched product, also in the form of uranium hexafluoride, is generally collected in International Standard "30B" transport cylinders (Fig. 12.8). Depleted tails, again in the form of uranium hexafluoride, is collected in 48Y-type cylinders.

Isotope separation in a gas centrifuge is a purely physical process, no chemical transformation occurs; the feed material is uranium hexafluoride, the product and tails are also uranium hexafluoride.

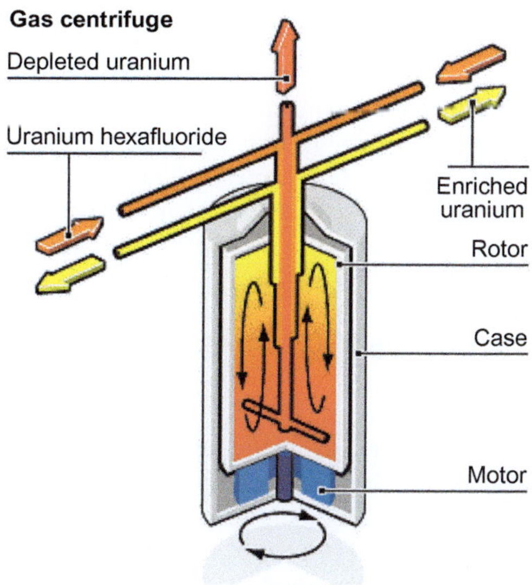

Figure 12.4 Gas centrifuge.
Source: uraniumworld.

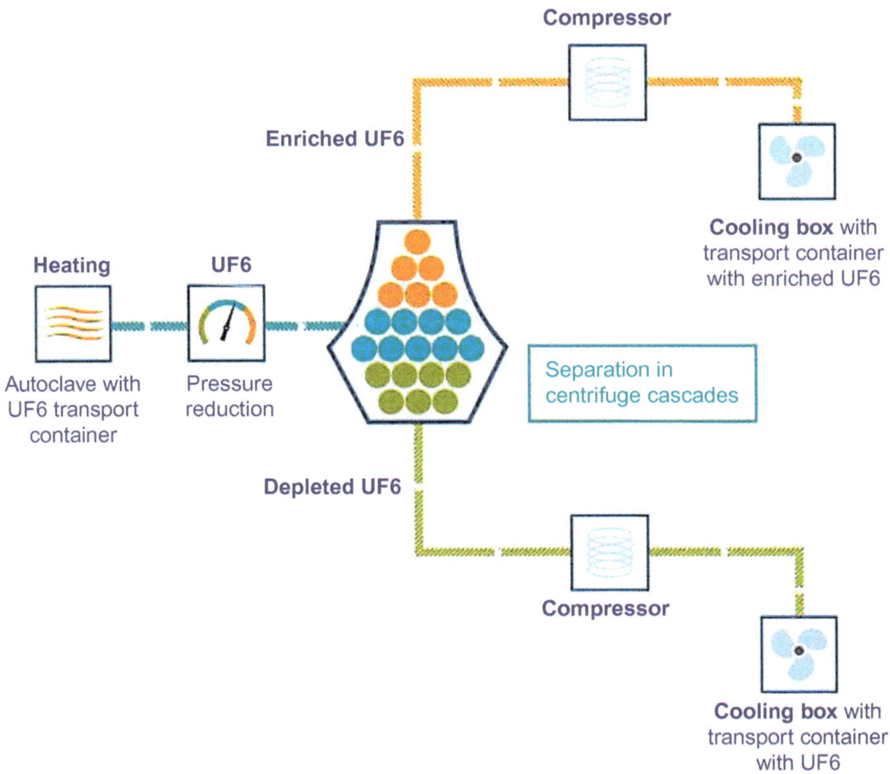

Figure 12.5 Representation of a cascade.
Source: URENCO.

Initial conversion of mined uranium to uranium hexafluoride is carried out in a variety of facilities prior to enrichment. Enriched product is subsequently deconverted, for example, to uranium dioxide, in a variety of facilities ahead of being fabricated into various designs of nuclear fuel elements for nuclear power plants.

Fig. 12.9 shows where uranium enrichment fits into the nuclear fuel cycle.

12.2.3 Current enrichment facilities: The main primary suppliers

12.2.3.1 URENCO

URENCO is a British/Dutch/German owned uranium enrichment company with its headquarters near London. URENCO operates four enrichment plants: Capenhurst, near Chester in the United Kingdom, Almelo in the Netherlands, Gronau in Germany, and Eunice, near Hobbs, New Mexico, United States. The combined reported enrichment capacity of the plants is 18,000 tSWU pa, (Fig. 12.10).

All the plants use gas centrifuge technology. The majority of the centrifuges are TC12-type machines. The most advanced TC21 machines are installed in a proportion of the cascades in the Gronau, Germany, and Eunice, US facilities.

Figure 12.6 Picture of cascade.
Source: URENCO.

Figure 12.7 Picture of 48Y-type cylinders.
Source: NNSA.

Figure 12.8 Picture of 30B-type cylinder.
Source: NNSA.

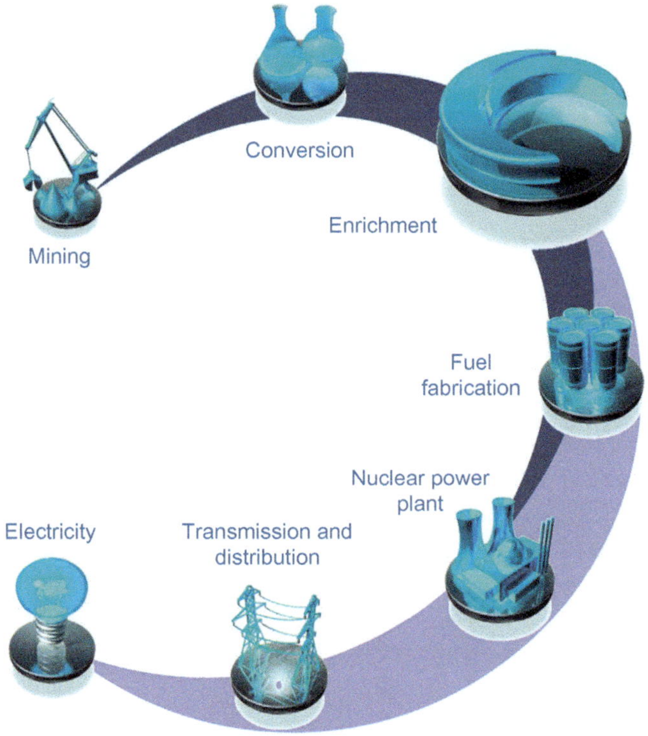

Figure 12.9 The nuclear supply chain.
Source: URENCO.

Figure 12.10 Pictures of URENCO enrichment facilities.
Source: URENCO.

URENCO owns a 50% share of Enrichment Technology Company with AREVA of France holding the remaining 50% share. Enrichment Technology Company is responsible for building and supplying gas centrifuge machines to both URENCO and AREVA.

Commercial gas centrifuge operations commenced at both Capenhurst and Almelo in 1976. The Gronau plant started up in 1985 and the first cascades at Eunice came online in 2010.

URENCO's major expansion of capacity was over the 15 year period 2000–15 when its gas centrifuges replaced tranches of out-dated gas diffusion technology in the United States and France. Going forward, URENCO's focus is expected to be on maintaining current capacity. See also URENCO Annual Report and Accounts (2014).

12.2.3.2 AREVA

AREVA is a French full service nuclear company, involved in all parts of the fuel cycle from uranium mining to fuel fabrication together with nuclear power plant supply. Its headquarters are in Paris.

AREVA operate gas centrifuge facilities (the Georges Besse II Enrichment Plant) at Tricastin in Provence, France. The South Unit began operations in 2011, the North Unit in 2013. By 2016, the combined capacity of the two units will reach 7500 tSWU pa.

AREVA bought a 50% share in Enrichment Technology Company from URENCO in 2006. The centrifuges installed in the Georges Besse II Plant are

Figure 12.11 Picture of AREVA Tricastin enrichment facility.
Source: AREVA.

predominantly model TC12, with at least one cascade of an uprated TC12 + variant (Fig. 12.11). See also http://www.areva.com/EN/operations-887/tricastin-site-the-georges-besse-ii-enrichment-plant.html.

12.2.3.3 Russia

Russia has four operating enrichment plants with a total capacity of 24,000 tSWU pa: (1) Novouralsk, near Yekaterinburg in the Urals, (2) Zelenogorsk (previously known as Krasnoyask—45), (3) Seversk near Tomsk, and (4) Angarsk near Irkutsk, the last three all in Siberia, (Fig. 12.12, Table 12.1).

All plants are equipped with various generations of Russian developed gas centrifuges. Machines are currently manufactured at Kovrov Mechanical Plant in Vladimir region and Urals Gas Centrifuge Plant in Novouralsk prior to shipment to and installation at the plant of use. Machines were also made at Tocmash up to 2012. The last sixth and seventh generation machines were installed in 2005. Eighth generation equipment was installed 2004–12. Ninth generation machines have been installed from 2013.

For more detailed information, see Bukharin (January 2004). See also http://www.world-nuclear.org/info/Country-Profiles/Countries-O-S/Russia--Nuclear-Fuel-Cycle/#Enrichment.

Figure 12.12 Picture of Russian gas centrifuges.
Source: Tenex.

Table 12.1 Individual Russian Enrichment Plant Capacities

Plant	Capacity (1000 tSWU pa)
Novouralsk	10.0
Zelenogorsk	8.7 (expanding to 12)
Seversk	3.0
Angarsk	2.6
Total	24.3

In 2013, a TVEL-Kazakh Joint Venture acquired a 25% share of the Novouralsk facility with an entitlement to half the output. This means Kazatomprom in Kazakhstan can access up to 2500 tSWU pa enrichment capacity.

12.2.3.4 China

China has three operating gas centrifuge factories; Hanzhun in Shaanxi province, (Fig. 12.13), Langzhou in Gansu province, (Fig. 12.14), and Emeishan in Sichuan province.

China's first gas centrifuge facilities were supplied by Russia with sixth generation Russian machines installed. The first Russian supplied centrifuge plant of 500 tSWU pa capacity started operation at Langzhou in 2001. Two modules came into operation at Hanzhun 1997–2000, providing addition of 500 tSWU pa capacity. A further 500 tSWU pa of Russian supplied capacity was commissioned in 2011.

Figure 12.13 Picture of Hanzhun enrichment facility.
Source: http://lewis.armscontrolwonk.com/archive/6826/chinas-new-centrifuge-plants.

Figure 12.14 Google Earth image of Langzhou facility.

China has and continues to develop its own indigenous gas centrifuge technology. The first plant with Chinese manufactured machines installed was brought into operation in 2013. And is thought to have a capacity of about 500 tSWU pa. Therefore total Chinese capacity at the end of 2014 was thought to be about 2000 tSWU pa and have reached around 2800 tSWU pa by the end of 2015.

With its ambitious nuclear reactor construction program, of all the enrichers, China is expected to continue to expand its enrichment capacity, based on its own indigenous technology, to at least meet its own internal needs and sell some surplus enrichment into the global market. China aims to be fully independent in enrichment capability including R and D, engineering, manufacturing and operating. See also http://www.world-nuclear.org/info/Country-Profiles/Countries-A-F/China--Nuclear-Fuel-Cycle/.

12.2.4 Other countries with operating civil enrichment facilities

12.2.4.1 Brazil

Brazil has a civil enrichment facility at Resende. The facility uses gas centrifuge technology developed by the Brazilian navy at Aramar in the 1980s for a naval propulsion program.

The facility at Resende is operated by Industrias Nucleares do Brazil (INB) and is intended to serve much of the enrichment needs of Brazil's Angra reactors. The Resende plant was officially opened in 2006 and Stage 1 has seen ramp up to four modules totaling 115 tSWU pa capacity. INB plans to reach 200 tSWU pa capacity by 2018. See also http//www.world-nuclear.org/info/Country-Profiles/Countries-A-F/Brazil/.

12.2.4.2 Japan

Uranium enrichment in Japan is undertaken by Japan Nuclear Fuel Ltd (JNFL) at Rokkasho, Aomori prefecture. The facility has used Japanese indigenous gas centrifuges of various generations.

The first generation plant operated 1992–2010 with a capacity of up to 1050 tSWU pa. A second generation plant using machines with carbon fiber rotors started operating in 2011. The ultimate aim is to provide a capacity of 1500 tSWU pa. See also http//www.jnfl.co.jp/English/business/uran.html.

12.2.5 Other countries with uranium enrichment capabilities

12.2.5.1 Argentina

INVAP operated a small (20 tSWU pa) gas diffusion enrichment facility in Argentina 1983–89. The plant was unreliable and produced very little LEU. In 2006, Comision Nacional de Energia Atomica (CNEA) said it wanted to recommission the plant using its own advanced diffusion technology, which was said to be competitive, and develop up to 3000 tSWU pa capacity. The facility was officially inaugurated on 1 December 2015. See also http://www.world-nuclear.org/info/Country-Profiles/Countries-A-F/Argentina/.

12.2.5.2 Pakistan

A small (15 tSWU pa) uranium centrifuge plant at Kakuta has been operated since 1984. It does not have any apparent civil use and was expanded threefold in 1991. Another plant is reported to be at Gadwai. It is not under civil safeguards. See also http://www.world-nuclear.org/info/Country-Profiles/Countries-O-S/Pakistan/.

12.2.5.3 India

A small centrifuge enrichment plant, is operated by the Department of Atomic Energy (DAE) Rare Materials Plant (RMP) at Ratnahalli, near Mysore, with some supply to research reactors. It started up in 1992 and is apparently being expanded to 25 tSWU pa. DAE in 2011 announced that it would build an industrial-scale centrifuge complex in Karnataka having both civil and naval purposes. India's enrichment plants are not under civil safeguards. See also http://www.world-nuclear.org/info/Country-Profiles/Counties-G-N/India.

12.2.5.4 Iran

Iran's uranium enrichment capability and capacity has been a topic of international controversy and negotiations for some years, the debate being potential military versus civil purpose to Iran's program. More detail can be read in http://www.world-nuclear.org/info/Country-Profiles/Countries-G-N/Iran/.

Table 12.2 summarizes what is publically known about Iran's uranium enrichment capability.

The Iranian capacity is comprised mainly of IR-1-type gas centrifuges (which are based on Pakistan's P1 design) with variants such as IR-2 to IR-6 reported as developments.

The first facility, PFEP, started operations in 2003. Operations at PFEP and FEP are under international civil safeguards, though monitoring is constrained.

Table 12.2 Iranian Enrichment Plant Capacities

Facility	Capacity (tSWU pa)
Natanz Pilot Fuel Enrichment Plant (PFEP)—above ground	<1
Natanz Fuel Enrichment Plant (FEP)—below ground	Approx. 6.5 (Nov. 2013)
Fordow Fuel Enrichment Plant (FFEP)—below ground	<1

12.2.6 Civil enrichment services demand/supply balance

Section 12.1.3 detailed the forward demand for uranium enrichment services.

In Section 12.2, plants and their capacities to provide civil nuclear enrichment services have been described. This allows overlay of available supply against demand for the period 2015–24. However, if this is done simplistically, without some key additional considerations, it would immediately appear that there is a significant excess of enrichment services capacity over demand. There are three other main factors to take into account in considering the real demand/supply balance: (1) tails re-enrichment, (2) enriched product inventories, and (3) civil utilization of ex-military materials. Each of these factors is discussed in turn.

12.2.6.1 Tails re-enrichment

In most commercial enrichment contracts, the customer agrees to supply the enricher with a given quantity of feed material in return for a given quantity of enriched product and pays for the services provided by the enricher. The tails arising from the enrichment performed are owned by the enricher. The enrichment contract will specify the tails assay, quite typically in the range 0.2–0.3% ^{235}U. Clearly the tails still contain residual ^{235}U and it is possible to use such tails as feedstock in a further enrichment cycle, producing additional enriched product and "tails of the tails" or "secondary tails." Such two-stage enrichment is illustrated in Fig. 12.15.

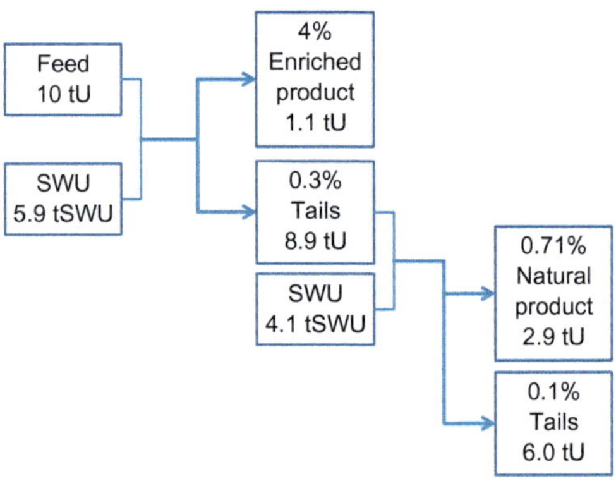

Figure 12.15 Re-enrichment of tails.
Source: URENCO.

The economics of tails re-enrichment depend, among other factors, on the price of natural uranium feedstock, the price of enrichment services, and the marginal costs of operating enrichment facilities. However, in the case of any surplus of enrichment capacity, there are drivers to devote some enrichment capacity to the re-enrichment of tails to regenerate natural uranium feedstock or additional enriched product for sale.

An alternative to re-enriching tails is a mode of operating enrichment plants known as "underfeeding." In this mode, an enricher can choose to use more SWU than specified in the contact to provide the required quantity of enriched product to the customer, but strip the tails to a lower assay than specified in the contract and thereby consume less feedstock than is provided by the contract. The saved feedstock is then owned by the enricher and can be sold on the market. The principle of underfeeding is illustrated in Fig. 12.16.

Underfeeding and tails re-enrichment are somewhat interchangeable in terms of their utilization of enrichment capacity and both are labeled tails re-enrichment in assessing demand/supply balance. Because it is desirable to keep capital intensive gas centrifuge plants in continuous operation, enrichers tend to underfeed/re-enrich tails at times when demand for primary enrichment is suppressed, such as the post-Fukushima period (see Section 12.2.6.2).

In the case that natural uranium was cheap compared to enrichment services, based on economics, enrichers could choose to "overfeed" plants. This involves enrichers acquiring additional feedstock, compared to that supplied in a contract, and using less SWU than specified in the contract to provide the customer with the required quantity of enriched product. Overfeeding has only been undertaken when

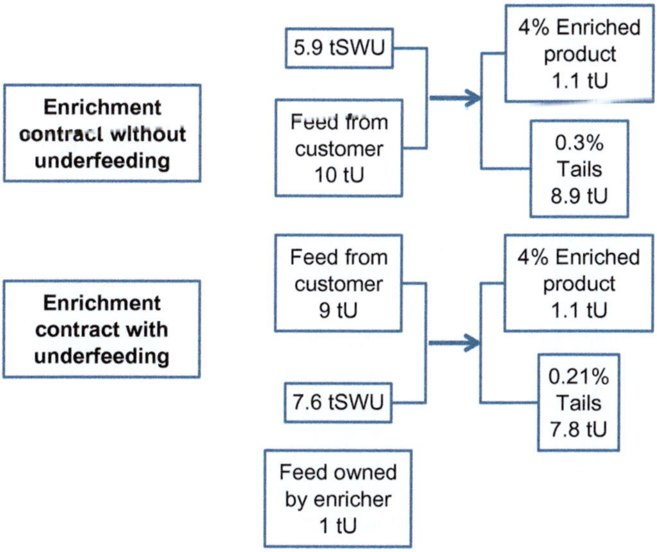

Figure 12.16 Explanation of underfeeding.
Source: URENCO.

uranium is cheap and/or enrichment capacity is in short supply compared to demand, or if, as with gas diffusion in the past, the enrichment process is expensive to operate.

12.2.6.2 Enriched product inventories

Historically, for various reasons, a number of parties, including governments and utilities, have held varying inventories of enriched product (to provide a buffer against supply chain interruption, for example).

Furthermore, post-Fukushima enrichment services contracted for Japanese reactors, which have been shutdown for at least 4 years, have continued to some extent with consequent buildup of product stocks.

Part of the forward use of such inventories to fulfill demand has to be added on the supply part of the equation; obviously dampening the demand for enrichment services from primary production.

12.2.6.3 Civil utilization of ex-military materials

A further source of existing enrichment inventory that can be utilized in the civil fuel cycle is material derived from downblending HEU surplus to requirements because of the decommissioning of nuclear weapons.

12.2.6.3.1 "Megatons to Megawatts": The Russian HEU deal

A celebrated initiative called the “Megatons to Megawatts” Programme from 1993 to 2013 saw 500 t HEU from decommissioned Russian weapons diluted in Russia to 14,500 t LEU and shipped to the United States for use in commercial power reactors.

The agreement between the United States and Russian governments was made in 1993 as part of arms reduction initiatives. In 1994, the United States Enrichment Corporation (USEC) and Tenex were made executive agents, respectively, for the US and Russian governments to progress the program.

The agreement was essentially an enrichment deal totaling about 90,000 tSWU. About 150,000 t natural uranium was exchanged for the LEU of which 112,000 t was sold on the world market and 38,000 t was sent to Russia.

The last of the 500 t HEU was downblended at Zelenogorsk in Aug. 2013 and shipped to the United States in Nov. 2103. This marked the end of the “Megatons to Megawatts” program.

Russia has other surplus military HEU which it may downblend for its own civil use. See also http://world-nuclear.org/info/Nuclear-Fuel-Cycle/Uranium-Resources/Military-Warheads-as-a-Source-of-Nuclear-Fuel/.

12.2.6.3.2 US HEU disposition

For its part, the US government initially declared just over 174 t HEU surplus from military stockpiles. Downblending and release to the civil market has proceeded in batches and whereas the Russian HEU deal is now part of history, release of US government ex-weapons inventory into the civil market needs to be factored into the forward supply/demand balance.

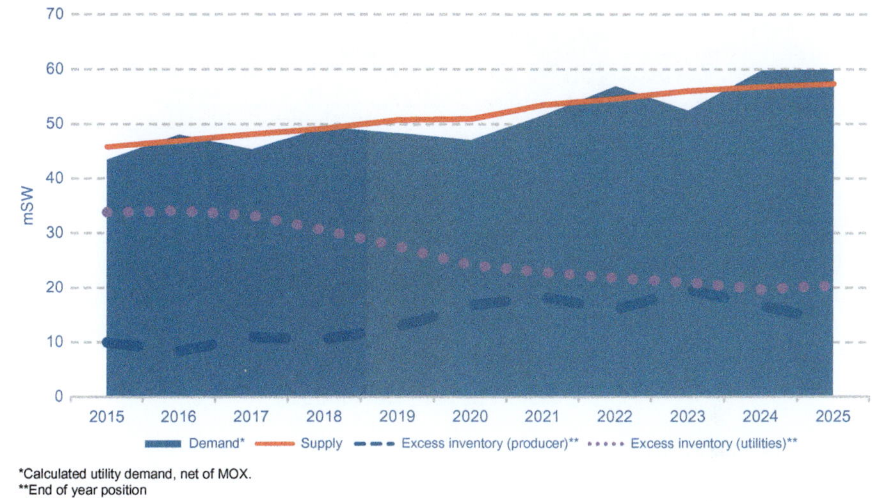

Figure 12.17 A view of the demand/supply balance 2015–25.
Source: URENCO.

12.2.6.4 Overall demand/supply balance forecast 2015–25

Taking into account primary demand, primary enrichment services supply, and the key modifiers discussed previously; tails re-enrichment, inventory use, and civil utilization of ex-military materials, one view of the demand/supply balance for enrichment services over the period 2015–25 is shown in Fig. 12.17.

Tails re-enrichment essentially brings enrichment services supply in balance with demand.

12.3 Historic technologies

The focus of this chapter is to consider uranium enrichment now and in the future, but for completeness, this section briefly describes historic uranium isotope separation technologies.

12.3.1 Gas diffusion

The principle of gas diffusion to separate ^{235}U and ^{238}U is based on forcing uranium (again in the form of uranium hexafluoride) through a semipermeable barrier. The lighter $^{235}UF_6$ will tend to travel slightly more easily (faster) through the barrier than the heavier $^{238}UF_6$. As with gas centrifuges, the degree of separation by passage through a single barrier is small, hence the gas diffusion process needs to be repeated through many stages to achieve the enrichments required for nuclear power plants.

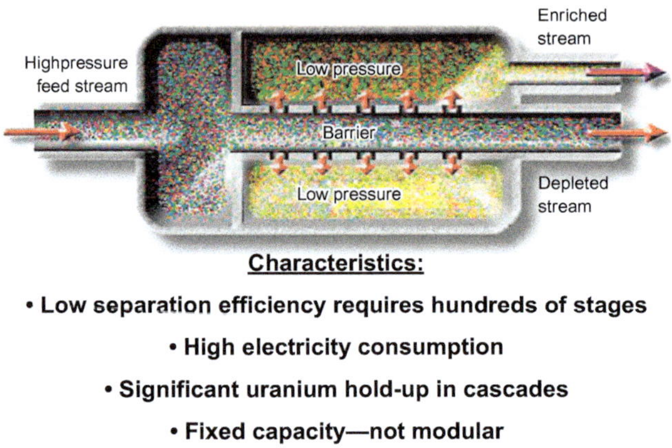

Figure 12.18 Gas diffusion.
Source: Portsmouth Virtual Museum.

Fig. 12.18 represents the principle of the gaseous diffusion process.

Gas diffusion was the first uranium isotope separation technology deployed on an industrial scale, reaching back to the United States in the 1940s. In addition, the United Kingdom, France, Russia, and China have all utilized gas diffusion for both military and civil enrichment. (Argentina has also used gas diffusion on a small scale, see Section 12.2.5.1.)

Gas diffusion plants use large compressors to drive uranium hexafluoride through the diffusion barriers at pressure. They consume significantly more electricity than even the early generation equivalent capacity gas centrifuge plants. Gas diffusion plants are monolithic rather than modular in construction and have high in-process inventories (many tonnes) of material again in sharp contrast to gas centrifuge plants, which operating under high vacuum and have very low (kilogram) quantities of material held up in-plant. By 2013, all commercial gas diffusion plants in the world had shutdown, being wholly superseded by the superior economic and flexible gas centrifuge technology on a global basis. See also http://world-nuclear.org/info/Nuclear-Fuel-Cycle/Conversion-Enrichment-and-Fabrication/Uranium-Enrichment/.

12.3.2 Electromagnetic process

A very early technology was electromagnetic isotope separation (EMIS), using devices called calutrons (in effect large-scale mass spectrometers).

In calutrons, ions of ^{235}U and ^{238}U are separated because they follow different curved paths when they move through a magnetic field. The process is very energy intensive, about 10 times that of diffusion.

EMIS was developed in the 1940s as part of the US Manhattan Project, but was abandoned in favor of gas diffusion.

Calutrons were found in Iraq during the Gulf War in 1992, indicating the existence of a military program. See also http://world-nuclear.org/info/Nuclear-Fuel-Cycle/Conversion-Enrichment-and-Fabrication/Uranium-Enrichment/.

12.3.3 Aerodynamic processes

Two aerodynamic processes were brought to demonstration stage around the 1970s. One is the jet nozzle process, with a demonstration plant built in Brazil, and the other is the Helikon vortex tube process developed in South Africa. Neither is in use now.

The processes depend on a high-speed gas stream bearing UF_6 being made to turn through a very small radius, causing a pressure gradient similar to that in a centrifuge. The light fraction can be extracted towards the center and the heavy fraction towards the outside. Thousands of stages are required to produce enriched product for a reactor. Both processes are energy intensive. See also http://world-nuclear.org/info/Nuclear-Fuel-Cycle/Conversion-Enrichment-and-Fabrication/Uranium-Enrichment/.

12.3.4 Chemical process

One chemical process has been demonstrated to pilot stage but not used. The French Chemex process exploited a very slight difference between the ^{235}U and ^{238}U isotopes valence change behavior in oxidation/reduction using aqueous (III valence), and organic (IV valence) phases. See also http://world-nuclear.org/info/Nuclear-Fuel-Cycle/Conversion-Enrichment-and-Fabrication/Uranium-Enrichment/.

12.4 Future technologies

12.4.1 The American centrifuge (USEC and Centrus)

The American centrifuge is a large gas centrifuge, which has been under development/demonstration in the United States over several decades.

Historically, uranium enrichment in the United States was carried out by the Department of Energy (DOE). The DOE operated gas diffusion plants at Oak Ridge and Paducah in Tennessee, and Portsmouth (also called Piketon) in Ohio. Uranium enrichment operations ceased at Oak Ridge in 1987. The remaining gas diffusion plants were taken over by USEC in 1993 and USEC was privatized through an initial public offering in 1998. The Portsmouth gas diffusion plant stopped enriching uranium in 2001 and the Paducah gas diffusion plant ceased uranium enrichment in 2013.

In Dec. 2013, USEC announced that it had reached agreement with a majority of its debt holders to file a prearranged and voluntary Chapter 11 bankruptcy restructuring in the first quarter of 2014. In Sep. 2014, it was announced that the

Figure 12.19 Picture of American Centrifuge Project (ACP) cascade.
Source: NEI.

company had emerged from bankruptcy proceedings with a new name, Centrus Energy Corporation.

Centrus is now a contractor to US DOE for the development and deployment of the American centrifuge, when market conditions are appropriate.

The American centrifuge was developed by US DOE in the 1980s. The DOE invested $3 billion over 10 years to develop the technology, built approximately 1500 machines and accumulated more than 10 million machine hours of run time.

US DOE and USEC then sought to develop the alternative laser enrichment Atomic Vapour Laser Isotope Separation (AVLIS) technology (see Section 12.4.2.1), with some $2 billion spent over a 26 year period, before AVLIS was abandoned in 1999 in favor of a return to the American Centrifuge Program.

The current production machine, known as the AC100, has been installed at a test/pilot facility in Piketon. Demonstration cascades with a hundred or so machines have operated since 2007, (Fig. 12.19).

It is possible that an industrial-scale American centrifuge plant will be built in the future, hence its inclusion as a future technology in this chapter. See also http://www.centrusenergy.com/american-centrifuge.

12.4.2 Laser enrichment

Laser enrichment processes fall into two categories: atomic or molecular.

12.4.2.1 Atomic Vapour Laser Isotope Separation (AVLIS or SILVA in France)

The feedstock for the AVLIS process is uranium metal. The metal is heated to form a vapor. Laser light at a very specific frequency is then directed at the vapor such that ^{235}U atoms only are ionized by the ejection of an electron. Positively charged $^{235}U^{+}$ ions are attracted to a negatively charged collector plate. The ^{238}U atoms in the vapor are not ionized by the impact of the laser light and pass unaffected through the process (Figs. 12.20 and 12.21).

Like all laser isotope separation processes, AVLIS offers the possibility of lower-energy inputs, lower capital costs and high separation factors; but after

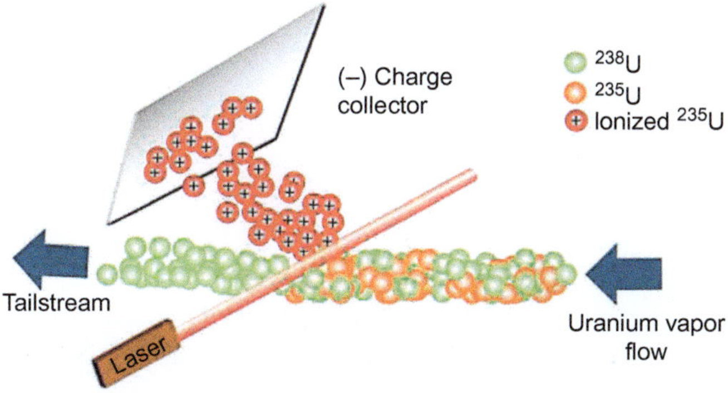

Figure 12.20 AVLIS process.
Source: NRC.

Figure 12.21 Picture of AVLIS.
Source: Wikepedia.

research, not least by URENCO, France, and the United States, these advantages have not proved commercially realizable to date.

Problems thrown up by research include the complexity and reliability of the required laser systems and the robustness of collector systems. Furthermore, because it uses metal feedstock, modification to the global fuel cycle, which is oriented around UF_6 as the feedstock for enrichment processes, would be required. See also http://world-nuclear.org/info/Nuclear-Fuel-Cycle/Conversion-Enrichment-and-Fabrication/Uranium-Enrichment/.

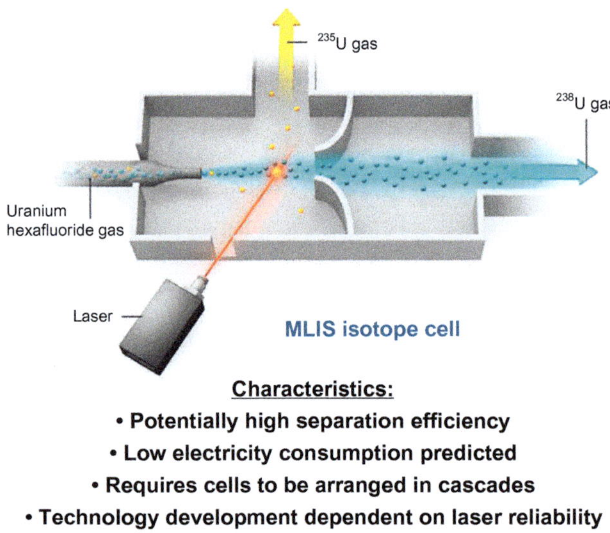

Figure 12.22 Molecular Laser Isotope Separation (MLIS) process.
Source: spectrum.ieee.org.

12.4.2.2 Molecular Laser Isotope Separation (MLIS)

Most molecular processes that have been researched work on the principle of photodissociation of UF_6 to UF_5^+ using tuned laser radiation to break the molecular bond holding one of the six fluorine atoms to a ^{235}U atom. This enables the ionized UF_5^+ to be separated from the unaffected UF_6 molecules containing ^{238}U atoms. Any process using UF_6 fits more readily into the conventional fuel cycle than the atomic process (Figs. 12.22 and 12.23). See also http://world-nuclear.org/info/Nuclear-Fuel-Cycle/Conversion-Enrichment-and-Fabrication/Uranium-Enrichment/.

12.4.2.2.1 SILEX/GLE

Separation of Isotopes by Laser EXcitation (SILEX) is a form of isotope separation developed in Australia, with SILEX Systems Ltd founded there in 1988.

In 1996, SILEX Systems Ltd. licensed its technology exclusively to the USEC for uranium enrichment. Following abandonment of SILEX by USEC in 2003, in favor of concentrating on development of the American centrifuge (see

Figure 12.23 Picture of MLIS.
Source: ic.pics.livejournal.com.

Section 12.4.1), SILEX Systems, in 2007, signed an exclusive commercialization and licensing agreement with General Electric Corporation (GE), in the United States. In 2008, GE spun off global laser enrichment (GLE), to commercialize the SILEX technology. A test loop program was undertaken at GE's facility in Wilmington, North Carolina and plans were announced for the first commercial uranium enrichment facility using SILEX.

Cameco of Canada bought into the project in 2008. In 2012, the US NRC issued a Combined Construction and Operating license for a plant up to 6000 tSWU pa capacity at Wilmington. However, due to the prevailing market conditions, GE decided to reduce funding to GLE and as a result, Cameco reported a full write-down of its investment.

It is possible that an industrial-scale SILEX plant will be built in the future when market conditions are appropriate, hence the inclusion of GLE/SILEX as a future technology in this chapter.

In 2013, GLE submitted a proposal to the US DOE to establish a laser enrichment plant at Paducah to re-enrich DOE-owned tails. The proposed plant capacity was 500–1000 tSWU pa. Negotiations with DOE continued through 2014/15. See also http://world-nuclear.org/info/Nuclear-Fuel-Cycle/Conversion-Enrichment-and-Fabrication/Uranium-Enrichment/.

12.4.2.2.2 CRISLA

Condensation Repression Isotope Separation by Laser Activation (CRISLA) is a variant of MLIS that employs low temperatures to cause UF_6 molecules to condense to form dimers, trimers, etc., and tuned lasers to selectively disrupt such condensation of ^{235}U containing species, thus achieving isotope separation.

For more detail, see Eerkens and Miller (2004).

12.5 Quality control of uranium hexafluoride in enrichment

12.5.1 Feed material

Natural uranium hexafluoride feedstock for enrichment is produced at a number of conversion facilities around the world. Currently the chief facilities in operation are Cameco's Port Hope facility in Canada, Converdyn's Metropolis facility in Illinois, United States, AREVA's Comurhex facilities in Malvesi and Tricastin in France, and conversion facilities in Russia, being consolidated at Seversk.

All natural uranium hexafluoride for enrichment should conform to the 2011 issued standard:

ASTM C787-11 Standard Specification for Uranium Hexafluoride for Enrichment.

Natural uranium hexafluoride feedstock is largely an undifferentiated commodity regardless of the plant of origin and enrichment facilities typically accept quality control documentation, including analysis results, from the supplying converter without resorting to additional sampling and analysis at the enrichment facility.

12.5.2 Enriched product

Enriched product produced at enrichment facilities, forms the feed material to nuclear fuel fabrication plants across the world. The quality standard, which must be met by enriched product, is the 2010 issued standard:

ASTM C996-10 Standard Specification for Uranium Hexafluoride for Enrichment to less than 5% ^{235}U.

The key measurement for enriched product is of course ^{235}U assay. Depending on feedstock, it is sometimes important to particularly check the concentrations of other isotopes such as ^{232}U, ^{234}U, and ^{236}U, to ensure the standard is met. ^{234}U and ^{236}U are neutron absorbers that depress the reactivity of nuclear fuel while the ^{232}U decay chain includes hard gamma emitter ^{208}Tl, which creates additional handling hazards (see Section 12.7).

A variety of sampling and analytical techniques may be deployed at enrichment facilities, including online mass spectrometry, product off-take stream sampling, and analysis, to postproduction homogenization and sampling of full product cylinders.

Figs. 12.24 and 12.25 show a full product cylinder sampling rig at URENCO.

The full cylinder sampling process is as follows:

The cylinder to be sampled is loaded into an autoclave with four sample bottles connected. The autoclave is closed and the cylinder is heated to liquefy and

Figure 12.24 30B cylinder inside sampling rig.
Source: URENCO.

Figure 12.25 Sampling rig in tipped position.
Source: URENCO.

homogenize its contents. One end of the autoclave is raised to allow liquid to flow out of the cylinder valve into the sample bottles. The cylinder is returned to the horizontal and allowed to cool so that the contents resolidify. The autoclave is then opened and the sample bottles removed, with one for the enrichment company, one for the customer, one for the umpire, and one spare. The total sampling cycle time is about 60 h.

12.5.3 Tails

A key point about quality control of tails is its potential to be re-fed for re-enrichment some time in the future. In this case, it becomes feed material and needs to conform to ASTM C787-11, as does natural feedstock.

If tails results from original feed conforming to ASTM C787-11, it is almost certain that the tails will also meet the specification. Online mass spectrometry and tails off-take stream sampling and analysis are periodically used to check the quality of tails.

12.6 Management of tails

The management of tails arising from primary enrichment is an important issue for enrichment service providers.

Three strategies have been deployed by primary enrichers to manage tails:

1. re-enrichment,
2. deconversion,
3. storage pending re-enrichment, deconversion, or some other use.

12.6.1 Re-enrichment of tails

As discussed in Section 12.2.6.1, ^{235}U assays in tails from primary enrichment are quite typically in the range 0.2–0.3% and are suitable for refeed to further strip out useful ^{235}U.

Historically, a significant proportion of enrichment capacity in Russia has been used for the re-enrichment of tails. Russia has not only re-enriched some of its own primary tails but it has also re-enriched primary tails from both URENCO and AREVA enrichment operations. The low ^{235}U assay secondary tails arising from such re-enrichment operations in Russia have been used to dilute HEU derived from decommissioning of Russian nuclear weapons as part of the "Megatons to Megawatts Programme" (see Section 12.2.6.3.1).

Both URENCO and AREVA established conventional enrichment contracts with Russia in which tails from primary enrichment operations at URENCO and AREVA plants in Europe were transported to Russia to form feedstock for Russian plants. In return, URENCO and AREVA received back equivalent natural material or enriched product. Russia took ownership of the secondary tails.

URENCO's contract with Russia to re-enrich tails ran from 1996 to 2010. In total 100,000 tU of URENCO tails from its European plants were processed over this period.

AREVA's contract ran from 1999 to 2010 and in total 60,000 tU of AREVA tails were re-enriched.

Elsewhere, other campaigns of tails re-enrichment have been conducted, the most recent example being the refeed of approximately 9000 t of "high assay" US DOE tails through the Paducah gas diffusion plant immediately prior to its closure in 2013. See also http://www.world-nuclear-news.org/C-Tails_deal_gives_Paducah_another_year-1605127.html.

12.6.2 Tails deconversion

Another option to manage tails is to de-covert tails uranium hexafluoride back to uranium dioxide. This releases a valuable resource, hydrofluoric acid for reuse in chemical processes.

Uranium oxide is the most stable form in which to store tails for the long term and is also suitable for disposition (eg, back into an original uranium mine), should that be a future decision.

Deconversion of tails has been carried out in France since 1984. AREVA have operated four deconversion lines (the "W Plants"), at Tricastin with a capacity of 14,000 tU pa. Over 300,000 tU of tails have been deconverted in AREVA's facility since the start of operations. Most of the tails processed to date has been arisings from AREVA's now closed George Besse I gas diffusion plant. AREVA has also deconverted some URENCO tails. In contracts over the period 2003–14, about 46,000 tU of URENCO tails have been treated.

Deconverted tails in the form of U_3O_8 is stored in 10 t capacity cubic steel containers known as "DV70s." Deconverted French tails are stores at Bessines (Fig. 12.26) and Tricastin. Deconverted URENCO tails are stored by the Dutch national radioactive materials storage organization, COVRA, at Vlissingen in the

Figure 12.26 Picture of DV70 tails U3O8 containers in store at Bessines.
Source: AREVA.

Netherlands, and will be stored by URENCO at Capenhurst, United Kingdom and Gronau, Germany.

A tails deconversion facility based on AREVA technology has been built at Zelenogorsk in Russia for the treatment of Russian tails. The plant commenced operations in 2009 and has a capacity of 7000 tU pa.

URENCO, through its subsidiary URENCO ChemPlants, is also building a tails deconversion facility based on AREVA technology at Capenhurst in the United Kingdom. Again, with a capacity of 7000 tU pa, the plant is due to commence operation in 2017.

UF_6 is first vaporized in autoclaves and is reacted with steam to form uranyl fluoride. Uranyl fluoride is then reacted with hydrogen at 700°C to yield aqueous hydrofluoric acid for sale to the chemical industry and U_3O_8 powder that is packed into 10 t cubic DV70 containers for storage. The chemical reactions are:

$$UF_6 + 2H_2O \rightarrow UO_2F_2 + 4HF$$
$$3UO_2F_2 + 2H_2O_{(excess)} + H_2 \rightarrow U_3O_8 + 6HF_{(aq)}$$

Two tails deconversion plants have been built by Uranium Disposition Services (UDS), in the United States, one at Portsmouth with a target capacity of about 11,500 tU pa and the other at Paducah with a target capacity 15,500 tU pa. The Portsmouth facility began operations in 2010. The technology deployed is different from the AREVA technology deployed at Tricastin, in Russia and by URENCO. Instead, it is based on original Siemens technology (which is now also owned by AREVA).

Plans for plants to de-convert UF_6 to UF_4 have also been published by Russia (for a plant at Angarsk), and International Isotopes in the United States (for a plant in New Mexico). See also http://www.world-nuclear.org/info/Nuclear-Fuel-Cycle/Conversion-Enrichment-and-Fabrication/Conversion-and-Deconversion.

12.6.3 Tails storage

Today some quantities of depleted (tails), uranium are used as a carrier for plutonium oxide in MOX fuel and requirements to use tails would go up with any future increase in the use of MOX fuel. Another envisaged future use of some tails is as "blanket" material in fast neutron breeder reactors potentially to be commercially deployed from the mid-21st century. These are additional reasons for enrichers to hold stocks of tails as strategic assets.

Aside from re-enrichment or deconversion, tails depleted uranium hexafluoride may be safely stored for a long period of time in international standard cylinders such as the 48Y with periodic inspection and nondestructive testing (such as

Figure 12.27 Tails UF_6 storage pad.
Source: informaciondesenudo.com.

cylinder wall thickness checks), to confirm storage integrity. A picture of a typical tails hex storage pad is shown in Fig. 12.27.

12.7 Experience enriching reprocessed uranium

As well as freshly mined natural uranium, uranium arising from reprocessing can form feedstock for uranium enrichment operations.

Reprocessed uranium typically contains higher concentrations of the ^{232}U, ^{234}U, and ^{236}U isotopes and other species arising in irradiated material such as technetium (^{99}Tc), a fission product.

Enhanced ^{232}U concentrations are undesirable because its decay chain via ^{228}Th gives rise to species including ^{208}Tl which is a hard gamma emitter creating additional handling hazards. Enhanced ^{234}U and ^{236}U concentrations are also undesirable because the species are neutron absorbers and therefore decrease nuclear fuel efficiency, if present.

Clearly in today's commercial gas centrifuge plants (and in diffusion plants in the past), the process designed to increase the concentration of ^{235}U will also increase the concentration of any ^{232}U, ^{234}U, and ^{236}U present in feedstock relative to ^{238}U.

As described in Section 12.4.2, laser enrichment technologies may be suited to selective enrichment of ^{235}U without coenriching ^{232}U, ^{234}U, or ^{236}U.

Reprocessed uranium has been fed to gas diffusion plants in the past. Due to the large material hold-up, this results in contamination of the plant with reprocessed material, which takes a long time to "flush through." By contrast, with a very small material hold-up in process, gas centrifuge plants are particularly suited to campaigns to enrich reprocessed feedstock and with certain precautions and processes, can switch back to enrichment of un-irradiated natural material with little "memory" of reprocessed material contaminating the cascades.

Some examples of enrichment of reprocessed uranium are as follows:

- Until it shut down in 1982, some 16,000 tU of "magnox-depleted" reprocessed uranium from the UK's first generation magnox reactor fleet was fed to the UK diffusion plant at Capenhurst to increase the ^{235}U assay from about 0.4% to "pseudonatural" around 0.7% assay. The pseudonatural material was then fed through URENCO centrifuge plants at Capenhurst to form about 1650 tU of low-enriched material, which was fabricated into fuel for the UK's second generation advanced gas-cooled reactors fleet.
- Around 1000 tU of reprocessed uranium from La Hague and reconverted to uranium hexafluoride at Tricastin in France has been fed through cascades at URENCO's Almelo plant in the Netherlands to make reprocessed uranium fuel for Kansai in Japan, Synatom in Belgium, and EDF in France. Minor adaptations to the enrichment plant, for example, some remote handling and additional shielding were made to reduce the hazard from enhanced radiation from the feedstock.
- A number of utilities have sent reprocessed uranium arising from their reprocessing campaigns at Sellafield in the United Kingdom and La Hague in France to Russia for processing and fabrication into fuel. In Russia, blending techniques have been deployed to control the concentrations of ^{232}U, ^{234}U, and ^{236}U in finished fuel. The Seversk facility in Russia has been used to enrich some reprocessed uranium feedstock.

Currently, the Russian blending route is the only route available for recycle of reprocessed uranium into fuel. There are no conversion facilities in operation licensed to convert reprocessed uranium to hexafluoride ahead of the enrichment step. A dedicated French reprocessed uranium conversion facility was shut down in 1998. British Nuclear Fuels (BNFL), also in 1998 abandoned a project having partly constructed a dedicated "Line 3" reprocessed uranium conversion facility at Springfields in the United Kingdom.

Fluorides of ^{232}U decay chain elements, including the hard gamma emitting ^{208}Tl, are nonvolatile and may be separated out by feeding reprocessed uranium hexafluoride to a conversion process or by any gaseous transfer of reprocessed uranium hexafluoride from one cylinder to another, effectively "resetting the clock" for the in-growth of ^{232}U decay chain species. Fig. 12.28 shows how new radiation in-growth increases over time, reaching equilibrium after 10 years.

Accordingly, to practically fabricate nuclear fuel from reprocessed uranium, the conversion, enrichment, and fabrication steps need to be executed sequentially with minimum delays, typically within 24 months from conversion to enrichment, and within 6 months from enrichment to fabrication, to minimize radiation doses from handling the material.

Much detail on experience handling reprocessed uranium can be found in IAEA-TecDOC-CD-1630 (2007) available on the Internet.

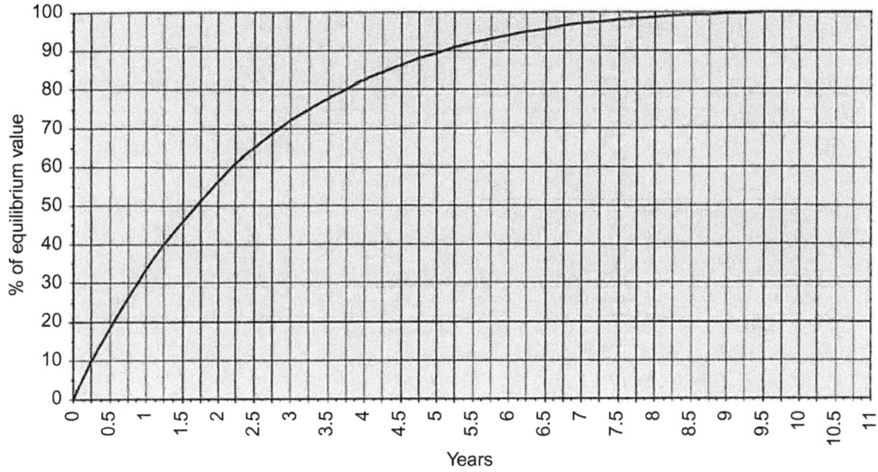

Figure 12.28 Radiation in-growth in newly converted reprocessed uranium.
Source: IAEA-TecDOC-CD-1630 (2007).

References

Bukharin, O., January 2004. Russian gaseous centrifuge technology and uranium enrichment complex. <www.partnershipforglobalsecurity-archive.org/Documents/bukharinrussianenrichmentcomplexjan2004.pdf>.

Cohen, K., 1951. The Theory of Isotope Separation as Applied to Large-scale Production of U235. McGraw-Hill, New York.

Eerkens, J.W., Miller, W.H., 2004. Laser isotope separation employing condensation repression. <www.osti.gov/scitech/servlets/purl/912774>.

IAEA-TecDOC-CD-1630, 2007. Use of Reprocessed Uranium. Proceedings of a Technical Meeting held in Vienna, 29–31 August 2007.

URENCO Annual Report and Accounts, 2014.

Nuclear fuel fabrication

13

Eileen Supko
Energy Resources International, Washington, DC, United States

13.1 Introduction

Fabrication of nuclear fuel assemblies for nuclear power plants (NPPs) is the step in the front end of the nuclear fuel cycle when nuclear fuel is transformed from a fungible commodity (eg, natural or enriched uranium, or other fissile material) into a highly engineered product, with designs that are tailored to a specific NPP's needs. That is, fuel assembly designs are specific to a given reactor type (generally based on the nuclear steam supply system (NSSS)) and a specific NPP's fuel management strategy (which considers cycle length, capacity factor, existing fuel in the reactor core, etc.) The nuclear fuel fabrication market for light water reactor (LWR) fuel assemblies is the largest—approximately 82% of the operating NPPs in the world are LWRs. Among LWR designs, there are two general categories of plants: (1) pressurized water reactors (PWR), including Russian VVER designs, and (2) boiling water reactors (BWR), which made up 63.4% and 18.4% of operating units at the end of 2014, respectively, as shown Fig. 13.1. Non-LWR NPP designs include gas-cooled reactors (GCR), pressurized heavy water reactors (PHWR), Russian graphite-moderated reactors (RBMK), and liquid-metal-cooled fast reactors (LMR). PHWRs are the second largest segment of operating NPPs after LWRs, with 11% of all operating reactors worldwide, as shown in Fig. 13.1 (ANS, 2015).

Fig. 13.1 provides a comparison of the percentage of operating NPPs in each category of reactor as well as the percentage of operating MWe for each reactor category. LWRs, both PWRs and BWRs, dominate both the number of operating NPPs and the amount of electricity produced by NPPs worldwide. This trend is expected to continue to through 2035 as LWRs grow from 81.8% of total nuclear generation to 88.6% (ERI, 2015).

13.1.1 Fuel fabrication service suppliers

As the LWR market is by far the largest segment of the nuclear fuel fabrication market, it is the most competitive. At the present time, there are three principal international nuclear fuel fabricators that supply a broad range of LWR fuel assemblies of various designs to owners and operators of NPPs in Asia, Europe, and North America: AREVA NP (AREVA), Global Nuclear Fuel (GNF), and Westinghouse Electric Company (Westinghouse). All three companies operate production facilities in the US and Europe, and GNF and Westinghouse also operate fuel production facilities in Japan. In addition to the principal fuel fabricators, there are a number of other

Uranium for Nuclear Power. DOI: http://dx.doi.org/10.1016/B978-0-08-100307-7.00013-2

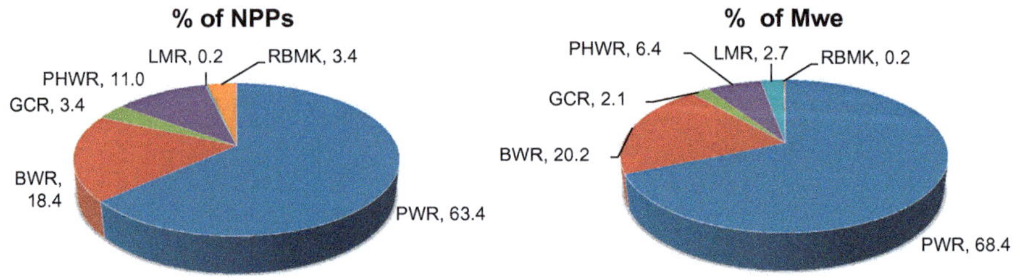

Figure 13.1 Types of NPPs, % of operating units and % of MWe.

companies that supply both LWR and non-LWR fuel to national and regional markets. The fuel fabricators are described next in alphabetical order.

AREVA NP is a wholly owned subsidiary of AREVA. AREVA is headquartered in Paris, France and has offices and manufacturing facilities around the world. AREVA's North American operations are administered by AREVA, Inc., which is currently headquartered in Charlotte, North Carolina. AREVA NP is a fully integrated fuel fabricator. In addition to fuel assemblies, it produces zirconium alloys, cladding, and other components through its subsidiary facilities in France and Germany.

AREVA operates two fuel fabrication facilities in Europe as well as several plants in France and Germany that produce zirconium alloys and fuel assembly components. AREVA's operating European plants for LWR fuel fabrication include the Romans, France plant that produces UO_2 powder and pellets as well as fabricated PWR fuel assemblies; and the Lingen, Germany fabrication facility that produces both BWR and PWR fuel assemblies. The Romans facility is rated at 1400 MTHM, or tonnes, per year with an annual dry UF_6 to UO_2 conversion capacity of 1800 MTHM and pelletization capability of 1400 MTHM/year. The capacity at Lingen is nominally 800 MTHM/year for conversion of UF_6 to UO_2, and 650 MTHM/year for pelletization and fuel assembly manufacturing. AREVA also operates a mixed oxide (MOX) fuel fabrication plant at Marcoule, in France, with an annual MOX fuel fabrication capacity of 195 MTHM.

AREVA's US fabrication operations are centered in Richland, Washington. The Richland plant manufactures both PWR and BWR fuel assemblies for US customers and BWR fuel for plants in Taiwan. Present capability at Richland for conversion of UF_6 to UO_2 powder is 1800 MTHM/year. Pellet manufacturing capacity is 1800 MTHM, and fuel assembly capacity is 750 MTHM/year. AREVA supplies UO_2 powder to Nuclear Fuel Industries of Japan from its Richland facility under a long-term supply contract. The Richland plant has also provided UO_2 powder in support of some of AREVA's European fabrication activities.

For almost two decades, AREVA NP and its predecessor companies have subcontracted with the Russian fabricator TVEL for the manufacture of fuel assemblies for NPPs in Germany, Sweden, and Switzerland. The fuel assemblies manufactured by TVEL contain reprocessed uranium (RepU) and are of standard AREVA designs using components supplied by AREVA. The RepU is blended with Russian HEU to the desired assays, processed, and fabricated by TVEL (ERI, 2015; WNA, 2013).

Cameco Corporation produces PHWR fuel for CANDU NPPs. It operates two manufacturing facilities: a fabrication facility in Cobourg, Ontario Province, for production of fuel bundle components and a fuel manufacturing plant in Port Hope, Ontario Province. One of the processes at the Port Hope conversion facility is to convert uranium trioxide (UO_3) to UO_2 powder, which the fuel manufacturing facility processes into UO_2 pellets for CANDU fuel assemblies. Capacity of the Port Hope fuel manufacturing facility is 1200 MTHM/year (Cameco, 2015a).

China National Nuclear Corporation (CNNC) has two nuclear fuel fabricators. The first is China Jianzhong Nuclear Fuel (CJNF), which through its subsidiary, China South Nuclear Fuel Co., Ltd (CSNFC), operates the fuel manufacturing facility at Yibin in Sichuan province. The second fabricator is China North Nuclear Fuel Co., Ltd (China Nuclear Northern, CNNFC), located in Baotou, Inner Mongolia. The CNNFC fuel fabrication facility in Baotou, also houses CNNC-Baotou Nuclear Fuel Company (CBNFC), which has been established to manufacture fuel for Westinghouse AP1000s built in China. CNNC has technology transfer agreements in place with AREVA, TVEL, and Westinghouse for PWR fuel supply (WNA, 2015a).

Precise fabrication capacity for various fuel designs being used in China is difficult to obtain. Data provided herein are estimates based on information from a variety of sources. In 2014, CSNFC announced that it expanded annual production capacity from 400 t to 800 MTHM. The facility is expected to continue to expand as fuel fabrication requirements grow. CNNFC fabricates fuel for two Canadian PHWRs in China, with an annual capacity of 200 MTHM/year and for Chinese-designed PWRs (including the Chasma PWRs in Pakistan), with estimated annual PWR fuel manufacturing capacity of 200 MTHM. It is also preparing to build a line to make pebble-bed fuel for high-temperature GCR. Baotou Nuclear Fuel Company has an initial capacity of 200 MTHM for production of AP1000 fuel assemblies (WNA, 2015a).

AREVA and CNNC have a joint venture in place to fabricate zirconium alloy in China and produce zirconium tubes for Chinese nuclear fuel production. The joint venture, CNNC AREVA Shanghai Tubing Company (CAST) entered production in 2012 (WNA, 2015a).

Enusa Industrias Avanzadas, SA (ENUSA) was created in 1972 with responsibility for the development of all commercial and industrial activities related to Spain's nuclear fuel cycle. The State Industrial Holding Company, SEPI, owns 60% of ENUSA and the Centro de Investigaciones Energeticas, Medioambientales y Technologiecas, CIEMAT, (Research Center for Energy, Environment, and Technology) owns the other 40%. ENUSA's headquarters are in Madrid, where it conducts fuel design and marketing activities. ENUSA's Fuel Division is responsible for fuel engineering and manufacturing and has been operating a PWR and BWR fuel fabrication facility in Juzbado since 1985. The facility has capacity of approximately 390 MTHM, with annual production of 300–350 MTHM/year for pellet and fuel assembly manufacturing. ENUSA has licensing agreements with Westinghouse (as the European Fuel Group (EFG)) and GNF (as GNF ENUSA

Nuclear Fuel SA (GENUSA)) for PWR and BWR fuel fabrication, respectively. With respect to PWR fuel, Westinghouse supplies fuel assembly components and tubing, and UO_2 powder. For BWRs, GNF supplies components and ENUSA has a contract with Westinghouse for the supply of UO_2 powder (ERI, 2015; ENUSA, 2014, 2015).

GNF was established in January 2000 as an international nuclear fuel joint venture limited liability company by General Electric Company (GE), Hitachi, Ltd and the Toshiba Corporation. Organizationally, 60% of GNF is held by GE, with the remaining 40% being shared equally between Hitachi and Toshiba. In establishing GNF, the three companies integrated their nuclear fuel marketing, design, development, and manufacturing functions. The US base for GNF is Global Nuclear Fuel-Americas, LLC (GNF-A), which is comprised of GE's former BWR fuel business, including marketing, design, development, and sales, together with the Wilmington, North Carolina manufacturing facility and the Joint Conversion Company (JCC). GNF Japan (GNF-J), formerly Japan Nuclear Fuel Company, Ltd (JNF), now includes all BWR marketing, design and development operations that GE, Hitachi, and Toshiba had previously been engaged in separately, together with the fuel manufacturing function of JNF, to serve the Japanese market (ERI, 2015).

The JCC was originally created to operate a new 1200 MTHM/year UF_6 to UO_2 powder production facility at GE's Wilmington, North Carolina site. Pellet and fuel assembly production at the Wilmington facility are 1000 MTHM annually. The JCC facility utilizes process equipment and dry conversion process technology licensed from AREVA. GNF-Canada, the maker of CANDU fuel, provides natural (eg, unenriched) UO_2 pellets used in the natural uranium ends (referred to as "blankets") in GNF BWR fuel assemblies. GNF-A also manufactures cladding and other fuel assembly components, fuel channels, control rod drives and blades in Wilmington. Because it does not have zirconium alloy production facilities of its own, GNF has purchased its zirconium alloy materials from Westinghouse, Wah Chang, and AREVA (ERI, 2015).

GNF-J supplies BWR fuel exclusively to the Japanese market. It operates a 750 MTHM/year fuel fabrication facility in the town of Kurihama of Yokosuka-shi in Kanagawa-ken, Japan. GNF-A supplies UO_2 powder and fuel assembly components from the Wilmington facility and GNF-J produces the fuel pellets and fabricates the fuel assemblies (ERI, 2015; NRA, 2013).

GNF-Canada operates a fabrication facility for PHWR fuel in Peterborough, Ontario. The facility has an annual capacity for pellet and fuel assembly production of 1500 MTHM (GNF, 2015b).

India's Department of Atomic Energy (DAE) Nuclear Fuel Complex (NFC) was established in 1971 and is responsible for the supply of nuclear fuel assemblies and reactor core components for all of the NPPs operating in India. The facility in Hyderabad fabricates both natural and enriched uranium fuel, zirconium alloy cladding and reactor core components. The facility has a 48 MTHM capacity for manufacture of LWR fuel, and a 435 MTHM capacity for manufacture of PHWR fuel. NFC will also manufacture FBR fuel for planned NPPs. NFC operates a 50 MTHM fabrication facility for MOX fuel in Tarapur (WNA, 2015b; NFC, 2015).

Industrias Nucleares do Brasil (INB) operates a PWR fuel fabrication plant in Resende, Brazil. The fuel fabrication facility, which as constructed with technology and support from Siemens-KWU, has operated since 1982. INB has technology transfer agreements with Westinghouse to produce fuel assemblies of Westinghouse design for Angra Unit 1 and with AREVA to manufacture fuel for use in Angra Unit 2, which is a Siemens-KWU PWR. According to INB, its fuel fabrication facility has 160 MTHM/year of powder production capability, 120 MTHM/year of pelletization capability, and 240 MTHM/year of assembly capability (INB, 2015).

Japan Nuclear Fuel Limited (JNFL) operates nuclear fuel cycle facilities at Rokkasho-Mura and is in the process of constructing a MOX fuel fabrication facility at the site. The facility will have a maximum MOX fuel fabrication capacity of 130 MTHM annually and is planned to begin operation in 2016 (NRC, 2013).

Kazatomprom operates the Ulba Metallurgical Plant (UMP) at Ust-Kamenogorsk, which provided UO_2 pellet production for both VVER and RBMK units for TVEL. UMP has a reported annual conversion capacity of 1200 MTHM and pelletization capacity of 1200 MTHM, but no assembly capability. Kazatomprom provides pellets for fuel manufacture in China, Japan, and India.

KEPCO Nuclear Fuel Company, Ltd (KNFC), a subsidiary of Korea Electric Power Corporation (KEPCO), began operating a fuel fabrication facility at Taejon, Republic of Korea in October 1988. The facility was built under a technology exchange agreement with Siemens-KWU. KNFC is responsible for all PHWR and PWR fuel fabrication activities and is presently supplying 100% of the Republic of Korea's PWR fuel fabrication requirements to reactors of Combustion Engineering, Framatome, Westinghouse and KEPCO design (Song et al., 2009).

KNFC has a capacity of 700 MTHM/year to convert UF_6 to UO_2 powder using a dry conversion process and for pelletization. Fuel assembly capacity is 550 MTHM annually. KNFC also provides nuclear fuel research and development activities, CANDU fuel fabrication, and all fuel and core design for both CANDU units and PWRs in the Republic of Korea. CANDU fuel fabrication capacity is 400 MTHM/year (KNFC, 2015; WNA, 2013).

KNFC's Techno Special Alloy (TSA) zircaloy tube mill began operating in November 2008. The fully operational plant has a production capacity of 1400 km of tubing per year. Current products include fuel tubes, thimble tubes, instrument tubes, as well as various sleeves and flanges. In addition, KNFC is building a new facility for the manufacture of zirconium alloy tubes. The facility will be capable of producing 600 km of nuclear fuel cladding tubes per year. Construction is expected to be complete in 2016, and the facility is expected to begin production in 2017 (ERI, 2015).

KNFC and Westinghouse formed a joint venture, KW Nuclear Components Company, Ltd (KWN), to manufacture Control Element Assemblies (CEA). Westinghouse holds 55% and KEPCO Nuclear Fuel Company (KNF) holds 45% of KWN. These CEAs would be for the Combustion Engineering (CE)-designed NPPs, including OPR1000 and APR1400 units in Korea and the United Arab Emirates (UAE). Manufacturing would take place at the KNFC fuel fabrication plant in Daejeon. Production of CEAs began in 2011 (ERI, 2015).

Mitsubishi Nuclear Fuel Co. Ltd (MNF) is owned by Mitsubishi Heavy Industries (MHI), 35%; Mitsubishi Materials Corporation (MMC), 30%; AREVA, 30%; and Mitsubishi Corporation, 5%. MNF operates a PWR fuel fabrication facility in Tokai Mura, Japan, including a 475 MTHM/year UF_6 to UO_2 conversion plant to produce powder for pellet manufacturing. The pelletization and fuel assembly manufacturing capacity is 440 MTHM/year. MNF provides fuel development, design, manufacturing, and sales (ERI, 2015; NRA, 2013).

Nuclear Fuel Industries, Ltd (NFI) is jointly owned by Westinghouse (52%), and NFI (48%—which is owned by Furukawa Electric Company, Ltd (50%) and Sumitomo Electric Industries, Ltd (50%)). NFI operates two fuel fabrication facilities, a 383 MTHM/year facility in Kumatori, Japan for PWR fuel, and a fuel plant in Tokai Mura that produces 250 MTHM/year of BWR fuel. Neither plant produces UO_2 powder. NFI has an agreement with Kazatomprom's UMP to supply UO_2 fuel pellets to NFI for its use in manufacturing fuel assemblies (ERI, 2015; NRA, 2013).

Joint Stock Company (JSC) TVEL, a Rosatom subsidiary, provides a wide range of services dealing with nuclear fuel design, development and manufacturing. All of Russia's uranium conversion, enrichment and fuel fabrication facilities operate within the TVEL organization. Fuel fabrication facilities include those operated by JSC Mashinostroitelny Zavod (MSZ) at Elektrostal outside of Moscow and JSC Novosibirsk Chemical Concentrates Plant (NCCP) that operates a fabrication facility at Novosibirsk in Western Siberia. The MSZ plant produces UO_2 powder and pellets, including pellet production using European RepU. MSZ fabricates fuel assemblies for VVER-440, VVER-1000 and for European reactors under agreement with AREVA. MSZ also has the capability to produce fuel assemblies for the RBMK units as well as LMR fuel for BN-600 and BN-800 fast reactors. The NCCP plant produces fuel for both the VVER-440 and VVER-1000 reactors. NCCP also has conversion and pelletization capacity. Reported TVEL fuel fabrication capacity varies widely from various sources, including the capacities at MSZ and NCCP. TVEL officials have reported LWR pellet manufacturing capability of 1760 MTHM and fuel assembly capacity of 2150 MTHM. Other organizations that report fuel fabrication capacity have shown TVEL's conversion capacity to be equal to its capacity for pellet manufacturing—thus, this report assumes a conversion capacity of 1760 in total at MSZ and NCCP. TVEL has reported a RMBK fuel manufacture capacity 460 MTHM and LMR fuel production capacity of 50 MTHM. TVEL is expected to increase its production capacity as needed to meet the requirements of new NPPs (TVEL, 2010; WNA, 2013).

Historically, UO_2 pellet production for both VVER and RBMK units took place at Kazatomprom's UMP. However, UO_2 powder and pellets are now being manufactured by TVEL to reduce Russian dependence on Kazakhstan. Zirconium production and manufacture of zirconium components is controlled by JSC Chepetsk Machine Building Plant in Glazov.

TVEL has historically provided almost 100% of fuel fabrication requirements to Russian-designed VVERs in Russia, Ukraine, and Eastern Europe. Recently TVEL developed a PWR fuel design for use in Western NPPs to expand its market, the TVS-K design (TVEL, 2014).

Westinghouse Electric Company is owned by Toshiba Corporation (87%), Kazatomprom (10%), and Japan's Ishikawajima-Harima Heavy Industries Co., Ltd (3%). Toshiba acquired Westinghouse from British Nuclear Fuels Limited (BNFL) in 2006. Westinghouse is headquartered in Cranberry Township, Pennsylvania, near Pittsburgh and operates a fuel fabrication facility in Columbia, South Carolina. Westinghouse is a fully integrated supplier. In addition to the Columbia fuel fabrication facility, Westinghouse operates the Western Zirconium Plant in Ogden, Utah, that produces zirconium and zirconium alloy materials for use in fuel components by Westinghouse and other fuel vendors. Tubing is produced by the Specialty Metals Plant (SMP) in Blairsville, Pennsylvania. Through earlier acquisitions by prior Westinghouse owners, the commercial nuclear power businesses of Asea Brown Boveri, Ltd (ABB), including ABB Atom, and CE, are owned by Westinghouse.

The Westinghouse fuel fabrication plant in Columbia, South Carolina has an annual capacity of 1500 MTHM for UF_6 to UO_2 conversion, pellet production and PWR and BWR fuel assembly manufacture. Westinghouse is the only Western fuel supplier that has produced fuel assemblies for Russian-designed VVER NPPs, having produced VVER-1000 fuel assemblies for the two Soviet-designed Temelin units in the Czech Republic until 2006 and for NPPs in Ukraine.

Westinghouse Electric Sweden operates a 600 MTHM/year fuel fabrication facility in Västerås, Sweden producing both PWR and BWR fuel. Westinghouse operates Springfields Fuels Limited in Lancashire, UK. The plant was has a capacity for UO_2 conversion of 730 MTHM, 380 MTHM pellet manufacturing capacity, and 640 MTHM assembly capacity. The facility also manufactures fuel for UK's Advanced GCR (AGR) NPPs with an annual production capacity 220 MTHM (WNA, 2013).

Other small fuel fabricators include the *Pakistan Atomic Energy Commission (PAEC)*, which produces PHWR fuel for Pakistan's Kanupp NPP. The facility is estimated to have a capacity of 20 MTHM/year. Argentina's *Combustibles Nucleares Argentinos (CONUAR) SA* operates a 160 MTHM/year fuel fabrication facility for fabrication of PHWR fuel for the Atucha and Embalse NPPs. Romania operates a 240 MTHM fabrication facility for PHWR fuel in Pitesti (WNA, 2015b).

13.1.2 Fabrication services market capacities

Nuclear fuel fabrication has several market segments as described previously. The largest segment is LWR fuel fabrication, which includes Russian VVER fuel fabrication. The other nuclear fuel fabrication markets include PHWR fuel, MOX nuclear fuel (which may be supplied to LWRs or fast reactors (LMR fuel), GCR fuel, RBMK fuel, and LMR fuel (which may include MOX fuel). Fuel fabrication facilities and their estimated capacities are summarized in the sections that follow.

13.1.2.1 LWR fuel fabrication

Fuel fabrication facilities and approximate capacities of the principal, national, and regional suppliers of LWR fuel are summarized in Table 13.1. The principal supplier facilities are arranged by their geographical location to provide a complete

Table 13.1 **Summary of all LWR fuel fabrication facilities and capacities (WNA, 2013; ERI, 2015)**

Country	Fuel fabricator	Location of fuel fabrication facility	Type of fuel	Production capacity, MTHM/year		
				Conversion	Pellet	Assembly
East Asia						
Japan	Global Nuclear Fuel—Japan	Yokosuka	BWR	0	750	750
	Mitsubishi Nuclear Fuel	Tokai Mura	PWR	475	440	440
	Nuclear Fuel Industries (W)	Kumatori	PWR	0	383	383
	Nuclear Fuel Industries (W)	Tokai Mura	BWR	0	250	250
Korea	KEPCO Nuclear Fuel Company	Taejon	PWR	700	700	550
China	CNNC-China South Nuclear Fuel Co.	Yibin, Sichuan	PWR	800	800	900
	CNNC-China Nuclear Northern Fuel Co.	Baotou, Inner Mongolia	PWR	200	200	200
	CNNC-Baotou Nuclear Fuel Co.	Baotou, Inner Mongolia	PWR	200	200	200
			Asia Subtotal	**2375**	**3723**	**3673**
Western Europe						
France	AREVA	Romans	PWR	1800	1400	1400
Germany	AREVA	Lingen	BWR/PWR	800	650	650
Spain	ENUSA	Juzbado	BWR/PWR	0	390	390
Sweden	Westinghouse Atom	Vasteras (a)	BWR/PWR	600	600	600
UK	Westinghouse	Springfields	PWR	730	380	640
			Europe Subtotal	**3930**	**3420**	**3680**

United States						
US	AREVA	Richland, WA	BWR/PWR	1800	1800	750
US	Global Nuclear Fuel-Americas	Wilmington, NC	BWR	1200	1000	1000
US	Westinghouse Electric Company	Columbia, SC	PWR/BWR	1500	1500	1500
			US Subtotal	**4500**	**4300**	**3250**
Total installed fuel fabrication capacity for East Asia, W. Europe and North America				**10,805**	**11,443**	**10,603**
Other						
Brazil	Industrias Nucleares do Brasil	Resende	PWR	160	120	240
India	Nuclear Fuel Complex (DAE)	Hyderabad	BWR	48	48	48
Russia	TVEL, JSC Machine Building Plant (MSZ)	Elecktrostal	PWR	1100	1100	950
	TVEL, Novosibirsk Chem. Conc. Plant (NCCP)	Novosibirsk		660	660	1200
Kazakhstan	Ulba Metallurgical Plant	Ust-Kamenogorsk	PWR	1200	1200	0
			Other Subtotal	**3168**	**3128**	**2438**
Total Installed LWR Fuel Fabrication Capacity for World				**13,973**	**14,571**	**13,041**

picture of LWR fuel fabrication supply capability on a world and regional basis. As shown in Table 13.1, present world annual LWR fuel fabrication capacity—based on UF_6 to UO_2 conversion capacity—from all identified sources is approximately 13,970 MTHM. While total assembly capacity is lower at approximately 13,000 MTHM, assembly capacity is widely recognized to be expandable, as it is generally limited only by the number of shifts being run at the plant.

In East Asia, total UF_6 to UO_2 conversion capacity is approximately 2375 MTHM—significantly lower than the capacity to manufacture pellets and LWR fuel assemblies, which is an estimated 3700 MTHM. Japanese fuel fabricators, which do not have conversion capability, import powder or pellets from other suppliers. China is expected to increase LWR fuel fabrication capacities significantly over the next decade to meet its growing requirements.

In Western Europe, annual conversion capacity for LWR fuel is approximately 3900 MTHM/year, with pelletization and assembly capacity for LWR fuel of 3420 MTHM and 3680 MTHM, respectively.

In the US, the annual conversion capacity is about 4500 MTHM, pelletization is about 4300 MTHM, and assembly capacity is 3250 MTHM. In the past, a substantial portion of UO_2 powder has been exported from the US to Japan to make up for the imbalance between UO_2 conversion capacity and fuel assembly capacity. This is expected to resume as fuel fabrication restarts to support NPP restarts in Japan.

LWR capacity in Russia is an estimated 1760 MTHM for UO_2 conversion and pellet manufacture and 2150 MTHM for fuel assembly production. As shown in Table 13.1, this capability is split between MSZ and NCCP. Kazatomprom's UMP has an annual production capacity of 1200 MTHM for UO_2 conversion and pellet manufacture. India and Brazil have small LWR fuel fabrication capacities to meet indigenous requirements.

13.1.2.2 PHWR fuel fabrication

As shown in Table 13.2, PHWR fuel fabrication is done on a national basis to support PHWR operations in the country in which the fabrication facility is located. In Korea, KNFC operates a 400 MTHM facility operated by KNFC in Taejon. CNNFC operates a 200 MTHM facility at Baotou, China. In Canada, two PHWR facilities are operational: Cameco's 1200 MTHM facility in Port Hope, Ontario and GNF-Canada's 1500 facility in Peterborough, Ontario. In Argentina, CONUAR operates a 160 MTHM facility in Cordoba and Ezeiza. In India, NFC operates a 435 MTHM facility in Hyderabad, A small 20 MTHM facility has been constructed in Pakistan, and a SNN operates a 240 MTHM facility in Pitesti, Romania. Total PHWR fuel production capacity is 4155 MTHM.

13.1.2.3 MOX fuel fabrication

MOX fuel has been utilized in LWRs in four European countries (Belgium, France, Germany, and Switzerland). In Japan, 12 NPPs are also licensed to utilize MOX fuel and a facility for MOX fuel fabrication is undergoing licensing.

Table 13.2 Summary of PHWR fuel fabrication facilities and capacities (WNA, 2015b)

Country	Fuel fabricator	Location of fuel fabrication facility	Type of fuel	Production capacity MTHM
East Asia				
Korea	KEPCO Nuclear Fuel Company	Taejon	PHWR	400
China	CNNC-China Nuclear Northern	Baotou, Inner Mongolia	PHWR	200
			Asia Subtotal	**600**
North America				
Canada	Cameco Corporation	Port Hope, Ontario	PHWR	1200
Canada	Global Nuclear Fuel—Canada	Peterborough, Ontario	PHWR	1500
			North America Subtotal	**2700**
Other				
Argentina	CONUAR	Cordoba and Ezeiza	PHWR	160
India	Nuclear Fuel Complex (DAE)	Hyderabad	PHWR	435
Pakistan	PAEC	Chashma	PHWR	20
Romania	SNN	Pitesti	PHWR	240
			Other Subtotal	**855**
Total Installed PHWR Fuel Fabrication Capacity for World				**4155**

MOX fuel has also been developed for other reactor types such as PHWRs and LMRs. Table 13.3 provides a summary of worldwide MOX fuel fabrication capacity.

In Japan, the MOX fuel fabrication facility at Tokai Mura, operated by Japan Atomic Energy Agency, is a demonstration facility that had a capacity of 5 MTHM annually. In 2014, JAEA announced that it would decommission the facility rather than expend capital to bring the plant up to the new regulatory standards. The larger JNFL MOX fabrication facility at Rokkasho-Mura, which will have a capacity of 130 MTHM, is undergoing review by Japan's Nuclear Regulatory Authority (NRA) and is currently expected to begin operation in 2017 (WNA, 2015d; JNFL, 2015; JAEA, 2015).

Table 13.3 Summary of MOX fuel fabrication facilities and capacities

Country	Fuel fabricator	Location of fuel fabrication facility	Type of fuel	Production capacity	
				Pellet	**Assembly**
East Asia					
Japan	Japan Nuclear Fuel Limited	Rokkasho-Mura	MOX (LWR)	130	130
			Asia Subtotal	**130**	**130**
Western Europe					
France	AREVA	Marcoule	MOX (LWR)	195	195
			Europe Subtotal	**195**	**195**
Other					
India	Advanced Fuel Fabrication Facility (DAE)	Tarapur	MOX (LWR, PHWR)	50	50
Russia	TVEL, Mining & Chemical Complex (MCC)	Zheleznogorsk	MOX (LMR)	60	60
			Other Subtotal	**110**	**110**
Total Installed MOX Fuel Fabrication Capacity for World				**435**	**435**

AREVA operates the 195 MTHM MOX fuel fabrication facility, Melox, at Marcoule. Melox began initial operation in 1995 and the facility achieved an annual production rate of 100 MTHM in 1997. AREVA, in partnership with the Shaw Group, as AREVA MOX Services, is building the MOX fuel fabrication facility (MFFF) at the Savannah River Site, in South Carolina, under contract to the US Department of Energy (DOE). If completed, the facility would manufacture MOX fuel from excess weapons plutonium for commercial NPPs. It is unclear when or if construction of this facility will be completed due to the rising cost of construction (AREVA, 2015a).

In India, DAE operates a MOX fuel manufacturing facility, Advanced Fuel Fabrication Facility (AFFF), at Tarapur. The facility has fabricated lead MOX fuel assemblies for irradiation in both BWRs and PHWRs to gain experience in MOX fuel fabrication for India's planned fast reactor program (Kamath, 2010; WNA, 2015e).

TVEL's Mining and Chemical Complex (MCC) at Zheleznogorsk operates a 60 MTHM/year MFFF that began operation in 2014. The MOX fuel will be used in Russian fast reactors the BN-800 and future BN-1200 reactors. First production of fuel assemblies for Beloyarsk 4 (BN-800) began in 2014, and full capacity is expected in 2016 (WNA, 2015c; TVEL, 2010).

13.1.2.4 Fuel fabrication capacity for other types of NPPs

The facilities that fabricate fuel for GCRs, RBMKs, and LMRs are country-specific in that the facilities only build these fuel types for use in its home country's nuclear power program. In the UK, Westinghouse operates the former BNFL fabrication facility at Springfields, which fabricates fuel for the UK GCRs (specifically AGR). Springfields capacity for GCR fuel is 200 MTHM annually. Springfields also fabricated GCR fuel for magnox reactors; however, the last magnox fuel was shipped from the Springfields site in 2011. Wylfa is the last operational magnox reactor and it is currently licensed to generate electricity until the end of 2015, when it is expected to begin defueling and decommissioning (Magnox, 2015; Westinghouse, 2015).

India's AFFF at Tarapur fabricates advanced fuel designs to support India's planned fast reactor program. In addition to fabrication of MOX fuel discussed previously, the facility is performing prototype research for MOX LMR fuel as well as mixed carbide fuels and metallic fuels for the fast reactor program (Kamath, 2010).

In Russia, TVEL's MSZ at Eleckrostal includes fabrication capacity for LMR fuel for the BN-600 and BN-800 fast reactors. LMR fuel fabrication capacity is estimated to be 50 MTHM. MSZ also includes capacity to fabricate 460 MTHM of fuel for Russia's RBMK reactors. At MCC in Zheleznogorsk, the facility has a 60 MTHM capacity for fabrication of MOX LMR fuel for the BN-800 reactor as shown in Table 13.3 (TVEL, 2010; WNA, 2015c).

13.1.3 World fabrication requirements

As illustrated in Fig. 13.2, world fuel fabrication requirements for all NPP types are expected to increase from an estimated 10,900 MTHM in 2015 to approximately 12,650 MTHM by 2025—a 16% increase. LWR fuel requirements are projected to continue to dominate other fuel forms—with NPP requirements increasing from an estimated 6700 MTHM to approximately 8600 MTHM in 2025—a 28% increase over the next 10 years. Both RBMK and AGR fuel fabrication requirements are expected to decline over the next 10 years assuming that RMBK reactors in Russia and AGRs in the UK begin to reach the decommissioning stage. There is also minor growth in PHWR fuel requirements—an increase of about 1%—with requirements rising to 3610 MTHM by 2025. LMR fuel requirements, which are a small segment of the fabrication market, increase by 400% to 32 MTHM as fast reactor development in Russia, China, and India increases and new fast reactors begin operation over the next 10 years (ERI, 2015) (Table 13.4).

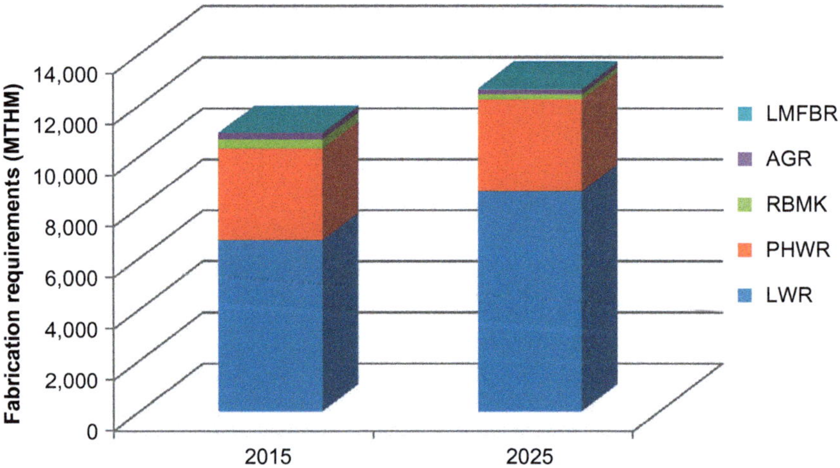

Figure 13.2 Fuel fabrication requirements by reactor type, 2015 and 2025.

Table 13.4 Summary of fuel fabrication facilities and capacities for other fuel types

Country	Fuel fabricator	Location of fuel fabrication facility	Type of fuel	Production capacity assembly
Western Europe				
United Kingdom	Westinghouse	Springfields	GCR	200
			Europe Subtotal	**200**
Other				
India	Advanced Fuel Fabrication Facility (DAE)	Tarapur	LMR	Prototype
Russia	TVEL, JSC Machine Building Plant (MSZ)	Elecktrostal	LMR (BN-600)	50
	TVEL, JSC Machine Building Plant (MSZ)	Elecktrostal	GMR (RBMK)	460
	TVEL, Mining & Chemical Complex (MCC)	Zheleznogorsk	MOX (LMR)	60
			Other Subtotal	**570**
Total Installed Other Fuel Fabrication Capacity for World				770

13.2 Fuel assembly design components

Nuclear fuel assembly designs are a function of the type of reactor in which the fuel will be used (PWR, BWR, VVER, PHWR, AGR, etc.), the fuel assembly lattice type, and the NSSS supplier of the NPP. LWR fuel assemblies typically consist of a fuel assembly skeleton, which is the hardware that provides structural support to the fuel, and fuel rods, which contain nuclear fuel, generally in the form of pellets. Fuel assembly components for non-LWR fuel are also discussed next.

13.2.1 PWR Fuel Assembly

The PWR fuel assembly skeleton includes spacer grids, top and bottom nozzles, guide tubes, and instrument tubes. Fuel rods are inserted into the skeleton to complete the fuel assembly. Spacer grids are structural elements along the length of the fuel assembly that provide structural and flow mixing functions. Spacer grids contain cells in a so-called "egg crate" design through which fuel rods are inserted and held into place in a square array. PWR assembly arrays range from 14 × 14 rods to 18 × 18 rods, depending upon the NSSS supplier of the NPP. The primary purpose of the top and bottom grids is to provide structure support for the fuel rods, while the spacer grids along the remaining length of the assembly provide not only structural support but also have a flow mixing function. Many modern PWR fuel designs also incorporate debris-resistant lower grids that are located between the bottom nozzle and bottom structural grid to prevent debris from damaging the fuel rod cladding. (NEI/ERI, 2008; AREVA, 2010; Westinghouse, 2005).

The top and bottom nozzles provide structural support for the fuel during fuel handling and operation. The top nozzle provides a grappling function for use during fuel handling operations and includes hold-down springs that offset the upward force of the coolant flow and prevent the assemblies from lifting off the bottom core plate during operation. The bottom nozzle directs coolant flow into the assembly and provides housing for debris filters in many of the modern PWR fuel designs. The debris filters are designed to minimize the amount and size of the debris particles entering the assemblies and hence minimize the potential for debris fretting failures. Current PWR fuel designs also incorporate easily removable top nozzle designs to facilitate fuel assembly reconstitution. The ability to reconstitute assemblies during refueling outages eliminates the necessity of premature assembly discharge due to minor fuel rod or structural skeleton damage (NEI/ERI, 2008).

Guide tubes are cylindrical metal tubes that provide axial positioning for the spacer grids, structural support to the fuel assembly, channels for insertion of control rods, and channels for burnable absorber rods, if used. Instrumentation tubes, which are generally a central guide tube in the fuel assembly, provide structural support for the fuel assembly as well as channels for the insertion of in-core monitoring instrumentation. Guide tubes and instrumentation tubes are typically fabricated of a zirconium alloy such as M5 or ZIRLO (NEI/ERI, 2008).

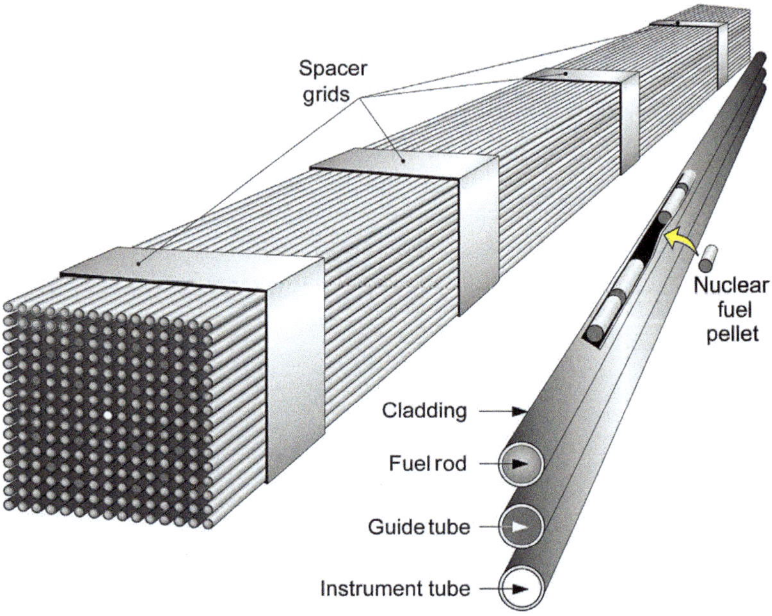

Figure 13.3 PWR fuel assembly (DOE, 2015).

In addition to the PWR fuel designs described previously that are used in Western-designed NPPs, VVER fuel assemblies are also a PWR fuel. However, in VVER fuel assemblies, the fuel rods are grouped in a hexagonal array rather than a square array. Depending upon the NSSS design, PWR fuel assemblies range in length from 4.06 to 4.80 m, and VVER fuel assemblies range from 3.20 to 4.57 m (NEI Magazine, 2014).

Fig. 13.3 provides a cutaway design of a typical fuel assembly showing the nuclear fuel pellets inside of fuel rods, fuel assembly array, and spacer grids. This particular graphic does not include the top and bottom nozzles (DOE, 2015).

13.2.2 BWR Fuel Assembly

The BWR fuel assembly skeleton includes the spacer grids, the upper and lower tie plates, the tie rods and the water rods or water channels. Fuel rods are inserted into the skeleton to form the fuel assembly. A fuel channel is placed over the fuel assembly and secured with a channel fastener. BWR spacer grids are similar in form and function to those in PWR fuel assemblies. However, rather than an "egg crate" design, some vendors use a grid in which cylinders are welded together to form the grid structure. A typical BWR fuel assembly may contain between six and eight spacer grids, depending on the manufacturer, and the grids support 9×9, 10×10 or 11×11 fuel rod arrays. BWR fuel assembly length ranges from 4.081 to

4.481 m depending upon the NSSS design of the NPP (NEI/ERI, 2008; NEI Magazine, 2014).

The lower and upper tie plates provide structural support for the fuel during fuel handling and operation. The lower tie plate of a BWR assembly provides structural support for the lower end of the fuel rods and also serves to distribute and direct coolant flow entering the assemblies. Debris filters are integral with the lower tie plate or may be attached to it. The upper tie plate provides structural support for the upper end of the fuel rods and contains a lifting fixture for use in fuel handling operations (NEI/ERI, 2008).

BWR tie rods are fuel rods that have specially designed end plugs that are screwed into the lower tie plate and also affixed to the upper tie plate to provide structural support for the fuel assembly. Water rods provide additional moderation to BWR assemblies and provide axial positioning of the spacer grids. Fuel designers will utilize water rods to optimize the neutron flux shape within a fuel assembly. The actual configuration of water rods in a BWR fuel assembly will vary depending upon the specific BWR fuel design. AREVA employs a square water channel that displaces nine fuel rods in its ATRIUM 10 and ATRIUM 11 designs. GNF utilizes two large cylindrical water rods displacing eight fuel rods in its GNF2 design. The Westinghouse BWR design consists of four 5×5 mini-assemblies in its SVEA-96 Optima3 design. To form the water channel, one fuel rod (in the center of the larger fuel assembly) is removed from each of the mini-assemblies to form a water cross (NEI/ERI, 2008; AREVA, 2015b; GNF, 2015a; Westinghouse, 2013).

BWR fuel channels are four-sided structural elements, which are the length of the fuel rods and are typically fabricated of Zr-4 or another zirconium alloy. The fuel channel encloses a BWR fuel assembly and performs three primary functions of providing a channel for through-bundle water flow path, acting as a guide for BWR control blades, and providing structural support to the fuel assembly. At one time BWR fuel channel designs were sold separately from fuel assemblies. However, today, BWR fuel channels complement specific fuel assembly designs to enhance flow mixing in the assembly and improve fuel performance (NEI/ERI, 2008).

13.2.3 PHWR Fuel Bundle

PHWR fuel assemblies include a circular zircaloy end support plate, that provides support to the fuel tubes, fuel tubes (referred to as fuel sheath in Fig. 13.4), zircaloy bearing pads, interelement spacers, and a pressure tube that surrounds the fuel assembly. The bearing pads maintain the spacing between the fuel assembly and the pressure tube. In addition, the interelement spacers maintain the separation of the fuel tubes from one other without the need to utilize spacer grids. The fuel tubes, which contain UO_2 pellets, are closed with zircaloy end caps and have a graphite inner layer (referred to as "Canlub") that is used to reduce the stresses within the fuel tube. The fuel tube end caps not only provide a seal for the fuel tubes, but they also provide a means of attaching the tubes to the end support plate.

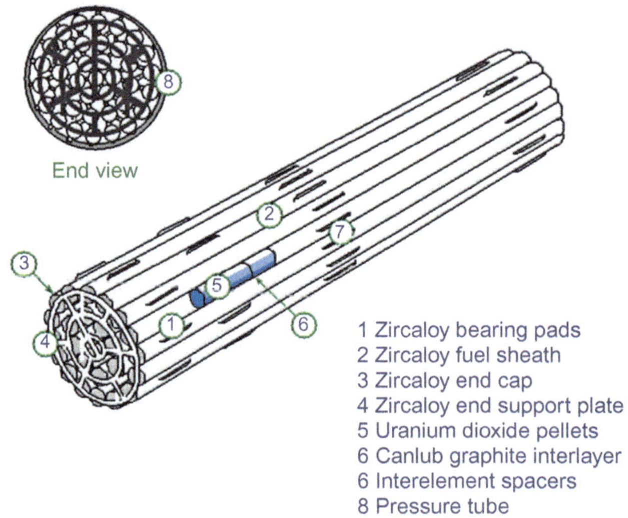

Figure 13.4 CANDU fuel assembly (CANDU.ORG, 2015) waiting for response regarding permission.

CANDU PHWR fuel assemblies utilize 28, 37, or 43 fuel tubes, which are typically made of a zirconium alloy and are grouped into a cylindrical fuel bundle roughly 10 cm in diameter and 50 cm in length around a central axis. Unlike LWR fuel which resides in the NPP core in a vertical position, PHWR fuel assemblies are loaded into horizontal channels or pressure tubes which penetrate the length of the reactor vessel (called the calandria) (Cameco, 2015a; CANDU.ORG, 2015; Ferenbach and Miller, 2009).

13.2.4 GCR Fuel Assemblies

GCR fuel assemblies, such as those used in AGRs in the UK, have a circular configuration, in which 36 stainless steel clad fuel pins are grouped. Each fuel pin contains 20 enriched UO_2 fuel pellets. The AGR fuel assemblies are covered with a graphite sheath that acts as a moderator during reactor operation. Fuel assemblies are stacked in a vertical configuration in a fuel channel, with eight assemblies per channel (Westinghouse, 2015; WNA, 2015e).

13.2.5 RBMK Fuel Assemblies

RBMK fuel assemblies, such as those used in RBMK NPPs in Russia, include two “subassemblies” that each contain 18 fuel rods and 1 carrier rod. The lower subassembly includes an end grid and ten spacer grids. The top subassembly has 10 spacer grids. The spacer grids provide structure support as well as improving heat transfer. The zircaloy fuel rods in each subassembly are arranged within two concentric rings around a central carrier rod. The two subassemblies are joined by a cylinder along the plane of the center carrier rod. Each fuel assembly in the reactor core is housed in an individual pressure tube. The total length of the fuel assembly is 10.025 m with 6.862 m being the active region. The fuel rods contain enriched UO_2 fuel pellets and may utilize a burnable absorber (INL, 2015).

13.2.6 Uranium dioxide production

There are three conversion processes used by fuel fabricators to convert UF_6 gas to UO_2 powder for the production of UO_2 pellets: (1) the "dry" process, (2) the ammonium diuranate (ADU) process, a wet process, and (3) the ammonium uranyl carbonate (AUC) process, also a wet process. The dry process results in significantly lower quantities of liquid waste than either of the wet processes and it is the most commonly used conversion process used today. In each of these processes, the first step is the heating of the UF_6 cylinders in an autoclave and the removal of the UF_6 in gaseous form. The remaining steps in the various UF_6 to UO_2 conversion processes are described briefly next (NEI/ERI, 2008).

The dry process was originally developed by Siemens, now part of AREVA, and has been licensed to several other fabricators. It consists of the hydrolyzation of the UF_6 with steam in a gas-phase reaction. This step is followed by the reduction of the resulting uranyl fluoride (UO_2F_2) with hydrogen and steam in a fluidized bed reactor to produce the UO_2. The initial powder product is then calcined in a rotary kiln with more steam and hydrogen to drive off any remaining fluoride and to dry the powder (NEI/ERI, 2008).

In the ADU process, the UF_6 gas is hydrolyzed by solution in water. Ammonia is added to the solution producing a precipitate of ammonium diuranate (ADU). The ADU is subsequently filtered, dried, calcined in the presence of steam and reduced in the presence of hydrogen to produce the UO_2 powder. In an earlier version of the wet process, ammonium carbonate was used instead of ammonia, producing AUC that was then treated in a similar manner to the ADU described previously (NEI/ERI, 2008).

In the *AUC process*, gaseous UF_6, carbon dioxide (CO_2) and ammonia (NH_3) are combined in water resulting in a precipitate of AUC. The AUC is then processed in a manner similar to ADU to produce the UO_2 powder (NEI/ERI, 2008).

13.2.7 Fuel pellet production

Following conversion of UF_6 gas into UO_2 powder, the next step in the fuel fabrication process pellet production of UO_2 pellets. The UO2 pellet manufacture involves the following steps: (1) mixing of the UO_2 powder with binding and lubricating agents, (2) compaction or cold pressing, (3) sintering, and (4) precision grinding. Binding and lubricating agents that are commonly employed in pellet manufacture are organic compounds such as aluminum or zinc stearate and stearic acid. These agents enhance the formation of pores to facilitate the increase in density during the sintering process. The binding agents provide additional strength to the cold pressed pellets and assist in reducing dust hazards associated with the handling of UO_2 powder. The lubricating agents assist in a uniform cold pressing operation (NEI/ERI, 2008).

Once the UO_2 powder has been mixed with the binding and lubricating agents, the next step is to cold press the resulting mixture to produce "green" pellets, which have a density of approximately 55–60% of theoretical density. The green pellets are then sintered in a high-temperature furnace to form a stable ceramic with the necessary heat transfer properties. The sintering process also drives off the remaining additives. Final pellet densities of about 96–97% of theoretical density are typically achieved. Once the sintering process is complete, the pellets are ground to their final dimensions in a grinder, inspected and stored for future fuel rod loading (NEI/ERI, 2008).

MOX fuel pellet manufacture can be accomplished via two methods: (1) dry mixing and (2) coprecipitation. In the dry mixing process, UO_2 powder and plutonium oxide powder (PuO_2) are ground together. Depleted UO_2 powder is typically used along with rejected UO2 pellets that have been ground into a powder. The mixture is then cold pressed into pellets, sintered, and ground to meet final specifications in similar processes used to produce UO_2 pellets. In the coprecipitation process, a mixture of uranyl nitrate and plutonium nitrate is converted by treatment with a base such as ammonia to form a mixture of ammonium diuranate and plutonium hydroxide. After the mixture undergoes a heating process, it will form a powder containing uranium dioxide and plutonium dioxide. The resulting powder is then pressed into pellets, sintered, and ground as in the formation of UO_2 pellets (Collins et al., 2011; AREVA, 2015c).

13.2.8 Use of burnable absorbers

Burnable absorbers are neutron-absorbing materials that are commonly incorporated in LWR fuel designs as a means of power shaping, local power-peaking control and overall long-term reactivity control. Burnable absorbers used for long-term reactivity control, such as gadolinium (GdO_2) in BWR fuel assemblies, depletes as the fuel assembly remains in the reactor core, thereby providing greater reactivity control during initial use of the fuel (when the new fuel assembly's reactivity is highest) and providing less reactivity control for the fuel assembly late in a cycle (when the fuel assembly's reactivity has decreased). The use of burnable absorbers varies among vendors and fuel designs, but can be generally grouped into three classifications: (1) discrete absorbers, (2) intimately mixed absorbers, and (3) surface coating absorbers (NEI/ERI, 2008).

The discrete absorbers, which are rarely used today, are only employed in PWR fuel designs and are typically incorporated in the form of an absorber-filled rod that is inserted into a vacant rod control cluster assembly (RCCA) guide tube. Typical absorber materials employed in these designs are Boron-10 (B^{10}) doped Pyrex glass or aluminum oxide-boron carbide (Al_2O_3-B_4C) pellets. Reactivity control is achieved through variation in the absorber loading per rod, the number of absorber rods per fuel assembly, and the total number of absorber-loaded fuel assemblies in the core (NEI/ERI, 2008; Westinghouse, 2010).

Intimately mixed absorbers, gadolinium (GdO_2) and erbium (ErO_2) oxides, are combined with the UO_2 powder prior to pelletization. This burnable absorber technology has been successfully employed in both PWR and BWR fuel designs by essentially all fuel fabricators. BWRs use exclusively gadolinia, while both gadolinia and erbia are employed in PWRs. Reactivity control is achieved through variation of absorber loading per fuel rod, absorber distribution within fuel rods, number of absorber-loaded fuel rods per assembly, and/or total number of absorber-loaded assemblies per core.

Surface coating absorbers, such as Westinghouse's Integral Fuel Burnable Absorber (IFBA), are applied as thin coatings of boron compounds such as zirconium diboride (ZrB_2) on the surface of individual pellets in PWR fuel rods. The coating is typically applied to specific pellets of like enrichment within a fuel rod, or to pellets spanning a wide region of a fuel rod to achieve specific power-peaking control (Westinghouse, 2010; NEI/ERI, 2008).

13.2.9 Fuel rod fabrication process

While the specifics of fuel rod design vary among individual fabricators and fuel type, a typical LWR fuel rod is composed of a zirconium alloy cladding tube (such as Zr-2, Zr-4, ZIRLO, M5, etc.), a UO_2 pellet column, two end plugs and an internal plenum spring. Each fuel rod has a unique identification number to provide the ability to trace the history of each rod manufactured. The internal plenum spring prevents pellet movement and possible damage during handling operations. A typical fuel rod manufacturing procedure consists of the following steps: (NEI/ERI, 2008).

- The bottom end plug is inserted into cladding tube and welded in place
- The UO_2 pellet column is loaded into the cladding tube. It may be pushed into horizontally positioned tubes or gravity loaded with the tubes tilted at an angle depending upon the manufacturing process
- The length of the pellet column is confirmed to be in accordance with manufacturing specification through insertion of a gauge is inserted into the cladding tube
- A plenum spring is inserted into the cladding tube on top of the pellet stack
- The top of the cladding tube is placed in a sealed chamber for pressurization. Helium is introduced into the fuel rod to a specified internal fuel rod pressure. If it had not been done previously, the top end plug is inserted into the cladding tube once rod pressurization has been achieved. The end plug is then welded to the tube. If the plug contained a hole to achieve rod pressurization, the hole is seal welded as well
- Inspections of the completed fuel rod are performed

Fuel rod weld integrity is generally verified through visual, ultrasonic, and/or X-ray inspection techniques. Each fuel rod may be weighed for gross loading verification and/or gamma scanned to provide verification of pellet enrichments and orientation (Kok, 2009; NEI/ERI, 2008).

13.3 Current and future trends

A number of fuel cycle trends are worth noting including the continued emphasis on fuel reliability and increases in NPP cycle length, capacity factor, and fuel assembly burnup. The development of fast reactors in Russia, China, and India, along with fuel for these plants is also expected to experience continued growth. Each of these topics is discussed briefly next with references to other resources.

13.3.1 Fuel reliability

The nuclear industry continues to make progress in the reduction of fuel failures. In the US, there was a concerted effort to meet the Institute of Nuclear Power Operation's (INPO) goal for zero fuel defects by 2010. This *Zero by Ten Initiative* is now called *Driving to Zero*. The Electric Power Research Institute (EPRI), INPO, NPP operators and fuel fabricators worldwide have worked together to develop guidelines to address the known failure mechanisms, which include: PWR corrosion and crud, BWR corrosion and crud, grid-to-rod fretting (GTRF), pellet-cladding interaction (PCI), and foreign material induced failures. EPRI published an updated Fuel Surveillance and Inspection Guidelines in 2012 and continues to review its five fuel reliability guidelines periodically to reflect the most recent knowledge gained through fuel surveillance and inspection programs. According to EPRI, when the industry began the *Zero by Ten Initiative* in 2007, 30% of US reactors were experiencing fuel failures. By the end of 2010, this figure was reduced to 6% and has remained constant (EPRI, 2005, 2008, 2013a).

Even though fuel failures have been significantly reduced over the last decade, fuel reliability remains a major industry focus. Improved debris-resistant fuel designs and better in-plant housekeeping have contributed to reductions in debris-related fuel failures. Vendor research and development have been focused on eliminating or at least significantly reducing other types of failures while continuing to increase fuel assembly discharge burnup. However, fuel failures resulting from materials corrosion and hydriding, grid-rod fretting, pellet-clad interaction, etc., have not yet been totally overcome in either BWRs or PWRs, and axial growth and distortion of PWR fuel assemblies and distortion of BWR channels remain major problems for operators at some plants.

EPRI established a fuel reliability database (FRED) in 2004 to better share fuel reliability information across the industry. FRED contains information on the fuel types in core, the reliability of the fuel during operation, and other fuel-related issues that affect operation. For PWRs, GTRF remains the dominant failure mechanism in US reactors. For US BWRs, debris-related failures have been the dominant failure mechanism since 2008. With most PWRs having transitioned to fuel designs with grid fretting resistance, debris-related failures are expected to become the dominant failure mechanism in US PWRs in the future (EPRI, 2006, 2014c).

Higher fuel duty and longer cycle lengths have increased the severity of BWR channel distortion problems in recent years. Channels may bow, bulge or twist, altering the clearance that allows the control rod to move freely, resulting in potential safety implications due to degraded control rod performance. According to EPRI, 17 out of 35 BWRs in the US have reported control blade interference due to channel distortion in the last decade. Affected fuel designs include Zircaloy-2 channels manufactured by all three US fuel vendors. EPRI, US utilities, the BWR Owners Group, fuel vendors, and INPO developed a Channel Distortion Industry Action Plan (CDIAP) to coordinate research in this area to better understand the mechanisms associated with channel distortion. Lead Channel Test programs have been initiated to evaluate alternative channel materials and fuel vendors have introduced enhanced BWR fuel channel designs to enhance bundle flow characteristics (EPRI, 2012, 2014a,d).

13.3.2 Fuel cycle trends

Worldwide NPP operating/refueling cycle lengths are generally increased over the past two decades—from nominal 12-month cycles up to 24-month cycles. Regarding operating/refueling cycle lengths, almost all US NPPs are on nominal 18- or 24-month cycles. However, one US plant has found it economical to return to an annual cycle. In Europe, PWRs generally operate on 12- to 18-month refueling cycles, depending upon the country and BWRs generally operate on 24-month cycles. NPPs in Eastern Europe, Russia, and former Soviet countries continue to operate on 12-month refueling cycles. In Korea and Taiwan, NPPs typically operate on 18-month cycles and China is moving a number of its NPPs in this direction. Prior to the Fukushima Daiichi accident in 2011, Japanese utilities received approval to increase the allowed operating period from 13 months to as much as 18 months, equivalent to cycle lengths of 21 months when the refueling outage is included, but these plans have not been implemented. Longer operating cycles allow NPP operators to maximize power generation, reduce the number of outages, lower outage costs, and make more efficient use of outage personnel. As cycle length increases, the fraction of fuel assemblies replaced at each outage also increases. As a result, longer cycles require higher enrichments even with the same design burnups, resulting in an increase in fuel cycle requirements. Generally, non-fuel operating cost factors, and not the fuel cost, determine a NPP's cycle length (ERI, 2015).

In 2014, the worldwide NPP capacity factor continued to be negatively impacted by the ongoing reactor outages in Japan, with a world average capacity factor of 73.9%. Excluding the reactor outages in Japan, the worldwide capacity factor was 83% in 2014. US capacity factors increased in 2014 to reach a record 91.7% as no plants were in long-term outages. The average world capacity factor is expected to improve slowly to an estimated 85%, as Japanese reactors resume operation or decisions are made regarding further plant closures (ERI, 2015).

Average design discharge exposure (also called fuel assembly burnup) for fuel loaded in LWRs is estimated at 49.5 GWD/t for fuel loaded in 2014. The actual discharge exposures are often slightly below the design discharge exposures. While new "Gen-3" reactors such as AREVA's EPR and Westinghouse's AP1000 are expected to use design discharge exposures of 60 GWD/t, most reactors are expected to remain near their current burnup levels (ERI, 2015). According to research conducted by ANT International, batch average discharge burnups for PWRs in the US are an estimated 43–58 GWD/t and for US BWRs are 43–52 GWD/t. In Western Europe, batch average burnups range from 45 to 65 GWD/t, in Asian countries—from 45 to 50 GWD/t, and in Russia and Eastern Europe—from 50 to 60 GWD/t (ANT International, 2014).

Interest in high burnup fuel is driven by longer cycle lengths, reductions in spent fuel generation, more efficient use of uranium and enrichment services on a unit energy production basis, and improved fuel cycle costs. While requiring higher initial enrichment assays, higher burnup designs result in fewer assemblies in each reload, and a decline in the nuclear fuel cycle requirements in the range of 4–8%, compared to a situation where no advances in fuel design take place and the same cycle length is maintained. Ongoing concerns about fuel performance at high discharge burnup and the industry-wide push for zero fuel failures are likely to restrict increases in discharge burnup beyond the current levels in the near term. If discharge exposures were pushed higher than 60 GWD/t, enrichment assays would be required to increase above the current licensed limit of 5.0 w/o ^{235}U for most fuel facilities (ERI, 2015).

13.3.3 Fast reactor fuel

Fast reactors can utilize a wide range of fuel types, a mixture of transuranic elements as fuel, and various chemical forms. Fast reactor fuel can be made from UO_2, MOX, single or mixed nitride ceramics, and metallic fuels. Fast reactor fuel can be made from pellets, in a manufacturing process similar to that for LWR fuel described previously, or it can be manufactured using a method called vibro-packing. Fast reactor development is expected to continue to grow as certain countries look to close the fuel cycle either to address fuel resource needs or to reduce growing quantities of spent nuclear fuel.

Vibro-packed MOX fuel is made by agitating a mixture of granulated uranium and plutonium oxides with uranium powder, with binding oxygen and other gases added during the agitation process. Vibro-packed MOX fuel can be more easily recycled than pelletized MOX fuel and there are reportedly fewer interactions between the fuel and cladding (WNN, 2014a).

A reactor core in a fast reactor is much smaller than that of an LWR and the cores will typically have two different fuel regions—the seed fuel uses fuel with a high fissile content and a higher power level, and the blanket fuel has a low fissile content but utilizes fuel material with high neutron absorption cross sections (in a "breeder" fast reactor) or actinide material to be transmuted (in a "burner" fast reactor) (WNA, 2015e).

The BN-600 at Beloyarsk operates as "breeder" with a central bundle that includes 127 rods (2.4 m in length and 7 mm in diameter). The rods contain ceramic pellets in three uranium enrichment levels: 17%, 21%, and 26%. Blanket fuel bundles have 37 rods containing depleted uranium. The BN-600 has utilized mostly

UO_2 fuel, but vibro-packed MOX fuel has been successfully used in the BN-600 reactor for the past decade. Vibro-packed MOX fuel assemblies have also been fabricated for the BN-800 NPP at Beloyarsk 4, which began operation in 2014. The assemblies were produced at the Research Institute of Atomic Reactors (NIIAR) in Dimitrovgrad (NIIAR, 2010; WNA, 2015e; WNN, 2014).

Many of the early fast reactors, such as EBR-II in the US, utilized metallic fuel and some fast reactor designers, such as GEH, are developing metallic fuel designs. In metallic fuel, a metallic fuel slug is loaded into the fuel cladding and the gap between the fuel slug and cladding is filled with sodium. The sodium acts as a thermal bond until the fuel swells to meet the cladding. Fuel slugs can be the full length of the fuel cladding tube or multiple slugs can be stacked—much in the way that ceramic UO_2 pellets are stacked in LWR fuel. The Mark-I and Mark II fuel utilized in EBR-II was made with 95% uranium metal and a 5% fissium alloy. Fissium is a mixture of fission products. Subsequent fuel was made from recycling the metal fission products along with recovered uranium metal. The General Electric-Hitachi PRISM fast reactor design, which is based on the EBR-II design, would utilize metallic fuel such as an alloy of zirconium, uranium, and plutonium (Chang, 2007; GEH, 2015).

A new Russian fast reactor that is under development, the BREST fast neutron reactor, will utilize lead as the primary coolant. The fuel type considered for the first core of the BREST fast reactor is a nitride of depleted uranium mixed with plutonium and minor actinides (MA). Reprocessing is limited to the removal of fission products without separating plutonium and MA from the mix (U-Pu-MA). One of the notable characteristics of the BREST plant and other planned fast reactors is that a reprocessing plant is colocated with the reactor, eliminating in principle any accident or problem due to fuel transportation (Alemberti et al., 2014; WNN, 2014b).

13.3.4 Expansion of fuel fabrication markets and new market participants

The past two decades saw consolidation of fuel suppliers in the US and Western Europe resulting in the three major Western fuel suppliers that exist today: AREVA, GEH, and Westinghouse. The next decade may be characterized as one of expansion of existing fabricators into new regions or the introduction of new fuel products from new market entrants.

TVEL has historically provided almost 100% of fuel fabrication requirements to Russian-designed VVERs in Russia, Ukraine, and Eastern Europe but it did not provide fuel to Western-designed NPPs. Recently TVEL developed a PWR fuel design for use in Western NPPs to expand its market. In February 2012, TVEL announced that it had signed a contract with Vattenfall Nuclear Fuel of Sweden that covers Lead Test Assemblies (LTAs) of square 17×17 lattice TVS-Kvadrat. Conversely, Westinghouse is the only Western fuel supplier that has produced fuel assemblies for Russian-designed VVER NPPs, having produced VVER-1000 fuel assemblies for the two Soviet-designed Temelin units in the Czech Republic until 2006. Under an the initial 2008 contract between Westinghouse and Energoatom of Ukraine, the first Westinghouse fuel assemblies were loaded into the South Ukraine NPP in 2010. In December 2014, Westinghouse and Ukraine's Energoatom agreed to significantly increase future deliveries of Westinghouse-supplied fuel assemblies to Ukrainian NPPs through 2020.

There is also expansion of fabrication capacity in China for Western-designed NPPs that are under construction in China. CNNC has put in place technology transfer agreements with AREVA, TVEL and Westinghouse so that China will be capable of fabricating fuel for its Western-designed NPPs. Kazatomprom is also exploring joint ventures with a number of fuel manufacturers that would allow fabrication of fuel assemblies in Kazakhstan. Kazatomprom already exports fuel pellets to a number of countries including China and India.

In April 2015, the Nuclear Utility Fuel Advisory Board (NUFAB), whose members include several of the largest US nuclear operators—Dominion, Duke Energy, Exelon, and Southern Nuclear, sent a letter to the US Nuclear Regulatory Commission (NRC) regarding a new fuel product under development by Lightbridge Corporation (Lightbridge). According to Lightbridge, its metallic fuel rod design includes three components: a- central displacer of zirconium (Zr), which serves to reduce centerline temperatures and allows for the incorporation of burnable poison material within the rod; a four-lobed fuel core composed of a Zr-U alloy, and a Zr-1Nb cladding alloy that is metallurgically bonded to the fuel core. The metallurgical bonding that takes place during the fuel rod fabrication process results in each fuel rod being a monolithic form composed entirely of metal. According to Lightbridge, the fuel rod design is more robust than current tubes that utilize ceramic pellets. NUFAB members are working with Lightbridge to submit an application to the NRC in 2017 for the use of LTA, with insertion of LTA's into a US PWR as early as 2020 (NUFAB, 2015; WNA, 2015e).

13.4 Sources of further information and advice

There are a wide variety of sources that provide information on fuel fabrication processes, capacity, and fuel cycle trends. Documents from many of these sources are cited herein and are included in the references to this chapter. Additional resources include:

- Addition information regarding fuel fabrication can be found at the various web sites for the fuel fabricators.
- The World Nuclear Association also provides a high-level overview of the fuel fabrication process and fuel cycle facilities in countries with nuclear power programs.
- ANT International, which provides training in the area of nuclear fuel, has very informative information available on its web site, for purchase or through various seminars that the company holds periodically.

- EPRI's Fuel Reliability Program provides information regarding fuel assembly failure mechanism, ongoing research programs, and plant specific experience.
- The proceedings of technical meetings sponsored by the International Atomic Energy Agency, TopFuel, and other conferences also provide a wealth of information on fuel for existing NPPs, fuel cycle innovations, and research into new fuel designs and new reactor types.

References

Alemberti, A., Frogheri, M.L., Hermsmeyer, S., Ammirabile, L., Smirnov, V., Takahashi, M., Smith, C.F., Wu, Y., Hwang, I.S., 2014. Lead-cooled Fast Reactor (LFR) Risk and Safety Assessment, White Paper, Revision 8. Gen IV Int. Forum, April 2014.

ANS, 2015. 2015 Nuclear News, Reference Special Section, Power Reactors by Type, Worldwide. Nuclear News, March 2015.

ANT International, 2014. Peter Rudling, Advanced Nuclear Technology International, et al. Annual Report, December 2014.

AREVA, 2010. HTP: Robust Technology for PWR Fuel Assemblies. Available from: <http://us.areva.com/home/liblocal/docs/Catalog/PWR/ANP_U_328_V1_10_ENG_HTP.pdf> (accessed 12.07.15.).

AREVA, 2015a. Production of MOX Fuel. Available from: <http://www.areva.com/EN/operations-1095/melox-operations-production-of-mox-fuel-assemblies.html> (accessed 10.07.15.).

AREVA, 2015b. Available from: <http://us.areva.com/EN/home-793/areva-inc-flexible-fuel-cycle-solutions.html> (accessed 12.07.15.).

AREVA, 2015c. Fabrication of Recycled Fuel. Available from: <http://www.areva.com/EN/operations-1173/the-fabrication-of-mox-fuel-recycled-nuclear-power.html> (accessed 10.07.15.).

Cameco, 2015a. Cameco Corporation, 2014 Annual Information Form, 6 March 2015.

Cameco, 2015b. Available from: <http://www.cameco.com/uranium_101/fuel-processing/fuel-manufacturing/> (accessed 10.07.15.).

CANDU.ORG, 2015. CANTEACH at Candu.org. Available from: <https://canteach.candu.org/Image%20Library1/19980101-%20Intro_to_CANDU6_China/19980101_021.jpg> (accessed 12.07.15.).

Chang, Y.I., 2007. Argonne national laboratory, technical rationale for metal fuel in fast reactors. Nucl. Eng. Technol. 39 (3), 161–170, June 2007.

Collins, E., Voit, S., Vedder, R., 2011. Oak Ridge National Laboratory, Milestone Report, Evaluation of Co-precipitation Processes for the Synthesis of Mixed-Oxide Fuel Feedstock Material, 23 June 2011.

DOE, 2015. Fuel Assembly for Production of Nuclear Power. Available from: <http://www.energy.gov/photos/yucca-mountain> (accessed 13.07.15.).

ENUSA, 2014. Annual Report 2014. Available from: <http://www.enusa.es/wp-content/uploads/2015/06/ANNUAL-REPORT-GRUPO-ENUSA-20141.pdf>

ENUSA, 2015, Manufacturing. Available from: <http://www.enusa.es/en/areas-de-negocio/nuclear/fabricacion/> (accessed 06.07.2015.).

EPRI, 2005. EPRI, the challenge of nuclear fuel reliability. EPRI J. Fall.

EPRI, 2008. EPRI, the Path to Zero Defects: EPRI Fuel Reliability Guidelines.

EPRI, 2012. EPRI. BWR Channel Distortion. Available from: <http://mydocs.epri.com/docs/Portfolio/PDF/2012_APC2_BWR-Channel-Distortion.pdf> (accessed 10.07.15.).

EPRI, 2013a. EPRI, Fuel Reliability Trends Strongly Move in Right Direction. Available from: <http://www.epri.com/Our-Work/Documents/Nuclear/Fuel%20failure%20trends_Mar2013.pdf> (accessed 10.07.15.).

EPRI, 2013b. EPRI, EPRI Fuel Reliability (QA) Program Overview, 201 Research Portfolio.

EPRI, 2013c. EPRI, in Use: Mitigation of PWR Fuel Failures Caused by Corrosion and Crud, Fuel Reliability, August 2013.

EPRI, 2014a. EPRI Fuel Reliability (QA) Program Overview, 2014 Research Portfolio.

EPRI, 2014b. EPRI, in Use: Mitigation of BWR Fuel Failures Caused by Corrosion and Crud, Fuel Reliability, January 2014.

EPRI, 2014c. EPRI, in Use: Mitigation of Fuel Failures Caused by Foreign Material, January 2014.

EPRI, 2014d. EPRI, in Use: Silicon Carbide Composite BWR Channel Development, Nuclear Sector Roadmaps, January 2014.

ERI, 2015. Energy Resources International, Inc. (ERI), 2015 Nuclear Fuel Cycle Supply and Price Report, Fabrication Services Market, June 2015.

EPRI, Yang, R., Cheng, B., Deshon, J., Edsinger, K., Ozer, O., 2006. Fuel R&D to improve fuel reliability. J. Nucl. Sci. Technol. 43 (9), 951–959, <http://www.tandfonline.com/doi/pdf/10.1080/18811248.2006.9711181>.

Ferenbach, P.J., Miller, A., 2009. Atomic Energy of Canada Limited, Chapter 15, HWR Fuel Cycles, Nuclear Engineering Handbook, CRC Press, 2009.

GEH, 2015. GE Hitachi, PRISM Reactor. Available from: <http://gehitachiprism.com/what-is-prism/how-prism-works/> (accessed 12.07.2015.).

GNF, 2015a. GNF2 Brochure, GNF2: High Performance Fuel Assembly. Available from: <https://nuclear.gepower.com/fuel-a-plant/products/gnf2-advantage.html> (accessed 10.07.2015.).

GNF, 2015b. CANDU: Fueling the Canadian Nuclear Industry for Nearly 60 Years. Available from: <https://nuclear.gepower.com/fuel-a-plant/products/candu-fuels.html> (accessed 12.07.2015.).

IAEA, 2003. International Atomic Energy Agency, Status and Advanced in MOX Fuel Technology, Technical Report Series 145, May 2003.

INB, 2015. FCN Components and Assemblies. Available from: <http://www.inb.gov.br/eng/WebForms/Interna2.aspx?secao_id = 56>.

INL, 2009. RBMK Fuel Assembly. Available from: <http://web.archive.org/web/20090908013425/http://www.insc.anl.gov/rbmk/reactor/assembly.html> (accessed 10.07.2015.).

JNFL, 2015. MOX Fuel Fabrication Plant. Available from: <http://www.jnfl.co.jp/english/business/mox.html>.

Kamath, H.S., Nuclear Fuels Group, 2010. Bhabha Atomic Research Centre, Recycle Fuel Fabrication for Closed Fuel Cycle in India. Asian Nuclear Prospects 2010.

KNFC, 2015. KEPCO Nuclear Fuel Company, Nuclear Fuel Production Capacity. Available from: <http://www.knfc.co.kr/eng/production_capacity> (accessed 06.07.2015.).

Kok, K.D., 2009. URS Washington Division, Chapter 9, Nuclear Fuel Fabrication. Nuclear Engineering Handbook, CRC Press, 2009.

Magnox, 2015. Available from: <http://www.magnoxsites.co.uk/> (accessed 10.07.2015.).

NEI/ERI, 2008, Energy Resources International, Inc. Michael Schwartz, Eileen Supko, et al., NEI Nuclear Fuel Fabrication Handbook, March 2008.

NEI Magazine, 2014. Nuclear Engineering International, Fuel Design Data, September 2014.

NIIAR Bychkov, A. et al. 2009. Research Institute of Atomic Reactors (NIIA), Vibropac MOX fuel for fast reactors—experience and prospects. In: International Conference on Fast Reactors and Related Fuel Cycles, International Atomic Energy Agency, 2009.

NFC, 2015. DAE Nuclear Fuel Complex. Available from: <http://www.nfc.gov.in/html-products.htm> (accessed 06.07.2015.).

NRA, 2013. Japan Nuclear Energy Safety Organization, Operational Status of Nuclear Facilities in Japan, 2013 Edition, IX-2 List of Operation and Construction Status of Fabrication Facilities.

NUFAB, 2015. Nuclear Utility Fuel Advisory Board, Letter to U.S. Nuclear Regulatory Commission, Lead Test Assemblies of Lightbridge Designed Fuel, 14 April 2015. Available from: <http://pbadupws.nrc.gov/docs/ML1513/ML15134A092.pdf>.

Song, K.W., et al., 2009. Progress in nuclear fuel technology in Korea, Korea atomic energy research institute. Nucl. Eng. Technol. 41 (4), May 2009, Special Issue in Celebration of the 40th Anniversary of the Korean Nuclear Society, March 2009. <http://www.kns.org/jknsfile/v41/JK0410493.pdf>.

TVEL, 2010. JSC TVEL, Nuclear Fuel Cycle in Russia, IAEA TCM on Nuclear Fuel Cycle Information System and Leveraging Synergies for Sustainability, 16 December 2010.

TVEL, 2014. Fuel Company TVEL—reliable supplier of nuclear fuel. State Atomic Energy Corporation Rosatom, 19 November 2014. Available from: <http://www.atomeks.ru/mediafiles/u/files/Asia_2014/materials/02_Fuel_Company_TVEL_reliable_supplier_of_nuclear_fuel.pdf> (accessed 12.07.15.).

TVEL, Molchanov, V., 2003. TVEL, Perspective Decisions of WWER Nuclear Fuel: Implementation at Russian NPPs, September 2003. Available from: <http://www.iaea.org/inis/collection/NCLCollectionStore/_Public/36/040/36040491.pdf>.

Westinghouse, 2005. Westinghouse Fuel for Combustion Engineering-Designed Reactors. Available from: <http://www.wecnq.com/Products_&_Services/docs/flysheets/NF-FE-0012.pdf> (accessed 12.07.15.).

Westinghouse, 2013. Reference Fuel Design, SVEA-96 Optima3, WCAP-17769-NP, Revision 0, November 2013.

Westinghouse, 2015. Products. Available from: <http://www.westinghousenuclear.com/springfields/Products> (accessed 10.07.15.).

Westinghouse, Secker, J.R., Brown, J.A., 2010. Nuclear Fuel Division, Westinghouse Electric Company, Westinghouse PWR Burnable Absorber Evolution and Usage, Winter ANS Conference, 9 November 2010.

WNA, 2013. World Nuclear Association, The Global Nuclear Fuel Market—Supply and Demand 2013–2030, Fuel Fabrication Supply and Demand, 2013.

WNA, 2015a. World Nuclear Association, China's Nuclear Fuel Cycle. Available from: <http://www.world-nuclear.org/info/Country-Profiles/Countries-A-F/China--Nuclear-Fuel-Cycle/> (accessed 06.07.2015.).

WNA, 2015b. World Nuclear Association, Fuel Fabrication, Table 2, World PHWR Fuel Fabrication Capacity. Available from: <http://www.world-nuclear.org/info/Nuclear-Fuel-Cycle/Conversion-Enrichment-and-Fabrication/Fuel-Fabrication/> (accessed 06.07.2015.).

WNA, 2015c. World Nuclear Association, Russia's Nuclear Fuel Cycle. Available from: <http://www.world-nuclear.org/info/Country-Profiles/Countries-O-S/Russia--Nuclear-Fuel-Cycle/> (accessed 10.07.2015.).

WNA, 2015d. World Nuclear Association, Japan's Nuclear Fuel Cycle. Available from: <http://www.world-nuclear.org/info/Country-Profiles/Countries-G-N/Japan--Nuclear-Fuel-Cycle/> (accessed 10.07.2015.).

WNA, 2015e. World Nuclear Association, Fuel Fabrication, Table 3: World MOX Fuel Fabrication Capacity. Available from: <http://www.world-nuclear.org/info/Nuclear-Fuel-Cycle/Conversion-Enrichment-and-Fabrication/Fuel-Fabrication/> (accessed 06.07.2015.).
WNN, 2014a. World Nuclear News, First MOX for Beloyarsk 4, 6 March 2014. Available from: <http://www.world-nuclear-news.org/ENF-First-MOX-for-Beloyarsk-4-0603147.html> (accessed 10.07.2015.).
WNN, 2014b. World Nuclear News, Design Completed for Prototype Fast Reactor. Available from: <http://www.world-nuclear-news.org/NN-Design-completed-for-prototype-fast-reactor-0209147.html> (accessed 12.07.2015.).

Part IV

Environmental and Social Issues

Management for health, safety, environment, and community in uranium mining and processing

14

Frank Harris
Rio Tinto Uranium, Brisbane, Queensland, Australia

14.1 Framework for health, safety, environment, and community in uranium operation

One of the most critical areas in any modern uranium mine is protection of the workforce, the public, and the environment. In mining in general, and for uranium operations in particular, the mantra is safety before production. The establishment of an appropriate safety culture is seen as a mandatory requirement for any modern operation and this is obvious from the first moment when you visit a mine from the initial induction until you safely leave the site.

Historically, the importance on the various aspects of health, safety, environment, and community (HSEC) has not been as prevalent and not necessarily consistently applied across all aspects. The push for HSEC to become a critical component has come from many drivers including moral, social, regulatory, best practice, and pure economics. All these factors are incorporated into the social license to operate for a modern mine.

The terminology that "safety is king" on any mine site holds true with high emphasis on all aspects from minor first aid treatments to fatalities and catastrophic failures. Any injury is seen as significant and measures are implemented to prevent recurrence. The focus on safety extents from the head office, through the site management all the way down to individual employees and contractors.

The emphasis on health, environment and community issues has historically lagged behind the safety emphasis. But today, these aspects are often treated in the same manner as safety issues, and this has raised their profile in the mind of management, and to a lesser extent the employees and contractors.

14.2 HSEC across the uranium life cycle

Uranium operations go through a number of stages as they develop and each of these stages may have a different level of emphasis on HSEC requirements. The stages may be simplified into the initial exploration for the resource, the design of the operation (including pilot plants and testing of material and processes), the

Uranium for Nuclear Power. DOI: http://dx.doi.org/10.1016/B978-0-08-100307-7.00014-4

actual operational stage, rehabilitation, and remediation of the site, and the eventual handover and/or long-term surveillance of the site. Also of high importance is the type of operation. Common uranium operations can be separated by mining methods such as in situ recovery (ISR), open cut, or underground operations.

During exploration, there often is not as high an emphasis on HSEC requirements, often due to the exploration being undertaken by either smaller companies or contractors who may not have the same degree of HSEC focus as an operational mine. Often exploration is often undertaken in remote localities and hence operators are away from the more intensive scrutiny that a production facility would have. The potential for a large incident is also smaller although the remote location of exploration may increase the consequence due to the reduced ability to undertake remedial actions (for example, an injury which could be quickly treated at a operation may require on-site stabilization and considerable delay before full medical attention is obtained). Due to perceived reduction in risk at remote sites, correct emphasis on the aspects beyond safety is often difficult. Community aspects have historically been an issue, with the exploration activities often setting the scene for future community interactions. There has been a marked improvement in HSEC awareness during exploration, but in general, there remains a gap.

The design stage is where the biggest potential HSEC risks can be identified and addressed. In particular, environmental risks and consequences can be engineered out of a facility by appropriate design (ie, proper bunding, tailings retention system design, dust control, etc.). Similarly, community issues can be addressed in a proactive manner (ie, community consultation, identification of significant sites, indigenous and local employment opportunities, etc.). Safety and health issues should also be considered (such as appropriate access, ventilation systems, etc.) but this is often more a standard approach to the design of operational structures. Some mining methods will require additional emphasis on one aspect of HSEC (eg, ground control in underground, airborne dust in open cut, groundwater in ISR) due to the inherent risks for that specific mining methodology. It is also in the design stage where detailed knowledge about the preexisting background conditions should be obtained, as this will be critical information for the final remediation of the site post operation (this has historically been a major weakness for many uranium operations globally).

During production is the stage where you can see the highest emphasis on HSEC requirements. It is often the time where there is the highest level of scrutiny (both internal and external) and significant investment in HSEC is part of the operational approach. Most uranium operations will have specialist teams for each aspect of HSEC. The highest focus will always remain safety although some of the other aspects will gain a similar emphasis due to site-specific factors (eg, environment if in an ecologically sensitive area). In the production stage, the management systems will continually monitor HSEC performance and any issue can be corrected rapidly and appropriately. Government regulatory authorities will also be involved and any breach will generally be investigated and acted on rapidly. Public scrutiny is also likely to be high and unresolved issues are potentially ground for community concern and conflict. The emphasis on HSEC is likely to be at a higher level than a conventional mine due to the greater regulatory attention on uranium mining.

Radiation management is likely to be given a high emphasis (potentially at a far higher level than qualitative risk assessment would warrant) from both the health and environmental perspective.

During rehabilitation and remediation, the emphasis would shift to a higher emphasis on environment and community aspects. Safety would remain an immediate focus but, due to both the comparative simplicity of the rehabilitation tasks and the presence of preexisting control systems from the operations phase, would comparatively have reduced importance. Due to the long half-life of residual radionuclides, this stage will also have to allow for long-term containment and mitigative structures. Negotiation with both government and community on what are the short and long-term targets for remediation success also have to be completed and implemented in the rehabilitation structure. A particular aspect for consideration with uranium mines is that radiological aspects for rehabilitation should consider the preexisting background conditions. Closure criteria should be those above the premining values and not the total radiation dose (critical for surface uranium deposits where it is possible to have a lower post closure radiological exposure than was in place prior to mining).

The final stage is the handover of the site and any residual long-term monitoring and surveillance needed. Full handover is the generally preferred result at the close of mining but this is not always possible in all jurisdictions. Irrespective of the degree of handover there is likely to be a requirement for some form if monitoring and surveillance undertaken by the government and/or the mining company. The emphasis will be strongly directed at the environmental aspects but there may be some community issues also being considered (eg, final land use, unauthorized entry, and evacuations). Table 14.1 shows the relative importance of the HSEC parameters for the various types and phases of uranium mining and processing. The relative importance is generic only and actual importance will also be driven by site specific factors for an individual operation.

14.3 Managing health in uranium operations

14.3.1 Introduction

14.3.1.1 Radiological health aspects

The management of health is similar to conventional mining operations with the exception of the need to manage radiological risks. Due to the nature of uranium and its more radioactive decay products, radiation protection will be a priority throughout all stages of the mine although most critical during the operational phase. An objective review of the potential risks associated with most modern uranium operations is that the radiological risk is actually one of the lesser "true" risks in comparison with other conventional health risks. However, most uranium mines will require extensive oversight of the radiological aspects as this is seen as a key component of the management of uranium mines and this will generally be undertaken irrespective of the level of actual risk.

Table 14.1 Relative importance of HSEC parameters for various types and phases of uranium mining and processing

HSEC parameter	Exploration	Design			Production			Rehabilitation			Handover		
Exposure pathway		ISR	Open cut	U/Ground	ISR	Open cut	U/Ground	ISR	Open cut	U/Ground	ISR	Open cut	U/Ground
Health													
Radiological—Gamma	**	*	**	****	**	***	****	**	***	*	*	***	**
Radiological—RDP	*	***	**	*****	**	**	****	*	**	**	*	****	***
Radiological—Dust	**	*	***	**	**	**	**	*	****	***	*	**	**
Conventional health hazard	**	***	***	****	***	***	****	**	***	**	*	**	*
Disease prevention	**	**	**	**	****	****	****	**	**	**	*	*	*
Safety													
Incident	**	***	***	***	**	**	**	**	**	**	n/a	n/a	n/a
First Aid	***	***	***	***	***	***	***	***	***	***	n/a	n/a	n/a
Medical Treatment Injury	****	***	***	***	****	****	****	****	****	****	n/a	n/a	n/a
Restricted Work Case	****	***	***	***	****	****	****	****	****	****	n/	n/a	n/a
Lost Time Injury	****	***	***	***	****	****	****	****	****	****	n/a	n/a	n/a
Fatality or Total Permanent Disability	*****	***	***	*****	*****	*****	*****	*****	*****	*****	n/	n/a	n/a
Significant Potential Incident	*****	***	***	*****	*****	*****	*****	*****	*****	*****	n/a	n/a	n/a

Environment													
Gamma	**	*	**	**	**	**	**	*	****	***	*	***	***
Dust	***	*	****	***	*	****	**	*	**	**	*	***	**
Radon	*	**	****	****	***	****	****	**	****	****	**	****	***
Groundwater	*	*****	***	***	*****	***	***	****	***	***	*****	***	***
Surface water	**	**	***	***	***	***	***	**	***	***	**	****	***
Nonhuman Biota	*	***	***	***	***	***	***	**	***	***	**	***	***
Community													
Community consultation	***	****	*****	****	***	****	****	***	****	****	***	*****	****
Local Benefits and Detriments	**	***	****	****	****	****	****	***	***	***	***	****	****
Significant cultural sites	****	***	****	***	***	****	***	***	***	***	**	**	**
Community support	***	****	****	****	****	****	****	***	***	***	***	***	***
Closure criteria	**	**	***	***	***	***	***	***	****	****	****	*****	*****

n/a, Not applicable; *, Very low importance; **, Low importance; ***, Normal importance; ****, High importance; *****, Critical importance.
Note: The importance rating is generic only and individual aspects of the operation and site-specific factors need to be considered.

Uranium is a naturally occurring radioactive material (NORM) that undergoes a series of decays through the uranium decay series. Generally when mined, the uranium is in equilibrium with all the decay products, which means that radiation protection from all decay products must be considered. During uranium mining, there are three main exposure pathways to be considered: direct gamma, inhalation of radon decay products (RDP), and inhalation of radioactivity in airborne dust.

Direct gamma exposure occurs when personnel spend significant periods of time in proximity to large quantities of material containing uranium or a specific decay product, radium-226. In most cases, gamma dose rate is relatively stable over time so monitoring can easily be undertaken to identify areas with higher risk and appropriate corrective actions implemented. For underground mining operations gamma exposure is the largest exposure pathway for routine practices and for high-grade deposits can be a limiting factor for mining approaches and technology (eg, the high-grade Macarthur River and Cigar Lake mines). Protection mechanisms for external gamma radiation can be summarized as:

- Time: reduce the time spent in higher dose rate areas via appropriate planning, design and equipment. For example mining from outside the immediate ore body or restricting access from area of known high dose rate.
- Distance: maximize the distance from high dose rate areas. For example situating offices and workshops away from ore stockpiles or locating equipment controls from the active ore face.
- Shielding: placing benign material between active ore areas and the working area. For example, shotcrete underground can be used as shielding for the walls and ceiling and waste rock can be used as a road base in high-grade ore zones.

Inhalation of RDP are due to one of the uranium decay products, radon-222, being gaseous. As such, it can diffuse out of the primary material and enter the air in work areas. Radon itself is an inert, colorless and odorless gas with a half-life of 3.8 days and as such has only a limited direct effect on the exposure to an individual (basically it is breathed in and then breathed out giving only a short residency time in the lungs). However, radon decays into a series of short-lived (less than 27 min half-life) decay products. As these are particulates when breathed in, they remain within the lung and hence give rise to the exposure pathway. RDP are generally only a significant exposure pathway when there are significant volumes of uranium-bearing material in an area where ventilation is restricted. Generally, it is most significant in underground mining operations but can be significant during vessel entries, ore reclaim tunnels, or where degassing of process pregnant liquor occurs in ISR operations. Radon is not generally an issue for the storage of the final uranium product as by this stage in the process the parent of radon, radium, has been removed so no radon generation. The main methods for control of radon decay product exposure are:

- Reduce radon emanation: stop radon entering the work air by blocking off mined out zones, reduce the active radon generating surfaces in a mine, and reduce or seal groundwater ingress.
- Ventilation: provide active one pass ventilation with priority for fresh air to active work areas and prevent recirculation of air containing radon.

- Monitoring and time: Radon and RDP concentrations can change very rapidly so active monitoring programs may be required. In areas of enhanced RDP, restriction on access or working time can be implemented to reduce dose.

Inhalation of radioactivity in airborne dust arises when the inhaled air contains dust particles, which include uranium and/or its decay products. Generally, only the longer-lived decay products are significant: uranium-238, uranium-234, thorium-230, radium-226, lead-210, and polonium-210. In the mine, all the radionuclides can be assumed to be present but following processing disequilibrium occurs. The tailing material is reduced in the contribution from the two uranium radionuclides (^{238}U and ^{234}U). The final uranium product generally only has those two uranium radionuclides. In most operations, inhalation of radioactivity in dust only becomes the dominant radiological pathway for workers in the final product recovery area. However, if an operation has very high dust levels or specific dust generation activities than it can be a significant pathway in other areas. Protection methods for this pathway are similar for general dust control mechanisms. Specifically, dust-generating activities and processes should be identified and controlled (ideally using engineered solutions), access to areas of uncontrolled dust generation should be restricted, and personal protective equipment should be used in areas where other methods of dust control are not effective.

There are a number of other minor pathways of radiological exposure: ingestion of radioactivity, radioactivity through wounds, and adsorption of radioactivity through the skin. Normal occupational practices including personal hygiene, designated crib area, covering of wounds, and cleanliness should ensure that these pathways are not a significant source of exposure. Radioactivity through wounds (injection) is only potentially significant in the event of an emergency although generally the radiological risk is far smaller than the medical aspects of an injury.

Radiological aspects of mining are not necessarily restricted to just uranium mining. The mining of mineral sands are often integrally associated with exposure to natural radioactive material containing thorium-232 and its decay products. In addition, other underground mines can provide significant exposure to RDP particularly where ventilation is not a primary concern or where air is recirculated. For example, deep gold mines and some coal mines can have radiological exposures far in excess of those received in modern uranium mines. Dating back to the prenuclear era, the mining for other metals (such as silver in areas rich in pitchblende) were associated with high radiological exposures and resulting health impacts.

14.3.1.2 Conventional health risks

Although radiological risks are the major difference between uranium mines and other metal mines, there remain a number of conventional (nonradiological) health risks that require management. These risks will vary according to the mining methodology, processing methodology, ore composition, and local site factors. Common health risks that may need to be considered include the following:

- Physical health risks: thermal stress, vibration, noise, general airborne dust, ultraviolet radiation
- Chemical health risks: diesel particulates, silica, acids, alkalis, carcinogens, asbestos, heavy metals, ammonia, acid mist, chemical irritants (though the mechanism of harm of some of these is physical)
- Biological hazards: legionnaires disease, sewerage pathogens
- Ergonomic hazards: manual handling

From the health risk perspective, hearing loss from noise is perhaps the most common health impact, followed by manual handling related impacts. With respect to carcinogenic impacts, the highest risk is ultraviolet radiation from the sun. Specifically associated with uranium mining, there are some more significant health risks that are common. From the emergency perspective, the potential for a large release of ammonia gas ranks as one of the highest risks for a health impact (ammonia is a commonly used reagent for uranium precipitation). In underground mines, diesel particulates are a common health issue and, depending on the makeup of the ore body and surrounding rock, silica inhalation may be a concern.

It is always important to keep the risks in perspective. In absolute terms, the risk from radiological hazards in a modern uranium mine are far smaller than the risk from conventional health hazards. In developing an effective health regime, it is important to keep a balanced program that addresses the more significant conventional health hazards while maintaining control over the radiological hazards associated with uranium mining.

14.3.1.3 Disease prevention and health care

Although often not considered an integral part of the health aspects of a mine, the consideration of diseases and other health impacts should be a component of the overall HSEC program. This is not specific to uranium mining, although there may be some areas where additional work is required due to radiological concerns.

Mining often occurs in remote areas and utilizes a workforce that may not be indigenous to the area. This can significantly raise the risk for a range of health-related impacts and the affect of this can result in larger consequences than all other health (and for that matter safety) related impacts on the workforce. The hazard may simple be due to the displacement of the workforce from their normal support base, or may be more directly related to the operation such as the change in diet at the mine accommodation. Issues to be considered include:

- Disease: AIDS, malaria, sexually transmitted diseases, parasites (water borne or other)
- Lifestyle: obesity, mental illness, alcoholism, drug use
- Location: availability of health care, climatic extremes, dangerous wildlife, distance to hospitals and specialist treatment

The health impacts are not necessarily negative and in fact, there can be some very significant positive health impacts. These range from the availability of good food and health care (sometimes for the first time) to positive mental health effects. It is very common for there to be a "healthy worker" effect where the overall health impact of a mine is positive. In these cases, workers may have significantly better health than the general population or the population in the area where the worker originated.

With respect to uranium mines, the disease and general health issues are identical to those for a conventional mine. The one exception is some additional concern around radiation. Although this is not a direct effect of the radiation, it relates to fear or social stigma associated with working with uranium and can cause secondary health effects. Good worker communication on radiation and potentially the use of health screening can mitigate against these potential health impacts.

14.3.2 Current versus historical practices

The fact that in modern uranium mines radiological health risks are low is a testament to the improvement in the control of health hazards over time. When uranium mining commenced in earnest in the 1940 and 1950s, little was known about potential health risks. As such, there were few mitigative measures in place to reduce worker exposure and the doses received reflect this. In particular, the contribution of radon and RDP was very high due to a lack of ventilation and close proximity to the ore zones. Dust controls were either minimal or nonexistent and gamma monitoring was really only to identify the highest ore grades and not for safety purposes.

As a result of these historic practices, such as in very small-scale gouging operations, uranium mine workers received substantially higher doses than those permitted today. The combination of high doses and large workforce numbers has led to a number of epidemiology studies being undertaken on these historic mine workers. There were proven health effects from the uranium mining and specifically an increased risk of lung cancer primarily due to the inhalation of RDP.

In modern uranium mines, there are strict controls on radiation exposure and annual doses are in almost all cases well below the internationally accepted limit of 20 mSv/year (100 mSv in any 5-year period with no dose above 50 mSv in any single year). Average doses to uranium workers are typically similar to the natural background radiation (in the range of 1–4 mSv/year) and maximum doses for most operations are under half the limit (<10 mSv/year). At these levels, there is no direct indication of any short or long-term health effects. In fact, the doses are less than that experienced in a number of other professions not normally considered as radiation related (eg, international aircrew, deep gold miners, cave guides).

For modern uranium operations, nonradiological factors have far more potential for health impacts than radiological doses. The most common work related heath impact is hearing loss due to noise exposure and the highest fatal health effect is likely to be related to ultraviolet radiation from the sun (although this has significantly reduced recently due to the compulsory use of protective clothing). However, the impact of these nonradiological hazards is likely to be far smaller than the impact of other health factors such as disease and lifestyle factors.

14.4 Managing safety in uranium operations

14.4.1 Introduction

The prominence of safety as the first priority for a modern mining operation is one of the most significant historical changes within HSEC. This is irrespective of whether the operation is targeting uranium or some other commodity although there can be higher levels of scrutiny if it is a uranium operation.

The importance of safety is obvious from the first contact with an operation. Before you even arrive at the site entrance, you will almost certainly see signs advising of the importance of safety (and sometimes more general HSEC). This comes in many different forms (zero harm, target zero, safety is the first priority, etc.) but the clear message is that safety should be considered above all other aspects and the aim is for there to be no safety risk. This is reinforced by the induction process, which is generally required before access is allowed. This can range from a simple visitor induction to more workplace- and task-specific inductions for workers. These inductions generally are required periodically and a reinforced via regular safety meetings and toolbox talks. In the event of a safety issue, it is generally widely communicated to the workforce. Management will also have a strong commitment to safety and this will be reflected in management targets and often performance bonuses.

More advanced sites will often have visible measures to establish strong safety culture among the workforce. The safety culture is designed to ensure that at all levels of the workforce safety is actively encouraged and unsafe acts are not tolerated. This represents the move away from the policing type approach to a more "community" based approach, which reflects "the safe way is the way we work here." From the perspective of safety, this is a far more effective approach as it is self-fulfilling and the onus for safety is firmly with the individual.

The reason for the strong push for safety is simply that safety is good business and moral sense. Any injury is bad for business and larger-scale incidents (such as multiple injuries or fatalities) can be catastrophic to an operation as well as families concerned. A modern operation is legally required to provide a safe place of work and failure can do so can have heavy repercussions including legal action, fines, imprisonment, and loss of license to operate. However, equally of importance is that no one wants to be responsible for the injury or death of a workmate and this is a very strong motivator for a safe workplace.

There are many potential levels of safety impacts that are considered by a mining operation. A fundamental approach to safety is that all levels of impact should be reported and appropriate actions taken to prevent reoccurrence. Failure in reporting systems are often a first indicator of poor safety culture. Terminology can change from jurisdiction to jurisdiction but the approach is fundamentally the same. These levels of safety impacts are:

- Incident: No injury occurs, but there was potential for an injury. These can be minor but provide an early warning of potentially more serious safety issues. The management of incidents varies widely between operations and ranges from no action to recording and investigation.

- First aid case (FAC): An injury of some form occurs and it requires only minor treatment. This could be as simple as a paper cut or an insect bite. These are recorded both as an indicator of potential higher risks and as a precaution for the worker in case the injury deteriorates (it has occurred that a simple mosquito bite has become infected and given rise to more serious safety impacts).
- Medical treatment injury (MTI): A MTI is when the injury requires some form of treatment beyond normal first aid. For example, if the injury requires a stitch or if some form of medication needs to be prescribed, it is generally classified as a MTI.
- Restricted work case (RWC): A RWC is when an injury occurs and because of the injury, the worker cannot complete normal duties. The worker may be able to be reassigned to a different task so there is no lost time. For example, a manual laborer could sprain their ankle and not be able to do their normal task but could be assigned to the office for paperwork duties.
- Lost time incident (LTI): An LTI is when a worker cannot work for a whole shift as the result of an injury. For example, if a manual laborer seriously sprains their ankle and cannot access site safely and hence cannot complete the following day's shift.
- Fatality: A fatality is where one or more workers lose their life due to an accident in a workplace.
- Total permanent disability (TPD): A TPD is where a worker is injured to such an extent that they cannot work ever again as a result of their injuries.
- Significant potential incidents (SPIs): An SPI is where some incident occurs or is about to occur that could have resulted in a fatality or TPD. These can be seen as "free lessons," as they identify high-risk tasks where no one got hurt but the potential was there. Common SPIs include potential for electrocution and working at heights.

MTI, RWC, and LTI are often grouped together and are considered reportable injuries. The frequency of reportable injuries as a fraction of collective working hours is a common key metric for safety performance. For example, the All Injury Frequency Rate is defined as the number of reportable injuries per 20,000 working hours by a workforce. This is considered a lagging indicator of safety performance as it indicates previous performance. Leading indicators measure proactive safety initiatives including incident investigations and time in field for management.

For uranium operations specifically, there is no key distinguisher from other mining operations. Generally, the safety practices and indicators are identical and variations are related to the individual company's approach to safety. The only potential difference is that uranium operations often come under heavier government and public scrutiny than comparable conventional mines. As such, even higher levels of safety performance may be demanded and the consequences of safety impacts may be higher. Key public interest groups may use safety as a means of opposing the uranium operation and there may be additional regulatory bodies overseeing the safety aspects.

14.4.2 Current versus historical practices

From the historic perspective, the improvement in safety in mining has been a major success for the industry. Mining traditionally is considered a dangerous industry to work in and this was the case in the past. However, review of modern operations show that the safety performance of mining is now far better than most other heavy industries and often has better safety performance than industries considered safe (eg, safety performance of mining is far stronger than agriculture or heavy industry and can be better than such areas as office work). Because of the heavy scrutiny of uranium mining, safety performance is often either as good as or better than conventional mines.

14.5 Managing environmental impacts in uranium operations

14.5.1 Introduction

The protection of the environment is the dominant public concern associated with uranium mining. Uranium mining has all the conventional risks of mining with the additional potential risk associated with radiation. Due to the long half-lives of the radionuclides in the uranium series, the potential radiological risk is seen as very long-term and this has greatly raised the perception of risk around uranium mining.

In terms of real environmental impacts, conventional hazards are almost always the dominant risk. Often the biggest impact of a mine is the direct physical disturbance or the "footprint" of the operation. In this footprint the environment will be directly affected either by being removed (mined), covered (stockpile, plant areas, water/runoff storage and tailings retention structures), or modified to totally disrupt the preexisting environment (gardens, stabilization areas). Also very significant is the introduction of external biota (such as weeds and feral animals) which can provide both medium- and long-term impacts on an area (both on-site and along transport corridors). In terms of chemical impacts, once again conventional impacts generally dominate, including such impacts as salt deposition (from ventilation systems or road spraying) or the transport of soluble compounds in ground or surface waters (eg, fish kills from the transport of magnesium sulfate).

Radiological impacts are generally very small and do not pose any significant risk to the environment or surrounding human populations. However, it is of critical importance in the minds of the public and hence is given extensive attention at uranium operations. Opponents of uranium mining often quote the half-lives of the radionuclides (^{238}U 4.5 billion years, Th-230 75,000 years, Ra-226 1600 years) as a reason for not allowing the mining. However, this overlooks that conventional elements (such as arsenic, mercury, cadmium, lead, etc.) with immediate toxicity have an infinite physical half-life so mining with these impurities have similar long-term concerns. Uranium mining also has one unique environmental pathway, being the generation and emanation of the inert gas radon.

As with conventional hazards, knowledge of the chemical and physical properties of each long-lived radionuclide is required to determine their transport and impacts on the surrounding environment. Often the impacts on the environment are incorrectly assessed due to either consideration of only the parent radionuclide, uranium, without appropriate attention paid to the decay products. The behavior of all the key radionuclides in the uranium series is complex and the following is a summary, but care must be taken as the behavior can change according to site- and pathway-specific factors.

Uranium in equilibrium: For ore and waste rock, storage and handling the radionuclides will generally be in equilibrium that simplifies the consideration of environmental pathways. For the distribution of airborne material (fugitive dust) this type of material often dominates. Direct exposure pathways can be assessed using an equilibrium figure and individual determination of all radionuclides may not be required. Secondary pathways (such as being incorporated into aquatic pathways) will cause disequilibrium and therefore consideration of individual radionuclides may be required.

Uranium-238 and U-234: These two isotopes of uranium are physically and chemically identical and are almost always in equilibrium (some slight disequilibrium is possible due to alpha recoil of ^{234}U, which may increase its solubility.). The half-life of both uranium isotopes are very long (4.5 billion and 250,000 years, respectively) and as such the concentration can be measured using standard chemical techniques such as Inductively Coupled Plasma Mass Spectrometry (ICPMS). In the environment the solubility is very dependent on both pH and oxidization state. As a rule, uranium is most soluble at low and high pH and partially soluble at neutral pH.

Thorium-230: With a half-life of 75,000 years, thorium-230 dominates the half-life of tailings deposits. Thorium is insoluble in neutral pH and is unlikely to be significantly transported in ground or surface waters (aside from direct physical transport). In acid tailings systems there can be crusts enhanced in Th-230 that, when disturbed, can be aerially dispersed. Thorium-230 can be measured by radiochemical or gamma spectrometry and the chemical measure of natural thorium (thorium-232) may be used as a chemical proxy to determine where the thorium reports to in a process.

Radium-226: Has a half-life of 1600 years and is generally one of the most critical for environmental pathways due to its high solubility in near neutral pH. It also is the parent of radon-222 and hence is the source term for radon generating structures. Finally, because the majority of the gamma radiation from the uranium series arises from bismuth-214 (one of the RDP) the concentration of radium-226 will dominate the gamma signature of a structure. Radium-226 is relatively easy measured by gamma spectrometry (after sealing and decay to ensure equilibrium with the RDP). Chemically it is similar to barium, calcium, and magnesium.

Radon-222: Radon-222 and the subsequent inhalation of RDP dominate the airborne pathways associated with uranium mining. Radon is an inert gas with a half-life of 3.8 days. This time is sufficient for the radon to leave its host mineral, diffuse through the material, and enter the general atmosphere. It can then be dispersed by the prevailing winds and travel significant distances from the uranium operation. If the host material is in an area of reduced ventilation, high concentrations can be experienced leading to significant radiological exposure.

Lead-210: Lead-210 has a half-life of 22 years and is most important because its levels in the natural environment are often far higher than the other radionuclides (by up to an order of magnitude). This is because it is the long-lived decay product of the RDP. In the natural environment, radon escapes from the ground and decays in the atmosphere, producing a significant natural source of lead-210. The lead-210 preferentially accumulates on near atmosphere surfaces (eg, the top layer of soil, in foliage or specifically deposited in some plants like lichens). This natural enhancement in lead-210 concentration is often misinterpreted as a uranium operation related impact. It can be measured by beta radiochemistry or by gamma spectrometry.

Polonium-210: Polonium-210 has a half-life of 138 days and is a pure alpha emitter. As such, it can only be measured using radiochemistry and alpha spectrometry, which means it is often missed by normal measurements. As the decay product of lead-210 (via bismuth-210), it is also substantially naturally enhanced in environmental samples. It also has a very low volatilization temperature so will be preferentially emitted by operations involving heating of material. As such it often important as a NORM in products thought of as nonradioactive (eg, coal, iron ore) and may be important in operations where uranium is a byproduct (and part of the processing involves heating). When measuring polonium-210 care must be taken not to heat the sample (eg, ashing of vegetation samples), as otherwise the polonium-210 may be volatilized out of the sample.

Tailings material: Tailings material requires a special mention from the radiological perspective. Tailings will retain the majority of the radioactivity from the uranium series (only the uranium radionuclides and their short-lived decay products have been removed). The mobility of the radionuclides within the tailings structure will be heavily dependent on the pH of the material and also the underlying soil and rock may similarly change the solubility. For example, thorium-230 is mobile in acid tailings but rapidly precipitates out of solution as the water is neutralized by underlying soil and rock.

For both conventional and radiological hazards the common pathways of concern are:

- Gamma
- Dust
- Radon
- Surface water
- Groundwater
- Nonhuman biota

Gamma: Direct gamma is only important directly adjacent to areas associated with a uranium operation. The contribution decreases rapidly with distance and once more than a couple of hundred meters distant the contribution is virtually indistinguishable from the natural background. From the environmental standpoint, the only significant impact is post closure on the rehabilitated structure. From the public perspective, this pathway is important if there is any significant occupancy time directly on the rehabilitated structure. Gamma contribution can be decreased by using covering materials that have low or background levels of radionuclides.

Dust: Dusting is generally more significant during the operation of a project where there is significant dust raising operations. Stockpiles, crushing, ore transportation and tailings storage are all potential sources of dust. Active measures to reduce dusting can be taken during operations, and post operations passive measures, such as benign covering material, can be used to prevent dusting.

Radon: The emanation of radon-222 from tailings and stockpiles and its subsequent aerial distribution is generally the most significant environmental pathway during operations. With its 3.8-day half-life, radon-222 can travel reasonable distances but does become diluted as distance increases. Once away from the immediate area of the operation, the contribution of operational radon is generally far smaller than the natural background radon. Post closure radon is also a critical pathway, particularly if there are nearby populations. Some of the highest public doses from uranium mining have arisen from people either living over historical sites or using waste material as building products for dwellings. In these cases, the radon can concentrate within the restricted ventilation of a dwelling and contribute high doses, which may be several times the natural background dose.

Surface water: Discharge to surface water during operations can occur either intentionally or unintentionally. The impact from the transport of dissolved material is generally small but the potential for transport of solids can be more significant. During planned discharges, water quality can be monitored and if acceptable, liquid material can be discharged into a surface water body. In these cases, the impacts are almost always negligible because discharge limits are designed to protect the receiving environment. Unplanned discharges can occur for a variety of reasons, such as an accident or a severe weather event. This can lead to localized impacts (such as chemical induced fish kills or deposition of water dispersed solids) or more distant effects due to bulk transport of material. Post closure the risk is mainly associated with water-based erosion of rehabilitation structures. Water flow is a powerful erosion tool and design of rehabilitation structures should account both for local water courses and also for water flow off structures under high rainfall events.

Groundwater: Groundwater transport of dissolved material can be significant during both operations and also post closure. With respect to impact on the environment, groundwater dispersal is the most significant where the groundwater is shallow and interacts with the surface environment. From the radiological perspective, the most critical radionuclide is radium-226 due to its solubility in neutral pH waters. Uranium-238 and uranium-234 are also of importance. The consideration of the groundwater pathway is critical for ISR uranium operations.

Nonhuman Biota: Historically the radiological effect on nonhuman biota was only considered as it relates to the potential for exposure to humans under the premise that if individual humans are protected then the environment was too. However, this approach has significantly changed in the early 2000s with the release of International Commission on Radiological Protection (ICRP) recommendations on protection of the environment (ICRP103, ICRP 108). Species of biota now require protection even in the absence of human receptors and this has been a major shift in emphasis from the environmental protection standpoint. Due to the comparatively low level of radioactivity associated with uranium operations, the actual radiological impact on the environment is almost always very low and of far less significance that physical (mainly habitat removal) biological (introduction of non-native species) or chemical effects.

14.5.2 Current versus historical practices

Historically, notably pre-1970s in the west, environmental protection was not a major priority for uranium operations and there were and still are examples of poor environmental practice associated with uranium mining. During historic operations, these poor practices generally focused around poor handling of process materials and chemicals, poor waste management practices or failure of containment systems. Environmental impacts continue from legacy uranium operations where poor practices were utilized. These generally result from poor tailings and waste rock management, lack of control of erosion, release of secondary material due to chemical activity (acid rock drainage), external events or interactions, or physical risks. Proper rehabilitation of uranium sites was either not attempted or was poorly undertaken and this has provided numerous examples of legacy sites requiring corrective actions to prevent or reduce environmental impact. Examples of legacy sites can be found worldwide and remain a focus of many national and international organizations (eg, Wismuth in Germany, Mailuu Suu in Kyrgyzstan, UMTRA in United States, Rum Jungle, and South Alligator River in Australia. These legacy sites reinforce the need for both proper planning of waste disposal during operations as well as well considered and implemented rehabilitation plans. These impacts are not restricted to uranium mines but also apply for a wide range of conventional mining as well.

Modern uranium mines are generally far more proactive in preventing or mitigating environmental impacts. This commences well before they start operation and include often very prescriptive approval processes (often far in excess of conventional mines). Planning for the mine include consideration of environmental impacts and the incorporation of practices to prevent future environmental impact. During operations, environmental protection is incorporated into the ongoing management of the mine, and these activities and impacts are scrutinized by regulatory authorities.

Perhaps the biggest area of improvement of environmental protection is consideration of the rehabilitation and post closure phase of the operation. This is generally the most significant phase from the environmental perspective and has historically been the phase responsible for the most significant environmental impacts. Modern mines are generally required to consider rehabilitation during the planning and operational phases and it is common practice for costing for the rehabilitation phase to be guaranteed during operations. The physical aspects of the rehabilitation are to provide containment of waste materials from the general environment for long periods. One common criticism is the need to protect for long periods due to the long half-life of the radionuclides but this is not normally the real determining factor as impacts are likely to be dominated by physical and/or chemical effects not radiological impacts. In general, the requirements for rehabilitation and close out of uranium operations are far more stringent than for conventional mines and this means that the long-term environmental impact is likely to be significantly less than that from conventional mines.

14.6 Managing community impacts in uranium operations

14.6.1 Introduction

For modern mines, one of the most critical components is community support and often the future of a mine relies on obtaining a social license to operate. Although this is important for a conventional mine, for a uranium mine it gains far more importance because of the high potential for significant and organized opposition (on a national or international level). On the positive side, the community can gain significant benefits from a mining operation and these benefits can be one of the most significant factors influencing the approval process. However, the definition of the "community" is often difficult and open to (mis)interpretation. Most of what is expounded next applies to any mining operation today.

Community consultation: Community consultation is a necessary component at all stages of a mine. It is also inherently difficult and often difficult to quantify and control. It relies to a significant degree on the establishment of trust between the proponent of the mine and the communities. Trust requires a long-term commitment but can easily be lost and is far more difficult to reestablish. It also can be heavily influenced by historic and external forces and does not operate on a fixed timetable as required by mining projects.

One of the first and most critical aspects is defining what is the community (this can be a multiple answer question) and how it should be engaged. The incorrect choice can risk alienating important sections of the community. Community may require several levels of definition including indigenous communities, local communities, business, regional, national and international stakeholders. The level of involvement will alter with each definition as will the approach. Each community is likely to have different expectations and desires and they will sometimes be in opposition. The communities may also change over time either by shifts in attitude, movement or intergenerational change.

There is no one correct mechanism for undertaking community consultation. A process that may work extremely well with one group may be completely unacceptable for another sector of the community. For example, high-tech presentations may be effective when talking to government and media groups but may be frowned upon by indigenous groups who are more closely aligned with the land (eg, overhead photographs may be appropriate for some groups but ground level imagery will work better with others). It is vital to understand and relate to the group being consulted with and to adjust the communication to suit. "One size fits all" communication processes are unlikely to work as they rely on the concerned individuals to adjust, which is unlikely to happen.

Also important is the realization that not all in the community will be supportive and there will almost definitely be some opposition. This opposition can vary from ideological differences to local concerns to organized international action. Not all the opposition will be able to be solved, so a balance of community support is likely to be the best-case scenario for a uranium operation. The level of support and opposition is likely to ebb and flow over the mine's operational life. During initial phases, it is likely to be extremely high and then after a period of operation reach a low-level representative of the mine being "status quo." However, this can change in the event of either internal or external events such as an incident or an external campaign of opposition. The community concern can be expected to increase again once closure is approaching prior to reducing post closure.

Local benefits and detriments: One of the key components of community support or opposition is the local benefits and detriments. Mines are often a very important economic boost for a community and this can give a significant benefit for local people, a region or even a country. This boost can come in the form of direct and indirect employment, new facilities and infrastructure, regional development, improvement in support services (medical and educational are often very important), and new businesses. Often the mine becomes a major regional center providing a lifestyle and level of prosperity unknown premining. This can spread to affect a far larger region than just the mines immediate area and can lead to significant change in the overall standard of living. An example of this are the contribution of the Rössing Foundation to the Erongo region of Namibia such as the establishment of class leading hospital facilities (where Angelina Jolie gave birth to Shiloh) and other support services. Mines can also provide direct benefit to local indigenous people in the form of royalties and local employment opportunities as well as improved health and education.

However, the establishment of a mine can also have detrimental effects on the local community. The influx of people and new facilities can have a negative effect on communities. New diseases, use of alcohol and drugs, prostitution, disruption of social practices and increased economic division can become disruptive factors for a local community. These negative impacts should be identified and may require corrective action to mitigate the potential impacts.

Significant cultural sites: The siting of a mine is dependent on geology irrespective of the local community. As such, it is possible that the mine may be in an area of cultural significance to indigenous people and this is an important consideration for a mine. These significant cultural sites may become critical in whether a mine will proceed and will require close consultation with the indigenous people. This may require some areas to be totally annexed from mining or exploration and can require physical changes in how the mine is developed. This information may not be physically documented (may reside in verbal history) and requires an awareness that indigenous people may have a strong connection to the land that transcends economics or demand for minerals. It may also be that different segments of the indigenous community have different understanding of the cultural sites and this may require multigroup consultation. Obtaining and acting upon the requirements of the indigenous people should be obtained as early as possible to help facilitate mutually agreeable approaches.

Community support: Community support is not an absolute or measureable factor and can vary according to community segment and over time. For a mine, it is often a difficult component to manage because many of the factors governing the support are outside the control of the mine.

Closure criteria: Once mining is completed, the most desirable final solution is for the mine site to be returned to the local community with the mining company having no further involvement. In reality, this is often a difficult outcome and will be dependent on agreement on how the land is to be returned and under what conditions. Closure can mean the removal of a range of benefits and infrastructure, and can leave some detriments for local communities. The agreement on closure criteria is also vital for optimizing the rehabilitation phase as factors such as final land use and local demographics are essential for determining the rehabilitation standards. This can have significant economic effect, as the cost for rehabilitating for permanent habitation on-site can be significantly different from those for a transiting population or hunting and gathering, pastoral or forestry uses.

Closure criteria should be established such that it meets the community expectations and ensures human end environmental protection while being practical and reducing other impacts. Often the use of infrastructure and support facilities will be critical as with the completion of mining, basic necessities may become unavailable. For example, mining may have brought to an area water, electricity, and travel options that did not previously exist in an area. For a local community this is likely to be critical post closure, and may form a critical component of community consultation.

14.6.2 Current versus historical practices

Historically the involvement and success of community with uranium mining has been varied. In the early days of uranium mining, community involvement was at best an afterthought and was not a significant consideration in the start of mining. Often the mining, especially uranium mining, was undertaken as a matter of national interest and communities either adapted or were overridden in the development and operation of a mine. However, mines often developed to become a major regional center for the areas and provided significant benefits on the local and regional scale. These benefits can remain in place after the completion in mining and have established areas as regional centers providing benefits such as employment, medical and education support. Historically, the community has both positive and negative impacts and these vary from site to site and country to country. Even for a single site, the different segments of the community would have different opinions of the net impact based on their personal relationships and beliefs.

In modern times the role of the community is very much a consideration throughout the entire mine life cycle. Modern assessments of impacts require consideration of the community and often the approval process for a mine requires wide ranging community consultation. Often this is a limiting factor for the establishment of a mine and this commences from the first opportunity to explore an area through mine development and operation all the way to eventual rehabilitation and hand back to the local community. It remains an area which is often difficult for a mining enterprise to conduct as the definition of what is the community and how to measure acceptance involve many variables which are often not directly measureable. This is often made more difficult due to external concerns relating to the nature of uranium mining with organized opposition with extends into the international community. However, uranium mining also can provide the opportunity for significant community benefits that can greatly enhance the local and regional area. These benefits can raise an area in terms of critical factors such as education, medical support and employment, and can become a major economic driver to raise the standard of living.

14.7 Integration of HSEC and conclusion

The consideration of HSEC is now a critical component which will determine the success or otherwise of a uranium mining operation. HSEC is now integrated into the management of a mining operation and is no longer regarded as an addition or desirable component. Rather it is seen as a key enabler for a mine to commence and proceed, and is considered at all stages of the mining operation. Failure to consider and give due attention to HSEC will greatly reduce the potential for the mine to receive approval and will reduce its ability to be successful over time. Good management of HSEC and associated issues is seen as good business practice and a requirement for a modern mine.

Safe and secure packaging and transport of uranium materials*

15

Dylan Hem
RMIT University, Melbourne, VIC, Australia

15.1 Introduction

The transport of nuclear materials is strictly regulated and has an impressive safety record spanning six decades. No form of transport is subject to more stringent regulation. In transporting nuclear materials there has never been an incident resulting in a significant release of radioactivity.

There have been accidents over the years, but never one in which a container with highly radioactive material has been breached, or has leaked. This record is a tribute to the effectiveness of the regulatory framework as well as the collective competence of the entities performing packaging and transport activities.

Estimates from the UN's International Atomic Energy Agency (IAEA) and the World Nuclear Transport Institute (WNTI) indicate that about 20 million shipments of radioactive materials are transported annually. Each shipment is made up of either a single package or a number of packages moved from one location to another. The vast majority of these shipments, some 95%, relate to nonfuel cycle transports such as medical radioisotopes, smoke detectors, and cobalt sources for sterilization. Only 5% relate to fuel cycle transports (WNTI).

Despite being broadly classified with highly radioactive wastes and sterilization sources, uranium-containing materials in the front end of the nuclear fuel cycle, from mine to reactor via fuel cycle plants, have relatively low radioactivity and are less hazardous than many other materials, notably flammable fuels and industrial chemicals routinely carried by rail, road and ship.

15.2 Class 7 dangerous goods and regulatory practice/codes

Materials that pose a risk to people, property, and the environment are considered under the classification of "dangerous goods." To curb the potential hazards of these comprehensive regulatory frameworks are employed, which are dependent

* The majority of this chapter is taken from World Nuclear Transport Institute (WNTI) website and World Nuclear Association (WNA) information papers on transport. The author RMIT University has functioned as editor and collator, not as an authority, but acknowledges the kind assistance from Henry-Jacques Neau and staff at WNTI and Serge Gorlin at WNA in particular.

Uranium for Nuclear Power. DOI: http://dx.doi.org/10.1016/B978-0-08-100307-7.00015-6

on the characteristics of the material. Dangerous goods are defined in nine classes internationally,[1] with subdivisions under certain classes. Dangerous goods regulations define radioactive material (which has no subdivisions)—"Class 7" as any material containing radionuclides, where both the activity concentration and the total activity of consignment exceeds certain defined limits. The reason for regulation of radioactive material is because the radioactive decay of certain radionuclides emitting ionizing radiation. This unique characteristic of ionizing radiation may present a potential risk to people, property, and the environment.

Safety in transport is the responsibility of the consignor, but security is mainly the responsibility of the states concerned.

For the safe transport of Class 7 Radioactive Material, the IAEA has published advisory regulations since 1961. These regulations have come to be recognized throughout the world as the uniform basis for both national and international transport safety requirements in this class. Requirements based on the IAEA regulations have been adopted in about 60 countries, as well as by the modal organizations such as the International Maritime Organisation (IMO) and the International Civil Aviation Organisation (ICAO), and regional transport organizations. The IAEA has regularly issued revisions to the transport regulations to keep them up to date. The latest set of regulations is published as *Regulations for the Safe Transport of Radioactive Material*, 2012 Edition SSR-6.

The objective of the regulations is to protect people and the environment from the effects of radiation during the transport of radioactive material.

Protection is achieved by:

- Containment of radioactive contents
- Control of external radiation levels
- Prevention of criticality
- Prevention of damage caused by heat

The fundamental principle applied to the transport of radioactive material is that the protection comes from the design of the package, regardless of how the material is transported.

In 1993, the IMO introduced the *Code for the Safe Carriage of Irradiated Nuclear Fuel, Plutonium and High-Level Radioactive Wastes in Flasks on Board Ships* (INF Code), complementing the IAEA regulations. These provisions mainly cover ship design, construction, and equipment. The INF Code has been mandatory since January 2001 and it introduced advanced safety features for ships carrying mixed oxide fuel (MOX), used fuel, or vitrified high-level wastes (WNA, 2015).

15.3 Packaging of radioactive materials

The principal assurance of safety in the transport of nuclear materials is the design of the package, which must allow for foreseeable accidents. The consignor bears primary responsibility for this. Many different nuclear materials are transported,

[1] The *UN Recommendations on the Transport of Dangerous Goods* is the basis for most regional, national, and international regulatory schemes.

and the degree of potential hazard from these materials varies considerably. Different package standards have been developed by the IAEA according to the characteristics and potential hazard posed by the different types of nuclear material, regardless of the mode of transport.

Packages used for the transport of radioactive materials are designed to retain their integrity during the various conditions that may be encountered while they are being transported, thus ensuring that an accident will not have any major consequences. Packaging for radioactive materials includes, where appropriate, shielding to reduce potential radiation exposures. In the case of some materials, such as fresh uranium fuel assemblies or uranium hexafluoride, the radiation levels are negligible and no shielding is required.

Conditions that packages are tested to withstand include fire, impact, wetting, pressure, heat, and cold. It is important to reduce radiation doses to workers and the public to be as-low-as-reasonably-achievable principle by adopting best practice at the operating level. Packages of radioactive material are checked prior to shipping and, when it is found to be necessary, cleaned to remove surface contamination.

As with other hazardous materials being transported, packages of radioactive materials are labeled in accordance with the requirements of national and international regulations. These labels not only indicate that the material in the package is radioactive, by including the well-known trefoil, but also give an indication of the radiation field in the vicinity of the package (WNA, 2015).

Because safety depends primarily on the package, the regulations set out several performance standards in this area. They provide for five different primary packages (Excepted, Industrial, type A, type B and type C) and set the criteria for their design according to both the activity and the physical form of the radioactive material they may contain (WNTI).

In the front end of the fuel cycle, industrial, type A and in some cases type B packages can be used depending on the particular nuclear material. To prove their integrity under various conditions, they are subject to a series of tests during the design phase.

Industrial packaging is suited for low-level radioactive material, such as ore, oxide concentrate, uranium compounds, and low radioactivity wastes. The testing condition of this package (without fissile material) includes fall resistance from a maximum height of 1.2 m, which is variable depending on the mass of the packaging being stack tested (piling packages) (AREVA).

Type A packaging is suited for an average level of radioactive material, such as new nuclear fuels and medical radioisotopes. The testing condition of this package (without fissile material) includes:

- Water spray test to simulate heavy rain
- Fall resistance on a target[2] hard surface from a maximum height of 1.2 m for packages designed for solid materials, and 9 m for packages designed for liquid or gas materials
- Penetration test using a 6-km bar released from a height of 1 m

[2] In this context, IAEA *Regulations for the Safe Transport of Radioactive Material*, 2012 Edition SSR-6 refers to target as "The target for the drop test ... shall be a flat, horizontal surface of such a character that any increase in its resistance to displacement or deformation upon impact by the specimen would not significantly increase damage to the specimen."

Type A with fissile material:

- Fire test, impact resistance at 50 km/h (drop of 9 m) on a target hard surface
- Submission for accreditation from the Safety Authorities following an expert technical inspection by the regulator, such as IRSN (Institut de Radioprotection et de Sûreté Nucléaire) - *French institute of radiation protection and nuclear safety* in France (AREVA)

Type B packaging is suited for highly radioactive material, such as irradiated fuels, certain radioactive sources, plutonium, and vitrified nuclear waste. There are over 150 different kinds of type B packaging, which are tailored for varying material. The testing condition of this package includes:

- Impact resistance at 50 km/h (drop of 9 m) on a target hard surface
- Head-on fall resistance from a height of 1 m
- Fire resistance at 800°C for 30 min
- Immersion resistance up to 200 m for the most radioactive packages
- Submission for accreditation from the Safety Authorities following an expert technical inspection by the regulator

The IAEA regulations lay down these test procedures to demonstrate compliance with the required performance standards. The procedures, whereby the package integrity is related to the potential hazard, are important for efficient commercial transport operations. They also take into account the different conditions of transport characterized by the IAEA as follows:

- Conditions likely to be encountered in routine transport
- Normal conditions of transport (minor mishaps)
- Accident conditions

The regulations also detail marking and labelling provisions, requirements imposed on packages during transit, and prescriptions for their maintenance (WNTI), since most are reusable.

15.4 Security measures for transport of nuclear materials

In terms of security, shipments of radioactive materials must comply with relevant physical protection requirements developed by the IAEA, as well as the security requirements of the modal organizations such as the IMO and the ICAO. Security concerns since 2001 have extended from theft and diversion to terrorist scenarios.

In addition, shipments comply with the security requirements of the shipping states' governments (WNTI). A range of protection measures are employed during transport, ranging from the design of the package and the vehicles used as well as security forces, access control, employee screening, satellite tracking of shipments and coordination with local and national security authorities.

The physical protection of Class 7 materials during transport is assisted by minimizing both the total time the material remains in transit and the number and duration of transfers of the material, avoiding the use of regular movement schedules, and limiting the advance knowledge of transport information including date of departure, route and destination to designated officials having a need to know that information (WNTI).

In addition to the IAEA regulations for transport, there is a recommendation developed by the IAEA, *The Physical Protection of Nuclear Material and Nuclear Facilities*, INFCIRC 225. In this context, nuclear materials are those that carry a potential risk of being used in a nuclear explosive device. This requires states to take appropriate measures to ensure security and includes the physical protection requirements for nuclear material in use, storage and during transport. Three categories of security are defined depending on the nature of the material. The nuclear materials covered by INFCIRC 225 are principally plutonium and highly enriched ^{235}U and ^{233}U, for which the highest security category applies. INFCIRC 225 now extends to national as well as international transport (WNTI).

15.5 Current issues for the transport of nuclear materials

The difficulties associated with the transport of nuclear materials arise from several sources, each with varying degree of occurrence and consequence.

Most transport of Class 7 materials is radioisotopes for medical and industrial use (including some cobalt-60 sterilization sources in 4-t type B packages). But all of it requires some understanding of the particular regulations and hence training of people who handle the packages, so there may be cost and inconvenience to both shippers and others handling the packages, leading to occasional denial of shipment. Multiple layers of regulation with lack of international consistency provide disincentive to shippers. There are also problems of regulatory harmonization, with the competent authority on one country not being accepted in another. Occasionally there is *de facto* refusal to issue permits, and certain insurances for vessels carrying material with more than 1% fissile content may need to be taken out by the consignor or consignee.

Most reports of denial of shipment relate to nonfissile materials, either type B packages (mainly cobalt-60) or tantalum-niobium concentrates. For uranium concentrates, the main problem is limited ports that handle them, and few marine carriers that accept them. There has never been any accident in which a type B transport cask containing radioactive materials has been breached or has leaked.

Difference in regulatory structures can force delays and cost overruns, along with the lack of harmonization pervading throughout industry. A Euratom study in 2015 identified lack of harmonization and over-regulation in transport authorization for radioactive materials, particularly between countries, as a significant risk from a security of supply perspective. As previously mentioned, The IAEA *Regulations for the Safe Transport of Radioactive Material*, 2012 Edition SSR-6 provides guidance for international organizations and governments to alleviate these issues.

Also, particular reactor designs may have limited supply options for fabricated fuel, creating vulnerability to transport disruption through industrial action or geopolitical factors.

15.6 Transport of uranium concentrates and uranium hexafluoride

Nuclear materials have been transported since before the advent of nuclear power. The procedures employed are designed to ensure the protection of the public and the environment both routinely and when transport accidents occur. For the generation of a given quantity of electricity, the amount of nuclear fuel required is very much smaller than the amount of any other fuels. Therefore, the conventional risks and environmental impacts associated with fuel transport are greatly reduced with nuclear power.

Transport is an integral part of the nuclear fuel cycle. There are some 430 nuclear power reactors in operation in 32 countries but uranium mining occurs in only about 20, most of the production being from countries without nuclear power. Furthermore, in the course of 60 years of operation by the nuclear industry, a number of specialized facilities have been developed in various locations around the world to provide fuel cycle services. Hence, there is a need to transport nuclear fuel cycle materials to and from these facilities.[3] Indeed, most of the material used in nuclear fuel is transported several times during its progress through the fuel cycle—mine to conversion plant, to enrichment plant, to fuel fabrication plant, and finally to nuclear power plant. Transport is frequently international and often over large distances. Specialist companies organize the transport of any substantial quantities of radioactive materials.

When radioactive materials, including nuclear materials, are transported, it is important to ensure that radiation exposure of both those involved in the transport of such materials and the public along transport routes is limited. Personnel directly involved in the transport of radioactive materials are trained to take appropriate precautions and to respond in case of an emergency.

Prior to 2001 (adoption of the INF Code), although not required by transport regulations, the nuclear industry had already chosen to undertake some shipments of nuclear material using dedicated, purpose-built transport vehicles or vessels. (see later section)

Uranium ore concentrate is a material of low radioactivity and radiological hazard. Concentrate such as U_3O_8 is normally transported in ordinary industrial sealed 200-L drums in standard sea (ISO) freight containers (WNTI). No radiation protection is required beyond having the steel drums clean and within the shipping container. About 36 standard 200-L drums fit into a standard 6-m transport container. They are also used for low-level wastes within countries.

Uranium concentrates pose a minor risk due to the toxicity of the powder if it is released and ingested. In this respect, it is no different from most heavy metal compounds such as lead oxide. Uranium hexafluoride is also of low activity

[3] In the case of mining and milling, the facilities are often on the same site, reducing the requirement for transport.

and the radiological risk is minor. However, there would be a chemical hazard in the event of a release because it produces toxic byproducts on reaction with moist air (WNTI).

Natural uranium hexafluoride is usually shipped to enrichment plants in type 48Y cylinders, each 122 cm in diameter and holding about 12.5 tU hexafluoride (8.4 tU). These cylinders are then used for long-term storage of depleted uranium (DU) tails from the enrichment plant, typically at the enrichment site.

Enriched uranium hexafluoride is shipped to fuel fabricators in type 30B cylinders, each holding 2.27 t UF_6 (1.54 tU) (WNA, 2015). These cylinders are some 76 cm in diameter and are loaded in overpacks to enhance resistance to crashes, fires, and immersion. The cylinder design prevents criticality, and hence chain reactions. The loaded overpacks are generally transported using ISO flat rack containers for transport to fuel fabrication plants.

Depleted UF_6 or "Hex," the residual product from the enrichment process, has the same physical and chemical properties as natural Hex and is transported using the same type of type 48Y cylinders.

With very few exceptions, nuclear fuel cycle materials are transported in solid form.

15.7 Transport and packaging of plutonium

Reactor-grade plutonium is transported, following its separation in reprocessing, as an oxide powder because this is its most stable form. Most transports are to MOX fuel fabrication plants.

Plutonium oxide is insoluble in water and only harmful to humans if it enters the lungs, due to alpha activity (WNA, 2015). The primary risk is due to toxicity, except for criticality that is controlled by the package design (WNTI). Plutonium oxide is a chemically unreactive material. Plutonium metal, though its chemical properties are similar to those of other heavy metals like lead, is relatively insoluble, and dissolves to any extent only in acid or strong carbonate solution. Because plutonium-239 and plutonium-240 are not very radioactive, cleaning up a release of these substances would not be expected to cause much radiological hazard (Foulke and Weiner, 2002).

Plutonium oxide transport uses several different types of sealed packages and each can contain several kilograms of material. Criticality is prevented by the design of the package, limitations on the amount of material contained within the package, and on the number of packages carried on a transport vessel. Special physical protection measures apply to plutonium consignments. A typical transport consists of one truck carrying one protected shipping container. The container holds a number of packages with a total weight varying from 80 to 200 kg of plutonium oxide. A sea shipment may consist of several containers, each of them holding between 80 and 200 kg of plutonium in sealed packages (WNA, 2015).

In laboratories handling plutonium, a great deal of attention is paid to containment and accountability of it, and relatively few are licensed and equipped for this. (Foulke and Weiner, 2002).

15.8 Transport of fabricated fuel—uranium and MOX

Uranium oxide fuel assemblies are manufactured at fuel fabrication plants. They are made up of ceramic pellets formed from pressed uranium oxide that has been sintered at a high temperature (over 1400°C). The pellets are aligned within long, hollow, metal tubes, which in turn are arranged in the fuel assemblies, ready for introduction into the reactor. Different types of reactors require different types of fuel assembly, so when the fuel assemblies are transported from the fuel fabrication facility to the nuclear power reactor, the contents of the shipment will vary with the type of reactor receiving it.

In Western Europe, Asia, and the US, the most common means of transporting uranium fuel assemblies is by truck. A typical truckload supplying a light water reactor contains 6 t of fuel. In Russia and Eastern Europe, rail transport is most often used. Intercontinental transports are mostly by sea, though occasionally transport is by air. The annual operation of a 1000 MWe light water reactor requires an average fuel load of 27 t of uranium dioxide, containing 24 t of enriched uranium, which requires only a few trucks.

The precision-made fuel assemblies are transported in robust packages specially constructed to protect them from damage during transport. Uranium fuel assemblies, typically about 4 m long, have a low radioactivity level, and radiation shielding is not necessary (WNA, 2015). However, fuel assemblies do contain fissile material and criticality is prevented by the design of the package, (including the arrangement of the fuel assemblies within it, and limitations on the amount of material contained within the package), and on the number of packages carried in one shipment (WNTI, WNA, 2015).

All nuclear materials shipped by sea are in a highly stable form that is inherently safe and resistant to the effects of outside elements. The uranium oxide and the MOX fuel pellets are a hard, ceramic material that is so stable it can survive the high temperatures in the core of a nuclear power plant without significant degradation.

The transport casks for MOX are massive steel structures, such as the TN 12/2, which is made from 0.3-m thick forged steel and weighs about 100 t. The reusable casks contain about 5 t of MOX fuel, so the vast majority of the weight is the protective casing of the cask itself. The casks measure approximately 6 m long and 2.5 m in diameter.

15.9 Ships for MOX (and used fuel or high-level wastes)

British and French nuclear companies with the Overseas Reprocessing Committee, a consortium of 10 Japanese electric utilities, have regularly transported shipments of spent fuel from Japan to Europe for reprocessing and returned conditioned waste and recycled fuel back again in more than 180 round-trip voyages over more than 30 years. The safety record is second to none. These ships have transported more than 8000 t of nuclear material, and have traveled more than 7 million km, without a single incident involving the release of radioactivity.

The three largest ships belong to a British-based company Pacific Nuclear Transport Ltd (PNTL), a subsidiary of International Nuclear Services Ltd (INS). They all have double hulls with impact-resistant structures between the hulls, together with duplication and separation of all essential systems to provide high reliability and also survivability in the event of an accident. Twin engines operate independently. Each ship can carry up to 20 or 24 transport casks.

The three PNTL vessels now in service, *Pacific Heron, Pacific Egret,* and *Pacific Grebe*, were launched in Japan in 2008, 2010, and 2010 respectively. They are 4916 t deadweight and 104 m long. *Pacific Grebe* carries mainly wastes, the other two mainly MOX fuel. Earlier ships in the PNTL fleet mainly carried Japanese used fuel to Europe for reprocessing. The PNTL fleet has successfully completed more than 180 shipments with more than 2000 casks over some 40 years, covering about 10 million km, without any incident resulting in release of radioactivity (WNA, 2015).

The ships used to transport MOX fuel are among the world's safest. The fleet is certified to INF3 (Irradiated Nuclear Fuel class 3)—the highest safety category of the IMO for nuclear voyages. The vessels have been designed and built specifically to carry these nuclear materials as well as high-level nuclear wastes, and they employ a range of safety features far in excess of those found on conventional cargo vessels (Foulke and Weiner, 2002).

For the radioactive material in a large type B package in sea transit to become exposed, the ship's hold (inside double hulls, if INF3 vessels) would need to rupture, the 25 cm-thick steel cask would need to rupture, and the stainless steel flask or the fuel rods would need to be broken open. Either borosilicate glass (for reprocessed wastes) or ceramic fuel material would then be exposed, but in either case, these materials are very insoluble. However, these materials are outside the scope of this book.

The INF3 transport ships are designed to withstand a side-on collision with a large oil tanker. If the ship did sink, the casks will remain sound for many years and could be recovered because instrumentation, including location beacons, would activate and monitor the casks (WNA, 2015).

15.10 Sources of further information

Dangerous Goods International, 2014, The 9 classes of dangerous goods http://www.dgiglobal.com/classes#rad

University of Ottawa, 2011, Transportation of Dangerous Goods Class 7 Guide http://www.uottawa.ca/services/ehss/documents/TDGClass7_Final_SOP_April_29_11.pdf

WNTI, Security measures for nuclear fuel cycle transport http://www.wnti.co.uk/media/13807/4.pdf

WNTI, 2010, The INF Code and purpose-built vessels, Factsheet http://www.dounreay.com/UserFiles/File/Transports/INFshipWNTI.pdf

International Maritime Organisation (IMO) www.imo.org

International Civil Aviation Organisation (ICAO) www.icao.org

Transport Logistics International http://www.tliusa.com/

Edlow International http://www.edlow.com/

Pacific Nuclear Transport Ltd http://www.pntl.co.uk

References

AREVA, (n.d.). Transportation of Radioactive Materials: Regulations. Available from: <http://www.areva.com/EN/operations-1234/regulations-for-the-packaging-and-transportation-of-radioactive-materials.html>.

Foulke, L.R., Weiner, R.F., 2002. Plutonium Transportation—Risks and Benefits, American Nuclear Society (ANS), Latin American Symposium on Power Supply and Its Problems: The Nuclear Proposal,. Rio de Janerio, Brazil, <http://www3.ans.org/pi/sptopics/pdfs/pluttrans.pdf>.

World Nuclear Transport Institute (WNTI), 2012. Frequently Asked Questions About Nuclear Transport. Available from: <http://www.wnti.co.uk/nuclear-transport-facts/faqs.aspx>.

World Nuclear Transport Institute (WNTI), 2013. Fact Sheet: Package Types Used for Transporting Radioactive Materials. Available from: <http://www.wnti.co.uk/media/38413/FS2_EN_MAR13_V2.pdf>.

World Nuclear Transport Institute (WNTI), 2012. Nuclear Transport Facts: Front End. Available from: <http://www.wnti.co.uk/nuclear-transport-facts/what-is-transported-how/front-end.aspx>.

World Nuclear Association (WNA), 2015. Transport of Radioactive Materials. Available from: <http://www.world-nuclear.org/info/Nuclear-Fuel-Cycle/Transport/Transport-of-Radioactive-Materials/>.

Uranium mine and mill remediation and reclamation

16

Doug Collier
Australian Nuclear Science and Technology Organisation—ANSTO, Sydney, New South Wales, Australia

16.1 Introduction

Acceptable waste management and the rehabilitation of a uranium mine site are vital requirements for a uranium mining and milling project to be viable (Waggitt, 2007; WNA, 2008). Extensive information is available in the literature on waste management and uranium mine and mill site rehabilitation. Consideration of the remediation and reclamation required after the shutdown of a new uranium mine and mill site begins with an understanding of the method of mining, milling and waste disposal proposed to be used, the baseline environmental conditions existing before any project commences and the current regulations. Environmental impact and mine plans at the commencement of mining projects include plans for decommissioning and rehabilitation of the site that are acceptable to the regulator and to other stakeholders. This is consistent with broadly accepted best practice in the mining industry. Such a practice will be the overall most cost-effective practice in the long term and will be the most likely to be accepted by stakeholders and to have the best chance of succeeding.

The rehabilitation of uranium milling and mining sites requires many factors, generic to the broader mining industry, to be addressed. This chapter focuses primarily on radiological factors arising from proposed uranium ore mining and milling site rehabilitation, although references are provided in the "References" section to practices addressing the broader issues, which may simultaneously, partially, or completely also address any radiological risks. Further information on uranium mine and milling site rehabilitation can be obtained by reference to this literature.

Rehabilitation includes "remediation" of the specific aspects of a mine site as well as "reclamation" for long-term end use of the land and water resources. To achieve the objective of acceptable mining and milling site rehabilitation, future remediation and reclamation plans can utilize an extensive, existing knowledge base (OECD, 2014). Regulators and potential operators make use of this experience to ensure that the land, water and air, and therefore the public and the environment, are protected from radioactivity and other pollutants. Rehabilitation requirements are milling process specific as well as mine site specific.

Uranium for Nuclear Power. DOI: http://dx.doi.org/10.1016/B978-0-08-100307-7.00016-8

Figure 16.1 Legacy uranium mining site in Central Asia.

The objectives of mine site remediation and reclamation include the following:

- To remove hazards due to tailings, waste rock and contaminated water to acceptable values and standards
- To protect the environment in the long term
- To address any risks, possibly only perceived, of concern to the local population

The following sections of this chapter briefly discuss the different uranium mining methods and milling processes (which are described in more detail in Chapters 6 and 7), the radiological implications in regard to site remediation and reclamation and the methods and tools used to carry out site rehabilitation. They do not examine, in detail, site legacy and cases of poor past practice (see Fig. 16.1), but references are provided to enable the reader to further investigate that subject. Because the remediation of legacy sites has been extensively reviewed by the International Atomic Energy Agency (IAEA) and other organizations (OECD, 2014), this chapter on remediation focuses on best practice in the remediation of proposed new uranium mining sites. There is an opportunity to remediate legacy sites, if these existing sites are revisited for further development for uranium or other potentially valuable commodities. Examples of leading practice in uranium mine site rehabilitation, as reported in the literature, are available for a variety of specific site scenarios (mine type, milling processes used, waste storage facility design) to assist with planning going forward.

The IAEA, OECD/NEA, International Commission for Radiation Protection (ICRP), and the WNA, as well as national regulatory bodies and corporations, provide invaluable forums which gather data and provide tools for regulators and operators to best manage the mining and milling of uranium to minimize any detrimental impact of the industry (Woods, 2013), particularly in regards to the potential radiological impact. Industry has held regular meetings with the IAEA, as well as other international bodies, to provide up-to-date information on the subject of uranium mine site rehabilitation. One notable example is the UMREG/IAEA meeting based historically on the substantial recent experience gained in Germany on legacy uranium mine site remediation.

Recent requirements for stewardship in sourcing uranium for nuclear power plants has advanced the need to adopt leading practices in mine and milling site rehabilitation even where national regulatory requirements are less restrictive. It is expected that best practice principles will continue to develop into the future and that regulators and operators will be knowledgeable and will adapt to such developments (IAEA, 2010), as a requirement for sustainable development. Best practice can reduce the overall cost of a project particularly those costs associated with postmine and postmill closure.

Research in the uranium mining and milling industry continues for the placement, containment, and isolation of radioactivity in mine wastes to provide effective long-term containment without significant financial, health, or environmental liabilities placed on future generations.

Although the radiological impact of uranium mine wastes may specifically be small, the overall effects in combination with other toxic contaminants, such as acidity, alkalinity, heavy metals, turbidity, salinity, etc. increase the need to address the combined environmental impact.

There are risks associated with long-term waste disposal in the light of the requirement for the longer-term stability of tailings and for minimal environmental impact. Many of the risks and methods of addressing these risks have been identified in previous studies. Further developments are expected to continue to address these risks in the future, for example, the possibility for the removal of other radionuclides or radioactivity-containing minerals from tailings (which has proven difficult to achieve in past research for specific sites) to minimize the radioactivity present, in the long term, in large quantities of tailings deposited on the surface.

Postclosure management of mine sites is a complex subject of worldwide regulatory interest, as shown in recent Canadian guidelines (Cunningham et al., 2015). The Canadian uranium mining industry is a world leader in the adoption of best practice and has suggested a procedure, shown in Fig. 16.2, to assist in planning a uranium mine and mill site. This is an example of a procedure from mine and mill site operation up until the return of the site to the authorities or landowner, with a system of funding to achieve the necessary rehabilitation and monitoring.

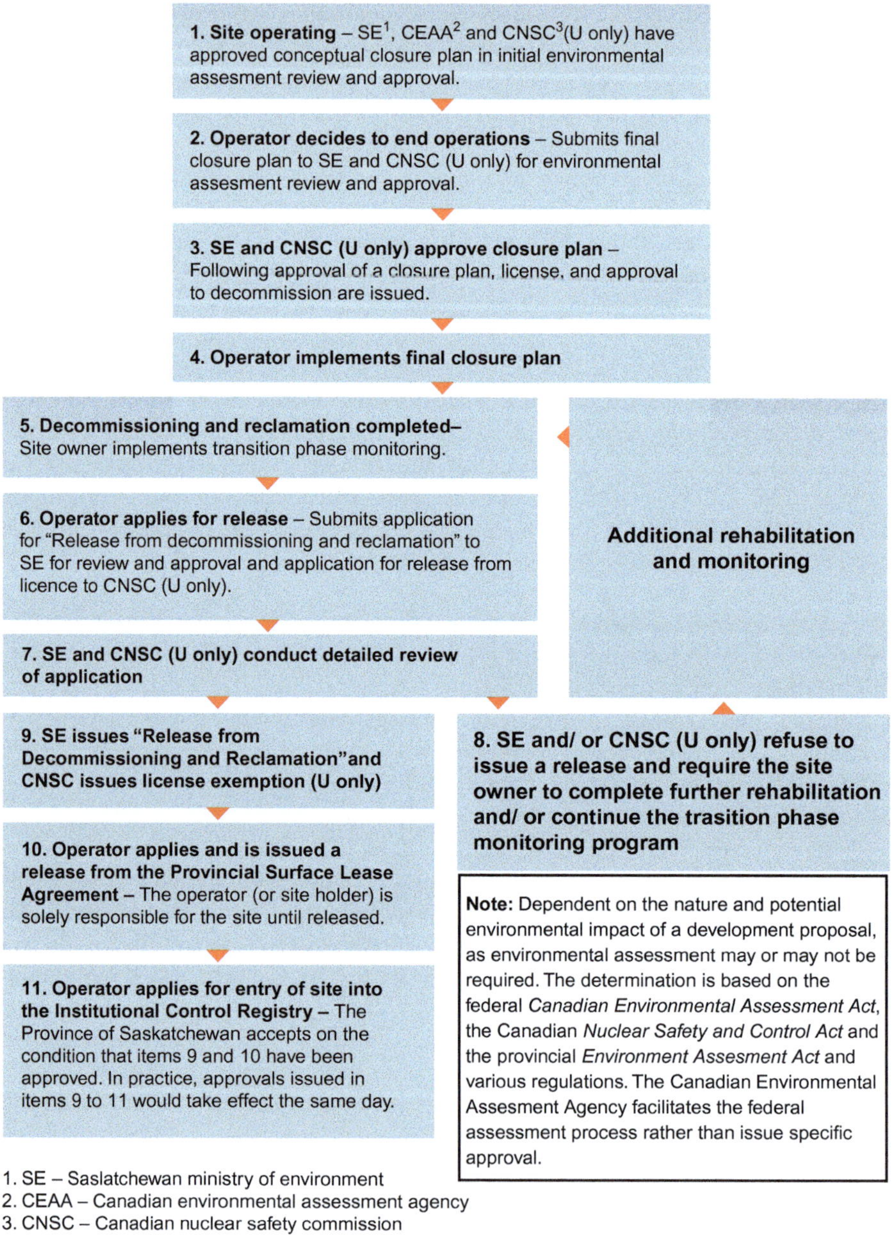

Figure 16.2 Progression of a mine/mill/waste management site closure and custodial transfer.
Source: Cunningham et al. (2015).

16.2 Uranium mine and milling sites

Mine types and milling processes are reviewed in Chapters 6 and 7. These methods are briefly revisited in this section as they affect uranium mine and milling site rehabilitation. All aspects of the mining, processing, and waste management at a uranium mine site must be planned and monitored in accordance with best practice. Elevated concentrations of radioactivity contained in site wastes must be managed in accordance with international standards and local regulations pertaining to the management of radioactivity.

The process of rehabilitation of a site begins with an environmental impact assessment and background, baseline study of the project before mining begins. The ultimate impact of a project is assessed during rehabilitation but the impact must be monitored continuously during the project from commencement of the project to mitigate adverse long-term effects and to alter the plan during operation, if the circumstances change. The environmental impact assessment is carried out to identify potential impacts of mining and milling and to put in place measures to address any identified potential risks.

Risk assessment, good management (including quality control (QC) and quality assurance (QA) systems) and sensitivity analysis of planned changes are essential components of the overall planning and management processes, from exploration to long-term, postclosure stewardship.

16.2.1 Introduction to mining and milling methods

Uranium ores are generally very large hard rock type deposits such as nonconformity, calcrete, or carbonaceous requiring mining, crushing and grinding, or porous sandstone deposits amenable to heap leaching, see Fig. 16.3, or in situ leaching (ISL). The orebody determines the mining method: surface open cut (see Fig. 16.4), underground, or in situ. The nature of the orebody, the ore grade (wt% U) and the mineralogy determine the ore mineral processing (beneficiation) and extraction methods used. These all, in turn, are determinants of the methods used for waste disposal and options for site rehabilitation. Closure will depend on stakeholder requirements depending on the location.

"Designing for closure" involves the development of closure objectives as a component of the overall project design. Sustainable remediation and closure objectives are site specific (IAEA, 2010) and involve issues including:

- Final or sequential land use
- Human health and safety
- Social impacts
- Ecosystem impacts
- Regulatory requirements
- Cost optimization

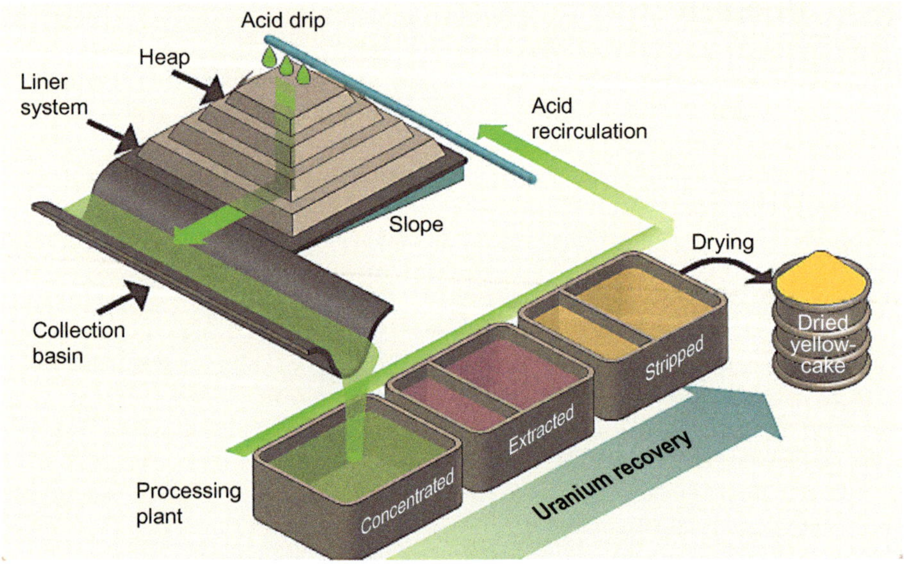

Figure 16.3 The Heap Leach Recovery Process (US NRC).

Figure 16.4 Open cut former uranium mine in Australia.

Although the evolution of standards and regulations might be regarded as potentially moving the goalposts and a source of risk, this must be transparently addressed in discussion with the regulators and other stakeholders. Regulatory requirements must be expected to evolve with increased knowledge and should not necessarily become more restrictive.

The additional potential risks from radioactivity must be addressed in rehabilitation. Pathways to exposure to radioactivity at uranium mining and milling sites include:

- Ingestion (contaminated plants and food and/or water)
- Direct exposure to gamma radiation
- Inhalation of dust, radon and radon progeny

With these potential pathways to exposure to radioactivity in mind, design options and optimization for successful decommissioning and remediation must iteratively consider and then reconsider:

- Mining options available
- Milling/processing options available
- Waste stream storage and disposal options
- Water management and treatment options
- Monitoring requirements
- Remediation options
- Mine and mill location options

The process to approve exploration and mining may specify the required site rehabilitation as indicated in the South Australian Royal Commission into the nuclear fuel cycle (South Australian government, 2015). The mining and extraction methods employed will determine the radionuclides present in wastes (gaseous, solid, and liquid) and the potential for their release into the environment. Baseline data collection (of local soils, waters, and plants) is carried out during the exploration drilling program to provide a reference point for future impact assessment.

16.2.2 Mining methods

Mining methods used include open cut, underground and ISL methods. Each of these has very different potential environmental impact advantages and disadvantages with regard to site remediation requirements. The deposition of solid wastes into a surface or underground waste management facility (WMF) is of fundamental importance in regard to the stability and permeability of the final facility and the longer-term environmental impact of its remediation.

The deposition of waste (tailings and other process wastes, contaminated equipment and mine waste) either onto a surface or into an underground facility must produce a waste structure with suitable characteristics for the long term. Backfill or cemented backfill of tailings as a paste into the WMF are techniques used to dispose of tailings and to stabilize the mine to improve mining extraction. These deposition methods can produce a structure with reduced permeability to surface and/or ground waters and to erosion and therefore the likelihood of impurity (including radioactivity) dispersion into the environment. The surface disposal of slimes containing higher radionuclide concentrations, with greater propensity to dispersion as dust and/or to erosion has led to the development of "whole-of-tailings" in paste techniques being used as backfill underground or into pits. These deposition methods are likely to be accompanied by barriers, on the surface or underground, to limit gaseous (radon) and solution migration through the waste.

16.2.3 Extraction processes

Uranium extraction processes include tank leaching, heap leaching, and ISL. These processes determine the nature of the wastes arising for disposal (solid and liquid), the particle size distribution (PSD) of the waste solids, the mineralogy and chemistry of the waste, the hydrological properties of the wastes and the physical properties of the waste. Pollutants from processing may be released (ARPANSA, 2014):

- As solids
- Into the atmosphere (as gases and dusts)
- Into the aquatic environment (surface waters)
- Into groundwater

The properties of the process wastes determine the rehabilitation risks, for example, the amenability of the site to decontamination and/or isolation during rehabilitation. The classification of wastes, see Fig. 16.5 and appropriate handling and disposal methods are discussed extensively in the literature (Williams, 1998).

The impact of extraction processes, as well as determining the solids tailings disposal methods used, will also determine the nature of short and long-term

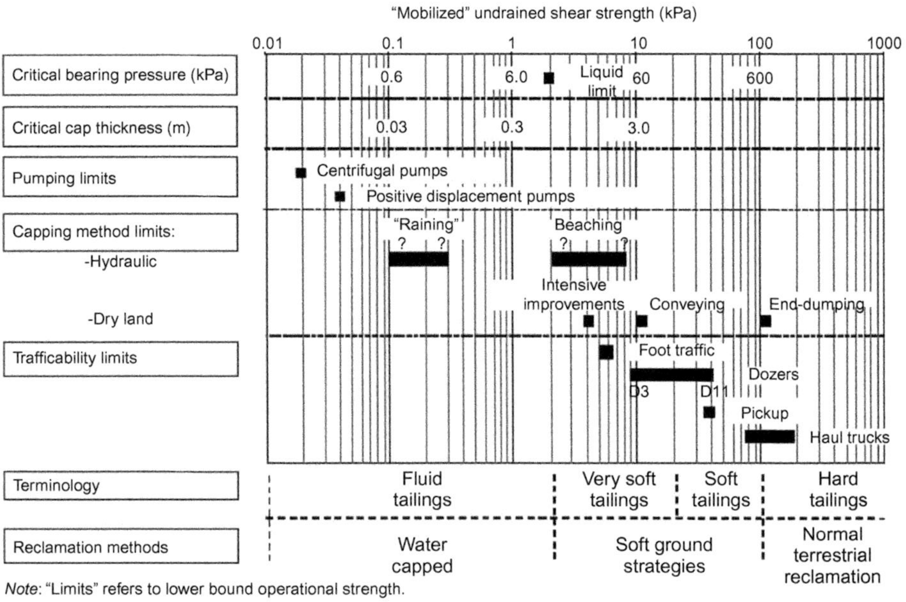

Figure 16.5 Proposed classification of tailings for remediation.
Source: From Jakubick et al. (2003).

wastewater treatment requirements. They also determine the longer-term monitoring and assessment required, as well as the likelihood of the need for appropriate disposal of plant and equipment during and after plant operation.

Time, distance, and the shielding provided by the waste placement method used and of covers determine the radiological dose received in the environment, as evidenced by the resilience of local bioavailability, bioaccumulation potential and the likely impacted species. Assessment of the impact from exposure to radiation (Larsson, 2004) is required for the short and long term.

Uranium extraction processes that facilitate mine closure must be adopted in the future.

16.3 Site remediation and reclamation

The expectation of society is that good practice, based on good science, will be used for site rehabilitation in adherence with regulations. In the US, the nuclear regulator (US Nuclear Regulatory Commission, 1997) focuses on the following for the regulation of uranium mine sites, and its regulations are frequently followed in other countries:

- Reducing radon emanation
- Preventing the spread of contamination through erosion
- Reducing contamination by seepage
- A thorough assessment of risk to the public and the environment
- Risk of groundwater contamination

The US Nuclear Regulatory Commission (NRC) has agreements in place, which authorize most of the individual states to regulate the sources of radiation that the NRC does not. This generally includes all naturally occurring radioactive materials (such as radium and radon) within their borders—see http://www.nrc.gov/about-nrc/radiation/protects-you/reg-matls.html.

The choice between alternative rehabilitation methods to minimize the environmental impact of a uranium mine site is very site specific, not only because of the differences in mining and milling methods used as discussed previously, but also because of the climate differences and the nature of the surrounding environment. Other local factors that must be considered in developing the necessary plans (Williams, 1998) include local regulatory requirements and local stakeholder requirements. Local water pathways are also critical in the evaluation of remediation options.

Guidelines have been developed by the geotechnical professions in many countries (IAEA, 2004) for the design, construction, operation, and rehabilitation of tailings dams for the mining industry. Efficient and cost-effective technologies for mine site rehabilitation of these areas (Saskatchewan Research Centre CLEANS project and the German Wismut (UMREG, 2013) rehabilitation projects) for use by future generations are required. Long-term rehabilitation measures are necessary with agreed objectives between stakeholders based on sound science and good management. Uranium Mill Tailings Remediation UMTRA in the USA (G. J. Rael Reference 18 in (IAEA, 2004)) involved extensive US government funding of designs for storage or disposal in robust waste management facilities for up to 1000 years. More recent research focused on the reduction of water infiltration through the facilities.

The design criteria for remediation are determined by:

- Remediation objectives and aesthetics, as required by the stakeholders
- Geotechnical, radiological, geochemical, ecological and hydrological factors as determined by the local site conditions

Implementation of best practice minimizes the potential for adverse environmental, social, and economic impacts. Best practice begins at the project conceptual stage with natural background baseline data collection and continues through to site remediation, closure, and site stewardship.

Other site-specific conditions such as local climate (wind, rain) and terrain (tropical, desert, etc.) and proximity to populations also determine the appropriate rehabilitation method options. Site decommissioning, remediation and land reuse are integral to the overall project plan and will require the application of principles such as the as-low-as-reasonably-achievable (ALARA) principle to meet environment, social, economic and governance objectives. Baseline data collection for local soil, plants, and water sources (IAEA, 2010) and the development of stakeholder relationships will establish the basis for future environmental and social impact assessment.

Remediation strategies must, to the maximum extent possible, produce a final landform structure that is maintenance free with minimal need for ongoing management or for future intervention. The commencement of remediation as early as possible during operations is the most cost-effective practice, as rehabilitation addressed as a legacy has proven expensive and uncertain to happen.

Compliance monitoring of parameters, as agreed with stakeholders, during and after decommissioning and monitoring to ensure closure objectives are met are part of the continual improvement process, particularly if the site facilities are classified radioactive and are regulated as such.

16.4 Risks addressed in rehabilitation

The presence of radioactivity at a uranium mine site increases public sensitivity and potentially increases the risk to the operation compared with that for the broader mining industry. Several studies have indicated a potential increased risk (Williams, 1998) arising from doses to humans and biota from uranium mill and mine wastes. Early planning will help identify those potential risks and provide remedies to address them as early as possible, which will also reduce the associated costs. This approach, together with gaining consensus with stakeholders as soon as possible, limits expensive data collection to only that necessary. The following sections discuss some of the risks faced by future uranium mine site operations in regard to site rehabilitation.

16.4.1 Regulatory risk

Best practice radiation protection is expected in the regulations (ARPANSA, 2005), however, changes in regulations are a significant risk to a project. Radiation protection philosophy and standards are continuously evolving. Regulatory reviews and changes can place unforeseeable risks on a project and can have a significant impact on site closure and release of the site and on the potential land end-use criteria (IAEA, 2004).

Codes and standards are available which provide the regulatory framework in which to manage and protect the environment. This includes the requirement for a radiation management plan (RMP) and a radioactive waste management plan using best practicable technology. This includes justification of practices and obtaining authorization. The regulator, which provides approvals, authorizations and reporting requirements, considers the potential sources of exposure to radiation.

Regulation changes will only be effective if they are realistic.

16.4.2 Effect of radiation on nonhuman biota

Previously, it was considered that in protecting humans from exposure to radiation, other organisms would also be protected (UNSCEAR, 2008), that is, that higher order organisms (such as humans) will experience the effects of radiation at lower doses than would lower order organisms. In recent years, the ICRP has suggested the need to also assess the impact of radioactivity on the biota in addition to its effect on people. Several reviews have identified the shortage of data on this subject for specific species and the limited modeling tools with which to make the assessments (Twining, 2012). A graded approach to dose limits for this impact is recommended by some regulators with "screening levels" of dose above background (ARPANSA, 2014).

16.4.3 Solids waste disposal

Waste management during operation and the proposed disposal methods are included in the justification of the practice. Waste rock and mill tailings may require both active and passive controls to limit the longer-term leaching and migration of pollutants arising from these wastes. An action plan is required to address the risk of failure of the installed systems to contain pollutants, as measured by site monitoring and affect assessments.

16.4.4 Acid mine drainage

Acid mine drainage (AMD) is an important aspect of uranium mine/milling site rehabilitation (Merkel et al., 2002), as it is in other mining industries, where contained sulfides can oxidize and produce acidic solution. AMD results in an increased potential to mobilize impurities, including radionuclides of uranium, radium, thorium, and lead. Ground and surface water treatment may be required continuously or intermittently in the short, medium, or long term depending on the concentrations of these elements, which are more likely to create concern as heavy toxic metals, rather than as being radiologically toxic.

16.4.5 Waste facility design

Above ground, below ground, and subaqueous disposal options have been used in the past for uranium mine waste disposal. The deposition of tailings above the ground may quarantine the land or limit its long-term use. These types of waste facility can also be more susceptible to catastrophic failure—see Fig. 16.6. Deposition may also make use of mined out pits. Designing a waste facility for the long term is challenging. The ability to validate models predicting long-term behavior is the subject of ongoing research and development.

Although underground placement of wastes has many obvious advantages over above ground deposition (more aesthetic and the wastes is less prone to erosion and wind dispersion), the contamination of groundwater, particularly if the wastes are

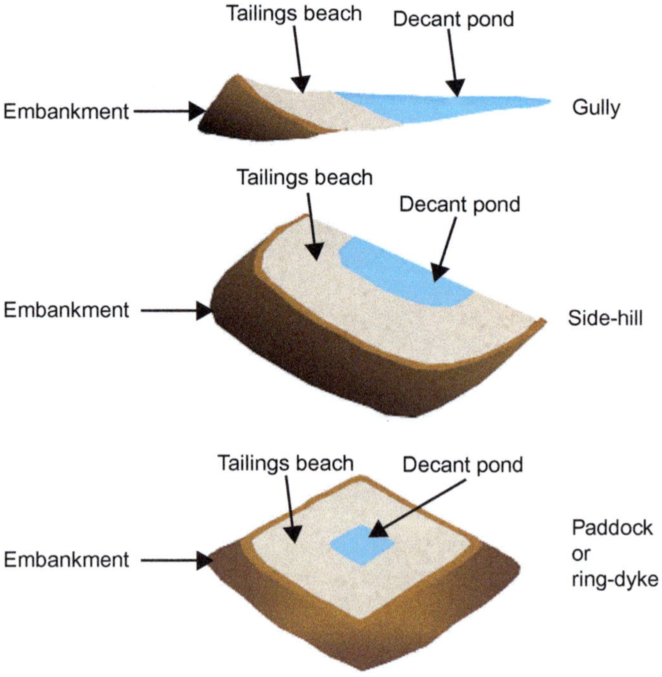

Figure 16.6 Types of above ground impoundment.
Source: From Williams (1998).

below the water table, requires special consideration and the effects will be site specific. Not all the tailings may fit back into the available underground mined space.

Maximized settled sludge densities increases the ability of an impoundment to accommodate wastes as well as reduces the permeability of above or below ground deposition (groundwater ingress is reduced). Sealing and injection of grouts are techniques used to help produce barriers to flow into and out of the structure. Cemented paste backfill or high-density backfilling using the whole-of-tailings PSD assist the facility to also accommodate troublesome slimes. Diversion of groundwater around waste masses by use of appropriate deposition and barrier techniques is also practiced.

A sound plan with a specific waste facility design, setting clear objectives and goals, a flexible project scope, with schedules, budget, project organization, setting roles, responsibilities, procedures and milestones will help to ensure that these risks are addressed using best current practice.

16.5 Tools used in rehabilitation

Concerns for the natural environment and land reuse at a uranium mining and milling site include:

- The risk of environmental degradation
- Contamination of water, plants and soil
- Reduced ecosystem viability and biodiversity
- Aesthetics
- Public amenity
- Access to land
- Quarantining of land for future beneficial land use

Uranium mill tailings are of particular environmental concern because:

- They contain most of the radioactivity originally in the ore
- The contained radioactivity is long lived
- They may also contain a range of biotoxic heavy metals and other compounds
- They may contain sulfidic minerals and be prone to produce AMD
- The particle size ranges from slime to approximately 500 μm making them susceptible to leaching, erosion, burrowing by animals and to having poor consolidation properties
- They often have large surfaces in contact with the elements thereby having high radon emanation, are prone to dust generation, interaction with surface water systems, etc. under adverse weather conditions
- They occupy large areas of relatively shallow depth, which sterilize otherwise valuable land from other future uses

Waste may also be in the form of low-grade ore or mine waste, as heap leached material or of other wastes from processing (scales, gypsum precipitates, etc.). The primary objectives of the isolation and the stabilization of mining and milling wastes is achievable using a systematic approach to decommissioning and rehabilitation (IAEA, 2004).

Design for the long term and impact assessment tools are discussed in the following.

16.5.1 Storage facility design for closure

The design criteria for a future uranium mine site are based on the characterization (mineralogy, chemistry, physical) of the tailings and waste rock, such as hydraulic conductivity, drainage, consolidation, etc. properties towards achieving an agreed plan for closure of the site. These criteria must be available at the environmental impact statement stage of a project. Factors taken into account in design include:

- The expected life of the facility to meet regulatory integrity requirements
- Cover design to prevent dispersion from waste
- A radon barrier to limit the dose from this source to agreed values
- Surface water runoff design to minimize erosion and permeation through the waste
- Erosion prevention measures
- Moist/wet tailings deposition options including under water (subaerial) disposal
- Geotechnical considerations
- Nonradiological as well as radiological contaminants
- Drainage from the waste
- Infiltration of water into the facility
- Spread of contamination by wind
- Control of runoff from contaminated areas
- Monitoring and periodic maintenance
- A monitoring plan and dewatering/monitoring wells

Many older designs such as "valley containment" and "marine disposal" are no longer considered best practice, depending on local conditions. Any design of a dam that allows uncontrolled discharge of tailings liquor is also not acceptable.

The failure of past waste storage facility designs has led to new dam designs, such as those using permeable reactive barriers and liners to sequester mobile contaminants. Designs have also used wicks to dewater the structure prior to covering to assist consolidation and long-term stability.

16.5.2 Modeling and model validation by monitoring

Models are used extensively to include hydrological, geochemical and geotechnical aspects into alternative site rehabilitation options for the prediction of contaminant transport (Merkel et al., 2002; Merkel and Hasche-Berger, 2006). Validation of the models provides confidence in the prediction of the long-term behavior at the site. The various models available to study the behavior of radioactivity (and their limitations) are closely compared in IAEA documentation (IAEA, 2007) and can be used to compare waste facility design options for the prediction of long-term impacts and to assess impacts at different stages of a project. Much of this is based on data available on the web such as that from the US, for example, the RESDRAD codes, which are available at https://web.evs.anl.gov/resrad/.

16.5.3 Engineering

A range of engineering measures are taken to address the potential adverse impacts of a uranium mine and mill site using containment, covers, water treatment, etc. to achieve the isolation of waste material and contaminants from the environment during both project operation and rehabilitation phases. The physical properties of the material are altered to enhance the strength of specific parts of the structure, while some parts of the facility are designed for water and slime retention during operation, which determines the facility ultimately requiring rehabilitation. Maximum physical stability of an impoundment reduces the need and cost for subsequent dewatering during both decommissioning and rehabilitation.

16.5.3.1 Covers

If above ground mining and mill waste disposal are unavoidable (completely or partly), the use of appropriately designed multicomponent/layered covers over the facility is best practice. Near surface disposal in the vicinity of a mine and mill can pose serious engineering challenges depending on the geomechanical and physical and chemical characteristics of the wastes. Covers are designed to:

- Minimize radon and dust dispersal
- Shield the environment from gamma radiation
- Minimize water and oxygen infiltration
- Control erosion
- Provide an acceptable landscape in the longer term

The integrity and time guarantee on the performance of covers for 200–1000 years has been studied by the US EPA, though the contained radioactivity takes longer to decay. Covers may need to be stable for a very long time depending on the concentration of radioactivity in the wastes and cover design. Care and maintenance is expensive but long term stewardship is expected. Covers should exploit the expected ecological changes in and on the cover rather than have a cover, which is purely a physical barrier vulnerable to erosion and other forms of deterioration. An appropriate cover design harmonizes natural forces and incorporates vegetation and landscape evolution. A cover's performance should ideally improve over a 1000-year period. Designs are site specific based on local data. Cover designs for different climatic conditions have been studied (Benson, 2002).

The thickness and nature (particle size, moisture content, etc.) of covers to abate radon flux to agreed satisfactory values can be calculated using the cover material characteristics, the predicted prevailing weather conditions, etc. Covers incorporating low concentration uranium mining wastes were examined in a French mine site study and the results compared with those obtained using modeling. The calculation included cover material characterization, hydrology, and meteorology (Ferry et al., 2002).

Australian radon flux measurements examined the effect of seasonal variation (Bollhöfer and Doering, 2015) on radon flux. The application of models for multi-layered covers to achieve a defined threshold radon emanation has also been reported (Dinis et al., 2011; IAEA, 2013). Work is largely based on earlier guidelines (US NRC, 1989) and research.

16.5.3.2 Barriers

Secure containment of pollutants in a waste facility is an essential objective and barriers assist in achieving this. This requires an understanding of the natural barriers available as well as designs for engineered barriers (Metcalfe and Rochelle, 1999).

Permeable structures surrounding a waste facility allow groundwater to flow around the waste and minimize the leaching of contaminants from the waste.

16.5.3.3 Dewatering

Excess water in a waste facility may be extracted and treated during operation to reduce the need for more expensive postdecommissioning dewatering prior to covering and reclamation. The disadvantages and advantages of various dewatering methods are described in IAEA publications (IAEA, 2004) and can have a significant impact on the time and cost for subsequent rehabilitation.

The final stage of site rehabilitation involves reclamation of the area for reuse. It involves top dressing of soil, profiling, and revegetation. This may be done once the objectives for pollution control have been achieved. It should make use of experience available from, for example, municipal landfill design and reclamation, but subsequent measures, such as the nature of replanting and construction on the reclaimed site, must not interfere with the controls previously placed to minimize radiation dose and the release of other pollutants.

16.5.4 Water treatment

Residual contaminated process water from the mine and mill and contaminated groundwater may need treatment during and after plant operation before release to the environment. The presence of excess carbonate containing neutralizing agents used to neutralize free acidity in contaminated water may increase the dissolution of uranium and its mobilization in the longer term. This has placed attention on plant neutralization process procedures and conditions and the examination of other (including lower cost) alternative neutralizing reagents. The use of processes such as high-density sludge (Merkel et al., 2002; Topp et al., 2003) and reverse osmosis are being used to treat excess process water before its discharge to the environment. Radium removal using barium chloride and ferric sulfate is also standard practice, when required.

The site water balance is closely monitored during decommissioning and postclosure.

Contaminated groundwater treatment alternatives include:

- Natural attenuation with monitoring
- In situ remediation

- Extraction of contaminated groundwater and treatment
- A combination of these processes

The objective is to eliminate seepage from the ground being remediated. Natural attenuation measures may not meet the clean-up schedule criteria. Collection and treatment may be necessary but ultimately (in the longer term) the aim is to have this done passively. However, initial extraction and treatment may speed up the remediation process. The costly need to treat AMD (IAEA, 2004) should be minimized by suitable facility design in regard to covers and barriers, surface water runoff and groundwater flow.

16.5.5 Communication and administration

An examination of the models used in the assessment of the environmental impact from radioactivity (IAEA, 2007) includes the statement that there should be a "strong emphasis on good communication between all stakeholders involved in a particular evaluation." Communication between all stakeholders in all aspects of the rehabilitation of a facility is essential. This must include a discussion of regulatory, legal, environmental, and economic issues and risks, and public concerns. The overall objective and the progress of a rehabilitation project must be clearly communicated. This ensures that all stakeholders understand and trust the process as being transparent.

Good communication and cooperation is also critical between the different groups of scientists involved, to ensure that the requirements of the different groups are clearly understood by everyone involved in the project.

Poor communication with stakeholders may result in the stakeholders not "accepting" the project.

16.5.6 Training

Training of staff is essential for the safety and effectiveness of any mining project. It is essential also for monitoring, modeling, impact assessment, reporting, etc., for closure.

Modelers, for example, need to be well trained in the use and application of the available models, and particularly in the interpretation of the results produced by the models. Poor training can lead to misinterpretation of model predictions.

There is the additional training requirement when processing and managing wastes from uranium mining and milling due to the presence of elevated concentrations of naturally occurring radioactivity. Specialist knowledge and training is required over a broad range of subjects to address this. They include professionals such as radio-analytical chemists, metallurgists and chemists with an understanding of the deportment of radionuclides in processing, modelers, and biologists with specialist understanding of the behavior of radionuclides in the environment, radiation safety specialists, etc. The overall workforce, from the operation phase to site rehabilitation and monitoring after closure, must be trained in various aspects of the additional risks arising from the presence of elevated concentrations of radioactivity, particularly in areas of health and safety awareness with qualified staff committed to ongoing knowledge transfer.

16.6 Funding for rehabilitation

It is unlikely that the presence of typical concentrations of radioactivity in uranium ore and in waste products will give rise to significantly higher mining costs than that required in other mining industries, as the same good practices are also demanded of the broader industry (CRC, 2008). Any proposed expenditure is carefully balanced against the likely benefit from such measures.

Some extra costs do arise for the management of radioactivity but most of the rehabilitation costs are required in any case for the management of other potential pollutants. The public perception of the uranium mining industry is such that good practice measures must be strictly adhered to and the practices be closely scrutinized. Forecast environmental (including radiological) impact must be compared before and after implementation of any measure. The objective is to minimize future remedial costs being borne by the taxpayer (see European Bank for Reconstruction and Development fund for rehabilitation of uranium mining legacy sites) and to minimize the chance of land being "sterile" for reuse postrehabilitation.

Organizations currently hold considerable liabilities for past practices and their impacts. The cost of not undertaking remedial work must be taken into account. The mine plan costs include financial assurance for closure and remediation (OECD, 2014; CNSC, 2012; Western Australian government, 2013). Monies are held as a bond. Full accounting and provision of future costs must be incorporated into the economics of a project to achieve sustainability over the long term.

The potential for a previously operated site to reopen after shutdown for the extraction of further uranium or other valuable constituents is one reason for delayed final closure and reclamation of a site.

16.7 Future trends

What are the long-term trends associated with rehabilitation of uranium mine and milling sites? These may be financial, health and/or those related to land reuse.

16.7.1 Regulations

There is an inescapable link between human health and the health of the environment. Regulations will change as the knowledge base relating to the impact of radioactivity on human and nonhuman biota increases. The development of a framework for radiological protection of the environment by national authorities has evolved from publications provided by the ICRP reports from 2007 to 2014). The Australian radiation protection authority (ARPANSA, 2014) and the authorities of other states involved in the rehabilitation of uranium mining and milling sites provide best practice guidance for the assessment of the impact of environmental exposures. Defined dose rate benchmarks are recommended to ensure that detrimental effects due to ionizing radiation are minimized. This framework is the basis for a best practice approach to environmental protection and provides models to assess exposure/dose to and the effect on humans and different types of flora and fauna.

Metal toxicity in mining has generally been accepted as having a greater impact than radioactivity, although it has been suggested that exposure to metal toxicity may make an organism more susceptible to radiological exposure, that is, there may be a synergistic effect of radiation dose and exposure to other agents. In many circumstances, the broader toxicity impacts can be more significant than radiological effects (Johnston et al., 2003). Many of these issues are being further studied. Radiological effects on biota will be better understood, and risks reduced, as data are obtained for specific flora and fauna and their dose exposure pathways, for example, the nature of their habitat.

16.7.2 Practice

A uranium mining and milling plan must have regulatory approval and include an acceptable closure and remediation plan, including rehabilitation costs. Radiological impact assessments (ARPANSA, 2014) must be carried out throughout the project. To ensure that effort and resources are not expended unnecessarily, these should be addressed as soon as possible within the environmental RMP. The establishment of a uranium mine and mill with the necessary waste disposal facilities is a "nuclear action" practice with subsequent integral site decommissioning and rehabilitation tasks. Real, potential, or perceived exposure of the environment and the associated risks must be addressed. A specific scenario at any time during a project will include a description of the current practice and the radiological issues (radioactivity sources, pathways to exposure, physicochemical properties of materials containing radioactivity, affected environment, etc.)—see Fig. 16.7.

The application of an environmental protection framework is recommended at the conceptual/planning, operational, and rehabilitation stages. This will assist in identifying the sources and nature of the radioactivity, the potential pathways to exposure, the relevant affected organisms, etc., as well as allow the assessment of predicted against actual impacts and determine the level of effort/cost for environmental protection.

Extra monitoring over the longer term may be needed to address local concerns and to demonstrate transparency of project management procedures.

16.7.3 Research

The following are areas of research that would assist an acceptable rehabilitation of a uranium mine site. Many of these issues are the subject of papers presented at

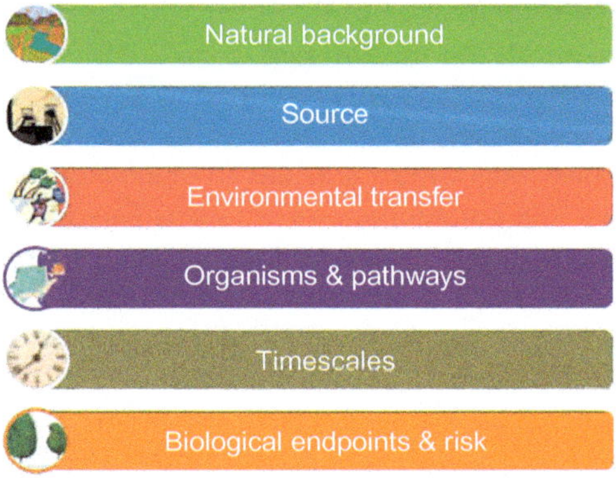

Figure 16.7 General aspects to consider when building a project scenario.
Source: From ARPANSA (2014).

meetings such as the IAEA-UMREG meeting in Bad Schlema, Germany, in August 2015. Much of the information from research to reduce the detrimental impacts of a mine site and to assess the effectiveness of rehabilitation is site specific. It may include the following topics:

- Up front beneficiation of uranium ore (crushing and milling, ore sorting, flotation, etc.) to reduce the quantity of finely milled material reporting as chemically contaminated waste tailings after mining and milling.
- The long-term stability of waste storage facilities remains of major concern. Emphasis is being placed on the development of innovative methods and techniques to increase settled densities of tailings to improve the integrity of waste facility structures and the biochemical and geochemical behavior of sealants to reduce the permeability and leachability of pollutants from rehabilitated areas.
- ISL practice is very site dependent and a thorough understanding of the site geology and hydrology is essential before approval to develop can be given.
- The effect of the migration of pollutants from uranium mining and milling waste on biota requires increased data to assess impact and to reduce the longer-term risk from its impact.
- Better data and improved, validated models are required. This includes a better understanding of the long-term migration of pollutants and the adsorption of radionuclides onto soils.

16.8 Conclusions

An improved understanding of the environmental impact of uranium mining and milling is required to address concerns that exist for the rehabilitation of sites used for uranium mining and milling. Waste management, for example, is a major source of risk in site rehabilitation from both a public perception and scientific viewpoint, as uncertainties still exist, such as:

- The effect of time and possibly a change in local site conditions on the rehabilitation methods proposed

- Long-term climate change effects
- Population density/demography uncertainties/changes
- Dose model/regulation changes, for example, linear nonthreshold dose response model unknowns

The risk is reduced as the effects on biota are better understood as further data are obtained and improvements are made to models. They are also reduced as waste disposal methods are adopted that are more practical in a remediation context.

Although ISL and underground mines (with underground disposal of tailings) have benefits in regard to site reclamation and land reuse, there are still risks associated with them. These risks include:

- Some process wastes may not be readily accommodated underground, such as some processing reagents, for example, SX reagents, and slimes;
- There are risks associated with radon emanation underground and specialized ventilation and mining techniques are required during operation to accommodate this. The longer-term benefits of an underground operation may outweigh this risk and allow underground mining, which will have a significant effect on the available waste management and site remediation options; and
- Uncontrolled flooding and water contamination are risks in underground operations, ISL sites and from heap leaching operations. This can have a significant effect on costly water treatment requirements in the short and long term during rehabilitation.

Future investment in knowledge gaps will reduce the risks particularly in regard to long-term pollutant migration. It will minimize the need for active maintenance in the longer term and provide solutions that are economically as well as technically feasible.

References

ARPANSA, 2005. Code of practice and safety guide on radiation protection and radioactive waste management in mining and mineral processing, RPS 9.

ARPANSA, 2014. Safety guide for radiation protection of the environment, Safety Guide, DRAFT. Commonwealth of Australia 2014 as represented by the Australian Radiation Protection and Nuclear Safety Agency (ARPANSA).

Benson, C.H., Albright, W.H., Roesler, A.C., Abichou, T., 2002. Evaluation of final cover performance: field data for the alternative cover assessment program (ACAP), Waste Management WM'02, 24–28 February, Tucson, Arizona, USA, Paper 521.

Bollhöfer, A., Doering, C., 2015. Long-Term temporal variability of the radon-222 exhalation flux from a landform covered by low uranium grade waste rock. J. Environ. Radioact. 127, 88–94.

CNSC, 2012. Uranium mining and milling: the facts of a well-regulated industry. Canadian Nuclear Safety Commission Fact Sheet, July.

CRC, 2008. Tailings and mine waste '08. In: 12th International Conference on Tailings and Mine Waste: Liners, Covers and Barriers, ISBN 978-0-415-48634-7.

Cunningham, K., Kristoff, D., Hovdebo, D., Webster, M., 2015. Post-closure management of mines in Saskatchewan, AusIMM Bulletin, June, pp. 84–86. Original document available from: <http://www.economy.gov.sk.ca/Institutional_Control-Decommissioned_Mines/Mills> (Institutional Control Program: Post Closure Management of Decommissioned Mine/Mill Properties Located on Crown Land in Saskatchewan).

Dinis, M. de L., Fiuza, A., 2011. Modeling of an optimised multilayer cover design for a uranium mill tailings disposal. In: WM2011 Conference, February 27–March 3, Phoenix, Arizona.

Ferry, C., Richon, P., Beneito, A., Robé, M.-C., 2002. Evaluation of the effect of a cover layer on radon exhalation from uranium mill tailings: transient radon flux analysis. J. Environ. Radioact. 63 (1), 49–64.

IAEA, 2004. The long term stabilization of uranium mill tailings, IAEA-TECDOC-1403, Vienna, August.

IAEA, 2007. Environmental modelling for radiation safety (EMRAS), IAEA-TECDOC-DRAFT, Vienna, September.

IAEA, 2010. Best practice in environmental management of uranium mining, TECDOC-1148, Vienna.

IAEA, 2013. Measurement and Calculation of Radon Release from NORM Residues. Tech, Vienna, Report Series No. 474.

Jakubick, A., McKenna, G., Robertson, A., 2003. Stabilisation of tailings deposits. International Experience, Proceedings of Mining and the Environment III, 25–28 May, Sudbury, ON, Canada.

Johnston, A., Humphrey, C., Martin, P., 2003. Protection of the environment from the effects of ionizing radiation associated with uranium mining. In: International Conference on the Protection of the Environment from the Effects of Ionising Radiation, 6–10 October, Stockholm, Sweden.

Larsson, C.-M., 2004. The Fasset framework for assessment of environmental impact of ionizing radiation in European ecosystems—an overview. J. Radiol. Prot. 24, A1–A12.

Merkel, B.J., Hasche-Berger, A. (Eds.), 2006. Uranium in the Environment, Mining Impact and Consequences. Springer, Berlin.

Merkel, B.J., Planer-Friedrich, B., Wolkersdorfer, C., 2002. Uranium in the aquatic environment. In: Proceedings of International Conference Uranium Mining and Hydrogeology, 15–21 September, Freiberg, Germany.

Metcalfe, R., Rochelle, C.A., 1999. Chemical Containment of Waste in the Geosphere. Geological Soc. Special Publication No. 157, London.

OECD, 2014. Managing environmental and health impacts of uranium mining. Uranium in the Environment, Mining Impact and Consequences, NEA No. 7062.

South Australian government, 2015. Nuclear Fuel Cycle Royal Commission. Issues paper 1, Royal Commission, Adelaide, South Australia. <nuclearrc.sa.gov.au/our-reports/>.

Topp, H., Russell, H., Davidson, J., Jones, D., Levy, V., Gilderdale, M., et al., 2003. Process water treatment at the Ranger uranium mine, Northern Australia. Water Sci. Technol. 47 (10), 155–162.

Twining, J., 2012. Tropical Radioecology, Volume 18 in the series 'Radioactivity in the Environment' (Series Editor, M.S. Baxter). Elsevier, Oxford, UK.

UMREG, 2013. Prevention and remediation of uranium production impacts. Available from: <UMREG@gmx.net>.

UNSCEAR, 2008. Effects of Ionising Radiation on Non-Human Biota, United Nations Scientific Committee on the Effects of Atomic Radiation. United Nations publications, New York.

US Nuclear Regulatory Commission, 1989. Calculation of radon flux attenuation by earthen uranium mill tailings covers. Regul. Guide. 3, 64.

US Nuclear Regulatory Commission, 1997. ff, Code of Federal Regulations Code 10, Energy. <http://www.nrc.gov/about-nrc/radiation/protects-you/reg-matls.html>.

Waggitt, P., 2007. Uranium mining legacy sites and remediation—a global perspective. In: IAEA, Namibia Meeting Presentation, October. <https://www.iaea.org/OurWork/ST/NE/NEFW/.../13%20Waggit4.PDF>.

Western Australian Government, 2013. Uranium in Western Australia, May.

Williams, D.J., 1998. Physical processes and capping of sub-aerially deposited tailings slurry. In: Asher, C.J., Bell, L.C. (Eds.), Workshop on Future Directions in Tailings Environmental Management, 14–15 September, Perth, Australia, Australian Centre for Mining Environmental Research, Brisbane.

Woods, P.H., 2013. International perspectives on uranium mining and environmental regulation. In: CSIRO Symposium, Determining the Hydrogeochemistry, Transport and Fate of Uranium in the Context of Environmental Outcomes, September, Perth, Western Australia, pp. 21–22.

WNA, 2008. Sustaining Global Best Practices in Uranium Mining and Processing. World Nuclear Association policy document, London.

Interregional technical cooperation for education and training in the uranium industry

17

Patricia Wieland
World Nuclear University, London, United Kingdom

17.1 Introduction

The interregional technical cooperation for education and training in the uranium industry increased substantially over the last decades. The major drives for cooperation are the need for competent human resources in regions where specific expertise is not available and the prevention of undue environmental impact, while assuring the uranium supply. The ultimate goal is to support the strengthening and maintenance of competencies for the safe, secure, and sustainable development of the uranium industry for energy production.

With the global opportunities for new nuclear energy plants, an expanding demand for skilled nuclear workers along the entire nuclear fuel cycle emerged. Training methodology itself has equally evolved. The application of the systematic approach to training, whereby there is a clear definition of the knowledge and skills to be acquired based on real needs assessment, assures that the training program is necessary, valid, and effective. Even though some of the companies and governmental agencies have already a long-term human resources development program in place, the advantage of international networking with peers is incommensurable in a globalized supply chain market.

While the market is driven by economic motivations, the challenges for attending the nuclear fuel demand for the future years can be better faced when stakeholders work together. Intergovernmental organizations and international trade associations facilitate the share of views and leading practice. New technologies are being developed and implemented to optimize the uranium extraction, separation from other minerals, isotopic enrichment, and fabrication of fuel elements. Although the innovative methods are subject to industrial secrecy and propriety process, surely the challenges are better solved by experts with an international experience and global perspective.

With due consideration to the diversity of cultures and context, as well as to fair competition, knowledge sharing among peers contributes to the sustainable development of mining sites in different countries. This is useful not only for large uranium providers like Kazakhstan, Canada, and Australia, but also for state-owned companies for domestic use only and for multistakeholder-owner industries.

Uranium for Nuclear Power. DOI: http://dx.doi.org/10.1016/B978-0-08-100307-7.00017-X

This chapter describes the mechanisms for interregional technical cooperation for education and training in the uranium industry, providing briefly some examples of global and regional networks; presents a summary of the core content of training programs, discusses the challenges, and presents future trends.

17.2 Relevant interregional initiatives in education and training

While the responsibility for training remains with national organizations, the role of interregional cooperation in education and training is recognizably relevant. This is mostly facilitated by intergovernmental organizations. However, bilateral agreements tend to increase with new and expanding nuclear programs.

17.2.1 Intergovernmental organizations

Intergovernmental organizations like the International Atomic Energy Agency (IAEA) and the OECD/Nuclear Energy Agency (NEA) provide support to member countries by diverse means.

The IAEA Statute (IAEA, 1956), clearly authorizes the Agency "To encourage the exchange of training of scientists and experts in the field of peaceful uses of atomic energy." The IAEA assistance for training is mainly provided by means of developing and promoting international standards and recommendations, organizing short training courses and fellowships, fostering specialized networks, facilitating peer review missions, and coordinating projects to enhance national capabilities.

The Logical Framework Approach (LFA) methodology helps countries efficiently plan and design national projects, identifying key problems, solutions to address them, and feasible alternatives for implementation. The LFA approach intends to ensure that training is considered as part of a long-term project, therefore, the participants should contribute significantly, once back to work after being trained.

When key issues are common to several countries, and there is a synergy among those countries to handle those issues together in a more efficient way, the technical cooperation project has a regional or interregional status. Examples of the interregional projects in place are the INT/2/15, which aims at supporting uranium exploration, resource augmentation, and production using advanced techniques, and INT/9/175, which aims at promoting safe and efficient clean-up of radioactively contaminated facilities and sites. Some projects are at the regional level, as they focus on the common issues that only that region faces. An example of regional level project is the RER/7/006, which aims at building capacity for developing and implementing integrated programs for remediation of the areas affected by uranium mining in the European region.

The IAEA facilitates peer reviews by organizing international team missions. The Uranium Production Site Appraisal Team (UPSAT) aims to improve the operational and safety performance of uranium production facilities through all phases of the uranium production cycle. The Safety Evaluation of Fuel Cycle Facilities during Operation (SEDO) aims to assist to promote the continuous development of operational safety by the dissemination of information on good safety practices to other member states. The facilities currently covered by SEDO are conversion and enrichment, fuel fabrication, spent fuel storage, reprocessing and associated waste conditioning, and fuel cycle research and development.

All these initiatives benefit from the IAEA Safety Standards applicable to the safety of nuclear fuel industry and other IAEA technical reports and reviews (IAEA, 2014). The information systems are another important resource, for example, the International Nuclear Information System (INIS) and the Integrated Nuclear Fuel Cycle Information System (INFCIS). The CONNECT network, launched in 2014, is a web platform with a professional networking system that acts as a gateway for interconnecting existing and planned IAEA networks.

IAEA assigns high priority to training, not only using the technical cooperation funds. The 2014 regular budget dedicated some 30% of the major program on nuclear power, fuel cycle and nuclear sciences to capacity building and nuclear knowledge maintenance for sustainable energy (IAEA, 2013). The nuclear knowledge management service offers several means for assistance, among missions, self-assessment tools, training events and networks. One interesting tool is the education and training portals. The cyber learning platform for nuclear education and training (CLP4NET) intends to help Member States and partners to ensure high standards of nuclear education and training in a modular object-oriented dynamic learning environment (MOODLE) learning management system.

The OECD/NEA also contributes to training. An international expert group under the auspices of OECD/NEA carried out a study on the evolution in nuclear education and training since the publication of "Nuclear Education: Cause for Concern" in 2002. Its findings identified the achievements and the remaining gaps and actions required to the OECD countries (OECD/NEA, 2012). Although the study focuses on the nuclear energy plants, it is a good basis for human resources planning for any stage of the uranium industry. It presents the results of the survey on education and training centers in its Member States and highlights the importance of strategic approach drawing from a common job taxonomy framework.

The Joint IAEA/OECD-NEA Uranium Group aims primarily to foster the exchange of technical information in the fields of uranium resources, exploration and production technology and produces since 1965 the biennial publication: Uranium Resources, Production and Demand, also familiarly known as the "Red book" (OECD/NEA, 2014). Members of the Uranium Group are experts from government agencies and from industrial and research organizations with interests in uranium geology, mining, and processing. The experts are nominated by governments to provide updated country reports on uranium resources and exploration.

17.2.2 Global and regional networks committed to education and training

17.2.2.1 World Nuclear University

The World Nuclear University (WNU) is a global network and the four major founding supporters are the IAEA, the OECD's NEA, the World Nuclear Association, and the World Association of Nuclear Operators. Inaugurated in 2003 and encompassing key institutions of nuclear learning in more than 30 nations, since its inception, the WNU network has expanded significantly globally.

WNU offers a range of unique nuclear educational and training programs around the world, organized by the WNU Coordinating Centre, located in London, UK, in joint collaboration with members of the WNU global network. By drawing on the support of industry, governments, and academia, these programs are designed to meet the training requirements of international nuclear professionals, including leadership development. As of December 2014, over 4000 participants from over 79 countries have attended WNU programs (Wieland and Davies, 2015). The flagship program is the WNU Summer Institute, an intensive 6-week program for future nuclear leaders held annually in different locations. The WNU Summer Institute offers a comprehensive set of dynamic lectures, workshops, field trips to nuclear and industrial facilities, and team projects led by some of the world's foremost experts. These cover the global environment and sustainable development, nuclear-related technology innovations, nuclear policy and safety, security and safeguards.

Of special interest for the uranium industry, the WNU School of Uranium Production (WNU-SUP), created in 2006, combines the infrastructure of Diamo state enterprise in Czech Republic and expertise from industry and academy. Its programs focus on professional training in all aspects of uranium production: from surveying, exploration, extraction by different mining methods and ore processing, environmental and radiation protection to the decommissioning and rehabilitation of mining areas, following the termination of extraction. Until January 2014 232 participants from 18 countries, successfully passed 27 different training programs of all uranium mining life cycle aspects (Trojacek, 2014).

17.2.2.2 European Nuclear Education Network

The European Commission supports human resource development in Europe and one the successful outcomes is the European Nuclear Education Network (ENEN), which was established in 2003. The ENEN mission is the preservation and the further development of expertise in the nuclear fields by higher education and training. This objective is realized through the cooperation between universities, research organizations, regulatory bodies, the industry, and any other organizations involved in the application of nuclear science and ionizing radiation in Europe. One of the programs is the Certification European Master of Science in Nuclear Engineering, whereby students who have obtained a master degree in nuclear engineering, or equivalent, from an ENEN member institution can apply for the certification, on the condition that their master program fulfils the ENEN requirements. Another interesting program to attract and recognize the potential of young professionals is the PhD event and Prize. The objectives of which are: to provide a forum for PhD students to present research work to their fellows and colleagues in a friendly but competitive spirit to promote the research work of PhD students in the nuclear field and to set up a bridge between PhD students and professionals in the nuclear field.

17.2.2.3 Other networks

Other networks were created to assist the regional preservation, promotion, and transference of nuclear knowledge in the Latin American, Africa, and Asia regions, respectively: *Red Latinoamericana para la Educación y la Capacitación en Tecnología Nuclear* (LANENT); AFRA Network for Education Science and Technology (AFRA-NEST); and Asian Network for Education in Nuclear Technology (ANENT). Interesting to note the growing trend of networks, mainly facilitated by social media. There are several national, regional and international networks organized by the young generation in the nuclear area. For example, the International Network of Emerging Nuclear Specialists (INENS) promotes next generation dialogue and develop practical policy recommendations on nuclear issues.

17.2.3 Relevant international discussion forum and trade association

While not aimed directly at training and education, the international discussion forums are relevant to exchange information and draw recommendations. As a global nuclear association, the World Nuclear Association supports and promotes the nuclear industry. Its website contains all relevant and updated information for the nuclear industry and they are access free (WNA, 2015). One of the exclusive membership benefits is the participation in the forums for global industry cooperation, which addresses topics of relevance to the nuclear industry to share best practice, conduct analysis on industry trends, and develop reports and positions on economic, safety and environmental issues. These working groups typically gather three times a year with meetings taking place in January, around April, typically at the World Nuclear Fuel Cycle Conference, and in September in the same week of the WNA Symposium in London. The Fuel Cycle Groups are focused on the fuel market report, transport, sustainable used fuel management, and security of the international fuel cycle. The Fuel Cycle Plenary receives relevant reports from the other working groups and attracts large audience; almost 100 representatives discuss the developments in nuclear fuel and trade. The reports produced by the working groups are exemplified by the Nuclear Fuel Report (WNA, 2015), that covers not only the mining industry, but also all the other stages up to the fuel fabrication.

The members of the World Nuclear Association support its public information mission. The World Nuclear News (WNN) disseminates daily, validated news classified in energy and environment, new nuclear, regulation and safety, nuclear policies, corporate, uranium and fuel, waste and recycling. This resource is free for all those interested in being informed about what is going on in the nuclear area.

17.2.4 Transnational education and training programs

Numerous transnational education and training programs have been initiated with a focus on qualification of personnel in industry and attraction of university students to fulfil the ambitious expansion of nuclear programs abroad. This section presents some of those initiatives.

To facilitate the organization of exchange programs involving undergraduate and graduate students and academics, agreements and memorandums of understanding between the countries are signed. A benchmark in international education is the Russian company Rosatom that has agreements in several countries. With traditional and renowned educational institutions, Russia is attracting several hundreds of students from Turkey, Vietnam, Bangladesh, Jordan and other countries where Russian design nuclear energy plants are expected to be built in coming years. For example, in 2014, 344 Vietnamese undergraduate and graduate students were studying in Russia and 150 engineers were helping with the construction of Rostov nuclear plant in Russia (WNA, 2015), preparing themselves for the construction of the Ninh Thuan 1.

While the choice of Russian technology for Ninh Thuan 1 has now been made, for the Ninh Thuan 2 plant consideration of possible technology options is ongoing. An intergovernmental agreement with Japan took effect from January 2012 and Japan has committed to train about 1000 staff for Ninh Thuan 2 (WNA, 2015).

From the point of view of the country, it not unusual to receive training from different providers in different countries, while the decision for a certain design or vendor is not taken. In many cases, the decision to build a nuclear energy plant can take decades. For such a long time, the training efforts might be lost, if there is no support from nuclear knowledge management from the beginning.

Training goes beyond the frontiers but to prepare competent personnel from an embarking country for new nuclear build, several local aspects of training need to be addressed, not only individual training. National infrastructure for research, development, and training is fundamental in a long run. The Centre for Nuclear Energy Science and Technology (CNEST) is being set up by Rosatom and the Vietnam Atomic Energy Institute (Vinatom), which will become the main coordinating body for nuclear research and development in Vietnam, including the Da Lat research reactor operation, commissioned in 1984. Rosatom also supported the establishment of a Nuclear Industry Information Centre (NIIC) at the Hanoi University of Science & Technology in 2012. This is an approach held in 20 Russian cities and in Turkey and Bangladesh as well, aiming at educating public, specially school students about nuclear energy (www.myatom.ru/en).

International training centers for national and international nuclear specialists are cleverly established by vendor companies. Rosatom established the Central Institute for Continuing Education and Training in Obninsk, Kaluga Region. The French company AREVA established international training centers in France, Germany, and US. In South Korea, the Korea Atomic Energy Research Institute (KAERI) Nuclear Training and Education Center (KNTC), provides domestic and international nuclear personnel in conjunction with international organizations on a wide range of training courses.

AREVA is developing uranium exploration and mining in several countries and has signed cooperation agreement to provide technical support and training. A joint venture between AREVA NC and the Centre for Advanced Studies of Mineral Resources (CESMAT) implements a training program for overseas students at the School of Mines in Ales, France. From 2008 to 2010, 35 participants from Namibia, Mongolia, South Africa, Gabon, Republic of Central Africa, and Senegal have participated in those training courses. (Ahmadzadeh and Petitclerc, 2009).

17.2.5 International conferences

International and regional conferences are opportunities to exchange information and to inspire new developments. Participating in conferences and international meetings is one of the preferred ways for researchers to gain new knowledge, along with becoming acquainted with the peer-reviewed publications. The uranium industry presents some challenges to optimize operational efficiency, and minimize waste generation, for which research is continually encouraged. Specialized international conferences are the best opportunities to learn from others' experience on how to deal with, for example, the separation of uranium from other minerals by complex metallurgy in unconventional mines, recover uranium as by product from phosphoric acid, deal efficiently with acid drainage in mines under decommissioning, and many other interesting issues.

Specifically on human resources development, the IAEA organizes every 4 to 5 years an International Conference on Human Resource Development. The last ones were in Abu-Dhabi in 2010 and in Vienna in 2014. The IAEA will organize the International Conference on Knowledge Management in Nuclear Energy—Challenges and Approaches from 7 to 11 November 2016 in Vienna, Austria. These Conferences are good opportunities to exchange information about practices, techniques, and methodologies with professionals from different areas, not only with nuclear experts, but also with information technologies and human resources specialists.

Examples of education and training conferences organized by nuclear societies are the NESTet organized by the European Nuclear Society and the Conference on Nuclear Training and Education (CONTE) by the American Nuclear Society.

17.3 Building competence in the uranium industry

17.3.1 Competence-based training

Education, training, and work experience are of primary importance for achieving competence in any area. Competence-based training differs from traditional training as it relates organizational goals and individual training needs. It establishes core competencies needed, which are the combination of knowledge, skills and attitudes for each job position. It also assesses existing competencies and identifies gaps to be filled with training, continually evaluating training effectiveness towards achieving these competencies. In addition, persons may need an appropriate level of work experience to fulfil particular responsibilities.

The identification of the competencies for each type of job role, function (operation, maintenance, safety, environment protection, supervision, control, etc.) and sector (mining, conversion, isotopic enrichment, reconversion, or fuel fabrication), is a challenging task. However, once done, the development of human resources and training implementation will became more efficient. International cooperation can be instrumental to build a consensus on key competencies for each job or role in the nuclear industry and for the dissemination of international guidelines for training design (IAEA, 1996).

Globalization and broadband internet are creating an increasingly diverse and interconnected world. Individuals need, for example, to master changing technologies, and to understand a large amount of available information. They also face collective challenges as societies—such as economy variations and environmental protection. They are often required to work with different languages and cultures and to foster social equity. In these contexts, the competencies that individuals need to meet their goals have become more complex, requiring more than a list of topics to be covered by a teacher in a classroom.

The design of a training program for the fuel cycle facilities depends on several aspects. Once the key competencies are defined, they can be converted into training objectives. The training objectives must be clear and relevant, and must be communicated with the persons that will receive the training. The objectives should describe the expected performance, under which conditions, and following certain technical standards and regulatory requirements. The training objective must contain an action verb that can be verified. To help verify if the training objectives are well defined, the following question should be asked to each objective: "At the end of the training, the person will be able to (training objective (example): understand the basic methods for radiation protection and apply them in typical workplace situations)." These objectives should be organized into a training plan from which the training material is drawn.

There is a wide range of competencies required to the nuclear fuel cycle related workers, being operators, managers, radiation protection and environmental protection officers, regulators, researchers, craft and technical workers, from different organizations. The range of competencies includes not only the capacity to apply a method for solving situations or a standard procedure to follow routinely. It should also include the ability to think critically, to learn from experience, to team work, to face and deal with changes and to benefit from the opportunities the global context provides.

17.3.2 Overview of training methodologies

Any methodology for training workers has to be objective and consider routine and accidental situations. With a systematic approach to training, each worker will have an individual training program that can be complied with in different methodologies, as for example, classroom lectures, mentoring, on-the-job-training, field visits, workshops, e-learning, case studies with real facts and lessons learned and participation in meetings and seminars. The selection of the best methodology, the training provider, and the timing to provide the training are to be defined jointly by the worker, the supervisor, and the human resources department training specialist.

International training events are very attractive and the participants should be wisely selected, according to the organization goals. The participants in these events are exposed to a multicultural learning environment. It presents advantages over national training programs, as for example the possibility to exchange information about innovations, to discuss cases and lessons learned, to visit nuclear installations in another country and to build a network of international peers.

It is unlikely that an organization can send several workers for a long training course at the same time and e-learning can be an important complement. Classroom formal lectures, when only informative, can be transformed in e-learning material with self-assessments. However, while learning from standardized and validated material and at the own pace and in any location presents several advantages, the e-learning still cannot compete with face to face lectures and interaction with peers, especially cannot compete with international training. Some people are not comfortable and disciplined enough to use computers for learning. On the other hand, the e-learning material format needs to be attractive. It is important to adapt the format to the background and interest of the target audience, or the e-learning will be not effective as expected. Some good examples of e-learning include features such as videos filmed in installations; discussion forums with encouragements for interactions, exercises with assessment after each module; simulation with virtual reality techniques.

17.3.3 Basic general training course content for the uranium industry

The content of any training program should be tailored to the target audience needs and background. Unless the trainees are from a nuclear engineering undergraduate course, it is unlikely that knowledge of nuclear topics has not received in-depth coverage in traditional educational curricula. Although the approaches are different for each stage of the fuel cycle, a holistic and global view of the uranium production towards the assurance of supply for nuclear power plants should be addressed, while addressing topical and local challenges.

Persons who work in the nuclear area should have an adequate level of understanding of concepts related to the technology as well as to safe and security. Environmental protection was only an afterthought in mining activities prior to the 1980s. Legacies from past mining and milling activities in several countries without due consideration for environmental sustainability (IAEA, 2011) remain to be solved, mostly by governmental organizations. Poor extraction efficiency sometimes led to high levels of uranium left in the tailings.

The occupational doses for mining workers is the highest among the other stages of the nuclear fuel cycle, in spite of remaining below the recommended limits of an average of 20 mSv per year over 5 years and 50 mSv in a single year. It is still dropping over the years thanks to regulatory control, training and automation. The average occupational radiation doses in mining dropped from 5.1 mSv per year in the years from 1980 to 1984 to 1.9 mSv per year from 2000 to 2002 (UNSCEAR, 2008). However, there is a potential for reaching higher doses, especially in new mining sites, if routine adequate protection is not in place and where automation, if needed, has not been installed. Therefore, international recommendations for radiological protection in mining are a fundamental subject for training.

Legislation from both nuclear and environmental agencies became more stringent for new mines and other nuclear installations. However, in general, complying with standards and regulatory requirements is not enough to operate an installation. A great deal of effort is devoted to communication with the population about the nuclear industry. In many countries, the public meetings are part of the licensing process. Communication with media and public requires dedicated and trained personnel at the installation site.

Finally, nuclear industry requires leaders and competent managers capable of making wise decisions and managing operational risks, especially those that may affect safety.

Table 17.1 presents a generic list of topics for a training course on uranium industry. It does not intend to be comprehensive and should be tailored for the specified target group.

17.3.4 Certification of personnel

To carry out certain functions or to assume certain responsibilities in the nuclear area, such as operators or supervisors, an internal qualification process is required. Qualification of personnel is the recognition of the adequacy of the combination of their educational level, training, and work experience. This may include consideration of the equivalence of certain elements of this combination. In some circumstances, such qualified persons may have to be certified by a regulatory body, or another recognized third party. The IAEA provides some recommendations for certification process on radiation safety (IAEA, 2001) and staff for nuclear plants (IAEA, 2002).

17.4 Future trends

Systematic planning for human resources in the nuclear industry depends on active monitoring of the global fuel market demand and supply capacities to assure the competent personnel in the right time and location. International organizations and world association publish authoritative information on current market and future scenarios. Training relies on the available technologies and should reflect the lifetime management approach, considering not only the environmental sustainability,

Table 17.1 **Generic topics for training course on uranium industry**

Topic	Content
Energy global setting	Global supply and demand for uranium as fuel; nuclear energy program in different countries; The fuel cycle for different types of reactors; Nuclear fuel as a cost key advantage; Role of international organizations; international conventions; International cooperation.
Mining and milling	Uranium deposit types and their origins; Uranium geochemistry; hydrology; Baseline data collection, Feasibility studies, economic evaluation; Environmental Impact Assessment (EIA); Exploration development, case studies of different mines; Mining processing methods in relation to uranium metallurgy; Water treatment; Closure; Sustainable development; Remediation and rehabilitation.
Uranium conversion	UO_3 refinery; Production of UO_2, UF_4 and UF_6; chemical characteristics of UF_6; Industrial safety; Overpacks cylinders for UF_6 storage and transportation.
Isotope enrichment	Uranium isotopic enrichment technologies; the separative work unit (SWU); Existing enrichment plants and supply; international low-enriched uranium (LEU) fuel bank; Management of the depleted uranium; Specific nonproliferation and safeguards aspects.
Fuel fabrication	Reconversion process; Quality control; Existing fuel fabrication plants; Fuel assemblies; fuel for CANDU reactor.
Radiological protection	Basic principles of radiation protection: justification, optimization and limitation; international recommendations; Quantity and units, basic dose calculations; Typical natural radiation doses in the world and occupational doses in the fuel cycle; Means for protection: reduce exposure time, limit the quantity, keep adequate distance, use adequate shielding, avoid contamination; Environmental monitoring.

(*Continued*)

Table 17.1 (Continued)

Topic	Content
Nuclear safety	Safety culture; Criticality safety; Defense in depth; IAEA specific safety standards; Safety assessment.
Nuclear law, international conventions	The international legal framework: conventions, treaties; country obligations to the conventions; Civil liability for nuclear damage; Export/import controls.
Regulatory control	Principles of nuclear regulation; Nuclear national legislation; Licensing process, inspection, enforcement actions.
Security	Internal and external threats; targets; security by design; Radioactive material control and accountancy; Human reliability; Coordination with other organizations in the country; Cybersecurity; Nuclear forensics.
Control of nuclear material, nonproliferation regime	Nuclear material control and accountancy; Agremeents; IAEA safeguards inspections.
Emergency planning and preparedness	Lessons learned from past accidents: Three Mile Island, Chernobyl, Fukushima; tailings dam ruptures, mine accidents. Emergency planning, Coordination with other organizations for emergeny response; Simulation exercises.
Transport of nuclear material	Basic requirements for transport of radioactive material in different forms; The role of international organizations.
Waste management and disposal techniques	Waste management at nuclear facilities; Waste Classification; Disposal options, technological, economical and socio-political aspects.
Stakeholders involvement and communication	Interested persons or stakeholders: operators, contractors, regulators, researchers, landowners, environmental organizations, academia, media, associations, public in general. Understanding public perception and its causes; Building public support; Effective participation in public meetings and press conferences.
Long-term project management	Economics; Local and international business; supply chain; vendor oversight, quality management; Project structuring, human resources, infrastructure and financing; Industry challenges; Leadership development.

but also the impact of the variation of uranium prices in the economic feasibility of the installation, especially for mining sites.

Mining as international business bounded by agreements and alliances calls for action across national borders. With the aim of fostering consistent quality of education and training, the interregional cooperation will increase as well as industry and universities partnerships.

A great deal is being spent on the remediation of legacy mine sites. Nowadays and in the future, exploration of uranium mines must avoid the generation of those legacies. For that, international cooperation is instrumental to assist countries in mining projects and to upgrade their regulatory framework on safety, security, and environment protection.

The decision making process is facilitated with managers and leaders with adequate nuclear knowledge and skills. Communication among stakeholders should be enhanced to allow discussion based on sound social and technical arguments. International networks and access to information facilitate the identification of common needs and the timely dissemination of practical information, especially for potential new comers in the nuclear industry.

In a globalized world, mobility of trained specialists among different countries will certainly rise. Therefore, quality control of education and training and common understanding of different nuclear job profiles and qualification may be an issue. Mutual competency recognition will call for accreditation, quality control, and peer reviews of training programs. The development of common job taxonomy and implementation of accreditation schemes are facilitators for transferable nuclear competencies (OECD/NEA, 2012).

Making the best out of modern technologies, improved language simultaneous translation tools will be a catalyst for translating training material. Virtual reality will be used for simulations of practical training. These features will greatly facilitate e-learning and training of multicultural personnel for company expansions abroad.

Still there is a need to fulfil the demand for craft and technical levels. This would be better dealt with locally, in local language and focusing in national regulations and procedures. However, international cooperation can assist in preparing the lecturers in a train-the-trainers program.

References

Ahmadzadeh, H., Petitclerc, J-L., 2009. Cooperation with emerging countries in advanced mining training programs involving an industrial partner. In: Int. Symp. Uranium Raw Material for the Nuclear Fuel Cycle: Exploration, Mining, Production, Supply and Demand, Economics and Environmental Issues, Vienna, pp. 22–26.

International Atomic Energy Agency, 1956. The Statute of the IAEA. Article III. Vienna.

International Atomic Energy Agency, 1996. Nuclear power plant personnel training and its evaluation. Technical Report Series 380. IAEA, Vienna.

International Atomic Energy Agency, 2001. Building Competence in Radiation Protection and the Safe Use of Radiation Sources. Safety Guide—IAEA Safety Standards Series RS-G-1.4. IAEA, Vienna.

International Atomic Energy Agency, 2002. Recruitment, Qualification and Training of Personnel for Nuclear Power Plants. IAEA SAFETY GUIDE No. NS-G-2.8. IAEA, Vienna.

International Atomic Energy Agency, 2011. The Uranium Mining Remediation Exchange Group. (UMREG). Selected Papers 1995–2007. STI/PUB/1524. IAEA, Vienna, ISBN 978–92–0–169310–5.

International Atomic Energy Agency, 2013. The Agency's Programme and Budget 2014–2015. GC(57)/2. IAEA, Vienna.

International Atomic Energy Agency, 2014. IAEA Nuclear Technology Review 2014. IAEA GC (58)/4. IAEA, Vienna.

OECD/Nuclear Energy Agency, 2012. Nuclear Education and Training: from Concern to Capability. Nuclear Development. NEA No. 6979. OECD/NEA, Paris, ISBN 978-92-64-17637-9.

OECD/Nuclear Energy Agency, 2014. Uranium 2014: Resources, Production and Demand. A Joint Report by the OECD Nuclear Energy Agency and the International Atomic Energy Agency. OECD/NEA, Paris.

Trojacek, J., 2014. World nuclear university school of uranium production: eight years' experience. In: Uranium Raw Material for the Nuclear Fuel Cycle: Exploration, Mining, Production, Supply and Demand, Economics and Environmental Issues, Vienna, Austria, 23–27 June 2014.

United Nations Scientific Committee on the Effects of Atomic Radiation, 2008. Report: Sources, Effects and Risks of Ionizing Radiation, vol. 1. UNSCEAR, Vienna.

Wieland, P., Davies, L., 2015. The World Nuclear University: exceeding expectations. Nucl. Future. vol. 11 (2), March/April. London.

World Nuclear Association, 2015. The Nuclear Fuel Report: Global Scenarios for Demand and Supply Availability 2015–2035. London. ISBN 97809931019-0-8.

World Nuclear Association, 2015. Information Library. Available from: <http://www.world-nuclear.org/Information-Library/> (accessed 05.04.15).

Index

Note: Page numbers followed by "*f*" and "*t*" refer to figures and tables, respectively.

A

Accessory minerals, 13–16, 275, 278–279, 283
Acid and alkaline uranium extraction and purification, 139–143
Acid mine drainage (AMD), 425–426
Acidic leaching, 139–140
Adits, 130
Adsorbtion, 16
Advanced heavy water reactors (AHWR), 253, 268–269
Advanced Spaceborne Thermal Emission and Reflection Radiometer, 62
Advanced Very High Resolution Radiometer, 62
Aerodynamic processes, 339
AFRA-NEST, 443
Alkaline igneous rocks, 273, 279
Alkaline lixiviant system, 177, 196–197, 200
Alkane Resources' Dubbo Zirconia Project, 247
Alta Mesa ISR project, 191*f*
Amenability, preliminary determination of, 177
American Assured Fuel Supply (AAFS), 223
American centrifuge (USEC and Centrus), 339–340
American Society for Testing and Materials (ASTM), 226
Americium, 256
AngloGold Ashanti, 146, 249
Appia Energy, 246
AREVA, 354, 445
  AREVA NP, 354
  as uranium enrichment primary supplier, 329–330
  and URENCO tails, 232
Argentina with uranium enrichment capabilities, 333
Arlit-In Azawa fault in Niger, 62
Asia, ISR operation in, 162–166
Asian Network for Education in Nuclear Technology (ANENT) ANENT, 443
As-low-as-reasonably-achievable (ALARA) principle, 424
Athabasca Basin in Canada, 57, 63, 88–89, 132–133
Atomic Energy Commission (AEC), 301, 312
Atomic Vapour Laser Isotope Separation (AVLIS) technology, 341–342
Aura Energy, 248
Australia
  Australasian Code for Reporting Exploration Results, 78–79
  ISR operation in, 160–161, 161*f*
  recoverable reserves of uranium in, 59–60
  uranium resources in, 86

B

Bag filter, 191–192
Basal-type uranium deposits (MWB), 31
Baseline water quality and in situ recovery technology, 180–181
Basin/basement redox-controlled deposits (HDbb), 30
Bastnäsite, 277–278
Batch-type rotary vacuum dryer systems, 191
Beharchuwa-Bokarda-Labed area of India, 63
Beryllium oxide (BeO), 262
Beverley, in South Australia, 160
Beverly ISR project, 158*f*, 161*f*, 173–174, 187*f*, 206*f*
Bhabha Atomic Research Center (BARC), 315
Black shale, 19
  deposits, 25
  uranium from, 247–249
Blend low-enriched uranium (BLEU) program, 227

Block leaching (underground), 145
Boiling water reactors (BWR), 261, 353
fuel assembly, 368–369
Brazil
with civil enrichment facility, 332
pilot-scale process, 315
Bulgaria, ISR operation in, 166
Burnable absorbers, 372–373
Byproduct, uranium as, 239
from black shales, 247–249
from coal ash, 250
from gold tailings, 249
from phosphate deposits, 239–241
processing phosphates, 241–242
production potential, 242–243
from rare earths deposits, 243–247
from seawater, 249

C

Calutrons, 338–339
Cameco Corporation, 355
Cameco uranium oxide concentrate refining to UO_3, 302–303, 302*f*
Canada
reactors in, 102
thorium as nuclear fuel, research and possibilities in, 268
Canada Deuterium Uranium (CANDU)-type reactors, 253
Canadian mines, 59–60
Carbonate deposits, 19, 24–25
Carbonatites, 277
Cash costs of operating mines, 112–113
Central processing plant (CPP), 198
Centre for Nuclear Energy Science and Technology (CNEST), 444
Centrus, 340
Chemical Plants Complex (CPC), 315
Chemical process, 339
Chernobyl accident, 9–10, 231
China, 244
ISR operation in, 162, 162*f*
reactors in, 102
thorium as nuclear fuel, research and possibilities in, 267–268
as uranium enrichment primary supplier, 331–332
uranium inventories in, 221
China National Nuclear Corporation (CNNC), 250, 314, 355
Cigar Lake mine, 59–60, 132–133
Civil nuclear power, 10
Class 7 dangerous goods and regulatory practice/codes, 405–406
Clay minerals, 70
Closed fuel cycle, 106
Coal ash, uranium from, 250
Coffinite, 13–16, 171
Collapse breccia pipe, 19, 23
Collapse breccia pipe deposits, 23
Combustibles Nucleares Argentinos (CONUAR), 359
Communication and administration, 431
Community impacts management, in uranium operations, 401–404
current versus historical practices, 403–404
Compact high-temperature reactor (CHTR), 269
Comurhex refining and conversion to UF_4, 303–304, 304*f*
Condensation repression isotope separation by laser activation (CRISLA), 343–344
CONNECT network, 441
Conventional deposits, 38–44, 39*t*
ConverDyn, 219–220
Conversion of natural uranium. *See* Natural uranium, conversion of
Converters, factors affecting, 315–316
Costs of production, 112–114
cash costs of operating mines, 112–113
full costs of planned and prospective mines, 113–114
future trends, 114
Covers, 429–430
Crow Butte ISR project in Nebraska, USA, 169*f*
Crushing and grinding, 136–138, 137*f*
Crystal fractionation, 26
Crystalchemical and redox trapping-related deposits (SCcr), 32
Curium, 256
Czech Republic, ISR operation in, 166, 167*f*

D
Dalmatovskoe/Klolovskoe deposit, 163–165
Declines, 130–131
Deep disposal well, 201–202
"Defense-in-depth" approach, 8
Definitive Feasibility Study (DFS), 242
Demand for uranium, 99
estimates of, 105
important reactor operating factors, 103–104
interplay with uranium enrichment sector, 104–105
national nuclear programs, 101–103
nuclear generating capacity, 100–101
Demand/supply balance forecast 2015–25, 337
Denmark, thorium utilization in, 272
Depleted uranium, 227–228, 234
Deposit models, critical factors in, 68–71
sedimentology, 70–71
source, transport, and fixing of the uranium, 69
sources of fluids, 69–70
timing of mineralization, 68–69
Deposits, uranium
classification of, 17–33
genetic classification and formation of uranium deposits, 25–33
IAEA geological classification, 17–25
future trends, 44–46
geology of, 11
IAEA UDEPO database, 33–34
new uranium deposits (2004–2014), 38–44
conventional deposits, 38–44
unconventional deposits and resources, 44
unconventional resources and deposits in UDEPO, 34–37
uranium minerals, 13–17
uranium provinces, 11–13, 14*t*
Desalination of seawater/brackish groundwater, 9
Designing for closure, 419–420
Dewatering, 430
Diagenetic-hydrothermal collapse breccia deposits (HDiaCb), 30
Discriminators, 135, 135*f*
Diversity of uranium deposits, 12*f*
DOE (Department of Energy), 219–220, 225, 227, 233
Dolni Rozinka uranium mine, 131*f*, 132, 139
Drill cuttings, 168–170
Drill-and-blast method, 129, 129*f*
Drilling and evaluation, 71
Dry process, 264, 300
Dust, 399
Dysprosium, 243–244

E
Eastern Europe, ISR operation in, 166–168
Electromagnetic (EM) energy, 62
Electromagnetic isotope separation (EMIS), 338
Electromagnetic process, 338–339
Electromobility frontier, 6–7
Elution, 141
and precipitation, 200–201
Emanometry, 63
End-member geological fluids, 25, 26*f*
Energy density, 4
Energy Information Administration (EIA), 218–219
Energy Northwest (ENW), 232–233
Engineering measures, 429–430
Enriched reprocessed uranium (ERU), 216, 230*f*
Enrichment, of uranium
American centrifuge (USEC and Centrus), 339–340
civil enrichment services demand/supply balance, 334–337
civil utilization of ex-military materials, 336
enriched product inventories, 336
tails reenrichment, 334–336
countries with operating civil enrichment facilities
Brazil, 332
Japan, 332–333
countries with uranium enrichment capabilities
Argentina, 333
India, 333
Iran, 333
Pakistan, 333
current technology and civil facilities, 324
gas centrifuge, principles of, 325–326

Enrichment, of uranium (*Continued*)
historic technologies, 337–339
aerodynamic processes, 339
chemical process, 339
electromagnetic process, 338–339
gas diffusion, 337–338
laser enrichment, 340–344
AVLIS, 341–342
molecular laser isotope separation, 342–344
main primary suppliers, 326–332
AREVA, 326–329
China, 331–332
Russia, 330–331
URENCO, 326–329
separative work units (SWUs), 321–322
tails, management of, 346–349
deconversion, 347–348
reenrichment of, 346–347
tails storage, 348–349
uranium, reprocessed, 349–350
uranium hexafluoride in enrichment, quality control of, 344–346
enriched product, 344–346
feed material, 344
tails, 346
Enrichment sector, 104–105
ENUSA, 355–356
Environmental impacts, managing in uranium operations, 396–401
current versus historical practices, 400–401
Environmental monitoring, 182, 183*t*
Equilibrium, uranium in, 397
EUP, 223, 232
Euratom Supply Agency (ESA), 220
Europe, thorium utilization in, 270
Europe
uranium inventories in, 220
European Nuclear Education Network (ENEN), 442–443
European Nuclear Energy Agency (ENEA), 77–78
Europium, 243–244
Evaporation ponds, 203
Evapotranspiration-related deposits, 32
Excess Uranium Transparency and Accountability Act, 220
Exploration, 83–85, 89–91
and delineation methods, 174–180
logging and subsurface evaluation, 177–180
preliminary determination of amenability, 177
Exploration for uranium, 53
critical factors in deposit models, 68–71
sedimentology, 70–71
source, transport, and fixing of the uranium, 69
sources of fluids, 69–70
timing of mineralization, 68–69
drilling and evaluation, 71
drivers of, 55–57
nuclear fuel cycle, 54–55
prospectivity, explorability, and exploration targeting, 57–60
synopsis, 72
techniques, 60–68
geochemical methods, 64–68
geological mapping, 62
geophysical methods, 63–64
property evaluation, 61
remote sensing, 62

F

Fast breeder reactors (FBR), 253, 261–262
Fast neutron reactors (FNRs), 117
Finland's Talvivaara Mining Company Plc, 247–248
Flotation, 138
Fluorination to UF_6, 305–306
Fluorine production, 311
Fossil fuels, 106
Four Mile leases, 160
France, reactors in, 102
"Free digging" material, 129
French nuclear sector, 110
Fuel cycles, alternative, 117–118
Fuel pellet production, 371–372
Fuel rod fabrication process, 373
Future, uranium in, 10

G

Gadolinium, 243–244
Gamma, 398
Gamma logs, 177
Gas centrifuge, principles of, 325–326

Gas diffusion, 337–338
Gas-cooled reactors (GCR), 353
fuel assemblies, 370
GE-Hitachi Global Laser Enrichment (GLE), 233
Generation III reactors, 117
Generation IV International Forum (GIF), 271
Generation IV nuclear reactors, implications of, 235–236
future trends, 236
Genetic classification and formation of uranium deposits, 25–33
hydrothermal deposits (H), 29–30
hydrothermal metasomatic systems (HM), 31
magmatic deposits (M), 26–28
other types (O), 33
synsedimentary deposits (S), 32
GENUSA, 355–356
Geochemical methods, 64–68
Geological mapping, 62
Geology, 61
Geophysical methods, 63–64
Germany, 261
ISR operation in, 168
reactors in, 102
thorium utilization in, 272
GIS capabilities, 72
Global laser enrichment (GLE), 342–343
Global Nuclear Fuel (GNF), 356, 369
Global uranium resources as of January 1, 2013, 85–91
requirements, 89–90
widespread and variable, 86–89
Gold tailings, uranium from, 249
Goscomgeo and China Guangdong Nuclear Uranium Corp. (CGN-URC), 248–249
Granite-hydrothermal systems (GH), 29
Granite-related deposits, 18, 21
Granites, 4
Greenland Minerals & Energy, 245
Groundwater conditions and in situ recovery (ISR) technology, 173–174
Groundwater monitoring, operational, 181
Groundwater restoration and well plugging, 209–210

H
Hard rock open cut mining, 129
Health, managing in uranium operations, 387–393
current versus historical practices, 393
radiological health aspects, 387–391
Health, safety, environment, and community (HSEC) in uranium operation
across uranium life cycle, 385–387
community impacts, managing, 401–404
current versus historical practices, 403–404
environmental impacts, managing, 396–401
current versus historical practices, 400–401
framework for, 385
health, managing, 387–393
current versus historical practices, 393
radiological health aspects, 387–391
integration of, 404
parameters for, 388*t*
safety, managing, 394–396
current versus historical practices, 395–396
Heap Leach Recovery Process, 434*f*
Heap leaching options, 143–145
Heavy water reactors, 261
Heavy-mineral sands, 283–285
Helikon vortex tube process, 339
Hexavalent uranium, 16, 139, 195
Highly enriched uranium (HEU), 92–93, 110, 215–216, 225–228
background, 225–226
blendstock, 232
future blend down of, 227–228
Russian–US HEU deal, 226–227
US HEU blend down, 227
High-temperature gas-cooled reactors (HTGR), 253, 262
History of uranium, 3–4
HolGoun's Uranium and Power Project, 250
Honeymoon ISR project, in South Australia, 160
Honeywell process, 301, 307
Host rock amenability, 170–172
deposit types, 171
uranium mineralogy, 171–172
Human Resource Development, 445

Husab mine, 107–108
Hydrofluorination to UF_4, 305
Hydrogen fluoride, 316
Hydrogen peroxide, 200
Hydrogeologic conditions and in situ recovery (ISR) technology, 173–174
  confinement, 173
  groundwater conditions, 173–174
Hydrological characterization, 182
Hydrothermal deposits (H), 29–30
Hydrothermal diagenetic systems (HD), 29
Hydrothermal metamorphic systems (HMp), 30
Hydrothermal metasomatic systems (HM), 31
Hydrothermal uranium deposits, 29

I

IAEA geological classification of uranium deposits, 17–25
  detailed geological classification, 18–20
  Type 1. Intrusive deposits, 21
  Type 2. Granite-related deposits, 21
  Type 3. Polymetallic ironoxide breccia complex deposits, 21
  Type 4. Volcanic-related deposits, 21–22
  Type 5. Metasomatite deposits, 22
  Type 6. Metamorphite deposits, 22
  Type 7. Proterozoic unconformity deposits, 22
  Type 8. Collapse breccia pipe deposits, 23
  Type 9. Sandstone deposits, 23–24
  Type 10. Paleo-quartz-pebble conglomerate deposits, 24
  Type 11. Surficial deposits, 24
  Type 12. Lignite-coal deposits, 24
  Type 13. Carbonate deposits, 24–25
  Type 14. Phosphate deposits, 25
  Type 15. Black shale deposits, 25
IAEA UDEPO database, 33–34
Identified Resources, 55–56, 80–81, 86
In situ leach (ISL) technique, 58, 126, 166
In situ recovery (ISR) technology, 58, 126, 157
  advantages of, 159–160
  in Asia, 162–166
  in Australia, 160–161, 161*f*
  automation and telemetries, 207–208
  block diagram, 158*f*
  in Bulgaria, 166
  in China, 162, 162*f*
  construction and installation, 204
  in Czech Republic, 166, 167*f*
  in Eastern Europe, 166–168
  environmental evaluations, 180–182
    baseline water quality, 180–181
    environmental monitoring, 182
    hydrological characterization, 182
    operational groundwater monitoring, 181
  exploration and delineation methods, 174–180
    logging and subsurface evaluation, 177–180
    preliminary determination of amenability, 177
  general process, 194*f*
  geology and hydrogeologic conditions, 170–180
  in Germany, 168
  host rock amenability, 170–172
    deposit types, 171
    uranium mineralogy, 171–172
  hydrogeologic conditions, 173–174
    confinement, 173
    groundwater conditions, 173–174
  ISR technique, 157–159
  in Kazakhstan, 162–163, 164*f*
  layout and patterns, 203
  lixiviant types, 193–197
    acid, 195–196
    alkaline, 196–197
  mine area site conditions, 180
  in Mongolia, 163
  plant configurations, 198–200
    central plant, 198
    elution and precipitation, 200–201
    remote ion exchange satellite plant, 198–200
  processing plant equipment, 185–192
    piping, 188
    process pads, 185–186
    retention ponds, 186–188, 187*f*
    tankage, at ISR facilities, 188
    weatherization, 188–189
    yellowcake dryer systems, 190–192

reclamation, 209–210
groundwater restoration and well plugging, 209–210
radiological decontamination, 209
reclamation and revegetation, 209
in Russia, 163–165, 165*f*
in Ukraine, 168, 168*f*
in United States, 168–170, 169*f*
in Uzbekistan, 165–166
waste management, 201–203
liquid radioactive byproduct waste, 201–203
solid radioactive byproduct waste, 203
well field instrumentation, 205–206
well fields, 182–185
well operation, 204–205
India
Department of Atomic Energy (DAE), 356
thorium as nuclear fuel, research and possibilities in, 268–269
with uranium enrichment capabilities, 333
Indian molten salt breeder reactors (IMSBRs), 269
Indicator mineral methods, 66
Indoor process facility at the Lost Creek ISR project in Wyoming, 190*f*
Industrial packaging, 407
Industrias Nucleares do Brasil (INB), 357
INFCIRC 225, 409
Inferred Resources (IR), 55–56, 80–81
global distribution of, 87*f*
Inkai ISR project Kazakhstan, central processing facility at, 198*f*
Integrated Nuclear Fuel Cycle Information System (INFCIS), 441
Interformational redox-controlled uranium deposits (HDir), 30
International Atomic Energy Agency (IAEA), 223, 239, 405, 440
International Atomic Energy Association (IAEA), 55–56, 77–78, 93–94, 406, 440
International Civil Aviation Organisation (ICAO), 406
International fuel reserves, 92–94
rationale, 92–93
status, 93–94
International Maritime Organisation (IMO), 406
International Nuclear Information System (INIS), 441
Interregional initiatives in education and training, in uranium industry, 440–445
basic general training course content for uranium industry, 447–448
building competence in uranium industry, 445–448
competence-based training, 445–446
training methodologies, 446–447
certification of personnel, 448
global and regional partnerships, 442–443
European Nuclear Education Network, 442–443
World Nuclear University (WNU), 442
intergovernmental organizations, 440–441
international conferences, 445
international discussion forum and trade association, 443
training course on uranium industry, 449*t*
transnational education and training programs, 444–445
Intrusive deposits, 18, 21
Ion exchange (IX), 141
Ionizing radiation, 9–10
Iran, with uranium enrichment capabilities, 333
Iron oxide deposits, 282
Iron oxide–copper–gold (IOCG) deposit, 55–56
Isotope separation process, 321, 325. *See also* Molecular Laser Isotope Separation
Itochu Corporation, 248

J

Japan
with civil enrichment facility, 332–333
thorium utilization in, 272
uranium inventories in, 220–221
Japan Nuclear Fuel Limited (JNFL), 357
Jet boring, 132–133, 133*f*
Jet nozzle process, 339
Joint Stock Company (JSC), 358
JORC code, 78–79

K

KAERI Nuclear Training and Education Center (KNTC), 444
Kazakhstan, 107
  ISR operation in, 162–163, 163*t*, 164*f*
  uranium production in, 107
  uranium resources in, 55–56, 59–60, 86, 88
Kazatomprom, 357
KEPCO Nuclear Fuel Company, 357
Kerr-McGee Corporation, 301
Khiagdinskoye deposit, 163–165
K-metasomatism related deposits (HMtK), 31
Korean nuclear program, 102
Kvanefjeld deposit in Greenland, 91, 244–245

L

Lakewater and sediment geochemistry, 67–68
Land position availability, 61
Landsat thematic mapper, 62
LANENT, 443
Langer Heinrich uranium mine, 138
Laser enrichment, 340–344
  AVLIS, 341–342
  molecular laser isotope separation, 342–344
    condensation repression isotope separation by laser activation (CRISLA), 343–344
    separation of isotopes by laser excitation (SILEX), 342–343
Lead isotopes, 67
Lead-210, 397–398
Legacy uranium mining site, 422*f*
Liabilities, 61
Lichtenberg open pit, 128
Light water reactor (LWR), 261, 353
  fuel fabrication, 359–362
    facilities and capacities, 360*t*
Lignite-coal deposits, 19, 24
Liquid radioactive byproduct waste, 201–203
  deep disposal well, 201–202
  evaporation ponds, 203
  reverse osmosis (R.O.), 201, 202*f*
  surface discharge and land application, 203
Liquid waste disposal well at the Alta Mesa ISR project, 202*f*
Liquid-fluoride thorium reactor (LFTR), 267, 271
Liquid-metal-cooled fast reactors (LMR), 353, 365
Lixiviant types, 193–197
  acid, 195–196
  alkaline, 196–197
Logging and subsurface evaluation, 177–180
Long-lived radioactive dust (LLRD), 134
Low-enriched uranium (LEU), 93, 110, 215–216

M

Magmatic deposits (M), 26–28
Magmatic-related deposits, 68
Magnetics, 63
Magnetotelluric methods, 63–64
Manual of Acid In Situ Leaching Uranium Mining Technology, 193
Mapping, geological, 62
Marble-hosted phosphate, 18
Market-related pricing, 119
Mary Kathleen, 247, 426*f*
McArthur River mine, 59–60, 132–134
Mechanical integrity test (MIT), 204
Mechanical sorting-related synsedimentary deposits, 32
Medium enriched uranium (MEU), 216
Megatons to megawatts, 336
Metal toxicity in mining, 433
Metallogeny, 59–60
Metamorphic rocks, 283
Metamorphite, 18
  deposits, 22
Metasomatic uranium deposits, 68
Metasomatite, 18
  deposits, 22
Meteoric water infiltration-related deposits (MW), 31
Milling and extraction, 135–145
  acid and alkaline uranium extraction and purification, 139–143
  block leaching (underground), 145
  conventional crushing and grinding, 136–138, 137*f*
  heap leaching options, 143–145
  ore beneficiation, 138–139

Mine and mill remediation and reclamation, 415
funding for rehabilitation, 432
future trends, 432–434
practice, 433
regulations, 432–433
research, 433–434
risks addressed in rehabilitation, 424–427
acid mine drainage (AMD), 425–426
effect of radiation on nonhuman biota, 425
regulatory risk, 425
solids waste disposal, 425
waste facility design, 426–427
site remediation and reclamation, 423–424
tools used in rehabilitation, 427–432
communication and administration, 431
engineering, 429–430
modeling and model validation by monitoring, 428
storage facility design for closure, 428
training, 431–432
water treatment, 430–431
uranium mine and milling sites, 419–423
extraction processes, 422–423
mining methods, 421–422
Mine area site conditions, 180
Mine Waste Solutions (MWS), 146
Mineral Resources and Ore Reserves, 78–79
Mineralogy, 67
of uranium, 171–172
Mining and milling of uranium, 123
books, proceedings, and technical reports, 148–149
first extensive uranium mining and production booms (World War II to 1980), 125–126
future trends, 146–148
considering the end at the beginning, 147–148
other trends and discussion, 148
technology developments, 146–147
historic mining for pigment and radium (to 1939), 125
milling and extraction, 135–145
acid and alkaline uranium extraction and purification, 139–143
block leaching (underground), 145
conventional crushing and grinding, 136–138, 137*f*
heap leaching options, 143–145
ore beneficiation, 138–139
modern era, 126
open cut mining, 126–130
hard rock open cut mining, 129
soft rock open cut mining, 128–129
trends and alternatives, 129–130
retreatment of mine tailings, 145–146
current considerations, 146
historic examples, 145–146
transition (1980–1995), 126
underground mining, 130–135
access options, 130–131
identification and placement of waste and mineralized material, 134–135
mining options, 131–133
ventilation considerations, 133–134
websites, 149–150
industry groups, 150
international organizations, 149
nongovernment organization, 150
top seven uranium producing companies, 150
Mitsubishi Nuclear Fuel (MNF), 358
Mixed oxide fuel (MOX), 110–111, 117, 216, 218, 228, 230*f*, 324
ships for, 412–413
and uranium, 412
Modeling and model validation by monitoring, 428
Modular object-oriented dynamic learning environment (MOODLE) learning management system, 441
Molecular laser isotope separation, 342–344
condensation repression isotope separation by laser activation (CRISLA), 343–344
separation of isotopes by laser excitation (SILEX), 342–343
Molten salt reactors (MSR), 253, 263–264, 269
Molycorp Inc's Mountain Pass deposit, 244
Monazite, 244, 246, 273–275, 289
Mongolia, ISR operation in, 163
Multilateral Enrichment Sanctuary Project (MESP), 93–94

N

Na-metasomatism, 31
Namibia, 26, 38–43, 46, 58, 88, 138
National Atomic Energy Commission of Argentina (CNEA), 315
National nuclear programs, 101–103
Natural and enriched uranium
 commercial inventories of, 218–222
  China, 221
  Europe, 220
  Japan, 220–221
  Russia, 221–222
  United States, 218–220
 fuel banks, 222–223
 future trends, 223–225
 Uranium Participation Corporation (UPC), 222
Natural uranium, 4, 6, 13, 105, 110–111, 216–217, 264
Natural uranium, conversion of, 297
 current status of conversion industry, 314–315
 environmental, 313
 factors that impact converters, 315–316
 history, 300–301
 needs, 299–300
 potential future developments, 318
 process outline, 300
 safety, 312–313
 UF_6 product, 311–312
  fluorine production, 311
  process comparisons, 312
 UO_2, conversion to, 316–318
 uranium oxide concentrate conversion to UF_6 prior to refining, 307–310
 uranium oxide concentrate refining/ conversion to UF_6 by wet process, 302–304
  Cameco uranium oxide concentrate refining to UO_3, 302–303, 302*f*
  Comurhex refining and conversion to UF_4, 303–304, 304*f*
  process byproducts and environmental aspects, 304
 uranium oxide conversion to UF_6 after refining, 304–306
  fluorination to UF_6, 305–306
  hydrofluorination to UF_4, 305
  reduction to UO_2, 304–305
Nb-Ta-Ti uranium oxides, 13–16
Neodymium, 243–244, 273
Neptunium, 256
New uranium deposits (2004–2014), 38–44
 conventional deposits, 38–44
 unconventional deposits and resources, 44
NI 43-101 code, 78–79
Nichols Ranch project in Wyoming, USA, 171*f*
 contained header house at, 207*f*
 IX columns at, 197, 197*f*
 pipe burial at, 189*f*
 RIX satellite connected to the resin transport at, 199*f*
Niger, 88
Nitrogen oxide, 303
Nolans Bore, 246–247, 281–282
Nonhuman biota, 399–400
 effect of radiation on, 425
Nonhydro renewable, 7
Nonpower uses, 9
Norway, nuclear power in, 270
Novosibirsk Chemical Concentrates Plant (NCCP), 358
Nuclear Energy Agency (NEA), 239, 440
Nuclear fission, 3
Nuclear fuel banks, 224*t*
Nuclear fuel fabrication
 fuel assembly design components, 367–373
  burnable absorbers, 372–373
  BWR fuel assembly, 368–369
  fuel pellet production, 371–372
  fuel rod fabrication process, 373
  GCR fuel assemblies, 370
  PHWR fuel bundle, 369–370
  PWR fuel assembly, 367–368
  RBMK fuel assemblies, 370–371
  uranium dioxide production, 371
 services market capacities, 359–365
  LWR fuel fabrication, 359–362
  MOX fuel fabrication, 362–365
  NPPs types, 365
  PHWR fuel fabrication, 362
  world fabrication requirements, 365–366
 service suppliers, 353–359
 trends, current and future, 374–378
  fast reactor fuel, 376–377
  fuel cycle trends, 375–376
  fuel reliability, 374–375

Nuclear Fuel Industries, Ltd (NFI), 358
Nuclear generating capacity, 99–101, 103*t*
Nuclear industry, 374, 410, 443
Nuclear materials, transport of
current issues for, 409
Nuclear Non-Proliferation Treaty, 10
Nuclear power plants (NPPs), 77, 101, 125, 322, 353, 365
Nuclear reactor designs
heterogeneous reactor, 255
homogenous reactor, 255
Nuclear reactors, 4–6, 106, 116
Nuclear renaissance, 109–110, 120
Nuclear steam supply system (NSSS), 353, 367–369

O

Oak Ridge National Laboratory (ORNL), 263
Olympic Dam, 21, 35, 88, 108
Open cut mining, 126–130
hard rock, 129
soft rock, 128–129
trends and alternatives, 129–130
Operating mines, cash costs of, 112–113
Operational groundwater monitoring, 181
Operations and maintenance (O&M) costs, 100
Ore beneficiation, 138–139
Organisation for Economic Co-operation and Development (OECD), 10, 55–56

P

Packaging of radioactive materials, 406–408
Pakistan, with uranium enrichment capabilities, 333
Pakistan Atomic Energy Commission (PAEC), 359
Pakistan Chemical Plants Complex (CPC), 315
Palangana/Hobson satellite in Texas, 192*f*
Paleo-quartz-pebble conglomerate, 19, 24
Partnerships committed to education and training, 442–443
Pele Mountain Resources Inc, 246
Peralkaline magmas, 29, 279
Peroxide, 142, 201
PhosEnergy process, 241–242
Phosphate deposits, 19, 25, 239–241
Phosphate rock (phosphorite), 91, 239, 241
Phosphate rock, 241–242
Phosphorous pentoxide (P_2O_5), 239–241, 243
Pipe burial, preparation for, 189*f*
Piping systems, in ISR operations, 188
Pitchblende, 22–23, 171
Plovdiv ISR facility, 166
Plutonium, 6, 228–229, 253–254, 364
recycling, 110–111
transport and packaging of, 411–412
Plutonium oxide, 411
Polonium-210, 398
Polymetallic iron oxide breccia complex, 18, 21, 44
Ponds, retention, 186–188, 187*f*
Pressurized heavy water reactors (PHWR), 253, 353
fuel bundle, 369–370
fuel fabrication, 362
facilities and capacities, 363*t*
Pressurized water reactors (PWR), 353
Price trends and outlook, 119–121
Primary supply, 56, 106–108
Process bleed, 204–205
Process pads, 185–186
Processing of uranium ore, flow sheet for, 136*f*
Prognosticated Resources (PR), 80–81
Proliferation, 9–10
Prompt fission neutron (PFN) logging, 64, 177, 180, 180*f*
Property evaluation, 61–62
Prospectivity, 57–60
Proterozoic unconformity, 18, 22
Prototype fast breeder reactor (PFBR), 261–262
Pu-239, 6, 117
PUREX process, 257, 264, 269

R

R&D programs, 117
Radiation, 9–10, 147, 386–387, 393, 406, 425
Radiation management plan (RMP), 425
Radioactive materials, packaging of, 406–408
Radiological decontamination, 209
Radiometric prospecting, 67

Radiometric sorting, 138
Radionuclides, 391, 396, 417, 431–432
Radium removal, 430
Radium-226, 390, 397, 399
Radon, 63, 67, 399
Radon emanometry, 63
Radon gas, 134
Radon-222, 390–391, 397
"Raise bore" technique, 132–133
Rare earth elements (REEs), 13–16, 80, 243, 253–254
Rare earth oxides (REO), 243–246
Rare earths deposits, uranium from, 243–247
Reasonably Assured Resources (RAR), 55–56, 80–81, 86
Reclamation, 209–210, 415, 430
  groundwater restoration and well plugging, 209–210
  radiological decontamination, 209
  and revegetation, 209
Recycled uranium and plutonium, 228–229
  future of, 229–230
Red Book, 78, 239, 240*t*, 247, 441
  unconventional uranium resources reported in, 240*t*
  uranium classification scheme, 80*f*
Redox trapping in continental environments, 32
Redox trapping in marine environments, 32
Reduction-oxidation (REDOX) processes, 170
Refueling, 117
Regulations, 8, 406, 423, 432–433
Rehabilitation of uranium milling and mining
  funding for, 432
  risks addressed in, 424–427
    acid mine drainage (AMD), 425–426
    effect of radiation on nonhuman biota, 425
    regulatory risk, 425
    solids waste disposal, 425
    waste facility design, 426–427
  tools used in, 427–432
    communication and administration, 431
    engineering, 429–430
    modeling and model validation by monitoring, 428
    storage facility design for closure, 428
    training, 431–432
    water treatment, 430–431
Remote ion exchange (RIX) satellite plant, 198–200, 199*f*
Remote sensing, 62
Reprocessed uranium (RepU), 110–111, 216, 218, 349–350, 354
Requirements of, uranium, 89–90, 104, 105*t*, 236–237
Resin-in-pulp (RIP) technique, 141
Resource situation, 4–5
Resources, 77
  classification, 80–83
  conventional and unconventional, 79–80
  dynamics, 83–85
  global uranium resources as of January 1, 2013, 85–91
    requirements, 89–90
    widespread and variable, 86–89
  history, 77–78
  international fuel reserves, 92–94
    rationale, 92–93
    status, 93–94
  "Red Book", 78
  and reserves, 78–85
  unconventional, 91–92
  United Nations Framework Classification, 94–95
Retention ponds, 186–188, 187*f*
Retreatment of mine tailings, 145–146
  current considerations, 146
  historic examples, 145–146
Revegetation, reclamation and, 209
Reverse osmosis (R.O.), 197, 201, 202*f*, 209–210
Rock forming minerals, uranium in, 17
Roll front deposit, 31, 171, 172*f*, 178*f*
"Room and pillar" mining, 132
Rössing deposit in Namibia, 26, 58
Rössing open pit uranium mine, 128*f*, 129–130
Rössing uranium mine, 141
Rotary vacuum dryer systems, 191, 193*f*
Rubber tire rotary drill rig and water truck, 174–177, 176*f*
Run-of-mine (ROM) ore, 136

Russia
ISR operation in, 163–165, 165*f*
reactors in, 102, 111
thorium utilization in, 271
as uranium enrichment primary supplier, 330–331
uranium inventories in, 221–222
Russian Federation, 93
Russian graphite-moderated reactors (RBMK), 353, 358
fuel assemblies, 370–371
Russian tails reenrichment activities, 231–232
AREVA and URENCO tails, 232
domestic consumption and export, 231–232
HEU blendstock, 232

S
Safe and secure packaging and transport of uranium materials, 405
Safety Assessment of the Molten Salt Fast Reactor (SAMOFAR), 270
Safety Evaluation of Fuel Cycle Facilities during Operation (SEDO), 440–441
Safety management in uranium operations, 394–396
current versus historical practices, 395–396
Samarium, 243–244
SAMREC code, 78–79
Sandstone, 19
Sandstone deposits, 23–24
Sandstone-hosted deposits, 13–16, 69–70
Sandstone-hosted uranium, 23, 170–171
Sandstone-related deposits, 58
Santa Quitéria mine, 91
Seawater, uranium from, 92, 249
Seawater desalination, 116
Secondary supplies of uranium and plutonium, 215, 216*f*
classification, 215–217
commercial inventories of natural and enriched uranium, 218–222
China, 221
Europe, 220
Japan, 220–221
Russia, 221–222
United States, 218–220
generation IV nuclear reactors, implications of, 235–236
future trends, 236
highly enriched uranium (HEU), 225–228
background, 225–226
future blend down of, 227–228
Russian–US HEU deal, 226–227
US HEU blend down, 227
natural and enriched inventories, 222–225
fuel banks, 222–223
future trends, 223–225
Uranium Participation Corporation (UPC), 222
nuclear fuel banks, 224*t*
recycled uranium and plutonium, 228–229
recycled uranium and plutonium, future of, 229–230
secondary supplies, continuing role of, 236–238
tails material, enrichment of, 230–234
future of, 233–234
Russian tails reenrichment activities, 231–232
US tails enrichment, 232–233
technology, role of, 217–218
underfeeding, 234–235
future underfeeding trends, 235
vs tails enrichment, 234–235
Secondary supply, 56, 84–85, 106, 110–112, 215, 216*f*
classification, 215–217
Security of energy supply, 87*f*, 96
Sedimentology, 70–71
Separation of isotopes by laser excitation (SILEX), 342–343
Separation of isotopes by laser excitation (SILEX), 342–343
Separative work units (SWUs), 105, 111, 321–322
Shafts, 130
Shipment, of radioactive materials, 405, 408
Short-wave infrared (SWIR) spectrometers, 67
Sibanye Gold, 249
Skarn-related deposits, 31
Smith Ranch/Highland project in Wyoming, USA, 170*f*, 173–174
Soft rock open cut mining, 128–129

Solid radioactive byproduct waste, 203
Solids waste disposal, 425
Solvent exchange (SX), 140
Solvent regeneration, 304
South Africa, thorium utilization in, 270
Speculative Resources (SR), 55–56, 80–81
Storage facility design for closure, 428
Sulfuric acid leach chemistry, 160, 163–165
Sulfuric acid storage at the Akdala ISR project, 195*f*
Supply of uranium, 106–112
  historical trends in meeting uranium demand, 108–110, 112
  primary supply from mines, 106–108
  secondary supplies, 110–111
Surface discharge and land application, 203
Surface water, 399
Surficial deposits, 19, 24
Synsedimentary deposits (S), 32

T

Tabular deposit, 171, 173*f*
Tabular intraformational redox-control deposits (HDiaTb), 29–30
Tailings material, 398
Tails, 346
  deconversion, 347–348
  management, 346–349
  reenrichment, 346–347
  storage, 348–349
Tails enrichment, underfeeding vs, 234–235
Tails material, enrichment of, 230–234
  future of, 233–234
  Russian tails reenrichment activities, 231–232
    AREVA and URENCO tails, 232
    domestic consumption and export, 231–232
    HEU blendstock, 232
  US tails enrichment, 232–233
Talvivaara Mining Company Plc, 247–248
Tankage, at ISR facilities, 188
Technological perspective, 5–6
Techsnabexport (TENEX), 226
Tectonolithologic intraformational redox-control deposits, 30
Tennessee Valley Authority (TVA), 227, 232–233
Terbium, 243–244
Tetravalent U-Th minerals, 13–16
Texas Rare Earth Resources Corp, 246
^{232}Th, 254, 256–257, 261–262
  decay schemes of, 54*f*
THOREX process, 269
Thorite, 275, 280–281
Thorium, 6
  advantages of, 253–254
  based nuclear power, 253
  current and recent thorium recovery, 289–290
  current research and future possibilities, 267–273
    Canada, 268
    China, 267–268
    Denmark, 272
    Europe, 270
    Generation IV International Forum (GIF), 271
    Germany, 272
    India, 268–269
    Japan, 272
    Norway, 270
    Russia, 271
    South Africa, 270
    thorium production for foreseeable future, 272–273
    United Kingdom, 271
    United States, 270–271
  fuel cycle, 254–258
    accelerator-driven systems, 258
    advantages of, 254–256
    disadvantages of, 256–257
    fast breeder reactors, 261–262
    fast reactor thorium utilization, 257
    heavy water reactors, 261
    high-temperature gas-cooled reactors, 262
    light water reactors, 261
    molten salt reactors, 263–264
    previous work on, 258–267
    resource requirements, 264–267, 265*t*
  geochemistry and mineralogy, 273–275
    thorium-bearing deposits, 275–285
    Thorium resources, 286–289
  geology and resources, 273–289
  -rich veins of uncertain origin, 280–281
  utilization in experimental and nuclear power reactors, 259*t*

Thorium dioxide (ThO_2), 255
Thorium high-temperature reactor (THTR), 262
Thorium-230, 397
Thorium-breeding molten salt reactor (TMSR), 267
Till geochemistry, 66
Titanium, 253–254
Trading of uranium, 118–119
Training
  competence-based, 445–446
  in mining project, 431–432
Tributyl phosphate (TBP), 301
Tumas calcrete-type deposit, 63
Type A packaging, 407–408
Type B packaging, 408

U
U-234, 397, 399
U-235, 5–6, 54–55, 105, 110, 262, 264, 321, 325, 337–338, 341, 346, 349, 409
  decay schemes of, 54*f*
  fission reaction, 117
U-238, 6, 54–55, 63, 255–257, 325, 341, 397, 399
  decay schemes of, 54*f*
UDEPO database, 33–34
UF_4, 301
  Comurhex refining and conversion to, 303–304, 304*f*
  hydrofluorination to, 305
UF_6, 299–301, 311–312, 348
  fluorination to, 305–306
  fluorine production, 311
  process comparisons, 312
UF_6, uranium oxide conversion to, 304–306
  after refining
    fluorination to UF_6, 305–306
    hydrofluorination to UF_4, 305
    reduction to UO_2, 304–305
  prior to refining, 307–310
Ukraine
  ISR operation in, 168, 168*f*
  reactors in, 102
Unconventional deposits and resources, 44, 45*t*
Unconventional uranium resources reported in Red Books, 240*t*
Underfeeding, 111, 234–235
  future trends, 235
  vs tails enrichment, 234–235
Underground mining, 130–135
  access options, 130–131
    adits, 130
    declines, 130–131
    shafts, 130
  identification and placement of waste and mineralized material, 134–135
  mining options, 131–133
  ventilation considerations, 133–134
Under-ream method, ISR well construction using, 205*f*
Undiscovered Resources, 80–81
United Kingdom
  reactors in, 102
  thorium utilization in, 271
United Nations Economic Commission for Europe (UNECE), 94–95
United Nations Framework Classification (UNFC), 94–95
United States
  environmental effluent monitoring in, 183*t*
  ISR operation in, 168–170, 169*f*
  thorium utilization in, 270–271
  uranium inventories in, 218–220
United States Enrichment Corporation (USEC), 111
UO_2, 300
  conversion to, 316–318
  reduction to, 304–305
UO_3
  Cameco uranium oxide concentrate refining to, 302–303, 302*f*
Uraninite, 171
Uranium cycle, 54, 255
Uranium dioxide production, 371
Uranium hexafluoride (UF_6), 299–300, 307, 312–313, 324–325, 410–411
  quality control of, 344–346
    enriched product, 344–346
    feed material, 344
Uranium hexafluoride (UF_6), uranium oxide concentrate refining/conversion to
  by wet process, 302–304
    Cameco uranium oxide concentrate refining to UO_3, 302–303, 302*f*

Uranium hexafluoride (UF_6), uranium oxide concentrate refining/conversion to (*Continued*)
  Comurhex refining and conversion to UF_4, 303–304, 304*f*
  process byproducts and environmental aspects, 304
Uranium market, 99, 102
Uranium Metal Plant, 315
Uranium minerals, 13–17, 125
Uranium ore concentrates (UOCs), 142, 299–300, 303
Uranium ore deposits, grade vs tonnage for, 59*f*
Uranium Production Site Appraisal Team (UPSAT), 440–441
Uranium resources by deposit types, 34*t*
Uranium supply and demand to 2020 and beyond, 114–118
  alternative fuel cycles, 117–118
  impact of prospective new technologies, 115–116
Uranyl anions, 192–193
Uranyl nitrate hexahydrate (UNH), 303
URENCO, 326–329, 348
U-rich granites, 66
US Atomic Energy Commission (AEC), 301
US Enrichment Corporation (USEC), 226
US HEU disposition, 336
US Nuclear Regulatory Commission (NRC), 423
US tails enrichment, 232–233
Uzbekistan, ISR operation in, 165–166

V
Volcanic-hydrothermal systems, 29
Volcanic-related deposits, 21–22

W
Waste facility design, 426–427
Waste management, 9–10, 201–203, 415, 434–435
  liquid radioactive byproduct waste, 201–203
    deep disposal well, 201–202
    evaporation ponds, 203
    R.O., 201, 202*f*
    surface discharge and land application, 203
  solid radioactive byproduct waste, 203
Waste management facility (WMF), 421–422
Water quality, baseline
  and in situ recovery (ISR) technology, 180–181
Water treatment, 430–431
Weatherization, 188–189
Weatherized wellhead enclosures in Wyoming, 190*f*
Well field instrumentation, 205–206
Well fields, 187*f*, 188*f*
  in ISR mines, 182–185
Well operation, 204–205
Well plugging, 209–210
Westinghouse Electric Company, 359
Wet process, 301–302
  uranium oxide concentrate refining/conversion to UF_6 by, 302–304
    Cameco uranium oxide concentrate refining to UO_3, 302–303, 302*f*
    Comurhex refining and conversion to UF_4, 303–304, 304*f*
    process byproducts and environmental aspects, 304
World Nuclear Association (WNA), 99, 322
World Nuclear News (WNN), 443
World Nuclear Transport Institute (WNTI), 405
World Nuclear University (WNU), 442
World uranium market and prices, 118–121
  price trends and outlook, 119–121
  trading of uranium, 118–119
World uranium production, 106–107, 107*t*, 109*f*

X
Xenotime, 275

Y
Yellowcake dryer systems, 190–192
Yttrium, 243–244
Yunnan Sparton New Environ-Tech Consulting Co, 250
Yuzhny fault zone, 62

Z
Zafarabad ISR project, 166*f*

CPI Antony Rowe
Chippenham, UK
2016-02-18 21:04